Mathematics for the Biological Sciences

Stanley I. Grossman

Department of Mathematics
University of Montana

James E. Turner

Department of Mathematics
McGill University

Macmillan Publishing Co., Inc.
New York
Collier Macmillan Publishers
London

Macmillan Publishing Co., Inc.
866 Third Avenue, New York, New York 10022

Collier-Macmillan Canada Ltd.

Library of Congress Cataloging in Publication Data

Grossman, Stanley I.
    Mathematics for the biological sciences.

    1. Biomathematics, I. Turner, James E., joint author. II. Title.
QH323.5.G76   510'.2'4574   73-4048
ISBN 0-02-348330-X

Printing: 1 2 3 4 5 6 7 8   Year: 4 5 6 7 8 9 0

Preface

Until recently, scientists in biology, medicine, and related areas felt little need for mathematics other than a standard course in statistics. The past decade, however, has seen a rapid rise in applications of varied mathematical tools to the study of many different kinds of biological phenomena. This has occurred to such an extent that it is no longer necessary to convince students of the life sciences that mathematics plays an important role. What is necessary is to make the relevant mathematics accessible in a reasonable time and to develop the student's ability to relate mathematics to problems in biology and medicine. These are the two aims of this book.

This text has developed from courses given to students majoring in all areas of biology and medicine at McGill University, Montreal, Canada. While some of the material contained here can be found in standard finite mathematics texts, we felt the need for a book that emphasized biological motivations at every step.

We have attempted to write a flexible book that will serve the extremely varied needs and backgrounds of the students for whom it is intended. To this end, prerequisites are kept to a minimum. The more difficult material in each chapter generally appears in the later sections and can be omitted without loss of continuity. On the other hand, the student with a better background can benefit from the more advanced material.

The principal subjects of the text are probability, vectors and matrices, linear programming, Markov chains, game theory, difference equations, and differential equations. It is our belief that, together with statistics, these are the most useful areas of mathematics in biological applications. We have not included any general discussion of statistical methods, because they have long been recognized as an essential part of biology. However, the chapters on probability theory here could effectively serve as a useful introduction to standard courses in statistics and biometrics.

The first six chapters assume only high school algebra. Only Chapters 7 and 8 and one section of Chapter 9 require some familiarity with calculus. The calculus needed for these chapters (basic differentiation and integration) is developed in Appendices B and C. These appendices could serve as a one-term course in calculus and were so designed. The calculus material also presupposes some knowledge of trigonometry; the needed material is developed in Appendix A.

Chapter 1 is introductory and is needed for the material that follows. It includes set theory, functions, permutations, combinations, and the binomial theorem.

Chapter 2, on discrete probability, includes all the probability theory that is standard in courses in finite mathematics. Such courses, however, often exclude such topics as the Poisson and geometric distributions, which are so useful in biological applications. These topics are included here.

Chapter 3, on vectors and matrices, completes the core of a basic finite mathematics course. Included in this chapter is a discussion of eigenvalues and eigenvectors. This material, the most difficult in the chapter, is needed in Chapter 5 for the discussion of Markov chains. It can be omitted at first reading.

After the first three chapters, there are a number of directions in which courses based on this book could continue. Chapters 4, 5, and 6 would complete a strong training in noncalculus mathematics. In Chapters 4 and 5, three of the most important applications of vectors and matrices are developed. Linear programming, Markov chains, and game theory have become important tools in ecology, medical diagnosis and treatment, genetics, and in many other areas. Chapter 4 depends only on Chapter 3, while Chapter 5 makes use of basic probability and therefore depends also on the early sections of Chapter 2. The two parts of Chapter 5, Markov chains and game theory, can be read independently. Chapter 6, on difference equations, is independent of the earlier chapters. Difference equations are an essential technique in describing processes of population growth and in many genetics applications.

Chapters 7 and 8 assume a basic knowledge of calculus. Chapter 7, on differential equations, gives the continuous analogue for the material developed in Chapter 6 (although it could be read independently of Chapter 6). The chapter does not depend on the material in the earlier part of the book. Chapter 8, on continuous probability, continues where Chapter 2 leaves off and develops the basic properties of continuous distributions. The normal distribution, undoubtedly the most important tool in probability theory for the biologist, is discussed in considerable detail.

Chapter 9 does not introduce any additional mathematics but instead develops, at some length, a number of mathematical models of biological processes. The models examined all relate to the basic problem of the growth, survival, and extinction of populations. All the mathematics required for these models is developed in the preceding chapters. The instructor is encouraged to discuss these models with his or her students as soon as the underlying mathematics has been developed. The examples discussed in the first eight chapters have often been simplified to facilitate the understanding of the mathematics. The longer models in Chapter 9 should prove to be especially useful in developing requisite model-building skills.

We would like to convey our thanks to our colleagues and students at McGill University and at the University of Montana for their many helpful suggestions for improving the material for this book. We are also grateful to Carl Allendoerfer and others who reviewed this text and offered much constructive criticism. Finally, special thanks to Everett Smethurst, our editor at Macmillan, Inc., for his continual help and encouragement, and to Maria Lam, who painstakingly and skillfully typed the manuscipt of this book.

S. I. G.

J. E. T.

Contents

Mathematics for the
Biological Sciences

Preliminaries 1

1.1 The Language of Sets

The idea of a "set" is a very general concept that comes up in every area of science and mathematics. A *set* is any well-defined collection of objects. The objects that make up the set are called the *elements* or *members* of the set. As examples, consider the set of all pages in this book (each page is an element of this set), the set of real numbers greater than 0 and less than 1, the set of patients in a certain hospital, the set of species that became extinct between 1900 and 1970.

A set is determined when we are able to decide whether any given object is a member of the set or not. This may be a difficult problem. For example, the collection of all "large" numbers is not a set unless we clearly define which numbers are large and which numbers are not large. The collection of all numbers greater than 1000 is a set, since we can obviously determine whether any given object is a member of this set or not.

There is a standard way to describe sets using a bracket notation. To illustrate this notation, suppose that S represents the set of positive even integers less than 10. In the bracket notation, we write

$$S = \{x : 0 < x < 10 \text{ and } x \text{ is an even integer}\}.$$

This should be read "S is the set of elements x such that $0 < x < 10$ and x is an even integer." More simply, we write $S = \{2, 4, 6, 8\}$.

Example 1.1.1   The following are examples of sets written in the bracket notation

1. $S_1 = \{x : 0 < x < 1 \text{ or } 2 < x < 3\}$.
2. $S_2 = \{x : x \text{ is a patient in a certain hospital}\}$.
3. $S_3 = \{x : x \text{ is a living human being}\}$.

The notation $x \in S$ will be used to mean that x is an element of S (or x belongs to S). The equation $S = \{x : x \in S\}$ expresses the obvious fact that the set S is equal to the set of elements x such that x belongs to S. Two sets S_1 and S_2 are said to be *equal* if they contain the same elements.

Example 1.1.2   The set $S_1 = \{x : 0 < x < 4, x \text{ is an integer}\}$ and the set $S_2 = \{x : x^3 - 6x^2 + 11x - 6 = 0\}$ are equal. The elements of both sets are the integers 1, 2, and 3. Therefore, we can write $S_1 = S_2 = \{1, 2, 3\}$.

A set may have an infinite number of elements. For example, $S = \{x : x \text{ is a positive integer}\} = \{1, 2, 3, \ldots\}$ is an infinite set. On the other hand, a set may contain no elements at all. The set $S = \{x : x \text{ is an odd integer divisible by 2}\}$ clearly contains no elements. Until 1969, the set of human beings who had walked on the moon contained no elements. If a set contains no elements, we will call it the *empty set* (or *null set*) and denote it by the symbol $\emptyset$. Note that $\emptyset$ is a set, since it is a very well-defined collection of objects; that is, no object belongs to S.

In applying this language to a particular problem, we usually form the set of all objects which are being studied in the problem. This is called the *universal set* of the problem. For example, a student preparing for an examination is interested in the set of all questions that can be asked on the examination. (This is a very large set.) The much smaller set of all questions that can reasonably be asked on the examination may be of more importance to the student than the universal set. As a second example, in a study of the incidence of diseases in a country, the universal set would be the set of all people in the country. For practical reasons, it may be necessary to work with a much smaller set of people chosen from the population. This leads naturally to the idea of subsets of a set.

Suppose that A is a given set of elements. Then B is a *subset* of A, written $B \subset A$, if every element of the set B is an element of the set A. Note that, if $A \subset B$ and $B \subset A$, then $A = B$. (Try to explain in words why this must be true.) We say that B is a *proper subset* of A if $B \subset A$ and $B \neq A$.

Example 1.1.3   Define $A = \{x:x \text{ is a human being}\}$ and $B = \{x:x \text{ is a female human being}\}$. Then, clearly, $B \subset A$ and B is a proper subset of A.

Example 1.1.4   Define $A_1 = \{x:x \text{ is an animal species}\}$, $A_2 = \{x:x \text{ is an insect species}\}$, and $A_3 = \{x:x \text{ is a mammalian species}\}$. Then A_2 and A_3 are proper subsets of A_1.

We should be careful to distinguish elements of a set and subsets of a set. For example, when we write $a \in \{a, b, c\}$, we say that the element a is a member of the set consisting of the three elements a, b, and c. When we write $\{a\} \subset \{a, b, c\}$, we say that the set consisting of the element a is a subset of the set consisting of the three elements a, b, and c.

Problems 1.1

1. Describe in words the following sets written in the bracket notation.
   (a) $S_1 = \{x:x^2 + 2x = 0\}$.
   (b) $S_2 = \{x:5 \leq x, x < 9\}$.
   (c) $S_3 = \{x:1 \leq x^2 \leq 9\}$.
   What are the elements of each of the sets S_1, S_2, and S_3?

2. Which of the following sets are equal and which are proper subsets of other sets in the following list?
   (a) $S_1 = \{x:0 \leq x \leq 1\}$.   (b) $S_2 = \{x:x > 0, 0 \leq x^2 \leq 4\}$.
   (c) $S_3 = \{x:0 \leq 3x^2 \leq 3\}$.   (d) $S_4 = \{x:x^2 + x = 0\}$.
   (e) $S_5 = \{x:x \geq 0, x < 2\}$.   (f) $S_6 = \{x:0 < x \leq 4 - x\}$.

3. Describe several ways that patients in a hospital are divided naturally into subsets. Describe the corresponding division of the hospital staff into subsets.

4. Consider the set of known symptoms for appendicitis. In general, a patient with appendicitis will have only a subset of these symptoms. If the known symptoms are s_1, s_2, and s_3, describe all possible subsets of symptoms. (Include the full set and the empty set as subsets.)

5. Suppose that S is a set containing n elements. Counting the empty set $\emptyset$

and the full set S as subsets, prove that there are 2^n subsets of S. (*Hint:* A given element of S is either in or not in a particular subset.)

6. Four drugs, D_1, D_2, D_3, and D_4, are available to treat a certain illness. The illness is treated by administering at least two of the drugs. If the order in which the drugs are administered is not relevant, write out the different ways of treating this illness with these drugs.

7. In a test of a weight-reducing diet, 300 volunteers follow the diet for two months. After 1 month, 240 of the volunteers have lost more than 10 pounds and 100 have lost more than 15 pounds. After 2 months, 260 have lost more than 10 pounds and 150 have lost more than 15 pounds. Assuming that no volunteer gains weight while following the diet, describe the relations among these four subsets of the set of volunteers.

1.2 Set Operations

In the previous section, we developed the concept of a set and its subsets. In this section, we will define several set operations or ways to combine sets to form other sets. For example, the set of all animal species combines with the set of all plant species to form the set of all species.

Definition 1.2.1   Union of Two Sets   *The union of the set A and the set B is the set composed of the elements of A, together with the elements of B. The union is written $A \cup B = \{x : x \in A \text{ or } x \in B \text{ or both}\}$.*

Example 1.2.1   Define $A = \{1, 3, 6, 8\}$ and $B = \{2, 4, 6, 8\}$. Then the union of A and B is $A \cup B = \{1, 2, 3, 4, 6, 8\}$. In this example, the elements 6 and 8 belong to both sets.

Example 1.2.2   Define A to be the set of all male smokers in a population and B to be the set of all fathers in the population. Then $A \cup B$ is the set of all males in the population who are either smokers or fathers or both.

In an obvious way, we can form the union of more than two sets. The union of the three sets A, B, and C is the set $A \cup B \cup C$ each of whose elements belongs to at least one of A, B, and C.

$$A \cup B \cup C = \{x : x \in A \text{ or } x \in B \text{ or } x \in C\}.$$

There is a simple way to represent sets and set unions by using *Venn diagrams*.

The set A is represented by the shaded area in Figure 1.1(a). Similarly, the sets $A \cup B$ and $A \cup B \cup C$ are represented in Figures 1.1(b) and 1.1(c) by the shaded areas.

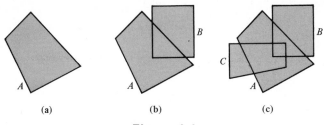

(a)   (b)   (c)

Figure 1.1

We are interested in a second way of combining sets to form new sets. For example, if A is the set of all men in a population and B is the set of all left-handed people in the population, we can form the set of all left-handed men in the population. This leads to the following definition.

Definition 1.2.2   Intersection of Two Sets   *The intersection of the set A and the set B is the set of elements which are contained in both A and B. The intersection is written $A \cap B = \{x : x \in A \text{ and } x \in B\}$.*

The intersection of more than two sets is defined in an obvious way. The intersection $A \cap B \cap C$ of the three sets A, B, and C is the set of elements which belong to A, to B, and to C.

$$A \cap B \cap C = \{x : x \in A \text{ and } x \in B \text{ and } x \in C\}.$$

The Venn diagram representation of the intersection of any number of sets is the overlap of the areas representing the sets. The shaded areas in Figures 1.2(a) and 1.2(b) represent the intersections $A \cap B$ and $A \cap B \cap C$.

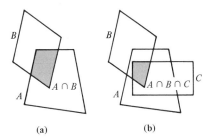

(a)   (b)

Figure 1.2

Example 1.2.3   Define $A = \{1, 3, 6, 8\}$ and $B = \{2, 4, 6, 8\}$. The intersection of A and B is $A \cap B = \{6, 8\}$.

Example 1.2.4   Define A to be the set of fruit flies in a given population that have a certain wing mutation and define B to be the set with a certain eye mutation. Then the intersection $A \cap B$ is the set of fruit flies in the population that have both the wing and eye mutations. (What is $A \cup B$ in this example?)

Theorem 1.2.1   *Consider any two sets A and B. Then the union $A \cup B$ and the intersection $A \cap B$ satisfy the relations*
  1. $A \cap B \subset A$, $A \cap B \subset B$.
  2. $A \subset A \cup B$, $B \subset A \cup B$.

Proof: Every element of $A \cap B$ is an element of A and of B. This is precisely what is meant by $A \cap B \subset A$ and $A \cap B \subset B$. Similarly, the set $A \cup B$ contains all the elements of A and B. Therefore, $A \subset A \cup B$ and $B \subset A \cup B$.

If often happens that two sets A and B have no elements in common. To describe this possibility, we introduce the following definition.

Definition 1.2.3   Two Disjoint Sets   *Two sets A and B are disjoint if they contain no elements in common. In other words, A and B are disjoint if $A \cap B = \varnothing$.*

Example 1.2.5   Define A to be the set of positive integers and B to be the set of negative integers. Then A and B are disjoint sets, since there is no integer that is both positive and negative.

Example 1.2.6   Define A to be the set of people in a population over 20 years of age and B to be the set of people under 10 years of age. Then A and B are disjoint sets; that is, $A \cap B = \varnothing$.

Definition 1.2.3 can be extended to more than two sets.

Definition 1.2.4   Mutually Disjoint Sets   *The n sets $A_1, A_2, \ldots, A_n$ are said to be mutually disjoint (or pairwise disjoint) if no two of the sets have* *elements in common. In other words, $A_1, A_2, \ldots, A_n$ are mutually disjoint if $A_i \cap A_j = \emptyset$ when $i \neq j$ for $i, j = 1, 2, \ldots, n$.*

Example 1.2.7   In a certain population, define $A = \{$people under 10 years old$\}$, $B = \{$people between 10 and 20$\}$, and $C = \{$people over 20$\}$. Then the sets A, B, and C are mutually disjoint, since no person belongs to two of the sets.

There is a third set operation that is extremely useful. Suppose that we are studying the incidence of tuberculosis in a population. The universal set of this problem is made up of all the individuals of the population. We are interested in the set of individuals who have tuberculosis and also in the set of individuals who do not have tuberculosis. This leads to the following definition.

Definition 1.2.5   Complement of a Set   *The complement of a set A relative to the universal set S is the set $\overline{A}$ composed of all the elements in S which are not in A. The complement $\overline{A}$ is also written A^c or $S \setminus A$ (read "S minus A"). In the bracket notation,*

$$\overline{A} = A^c = S \setminus A = \{x : x \in S \text{ and } x \notin A\}.$$

The symbol $\notin$ stands for "does not belong to."

More generally, the set $B \setminus A$ is the set of elements of B that are not in A. Note that A and $B \setminus A$ are disjoint sets. (Why?) The sets $A \setminus B$ and $B \setminus A$ are illustrated in the Venn diagram, Figure 1.3. From the Venn diagram, we

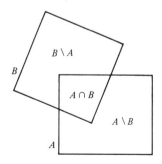

Figure 1.3

have the general result

$$A \cup B = (A \setminus B) \cup (A \cap B) \cup (B \setminus A).$$

This is a decomposition of the union of two sets into three disjoint subsets.

Example 1.2.8   For a certain population, define M to be the set of males and define T to be the set of people who have tuberculosis. The universal set S is the set of all people in the population. Then, $\overline{M} = \{$females$\}$, $\overline{T} = \{$people who do not have tuberculosis$\}$, $M \setminus T = \{$males who do not have tuberculosis$\}$, $T \setminus M = \{$females with tuberculosis$\}$. The reader should verify that $M \cup T = (M \setminus T) \cup (M \cap T) \cup (T \setminus M)$.

Example 1.2.9   In a certain environment, there are 100 coexisting animal species. Define A to be the set of species that feed by day and B to be the set of species that feed by night. Describe the sets $A \cup B$, $A \cap B$, $\overline{A}$, and $A \setminus B$. If 80 species feed by day and 30 species feed by night, how many species feed only by day? How many feed both by day and by night?

Solution: The set $A \cup B$ is the set of species that feed either by day or by night. Clearly, this is the set of all 100 species. Similarly,

$$A \cap B = \{\text{species that feed both by day and by night}\},$$
$$\overline{A} = \{\text{species that do not feed by day}\},$$
$$A \setminus B = \{\text{species that feed only by day}\}.$$

The species that feed only by day are the species that do not feed by night. There are 30 species that feed by night and, therefore, there are $100 - 30 = 70$ that feed only by day. Since 80 species feed by day, and 70 species feed only by day, there are $80 - 70 = 10$ species that feed both by day and by night.

   We are interested in the possible ways of dividing a set into disjoint subsets. For example, a collection of people can be divided into children and adults, males and females, and in many other ways. To formalize this idea, we introduce the following definition.

Definition 1.2.6   Partition of a Set   *A partition of a set A is a collection $A_1, A_2, \ldots, A_n$ of subsets of A that are mutually disjoint and whose union is A. This can be written $A_1 \cup A_2 \cup \cdots \cup A_n = A$ and $A_i \cap A_j = \emptyset$ if $i \neq j$ for $i, j = 1, 2, \ldots, n$.*

Figure 1.4 gives a Venn diagram representation of a partition of a set A.

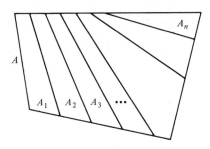

Figure 1.4

Suppose that $S_1, S_2, \ldots, S_n$ is a partition of the universal set S and suppose that A is any subset of S. Then the partition of S also gives rise to a partition of the subset A.

Theorem 1.2.2   *If $S_1, S_2, \ldots, S_n$ is a partition of the set S and if A is any subset of S, then $A \cap S_1, A \cap S_2, \ldots, A \cap S_n$ is a partition of A.*

Proof: We must prove that the sets $A \cap S_1$, $A \cap S_2, \ldots, A \cap S_n$ are mutually disjoint and that their union is A. Consider any element $x \in A$. Then x is an element of exactly one of the sets $S_1, S_2, \ldots, S_n$, since they form a partition of S. Therefore, x is an element of exactly one of the sets $A \cap S_1$, $A \cap S_2, \ldots, A \cap S_n$. Since these sets are all subsets of A, they are mutually disjoint and their union is A.

$$A = (A \cap S_1) \cup (A \cap S_2) \cup \cdots \cup (A \cap S_n).$$

Example 1.2.10   The division of the set of all species into the plant and animal kingdoms is a partition. A finer partition is given by the phyla. What other partitions of the set of all species are commonly used? (See Problem 1.2.5.)

Example 1.2.11   Among a group of 170 science students, 70 are taking at least one course in physics, 95 are taking biology, and 80 are taking mathematics. Suppose that 30 students are taking both mathematics and physics, 35 are taking both mathematics and biology, and 15 are taking both physics and biology. Suppose further that 5 students are taking all three subjects. How many students are taking exactly two of these three subjects?

Solution: Define P, B, and M to be the sets of students taking physics, biology, and mathematics courses, respectively. Then these sets have 70, 95, and 80 students. The sets $P \cap M$, $B \cap M$, and $P \cap B$ contain 30, 35, and 15 elements, respectively. The set $P \cap M \cap B$ contains 5 elements. The set $(P \cap M) \setminus (P \cap M \cap B)$ contains $30 - 5 = 25$ elements. These students study both mathematics and physics, but not biology. Similarly, $35 - 5 = 30$ students study only mathematics and biology and $15 - 5 = 10$ students study only physics and biology. We conclude that $25 + 30 + 10 = 65$ students are studying exactly two of the three subjects.

Problems 1.2

1. By means of a verbal argument and by using a Venn diagram, prove that $(A \cup B) \cup C = A \cup (B \cup C)$, where A, B, and C are any sets. If A is the set of adult males, B is the set of adult females, and C is the set of children in a population, what are the sets $A \cup B$, $B \cup C$, $(A \cup B) \cup C$, and $A \cup (B \cup C)$?

2. Prove that the complement $\overline{A}$ of a set A (relative to S) satisfies the following equations.
   (a) $A \cup \overline{A} = S$, $A \cap \overline{A} = \varnothing$.
   (b) $\overline{A \cap B} = \overline{A} \cup \overline{B}$, $\overline{A \cup B} = \overline{A} \cap \overline{B}$.
   (c) If $A \subset B$ and $C \subset B$, then $\overline{B} \subset \overline{A} \cap \overline{C}$.

3. By means of a Venn diagram or by a verbal argument, prove that $A \cap (B \cup C)$ is not in general equal to $(A \cap B) \cup C$.

4. Define I to be the set of all positive integers. Define A to be the set of even integers in I, B to be the set of integers in I divisible by 3, and C to be the set of integers in I divisible by 5.
   (a) What are the elements of $A \cap B$, $B \cap C$, and $A \cap B \cap C$?
   (b) Determine the elements of $D = \{x : x = n^2, n \in A \cap B \cap C, n < 20\}$.

5. Each living species is classified (according to one system of classification) into a kingdom, a phylum, a class, an order, a family, a genus, and a species. Draw a Venn diagram that illustrates this classification system.

6. Among a group of 1000 science students, 630 are taking at least one course in biology, 390 are taking chemistry, and 720 are taking mathematics. It is also known that 440 are taking both mathematics and biology, 250 both mathematics and chemistry, and 200 both biology and chemistry.

In addition, it is known that 130 students are taking all three subjects.
(a) Draw a Venn diagram that illustrates this problem.
(b) How many of the 1000 science students were taking no biology, chemistry, or mathematics?
(c) How many were taking only one of the three subjects?
(d) How many were taking exactly two of the subjects?

7. In a study of blood groups, 10,000 people were tested. Of these, 5500 were found to have the antigen A, 2500 the antigen B, and 3000 were found to have neither antigen. Define A, B, and O to be these three sets of people.
(a) Draw a Venn diagram that illustrates this problem.
(b) Describe in words the sets $A \cup B$, $A \cap B$, $A \cap O$, $(A \cup B) \cap O$, $\overline{A}$, and $\overline{A} \cap B$.
(c) How many people have both antigens A and B?

8. In a study of the effect of smoking on lung cancer, a large population of adults is divided into smokers and nonsmokers. The smokers are further divided into light, moderate, and heavy smokers. The population is also divided into those who have lung cancer and those who do not. Finally the population is divided into males and females. Draw a single Venn diagram that illustrates these three ways of partitioning the population.

1.3 Relations and Functions

The set operations defined in the previous section allow us to combine sets in various ways to form new sets. There is yet another way of combining sets to form a new set that is very useful in applications. Suppose that A and B are any two sets and suppose that a is an element of A and that b is an element of B. Then we can form the ordered pair (a, b) consisting of the element of A followed by the element of B. Two such ordered pairs are said to be *equal* if the elements of A in both pairs are the same and the elements of B in both pairs are the same. This means that $(a_1, b_1) = (a_2, b_2)$ if and only if $a_1 = a_2$ and $b_1 = b_2$. We now define the set of all such ordered pairs.

Definition 1.3.1   Cartesian Product of Two Sets   *Suppose that A and B are any two sets. The Cartesian product of A and B, written $A \times B$, is the set consisting of all ordered pairs (a, b), where $a \in A$ and $b \in B$.*

Example 1.3.1   Define $C = \{\text{state capitals}\}$ and $T = \{\text{two-digit numbers}\}$. Then the Cartesian product of these two sets is the set

$$C \times T = \{(c, t) : c \text{ is a state capital and } t \text{ is a two-digit number}\}.$$

One member of this set is (Boston, 75) and another is (Albany, 67). An interpretation of this example might be that the two-digit number represents the maximum temperature in the capital on a given day. Note that there are $5000 = 50 \times 100$ elements in the set $C \times T$. The actual maximum temperatures on a given day would give a 50-element subset of $C \times T$.

Example 1.3.2   The Cartesian Plane   Define $\mathbf{R}$ to be the set of real numbers. Then the Cartesian product $\mathbf{R} \times \mathbf{R}$ is the set of pairs (a, b) of real numbers. There is a well-known representation of the elements of $\mathbf{R} \times \mathbf{R}$ as points in the Cartesian plane. This representation is illustrated in Figure 1.5.

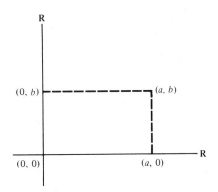

Figure 1.5

Example 1.3.3   Define M and W to be the set of all men and all women in a population. Then $M \times W$ is the set of all pairs (m, w), where m is a man and w is a woman. We may be particularly interested in those pairs in which there is some relationship between the man and the woman. For example, the set of of all husband–wife pairs and the set of all father–daughter pairs are well-defined subsets of the Cartesian product $M \times W$.

As in Examples 1.3.1 and 1.3.3, we are often interested in studying correspondences or relations between elements of A and elements of B.

Definition 1.3.2   Relation   *A relation R from a set A to a set B is a subset of A × B. The relation R satisfies R ⊂ A × B.*

Example 1.3.1 (continued)   Define $R = \{(c, t) : t$ is the maximum temperature in c on May 1 of this year$\}$. Then R is a relation from C to T. The relation R associates to each of the 50 capitals its temperature on May 1. For example, (Helena, 70) may be a typical element of R.

Example 1.3.2 (continued)   Define $L = \{(x, y) : y = 2x + 3\}$. Then L is a subset of **R** × **R** and it defines a relation from **R** to **R**. The points of L lie on a straight line in the Cartesian plane as drawn in Figure 1.6.

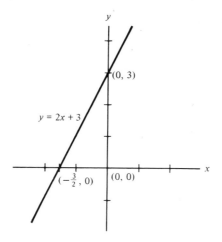

Figure 1.6

Example 1.3.3 (continued)   Define $R = \{(m, w) : m$ is the father of $w\}$. Then R is a subset of $M \times W$ that defines a relation from M to W. It associates to each male in the population the set of all his daughters. This set may be empty or it may have several elements. Typical elements of R may be (John Doe, Mary Doe) and (John Doe, Martha Doe).

The last example illustrates that a relation R can contain more than one pair corresponding to the same element of A. This means that, in general, a relation R does not associate to each element of A a unique element of B. The element $a \in A$ can appear in many pairs $(a, b_1), (a, b_2), \ldots, (a, b_k), \ldots$. On the other hand, in the first two examples of relations, specifying the element of the first set does determine the element of the second set uniquely. The maximum temperature of each capital on May 1 of this year is uniquely determined. In the second example, $y = 2x + 3$ is uniquely determined by x. This important special case of a relation is known as a function.

Definition 1.3.3   Function   *A function from a set A to a set B is a relation f from A to B with the property that, corresponding to each $a \in A$, there is a unique $b \in B$ such that $(a, b) \in f$.*

The element $b \in B$ associated to the element $a \in A$ by the function f is usually written $b = f(a)$. The set A on which the function f is defined is called the domain *of the function, written $D(f)$. The* range *$R(f)$ of the function f is the subset of B consisting of those elements b that can be written $b = f(a)$ for some $a \in A = D(f)$. We say that a function f is a* mapping *from the set A to the set B or from the domain of f to the range of f.*

How is this definition of functions related to the ordinary concept of functions? This is best illustrated by a number of examples.

Example 1.3.4   The subset of $\mathbf{R} \times \mathbf{R}$ denoted by $f = \{(x, y) : y = x^2\}$ defines a function that associates to every real number $x \in \mathbf{R}$ the real number $y = x^2 \in \mathbf{R}$. This function is more commonly written $f(x) = x^2$. The domain of f is $D(f) = \mathbf{R}$. The range of f is $R(f) = \{y : y \geq 0\}$, since x^2 is never negative.

Example 1.3.5   Describe the following functions as subsets of $\mathbf{R} \times \mathbf{R}$:

1. $f(x) = x^3$.

2. $g(x) = \dfrac{1}{1 + x^2}$.

3. $h(x) = \dfrac{2x}{1 + x^2}$.

Solution:
1. We write $f = \{x, y) : y = x^3\}$. As in Example 1.3.4, we have $D(f) = \mathbf{R}$. The range of f is $R(f) = \mathbf{R}$, since $f(x) = x^3$ takes on every real value for some x.

2. In this example, $g = \{(x, y) : y = 1/(1 + x^2)\}$. The domain of f is again $D(f) = \mathbf{R}$. The range of f is $R(f) = \{y : 0 < y \le 1\}$, since $0 < 1/(1 + x^2) \le 1$ for all real x.

3. In bracket notation, $h = \{(x, y) : y = 2x/(1 + x^2)\}$. The domain of f is $D(f) = \mathbf{R}$. The range of f is $R(f) = \{y : -1 \le y \le 1\}$, since $-1 \le 2x/(1 + x^2) \le 1$ for all real x. This can be shown by plotting the curve $y = 2x/(1 + x^2)$ in the Cartesian plane.

Example 1.3.6   Describe the functions (1) $f(x) = 1/(x - 1)$ and (2) $g(x) = \sqrt{x}$ as subsets of $\mathbf{R} \times \mathbf{R}$.

Solution:

1. We define f as the relation $f = \{(x, y) : x \in A \text{ and } y = 1/(x - 1)\}$. The function f is defined for all real numbers x except $x = 1$ (since we cannot divide by 0). Therefore, $D(f) = A = \mathbf{R} \setminus \{1\}$. The range of f is $R(f) = \mathbf{R} \setminus \{0\}$, the whole real line minus the point zero.

2. The square root of the real number x is not defined (as a real number) if x is negative. The domain of g is $D(g) = \mathbf{R}^+ = \{x : x \ge 0\}$, and the range of g is $R(g) = \mathbf{R}^+$. We adopt the convention that $\sqrt{x}$ denotes the positive square root of x. The negative square root is denoted by $-\sqrt{x}$. In bracket notation, therefore, the function $g = \{(x, y) : x \in \mathbf{R}^+, y = \sqrt{x}\}$.

The above examples illustrate that our definition of functions in terms of sets generalizes the ordinary concept of functions. The more general definition has many applications to biological problems in which the biological variable may not be a number or where the domain of the biological function contains elements that are not numbers.

Example 1.3.7   In a study of the spread of an infectious disease, 20 mice are kept in each of 10 compartments labeled 1 to 10. One mouse in each compartment is infected with the disease, and 2 days later the number of infected mice in each compartment is observed. Define the sets $I_1 = \{1, 2, 3, \ldots, 10\}$ and $I_2 = \{1, 2, 3, \ldots, 20\}$. Then a function $f \subset I_1 \times I_2$ is defined by the observations. If $a \in I_1$, define $b = f(a) \in I_2$ to be the number of mice in compartment a who are observed to be infected. For example, if 12 mice in the second compartment are infected, then $f(2) = 12$. The domain of f is I_1 and the range of f is a subset of I_2.

Example 1.3.8 In a study of bacterial growth, x units of a nutrient are supplied daily to a population of bacteria whose initial size is 10,000. This experiment is repeated six times with $x = 0, 1, 2, 3, 4,$ and 5. After 5 days, the bacteria populations are 0, 400, 4500, 9000, 25,000, and 70,000. This defines a function $f \subset A \times B$, where $A = \{0, 1, 2, 3, 4, 5\}$ and B is the set of real numbers. The range of f is $R(f) = \{0, 400, 4500, 9000, 25,000, 70,000\}$.

Problems 1.3

1. If $A = \{$men in a population$\}$, $B = \{$women$\}$, and $C = \{$children$\}$, describe the following sets.

   (a) $A \times B$.   (b) $A \times C$.   (c) $B \times C$.

2. For the sets of Problem 1, define relations corresponding to the ordinary family relations (husband–wife, father–child, mother–child) as subsets of $A \times B$, $A \times C$, and $B \times C$.

3. Nerve cells are connected in a network and function by sending impulses to other cells. Define N to be the set of all nerve cells in an animal. Define a nerve cell n_1 to be connected to a nerve cell n_2 if n_2 can receive impulses from n_1. Does this concept of connection define a relation in $N \times N$? Is this relation a function?

4. Define A to be the set of all living organisms. The classical species concept divides the set A into species consisting of organisms between whom an interchange of genes can occur (the interchanged genes occurring in offspring). Describe this concept of species as a relation in $A \times A$.

5. Describe the following functions as subsets of $\mathbf{R} \times \mathbf{R}$. Determine the domains and ranges of the functions.

   (a) $f(x) = \dfrac{1}{1 - x^2}$.   (b) $g(x) = 2^x$.   (c) $h(x) = 1 + \dfrac{2x}{1 + x^2}$.

6. If x and y are the temperature in Fahrenheit degrees and Celsius degrees, then $y = \frac{5}{9}(x - 32)$. Define this function as a subset of $\mathbf{R} \times \mathbf{R}$ and graph it as a subset of the Cartesian plane.

7. An infectious disease is introduced into a population P by one infected individual. Describe the spread of this disease as a relation R in the Cartesian product $P \times P$. Will this relation be a function in certain circumstances?

8. In a population P, define a relation $R \subset P \times P$ by $R = \{(a, b) : a \in P, b \in P,$ and b is of the same blood type as $a\}$. If $(a, b) \in R$, does $(b, a) \in R$? If $(a, b) \in R$ and $(b, c) \in R$, does $(a, c) \in R$?

9. Describe the following relations as subsets of $\mathbf{R} \times \mathbf{R}$. Draw the corresponding regions in the Cartesian plane. Which of these relations are functions?

   (a) $R_1 = \{(x, y) : y \leq x\}$.
   (b) $R_2 = \{(x, y) : x^2 + y^2 = 1\}$.
   (c) $R_3 = \{(x, y) : x^2 + y^2 \geq 1\}$.
   (d) $R_4 = \{(x, y) : y = x - x^2\}$.

1.4 The Mathematics of Counting: Permutations

The study of permutations (and combinations in the next section) is nothing more than a systematic study of the mathematics of counting. When we come to the study of probability, we will be very interested in the number of ways that an event can occur or in the number of ways that a set of objects can be arranged in order. For example, in how many ways can six people be arranged around a circular table? How many bridge hands are there? How many ways are there through a maze? If a gene in a diploid organism has four alleles, what is the corresponding number of genotypes?

The mathematics of counting is based on the following principle.

Theorem 1.4.1   Fundamental Principle of Counting   *Suppose that the set A_1 contains n_1 objects, the set A_2 contains n_2 objects, ..., and the set A_m contains n_m objects. Then the number of ways to choose one object from each of of the m sets is $n_1 n_2 \cdots n_m$.*

If the numbers $n_1, n_2, \ldots, n_m$ and m are not too large, it is possible to write out explicitly all the possible ways of choosing one object from each of the m sets. We illustrate the applications of the fundamental principle of counting by the following examples.

Example 1.4.1   In a certain environment, there are 14 species of fruit flies, 17 species of moths, and 13 species of mosquitoes. In how many ways can one species of each type be chosen?

Solution: In this example, $n_1 = 14$, $n_2 = 17$, $n_3 = 13$, and $m = 3$. By the fundamental principle of counting, there are $14 \cdot 17 \cdot 13 = 3094$ different ways of selecting one species of each type.

Example 1.4.2 Four classes of students each choose one representative on a committee. If the classes contain 47, 51, 54, and 55 students, in how many ways can the representatives be chosen?

Solution: The representative of the first class can be chosen in 47 ways. Similarly, for each of the other classes, the choices can be made in 51, 54, and 55 ways. By the fundamental principle of counting, the number of ways of choosing one representative from each class is $47 \cdot 51 \cdot 54 \cdot 55 = 7,119,090$ ways.

Suppose now that a set contains n objects. We are often interested in arranging the objects in a definite order. To order the n objects, we select one object to be the "first," then choose another to be the "second," and so on. Each ordering is called a permutation.

Definition 1.4.1   Permutation of n Objects *A permutation of n objects is an ordering of the n objects, that is, an arrangement of the n objects in a definite order.*

Example 1.4.3 List all the permutations of the four letters w, x, y, and z.

Solution: The possible permutations are $wxyz$, $wxzy$, $wyxz$, $wyzx$, $wzxy$, $wzyx$, $xwyz$, $xwzy$, $xywz$, $xyzw$, $xzwy$, $xzyw$, $ywxz$, $ywzx$, $yxwz$, $yxzw$, $yzxw$, $zwxy$, $zwyx$, $zxwy$, $zxyw$, $zywx$, and $zyxw$. By counting, we see that there are 24 permutations of the four letters.

Example 1.4.4 Three different chemicals, C_1, C_2, and C_3, are kept in three test tubes in a test tube rack. List the possible orders of the three chemicals in the rack.

Solution: The possible orders or permutations are $C_1 C_2 C_3$, $C_1 C_3 C_2$, $C_2 C_1 C_3$, $C_2 C_3 C_1$, $C_3 C_1 C_2$, and $C_3 C_2 C_1$. We conclude that there are six permutations of these three objects.

Before proving a general result about the number of permutations of n objects, we must introduce some notation. The symbol $n!$ (read n factorial) is used to represent the product $n(n-1)(n-2) \cdots 3 \cdot 2 \cdot 1$. This symbol is defined when n is a positive integer and, by convention, we will agree that $0! = 1$. Note that $1! = 1$, $2! = 2$, $3! = 6$, $4! = 24$, and so on. If n is a large positive integer, $n!$ is extremely large. The frequent occurrence of the factorial symbol in counting problems is a consequence of the following theorem.

Theorem 1.4.2   *The number of permutations of n objects is n!*

Proof: To arrange the n objects in a definite order, we choose one object to be the "first." There are n ways to make this choice. From the $(n - 1)$ remaining objects, we choose the second object. There are $(n - 1)$ ways of doing this. Repeating this process, the kth object can be chosen in $(n - k + 1)$ ways for $k = 1, 2, 3, \ldots, n$. By the fundamental principle of counting, there are $n! = n(n - 1)(n - 2) \cdots 3 \cdot 2 \cdot 1$ permutations of n objects.

Example 1.4.5   In how many ways can a group of six persons arrange themselves (1) in a row, (2) around a circular table?

Solution:
1. This is the number of permutations of six objects, or $6! = 720$.
2. To order the six persons in a circle, choose one person arbitrarily and order the remaining five persons relative to this person. This can be done in $5! = 120$ ways.

Example 1.4.6   Eight laboratory animals are to be ranked according to their abilities to perform certain tasks. Assuming that there are no ties, how many rankings are possible?

Solution: There are $8! = 40,320$ orderings or rankings according to abilities.

Very often, we are not interested in all possible orderings of the n objects but in the possible orderings of some subset of the n objects. For example, we may be interested in the number of ways of choosing k objects from n objects, where the k objects are to be chosen in a definite order.

Example 1.4.7   Three rats are to be chosen from a group of nine and placed in three cages, marked C_1, C_2, and C_3. In how many ways can this be done?

Solution: There are nine ways to choose the rat for cage C_1. There are then eight rats remaining, and one of these is chosen and placed in cage C_2. Finally, one of the seven remaining rats is placed in C_3. By the fundamental principle of counting, there are $9 \cdot 8 \cdot 7 = 504$ ways of doing this.

The preceding example suggests the following definition.

Definition 1.4.2   Permutation of n Objects Taken k at a Time
A permutation of n objects taken k at a time is any selection of k objects in a definite order from the n objects.

The following theorem generalizes the result of Example 1.4.7.

Theorem 1.4.3   *The number of permutations of n objects taken k at a time is*

$$n(n - 1)(n - 2) \cdots (n - k + 1) = \frac{n!}{(n - k)!}.$$

Proof: There are n ways to choose the first object, $(n - 1)$ ways to choose the second object, $\ldots$, $(n - k + 1)$ ways to choose the kth object. The result therefore follows from the fundamental principle of counting. Note that

$$n(n - 1)(n - 2) \cdots (n - k + 1) = n(n - 1)(n - 2) \cdots (n - k + 1)\frac{(n - k)!}{(n - k)!}$$

$$= \frac{n!}{(n - k)!}.$$

Example 1.4.8   How many four-letter "words" (not necessarily English) can be formed from the letters of the word "around"?

Solution: This problem involves permutation of six objects taken four at a time. By the previous theorem, the number of such permutations is

$$\frac{6!}{(6 - 4)!} = \frac{6!}{2!} = 6 \cdot 5 \cdot 4 \cdot 3 = 360.$$

Example 1.4.9   First, second, and third prizes are to be awarded in a competition among 20 persons. In how many ways can the prizes be distributed?

Solution: The number of ways is the number of permutations of 20 objects taken three at a time; that is,

$$\frac{20!}{(20 - 3)!} = \frac{20!}{17!} = 20 \cdot 19 \cdot 18 = 6840 \text{ ways.}$$

In some cases, not all the objects that are being permuted can be distinguished. For example, there are $3! = 6$ permutations of the letters ABB but, if the two B's are indistinguishable, the only distinct permutations are ABB, BAB, and BBA. We develop this idea by means of a more complicated example.

Example 1.4.10   A barn contains five stalls in a row for three cows and two horses. In how many different ways can the cows and horses be placed in the stalls if we do not distinguish among the cows and among the horses?

Solution: One way to solve this problem is simply to write out all the possibilities. If C stands for cow and H stands for horse; the possible orders are *CCCHH, CCHCH, CHCCH, HCCCH, HHCCC, HCHCC, HCCHC, CHHCC, CHCHC,* and *CCHHC.* By counting, we conclude that there are 10 such permutations. There is a more systematic way to solve problems of this type. There are $5! = 120$ permutations of five objects if all five objects are distinguished. If three of the objects cannot be distinguished, for each permutation of the five objects, there are a total of $3! = 6$ permutations which look exactly the same. To illustrate this, consider the ordering *CCCHH* and let us label the cows $C_1, C_2,$ and C_3. Then the six permutations are $C_1C_2C_3HH,$ $C_1C_3C_2HH, C_2C_1C_3HH, C_2C_3C_1HH, C_3C_1C_2HH,$ and $C_3C_2C_1HH$. These six permutations are not distinguished if we are only interested in the ordering of the cows relative to the horses. Therefore, the number of distinct permutations is $5!/3! = \frac{120}{6} = 20$. But the horses are also indistinguishable in the present problem; for example, $CCCH_1H_2$ is the same as $CCCH_2H_1$. We conclude that the number of distinct orderings is $\frac{20}{2} = 5!/3!2! = 10$. This agrees with the answer obtained by counting all possibilities.

Once the above reasoning is understood, the example can be easily generalized. Suppose that we have a set of n objects containing n_1 indistinguishable objects of type 1, n_2 indistinguishable objects of type 2,..., and n_k indistinguishable objects of type k. In Example 1.4.10, $n = 5, n_1 = 3,$ and $n_2 = 2$. In general, the set of n objects is partitioned into k subsets containing $n_1, n_2,$ ..., n_k objects with $n_1 + n_2 + \cdots + n_k = n$. If all objects were distinguished, there would be $n!$ permutations of the n objects. Let us define the *multinomial symbol* or *multinomial coefficient*

$$\binom{n}{n_1, n_2, \ldots, n_k}$$

to be the number of permutations of the n objects in which n_1 are indistinguishable, n_2 are indistinguishable, . . . , and n_k are indistinguishable. Then, by the reasoning of the example, we have the following result.

Theorem 1.4.4

$$\binom{n}{n_1, n_2, \ldots, n_k} = \frac{n!}{n_1! \, n_2! \cdots n_k!}.$$

Example 1.4.11   How many distinct permutations are there of the letters of the word "error"?

Solution: The five objects to be permuted consist of three (indistinguishable) r's, one e, and one o. The number of distinct permutations is therefore

$$\binom{5}{3, 1, 1} = \frac{5!}{3! \, 1! \, 1!} = 20.$$

Example 1.4.12   Three types of bacteria are cultured in nine test tubes. Three test tubes contain bacteria of the first type, four contain bacteria of the second type, and two contain bacteria of the third type. In how many distinct ways can the test tubes be arranged in a row in a test tube rack if we are interested only in the ordering of the bacteria types?

Solution: The set of nine test tubes is partitioned into three subsets containing three, four, and two indistinguishable objects, respectively. By Theorem 1.4.4, the number of distinct permutations is

$$\binom{9}{3, 4, 2} = \frac{9!}{3! \, 4! \, 2!} = 1260.$$

Problems 1.4

1. Consider the set of numbers 1, 3, 4, 6, 8, and 9. How many three-digit numbers can be formed from this set if repetition is not allowed? How many of these three-digit numbers are under 500? How many are odd?

2. Consider the set of numbers 1, 2, 3, 4, and 5. How many three-digit numbers can be formed from this set if repetition is allowed? How many of these numbers are under 500? How many are divisible by 111?

3. Determine integers n that satisfy the following equations.

   (a) $(n + 1)! = 72(n - 1)!$.   (b) $\begin{pmatrix} n \\ n - 2, 1, 1 \end{pmatrix} = 9n$.

   (c) $\begin{pmatrix} 3n \\ n, n, n \end{pmatrix} = 6$.

4. Evaluate the following multinomial coefficients.

   (a) $\begin{pmatrix} 5 \\ 2, 2, 1 \end{pmatrix}$.   (b) $\begin{pmatrix} 7 \\ 5, 2, 0 \end{pmatrix}$.   (c) $\begin{pmatrix} 7 \\ 4, 2, 1 \end{pmatrix}$.   (d) $\begin{pmatrix} 6 \\ 6, 0, 0 \end{pmatrix}$.

   (e) $\begin{pmatrix} 5 \\ 2, 1, 2 \end{pmatrix}$.   (f) $\begin{pmatrix} 17 \\ 15, 1, 1 \end{pmatrix}$.   (g) $\begin{pmatrix} 17 \\ 14, 2, 1 \end{pmatrix}$.   (h) $\begin{pmatrix} 17 \\ 13, 2, 2 \end{pmatrix}$.

5. How many distinct permutations are there of the letters of the following words?

   (a) letters.   (b) distinct.   (c) following.

6. Five drugs are used in the treatment of a disease. It is believed that the sequence in which the drugs are used may have a considerable effect on the results of treatment. In how many different orders can the drugs be administered?

7. Two chess players, A and B, play 12 games. In how many ways can the outcome be four wins for A, four wins for B, and four draws?

8. A biologist attempts to classify 46,200 species of insects by assigning to each species three initials from the alphabet. Will the classification be completed? What is the number of initials that should be used?

9. It is estimated that there are 2 million species of insects, 1 million species of plants, 20,000 species of fish, and 8700 species of birds. If one species from each of these four categories is to be chosen for a comparative study, in how many ways can this be done?

10. How many different "words" of three letters can be formed from the four letters A, U, G, and C if repetitions are allowed? (The words of the genetic

code are formed from triplets or codons of four bases: adenine, uracil, guanine, and cytosine.)

11. Ten laboratory rats are to be ranked according to their abilities to learn five different tasks. A rat is given a score of 3, 2, 1, or 0 points for each task if the task is learned in one, two, three, or more training periods.
    (a) How many different total scores are possible for a particular rat?
    (b) Assuming there are no ties, how many rankings are possible?

12. In an African religion, a professional diviner appeals to the god by reciting a verse. To choose the appropriate verse, 16 smooth palm nuts are grasped by the diviner between both hands. He then attempts to grasp them all in his right hand. Since palm nuts are relatively large, they are difficult to grasp in one hand. If one or two nuts remain in the left hand, this number is recorded. This procedure is repeated eight times, producing a sequence of eight numbers which determines the verse that the diviner recites. How many verses must the diviner know?

13. In a comparative study of bird species, five characteristics are to be examined. If there are six recognizable differences in each of three characteristics and five recognizable differences in each of the remaining two characteristics, what is the maximum number of different groups of species that could be distinguished by these five characteristics?

14. Suppose that, in Problem 13, there are 8000 bird species to be classified. Is it possible that each species can be uniquely distinguished by the above five characteristics?

1.5 The Mathematics of Counting: Combinations

In the counting problems we considered in Section 1.4, we were interested in the number of orderings of a set of objects. There is another type of counting problem in which the ordering of objects is not relevant. For example, we may be interested in choosing 100 people from a population of 1000 in order to conduct an experimental study. The order in which the people are chosen is probably not of interest. Instead we are interested in the number of possible ways that the group of 100 people may be chosen. To see how this can be calculated, consider the following simple example.

Example 1.5.1 From a group of five mice, three are to be chosen without regard to order. In how many ways can this be done?

Solution: If the order in which the mice were chosen was relevant, the number of ways would be the number of permutations of five objects taken three at a time. There are $5!/(5-3)! = 5!/2! = 60$ such permutations. But in many of these permutations, the same three mice have been chosen. If the mice are labeled M_1, M_2, M_3, M_4, and M_5, then the permutations $M_1M_2M_3$, $M_1M_3M_2$, $M_2M_1M_3$, $M_2M_3M_1$, $M_3M_1M_2$, and $M_3M_2M_1$ correspond to the same choice of three mice. Therefore, in order to calculate the number of different choices, we must divide the number of permutations $5!/2!$ by the number of permutations of three objects. The number of different choices of three mice is $5!/2!3! = 10$. These can be explicitly written $M_1M_2M_3$, $M_1M_2M_4$, $M_1M_2M_5$, $M_1M_3M_4$, $M_1M_3M_5$, $M_1M_4M_5$, $M_2M_3M_4$, $M_2M_3M_5$, $M_2M_4M_5$, and $M_3M_4M_5$. The reader should verify that these are the only distinct choices.

This leads to the following important definition.

Definition 1.5.1 Combination of n Objects Taken k at a Time
A combination of n objects taken k at a time is any selection of k of the n objects without regard to order.

The symbol $\binom{n}{k}$ *is used to denote the number of combinations of n objects taken k at a time. This symbol is usually called the* binomial symbol *or* binomial coefficient.

Theorem 1.5.1

$$\binom{n}{k} = \frac{n!}{k!(n-k)!}.$$

Proof: The proof is a generalization of the discussion in Example 1.5.1. From Theorem 1.4.2, there are $n!/(n-k)!$ permutations of n objects taken k at a time. Consider a particular permutation of this type. If we disregard order among the k objects, there are $k!$ permutations which cannot be distinguished from the original permutation. Therefore, the number of combinations of n objects taken k at a time is the number of permutations divided by $k!$. This implies that

$$\binom{n}{k} = \frac{n!}{k!(n-k)!},$$

and the proof is complete.

Example 1.5.2 Four strains of bacteria are chosen from eight available strains for a growth experiment. In how many ways can this be done?

Solution: This is the number of ways of choosing four objects from eight objects without regard to order. This number is

$$\binom{8}{4} = \frac{8!}{4!4!} = 70 \text{ ways.}$$

Example 1.5.3 Six boys and 11 girls in a class are suspected to have an infectious disease. Blood samples are to be taken from 2 of the boys and 2 of the girls to test for the disease. In how many ways can this be done?

Solution: There are

$$\binom{6}{2} = \frac{6!}{2!4!} = 15$$

ways to choose the two boys and

$$\binom{11}{2} = \frac{11!}{2!9!} = 55$$

ways to choose the two girls. By the fundamental principle of counting, there are $15 \cdot 55 = 825$ ways to choose two boys and two girls.

Example 1.5.4 A committe contains 12 members. A minimum quorum at meetings of this committee consists of 8 members.

1. In how many ways can a minimum quorum occur?
2. In how many ways can a quorum occur?

Solution: Since order is obviously irrelevant, this is a problem involving combinations.

1. There are

$$\binom{12}{8} = \frac{12!}{8!4!} = 495 \text{ ways}$$

of choosing eight members.

2. A quorum is 8, 9, 10, 11, or 12 members. The number of ways that a quorum can occur is

$$\binom{12}{8} + \binom{12}{9} + \binom{12}{10} + \binom{12}{11} + \binom{12}{12} = 495 + 220 + 66 + 12 + 1$$

$$= 794 \text{ ways.}$$

Example 1.5.5   A laboratory cage contains eight white mice and six brown mice. Find the number of ways of choosing five mice from the cage if (1) they can be of either color, (2) three must be white and two must be brown, and (3) they must be of the same color.

Solution:

1. In this problem, color is irrelevant. There are

$$\binom{14}{5} = \frac{14!}{5!9!} = 2002 \text{ ways}$$

that 5 of the 14 mice can be chosen.

2. There are

$$\binom{8}{3} = 56 \text{ ways}$$

to choose three white mice and

$$\binom{6}{2} = 15 \text{ ways}$$

to choose two brown mice. Therefore, there are $56 \cdot 15 = 840$ ways to choose three white and two brown mice.

3. There are

$$\binom{8}{5} = 56 \text{ ways}$$

to choose five white mice and

$$\binom{6}{5} = 6 \text{ ways}$$

to choose five brown mice. Therefore, there are $56 + 6 = 62$ ways to choose five mice of the same color.

The number of combinations of n objects taken k at a time can be thought of as the number of ways of partitioning a set of n objects into two subsets containing k and $n - k$ objects. A more general problem is to choose several subsets of a given set A of n objects. Can we count the number of ways of doing this? For example, in how many ways can nine persons be divided into three committees of four, three, and two persons, respectively? To answer this question, recall the definition of a partition of a set given in Section 1.2. Consider all possible partitions of A into subsets $A_1, A_2, \ldots, A_m$, where A_1 contains n_1 objects, A_2 contains n_2 objects, $\ldots$, and A_m contains n_m objects. By the definition of a partition, we must have $n_1 + n_2 + \cdots + n_m = n$. In how many ways can such a partition be formed?

First, there are

$$\binom{n}{n_1} \text{ ways}$$

to choose the n_1 objects in A_1 from the n objects in A. Once A_1 is chosen, there are

$$\binom{n - n_1}{n_2} \text{ ways}$$

to choose the n_2 objects in A_2 from the $n - n_1$ objects remaining. Similarly, there are

$$\binom{n - n_1 - n_2}{n_3} \text{ ways}$$

to form A_3, and so on. This process is continued until all the subsets have been chosen. The general result follows from the fundamental principle of counting and is summarized in the following theorem.

Theorem 1.5.2   *Let A be any set containing n elements and let $n_1, n_2, \ldots, n_m$ be positive integers with $n_1 + n_2 + \cdots + n_m = n$. Then there exist*

$$\binom{n}{n_1}\binom{n - n_1}{n_2}\binom{n - n_1 - n_2}{n_3} \cdots \binom{n - n_1 - n_2 - \cdots - n_{m-1}}{n_m}$$

different partitions of A of the form $A_1, A_2, \ldots, A_m$, *where* A_1 *is a subset of* A *containing* n_1 *elements,* A_2 *contains* n_2 *elements,* $\ldots$, *and* A_m *contains* n_m *elements.*

The expression in the above theorem for the number of partitions can be greatly simplified. In fact, we can prove that

$$\binom{n}{n_1}\binom{n-n_1}{n_2}\binom{n-n_1-n_2}{n_3}\cdots\binom{n-n_1-n_2-\cdots-n_{m-1}}{n_m}$$
$$= \binom{n}{n_1, n_2, \ldots, n_m},$$

where

$$\binom{n}{n_1, n_2, \ldots, n_m} = \frac{n!}{n_1! n_2! \cdots n_m!}$$

is the multinomial coefficient of Theorem 1.4.4. The proof of this identity is left as an exercise for the reader.

Example 1.5.6 In how many ways can nine persons be formed into three committees of four, three, and two persons, respectively?

Solution: This is the number of ways of partitioning a set A of nine objects into subsets A_1, A_2, and A_3 containing four, three, and two objects, respectively. This number is given by the multinomial coefficient

$$\binom{9}{4, 3, 2} = \frac{9!}{4! 3! 2!} = 1260.$$

Example 1.5.7 In a comparative study, 16 persons suffering from a thyroid condition are to be divided into three groups of 12, 2, and 2 persons. In how many ways can this be done?

Solution: This is the number of ways of partitioning a set of 16 objects into three subsets containing 12, 2, and 2 objects. This number is given by the multinomial coefficient

$$\binom{16}{12,2,2} = \frac{16!}{12!2!2!} = 10{,}920.$$

Problems 1.5

1. Evaluate the following binomial coefficients.

(a) $\binom{7}{4}$.   (b) $\binom{7}{3}$.   (c) $\binom{19}{1}$.   (d) $\binom{25}{23}$.

(e) $\binom{13}{2}$.   (f) $\binom{13}{11}$.   (g) $\binom{8}{6}$.   (h) $\binom{8}{8}$.

2. A group of 20 people is to be divided into one group of 10 and two groups of 5. In how many ways can this be done? Express the answer as a product of binomial coefficients and as a single multinomial coefficient.

3. In a genetics experiment, 4 white peas, 7 red peas, and 5 pink peas are chosen for pollination from a sample of 10 white, 10 red, and 10 pink peas. (The color of the peas refers to the color of their flowers.) In how many ways can this be done?

4. Suppose that there are five highways joining city A and city B and three highways joining city B and city C. How many different routes join cities A and C? In how many ways can the round trip from city A to city C be made (a) without traveling the same route twice and (b) without traveling the same highway twice?

5. Prove the following identities relating binomial coefficients.

(a) $\binom{n}{k} = \binom{n}{n-k}$.   (b) $n^2 = \binom{n}{2} + \binom{n+1}{2}$.

(c) $\binom{n+1}{k} = \binom{n}{k} + \binom{n}{k-1}$.

(d) $\binom{n+2}{k+2} = \binom{n}{k+2} + 2\binom{n}{k+1} + \binom{n}{k}$.

(e) $\binom{n}{k}\binom{k}{r} = \binom{n}{r}\binom{n-r}{k-r}$. (Assume that $r \le k \le n$.)

6. If n, r, s, and t are nonnegative integers such that $t \leq s \leq r \leq n$, prove that

$$\binom{n}{r}\binom{r}{s}\binom{s}{t} = \binom{n}{t}\binom{n-t}{s-t}\binom{n-s}{r-s}.$$

Give a set theory interpretation of this result.

7. A ranch has 20 brown horses, 15 black horses, and 10 white horses. How many of these horses can say that they are of the same color as 10 other horses on this ranch?

8. How many "words" can be formed from the symbols X and Y if each word must contain at least one X and if the maximum length of the words is three letters with the order of the letters not being relevant? (The chromosomes X and Y determine sex. Normal females and males are XX and XY, but nondisjunction of the sex chromosomes may give rise to X, XXX, XXY, and XYY individuals).

9. A certain course covers 10 topics in probability and 8 topics in other subjects. The final exam will have five questions with at most one from any topic. Three questions will be on probability and two will be on other subjects. In how many ways can the topics examined by chosen?

10. Interchanges may occur between any two of the n chromosomes of a cell.
    (a) In how many ways can exactly one interchange occur?
    (b) In how many ways can exactly k interchanges occur?

11. Continuing Problem 10, suppose that a cell has four chromosomes. In how many ways can four or fewer interchanges occur between pairs of chromosomes in this cell?

12. In a diploid organism, genes occur in pairs on paired chromosomes. (See Section 9.3 for an explanation of terminology.) A gene locus with one allele A produces a single genotype AA. A gene locus with two alleles A_1 and A_2 produces three genotypes, A_1A_1, A_1A_2, and A_2A_2, the genotypes A_1A_2 and A_2A_1 being identical. Write out the genotypes corresponding to a gene locus with (a) three alleles and (b) four alleles.

13. Continuing with Problem 12, prove that a gene locus with n alleles A_1, $A_2, \ldots, A_n$ produces $(n^2 + n)/2$ genotypes.

14. A single gene locus with two alleles in a diploid organism produces three genotypes. If there are n such loci, there are 3^n diploid genotypes corresponding to these loci. Suppose that there are n loci each with three alleles. What is the number of diploid genotypes corresponding to these loci?

15. Various types of anemia in humans are believed to be due to recessive genes. For example, sickle cell anemia, ovalocytosis, and thalassemia are believed to be due to abnormal alleles at three different gene loci. Denote the dominant normal alleles at these loci by A_1, B_1, and C_1 and the recessive abnormal alleles at these loci by A_2, B_2, and C_2.
    (a) What is the total number of diploid genotypes corresponding to these loci?
    (b) What is the number of normal (nonanemic) genotypes corresponding to these loci?

16. Ten persons are to be chosen from a group of 10 men and 10 women.
    (a) What is the number of ways that the 10 persons can be chosen?
    (b) What is the number of ways that more men than women can be chosen?
    (c) If at least 8 of the persons chosen must be women, what is the number of ways that the 10 persons can be chosen?

1.6 The Binomial and Multinomial Theorems

In many areas of mathematics, it frequently happens that we encounter expressions of the form $(a + b)^n$ and $(a + b + c)^n$, where a, b, and c are real numbers and n is a positive integer. Expressions of this type will play a fundamental role in our study of probability in Chapter 2. The binomial and multinomial theorems give a systematic method to write out these nth powers as a sum of terms.

In this section, we will make extensive use of the *sigma notation*. This gives a convenient way to write lengthy sums of terms. We define

$$\sum_{k=0}^{n} f_k = f_0 + f_1 + f_2 + \cdots + f_n.$$

The left-hand side is read "the sum of f_k from $k = 0$ to $k = n$." The symbol Σ is the Greek letter sigma, which we will always use to denote a sum.

Example 1.6.1   Evaluate the following sums.

1. $\sum_{k=0}^{3} k.$

2. $\sum_{k=1}^{5} k^2.$

3. $\displaystyle\sum_{k=0}^{4} 2^{-k}$.

Solution:

1. $\displaystyle\sum_{k=0}^{3} k = 0 + 1 + 2 + 3 = 6$. This is read "the sum of k from $k = 0$

   to $k = 3$ is 6."

2. $\displaystyle\sum_{k=1}^{5} k^2 = 1^2 + 2^2 + 3^2 + 4^2 + 5^2 = 1 + 4 + 9 + 16 + 25 = 55$.

3. $\displaystyle\sum_{k=0}^{4} 2^{-k} = 2^{-0} + 2^{-1} + 2^{-2} + 2^{-3} + 2^{-4} = 1 + \frac{1}{2} + \frac{1}{4} + \frac{1}{8} + \frac{1}{16} =$

   $\frac{31}{16}$.

Using the sigma notation we can now state the first main result of this section.

Theorem 1.6.1   Binomial Theorem   *If a and b are real numbers and n is a positive integer, then the product $(a + b)^n$ can be written*

$$(a + b)^n = \sum_{k=0}^{n} \binom{n}{k} a^{n-k} b^k$$

$$= a^n + na^{n-1}b + \frac{n(n-1)}{2} a^{n-2}b + \cdots$$

$$+ \frac{n(n-1)\cdots(n-k+1)}{k!} a^{n-k}b^k + \cdots + b^n.$$

Proof: Consider the product $(a + b)^n = (a + b)(a + b)\cdots(a + b)$, where there are n factors of $(a + b)$. To form the product, we multiply together one term from each of the n factors and then add all products of this type. [For example, $(a + b)^2 = (a + b)(a + b) = a^2 + ab + ba + b^2 = a^2 + 2ab + b^2$.] A typical term in the product is $a^{n-k}b^k$, where k is an integer between 0 and n. The problem now is to determine the number of times this term occurs. But to produce the term $a^{n-k}b^k$, we must choose b exactly k times from the n factors. This is a problem in combinations. The number of ways of choosing k objects from n objects without regard to order is

$$\binom{n}{k}.$$

Therefore, in the expansion of $(a + b)^n$, the term $a^{n-k}b^k$ appears with coefficient

$$\binom{n}{k}.$$

This is true for $k = 0, 1, 2, \ldots, n$, and the theorem is proved.

Example 1.6.2   Calculate $(x + y)^5$.

Solution: From the binomial theorem

$$(x + y)^5 = \sum_{k=0}^{5} \binom{5}{k} x^{5-k} y^k$$

$$= \binom{5}{0} x^5 y^0 + \binom{5}{1} x^4 y^1 + \binom{5}{2} x^3 y^2 + \binom{5}{3} x^2 y^3 + \binom{5}{4} xy^4 + \binom{5}{5} x^0 y^5$$

$$= x^5 + 5x^4 y + 10x^3 y^2 + 10x^2 y^3 + 5xy^4 + y^5.$$

Example 1.6.3   What is the coefficient of $x^3 y^4$ in the expansion of $(x + y)^7$?

Solution: A typical term in the expansion is

$$\binom{7}{k} x^{7-k} y^k.$$

The $x^3 y^4$ term occurs when $k = 4$ and the coefficient is

$$\binom{7}{4} = 35.$$

Example 1.6.4   As a special case of the binomial theorem, we have the result

$$2^n = (1 + 1)^n = \sum_{k=0}^{n} \binom{n}{k} 1^{n-k} 1^k$$

$$= \sum_{k=0}^{n} \binom{n}{k} = \binom{n}{0} + \binom{n}{1} + \binom{n}{2} + \cdots + \binom{n}{k} + \cdots + \binom{n}{n}.$$

This formula has an interesting interpretation in terms of sets. We know that

$$\binom{n}{k}$$

is the number of subsets containing k elements of a set of n elements. Therefore,

$$\binom{n}{0} + \binom{n}{1} + \binom{n}{2} + \cdots + \binom{n}{n}$$

represents the total number of subsets of a set of n elements. The total number
of subsets is therefore 2^n. This gives another solution of Problem 1.1.5.

To calculate a product of the form $(a + b + c)^n$, we would guess that a
theorem similar to the binomial theorem must be true. The following result
gives the appropriate generalization.

Theorem 1.6.2   Multinomial Theorem   *If $a_1, a_2, \ldots, a_m$ are real numbers
and n is a positive integer, then the product $(a_1 + a_2 + \cdots + a_m)^n$ can be written*

$$(a_1 + a_2 + \cdots + a_m)^n = \sum_{n_1 + n_2 + \cdots + n_m = n} \binom{n}{n_1, n_2, \ldots, n_m} a_1^{n_1} a_2^{n_2} \cdots a_m^{n_m}.$$

*The sum is taken over all possible combinations of nonnegative integers which
add up to n.*

Proof.   A proof very similar to the proof of the binomial theorem will be
given. In forming the product $(a_1 + a_2 + \cdots + a_m)^n$, we multiply together
one term from each of the n factors and add up all such products. A typical
term in the product is

$$a_1^{n_1} a_2^{n_2} \cdots a_m^{n_m},$$

where $n_1 + n_2 + \cdots + n_m = n$. The coefficient of this term is the number of
ways of partitioning n objects into m subsets containing $n_1, n_2, \ldots, n_m$

objects. From Theorem 1.5.2, this coefficient is

$$\binom{n}{n_1, n_2, \ldots, n_m},$$

and the theorem is proved.

Example 1.6.5   Calculate $(a + b + c)^3$.

Solution: In order to apply the multinomial theorem, we must write out all possible sets of nonnegative integers n_1, n_2, and n_3 that satisfy $n_1 + n_2 + n_3 = 3$. These are 300, 210, 201, 120, 102, 111, 030, 003, 021, and 012. Thus there are 10 terms in the expansion

$$(a + b + c)^3 = \sum \binom{3}{n_1, n_2, n_3} a^{n_1} b^{n_2} c^{n_3}.$$

Evaluating the 10 multinomial coefficients, we have

$$(a + b + c)^3 = a^3 + 3a^2b + 3a^2c + 3ab^2 + 3ac^2 + 6abc + b^3$$
$$+ c^3 + 3b^2c + 3bc^2.$$

Example 1.6.6   What is the coefficient of ab^2cd^2 in the expansion of $(a + b + c + d)^6$?

Solution: In this example, $n = 6$, and we require the coefficient of the term corresponding to $n_1 = 1, n_2 = 2, n_3 = 1$, and $n_4 = 2$. This coefficient is

$$\binom{6}{1, 2, 1, 2} = \frac{6!}{1!2!1!2!} = 180.$$

There is a remarkable extension of the binomial theorem to the case in which n is not a positive integer. When n is a positive integer and a and b are real numbers, the binomial expansion of $(a + b)^n$ is

$$(a + b)^n = a^n + na^{n-1}b + \frac{n(n-1)}{2!}a^{n-2}b^2 + \frac{n(n-1)(n-2)}{3!}$$
$$\times a^{n-3}b^3 + \cdots + b^n.$$

This is a finite sum with $(n + 1)$ terms. Note that we could write

$$a + b = a\left(1 + \frac{b}{a}\right) \quad \text{and} \quad (a + b)^n = a^n\left(1 + \frac{b}{a}\right)^n.$$

Therefore, to determine $(a + b)^n$, we need only to determine $(1 + b/a)^n$ and multiply by a^n.

For simplicity, define $x = b/a$ and consider the binomial expansion of $(1 + x)^n$.

$$(1 + x)^n = 1 + nx + \frac{n(n - 1)}{2!}x^2 + \frac{n(n - 1)(n - 2)}{3!}x^3 + \cdots.$$

We have proved this formula for the case when n is a positive integer. What happens when n is not a positive integer? For example, does the binomial expansion have any meaning when $n = \frac{1}{2}$ or $n = -1$? For $n = \frac{1}{2}$, the binomial expansion of $(1 + x)^n$ is

$$(1 + x)^{1/2} = 1 + \tfrac{1}{2}x + \frac{(\frac{1}{2})(-\frac{1}{2})}{2!}x^2 + \frac{(\frac{1}{2})(-\frac{1}{2})(-\frac{3}{2})}{3!}x^3 + \cdots.$$

For $n = -1$, the binomial expansion of $(1 + x)^n$ is

$$(1 + x)^{-1} = 1 + (-1)x + \frac{(-1)(-2)}{2!}x^2 + \frac{(-1)(-2)(-3)}{3!}x^3 + \cdots$$

$$= 1 - x + x^2 - x^3 + \cdots.$$

The binomial expansions for $(1 + x)^{1/2}$ and $(1 + x)^{-1}$ are now infinite sums of terms or *infinite series*. We will not develop the properties of such infinite series here, except to note the remarkable fact that, if $-1 < x < 1$, the series for $(1 + x)^n$ given above does have a finite sum (even if n is not a positive integer). This gives us a useful way of evaluating $(1 + x)^n$. The condition that x must lie between -1 and $+1$ ensures that the terms in the infinite sum such as

$$\frac{n(n - 1)\cdots(n - k + 1)}{k!}x^k$$

are small if k is large.

To illustrate the applications of this extension of the binomial theorem, consider the following examples.

Example 1.6.7   Evaluate (1) $(1.10)^{1/2}$, (2) $(1.05)^{1/5}$, (3) $\sqrt{5}$.

Solution:

1. To evaluate $(1.10)^{1/2}$, we define $x = \frac{1}{10}$ and we must evaluate the terms of the binomial expansion of $(1 + x)^{1/2}$.

$$(1 + \tfrac{1}{10})^{1/2} = 1 + \tfrac{1}{2}(\tfrac{1}{10}) + \tfrac{1}{2!}(\tfrac{1}{2})(-\tfrac{1}{2})(\tfrac{1}{10})^2 + \tfrac{1}{3!}(\tfrac{1}{2})(-\tfrac{1}{2})(-\tfrac{3}{2})(\tfrac{1}{10})^3 + \cdots$$

$$= 1 + .05 - .00125 + .0000625 - \cdots$$

$$= 1.04881\ldots.$$

It is clear that we could evaluate $(1.10)^{1/2}$ to any number of decimal places.

2. $(1.05)^{1/5} = (1 + \tfrac{1}{20})^{1/5} = 1 + \tfrac{1}{5}(\tfrac{1}{20}) + \tfrac{1}{2!}(\tfrac{1}{5})(-\tfrac{4}{5})(\tfrac{1}{20})^2 + \cdots$

$$= 1 + .01 - .0002 + \cdots$$

$$= 1.0098\ldots.$$

3. To evaluate $\sqrt{5}$, we write $5 = 4 + 1$ and $\sqrt{5} = \sqrt{4 + 1} = \sqrt{4(1 + \tfrac{1}{4})}$. Therefore, $\sqrt{5} = 2\sqrt{1 + \tfrac{1}{4}}$, and we evaluate $(1 + \tfrac{1}{4})^{1/2}$ by the binomial expansion.

$$(1 + \tfrac{1}{4})^{1/2} = 1 + \tfrac{1}{2}(\tfrac{1}{4}) + \tfrac{1}{2!}(\tfrac{1}{2})(-\tfrac{1}{2})(\tfrac{1}{4})^2 + \tfrac{1}{3!}(\tfrac{1}{2})(-\tfrac{1}{2})(-\tfrac{3}{2})(\tfrac{1}{4})^3 + \cdots$$

$$= 1 + .125 - .007801 + .000975 - \cdots$$

$$= 1.118174\ldots.$$

Therefore, $\sqrt{5} = 2(1.118174\ldots) = 2.236\ldots.$

Example 1.6.8   Geometric Series   The binomial expansion of $(1 - x)^{-1}$ is

$$(1 - x)^{-1} = 1 + (-1)(-x) + \frac{(-1)(-2)}{2!}(-x)^2$$

$$+ \frac{(-1)(-2)(-3)}{3!}(-x)^3 + \cdots$$

$$= 1 + x + x^2 + x^3 + \cdots.$$

This particular infinite series occurs so frequently in applications that it is given a special name, the *geometric series*. When x lies between -1 and $+1$, the geometric series has the finite sum $1/(1 - x) = (1 - x)^{-1}$. We will encounter this series in the study of probability in Chapter 2.

Example 1.6.9   Evaluate the sum of the geometric series $1 + x + x^2 + x^3 + \cdots$ when (1) $x = \frac{1}{2}$, (2) $x = \frac{1}{3}$, (3) $x = \frac{2}{3}$.

Solution:

1. We   could   evaluate $1 + \frac{1}{2} + (\frac{1}{2})^2 + (\frac{1}{2})^3 + \cdots = 1 + .5 + .025 +$ $.0125 + \cdots$, but we know that this is the binomial expansion of $1/(1 - \frac{1}{2}) = 2$. Therefore, 2 is the exact sum of the geometric series when $x = \frac{1}{2}$.
2. When $x = \frac{1}{3}$, the sum is $1/(1 - \frac{1}{3}) = \frac{3}{2}$.
3. When $x = \frac{2}{3}$, the sum is $1/(1 - \frac{2}{3}) = 3$.

Problems 1.6

1. Determine the coefficients of $x^3 y^4$ in the expansions of
   (a) $(x + y)^7$.   (b) $(x - y)^7$.   (c) $(x^3 + y^2)^3$.   (d) $(x + y^4)^4$.
2. Determine the coefficients of $a^2 b^5$ in the expansions of
   (a) $(a + b)^7$.   (b) $(a + b)^4$.   (c) $(ab + b^3)^3$.   (d) $(a^2 + b^5)^2$.
3. Expand the following binomial and multinomial expressions.
   (a) $(x + 2y)^4$.   (b) $(x - 2y)^4$.
   (c) $(x + y + z)^2$.   (d) $(1 + x + y + z)^2$.
4. Determine the coefficients of $x^2 y^3 z^4$ in the expansions of
   (a) $(x + y + z)^9$.   (b) $(x - y - z)^9$.   (c) $(1 + x + y + z)^{15}$.
5. By setting $x = -1$ in the binomial expansion of $(1 + x)^n$, prove that

$$\binom{n}{0} - \binom{n}{1} + \binom{n}{2} - \binom{n}{3} + \cdots + (-1)^n \binom{n}{n} = 0.$$

6. Suppose that S is a set containing n elements. Prove that the number of subsets of S containing an even number of elements is equal to the number of subsets containing an odd number of elements. (*Hint:* Refer to Problem 5.)

7. For a certain disease, there are n recognized symptoms. A victim of the disease may have all these symptoms, none, or any intermediate number. Prove that there are 2^n different possible combinations of symptoms.

8. Continuing Problem 7, suppose that there are six recognized symptoms for a certain disease. The disease will be diagnosed to be present if the patient shows four or more of the symptoms. How many different combinations of symptoms will give rise to this diagnosis?

9. By setting $a_1 = a_2 = \cdots = a_m = 1$ in the multinomial expansion of $(a_1 + a_2 + \cdots + a_m)^n$, prove that

$$\sum_{n_1 + n_2 + \cdots + n_m = n} \binom{n}{n_1, n_2, \ldots, n_m} = m^n,$$

where the sum is taken over all possible combinations of nonnegative integers which add up to n. Give a set theory interpretation of this result. (This is a generalization of Example 1.6.4.)

10. Prove that the number of terms in the multinomial expansion of

$$(a_1 + a_2 + \cdots + a_m)^n$$

is

$$\binom{n + m - 1}{n} = \frac{(n + m - 1)!}{n!(m - 1)!}.$$

(*Hint:* Observe that the number of terms in the expansion is equal to the number of ways of dividing n objects into m subsets, some of which may be empty. To do this, arrange the n objects in a row and insert $m - 1$ dividers between the objects.)

11. A laboratory animal has a choice of m different foods, each available in standard units. The animal is allowed to eat a total of n units of the m foods. Ignoring the order in which the n units are eaten, what is the number of different diets available to the animal? Calculate this number when $m = 5$ and $n = 10$.

12. By means of a binomial expansion, evaluate
    (a) $(1.04)^{1/4}$.   (b) $(9.18)^{1/2}$.   (c) $(28)^{1/3}$.   (d) $(33)^{1/5}$.

13. Evaluate the sum of the geometric series $1 + x + x^2 + x^3 + \cdots$ when
    (a) $x = \frac{1}{5}$.   (b) $x = \frac{1}{4}$.   (c) $x = \frac{3}{4}$.   (d) $x = \frac{19}{20}$.

14. The *finite geometric series* $S(x) = 1 + x + x^2 + \cdots + x^n$ can be evaluated by calculating $(1 - x)S(x)$. Prove that $S(x) = (1 - x^{n+1})/(1 - x)$ when $x \neq 1$, and that $S(1) = n + 1$.

15. Evaluate the following finite geometric series.

(a) $1 + (\frac{1}{2}) + (\frac{1}{2})^2 + \cdots + (\frac{1}{2})^9$.

(b) $1 + 3 + 3^2 + 3^3 + 3^4$.

(c) $1 + \dfrac{1}{3} + \dfrac{1}{3^2} + \dfrac{1}{3^3} + \cdots + \dfrac{1}{3^{20}}$.

(d) $1 + \dfrac{1}{10} + \dfrac{1}{100} + \dfrac{1}{1000} + \cdots + \dfrac{1}{1,000,000}$.

16. Yeast is growing in a sugar solution in such a way that the weight of the yeast increases by 10 per cent every hour. If the initial weight is 1 gram, the weight after n hours is $w(n) = (1.10)^n$. Calculate the weights after (a) 10 minutes, (b) 20 minutes (c) 30 minutes.

17. The weight of a patient suffering from a disease decreased from 160 pounds to 140 pounds after 25 days. The weight $w(t)$ after t days was given by the equation $w(t) = 160(\frac{140}{160})^{t/25} = 160(\frac{7}{8})^{t/25}$. Determine approximate values for $w(1)$, $w(2)$, and $w(10)$. [*Hint:* Expand $(1 - \frac{1}{8})^{t/25}$ by the general binomial theorem.]

Discrete Probability | 2

2.1 Introduction

Almost every human or natural event contains uncertainties that one attempts to analyze in terms of an intuitive concept of probability. Daily newspapers give "odds" on all sorts of sporting events.

What is the chance of rain on the next first of July? How can this question be answered meaningfully? One way to estimate this would be to equate the chance to the proportion of firsts of July in the past when it did rain. A different problem would be to estimate the chance of rain tomorrow. A reasonable method would be to determine from meteorological records those days in the past which had similar weather patterns to that of today. The chance of rain tomorrow would be the proportion of days with similar weather for which it rained on the next day.

Historically, probability developed from a mathematical analysis of gambling games. Games of chance have been played for more than 5000 years. David[1] describes an early dice game involving astragali: small, roughly rectangular bones from the ankle of a mammal. Each throw of such

[1] F. N. David, "Dicing and Gaming," *Biometrika*, 42:1–15 (1955).

a bone had four possible outcomes (since the ends are small, the probability that the bone lands on an end is neglected). The four outcomes were not equally likely due to the lack of symmetry of the bones. This must have made the calculation of probabilities very difficult.

Rabinovitch[2] gives examples of the use of probability in the Talmud, the book of Jewish law written mostly before the year 1000. For example, Rabbi Judah the Prince (second century) ruled: "[A mother] had one child circumsized and he died; a second one and he died; one must not circumsize the third." The first death may have been coincidence. The second indicated a high correlation between circumcision and death. Thus the Rabbi's ruling may be reworded as "there is a high probability that a male child (of this mother) who is circumsized, will die."

It is evident that the theory of probability has an extremely wide range of applications from predicting the weather to gambling to genetics. It is a fundamental concept which arises in every area of human activity. In fact, the modern theories of the structure of matter itself are formulated in terms of probability. To include such a diversity of areas of application, it is necessary to develop a very general theory.

2.2 Sample Spaces and Equiprobable Spaces

The common feature of every situation involving probabilities is an action or occurrence which can take place in several ways. It may rain tomorrow or it may not. We analyze these situations by comparing the likelihood of occurrences of the various alternatives.

The theory of probability is developed as a study of the outcomes of trials of an experiment. An *experiment* is a phenomenon to be observed according to a clearly defined procedure. It may be as simple as tossing a coin and observing the outcome or as complex as choosing 1000 people from a large population and testing them for a certain disease.

Definition 2.2.1   Trial of an Experiment   *A trial is a single performance of the experiment.*

Definition 2.2.2   Sample Space of an Experiment   *The sample space S of an experiment is the set of all possible outcomes of one trial of the experiment. If the experiment has a finite number of outcomes, the sample space is said to be* finite.

[2] N. L. Rabinovitch, "Probability in the Talmud," *Biometrika*, 56:437–441 (1969).

In a typical experiment with a finite sample space, there are N possible outcomes. For example, the experiment of throwing a single six-sided die has $N = 6$ possible outcomes. It is reasonable in this example to assume that each of the six possible outcomes is equally likely. This is expressed by saying that the probability of each of the outcomes is one in six, or $\frac{1}{6}$.

We now introduce some additional terms, which will be useful for defining probabilities in general.

Definition 2.2.3   Elementary Event   *An elementary event in the sample space S is any of the possible outcomes of the experiment. In other words, an elementary event is any element of the set S.*

Definition 2.2.4   Event   *An event E is any subset of the sample space S.*

Since an elementary event is a one-element subset of S, we can give the alternative definition that an event E is the union of elementary events.

For the example of throwing a single six-sided die, a typical event is the event $E = \{2, 4, 6\}$ that an even number occurs. The sample space $S = \{1, 2, 3, 4, 5, 6\}$ is the event that either 1 or 2 or 3 or 4 or 5 or 6 occurs on the throw of the die. Of course, this always happens and, for this reason, S is called the *certain event*. Another special event is $\emptyset$, the empty subset of S. This event corresponds to an outcome that is impossible on a trial of an experiment and is therefore called the *impossible event*.

Example 2.2.1   Consider the experiment of choosing an integer from 1 to 4.

1. What is the sample space of this experiment?
2. What are the elementary events?
3. List all the events.

Solution:

1. The possible outcomes of this experiment are the integers 1, 2, 3, and 4. The sample space is $S = \{1, 2, 3, 4\}$, the set of all possible outcomes.
2. The elementary events are the outcomes 1, 2, 3, and 4. (If the elementary events are thought of as subsets of S, they should be written $\{1\}$, $\{2\}$, $\{3\}$, and $\{4\}$.)
3. The events are the subsets of S. These are $\emptyset$, $\{1\}$, $\{2\}$, $\{3\}$, $\{4\}$, $\{1, 2\}$, $\{1, 3\}$, $\{1, 4\}$, $\{2, 3\}$, $\{2, 4\}$, $\{3, 4\}$, $\{1, 2, 3\}$, $\{1, 2, 4\}$, $\{1, 3, 4\}$, $\{2, 3, 4\}$, and $S = \{1, 2, 3, 4\}$. Note that there are $16 = 2^4$ subsets of the set S containing four elements.

Example 2.2.2   Describe the sample spaces of the following experiments.

1. Observe the order in which a deer, an elk, and a moose arrive at a lake.
2. Choose four blood cells from plasma containing red and white blood cells and observe the number of red cells chosen.
3. Test three persons for cancer and observe the number who have this disease.

Solution:

1. The deer, elk, and moose can arrive at the lake in $3! = 6$ possible orders. The sample space of this experiment is $S = \{DEM, DME, EDM, MDE, EMD, MED\}$.
2. The possible outcomes are 0, 1, 2, 3, and 4 (the possible numbers of red cells). The sample space is $S = \{0, 1, 2, 3, 4\}$.
3. The number of people who have cancer can be 0, 1, 2, or 3. The sample space of this experiment is $S = \{0, 1, 2, 3\}$.

The sample spaces which are easiest to analyze in terms of probabilities are the equiprobable spaces.

Definition 2.2.5   Equiprobable Space   *Suppose that in an experiment all the elementary events in its sample space S are equally likely to occur. Then S is said to be an equiprobable space. If the experiment has N outcomes, then the probability of each elementary event is $1/N$. This is written $P(A) = 1/N$, where A is any elementary event in S.*

Example 2.2.3   Consider the experiment of throwing a six-sided die and observing the number of dots on the upturned face. If we assume that the six possible outcomes are equally likely, then the probability of each element-ary outcome is $\frac{1}{6}$.

Example 2.2.4   A laboratory rat is placed in a maze and must choose one of five possible paths. Only one of these paths leads to a reward of food. Assum-ing that the rat is equally likely to choose any path, what is the probability that the rat chooses the path which leads to the food?

Solution: The sample space of this experiment is $S = \{$path 1, path 2, path 3, path 4, path 5$\}$ and the probability of choosing any one path is $\frac{1}{5}$. Since only one path leads to the food, we have P (the rat finds the food) $= \frac{1}{5}$.

Definition 2.2.6   Probability of Events in a Finite Equiprobable Space
Suppose that S is a finite equiprobable space. The probability of an event E in S,
written P(E), is defined to be the number n of elementary events in E divided by
the number N of elementary events in S:

$$P(E) = \frac{n}{N} = \frac{\text{number of elementary events in } E}{\text{number of elementary events in } S}.$$

Since n is a nonnegative integer which is less than or equal to N (E, being
a subset of S, cannot contain more elements than S), it follows that $0 \le$
$P(E) \le 1$. If $P(E) = 0$, the event E must be the impossible event $\varnothing$. If $P(E) = 1$,
the event E is the certain event S.

It is easy to think of examples of equiprobable spaces. If a fair coin is
tossed, the probabilities of the two elementary outcomes, heads or tails,
are equal. In fact, this is what is meant by a "fair" coin. If a card is chosen at
random from a deck of 52 cards, the 52 elementary outcomes have equal
probabilities. Thus the probability of choosing a heart is $\frac{13}{52} = \frac{1}{4}$, since the
event of choosing a heart contains 13 elementary events, each elementary
event having probability $\frac{1}{52}$.

The phrase "at random," when used to describe an experiment, will
mean, in the absence of other information, that we are dealing with an equiprobable space. This means that if the experiment consists of choosing
something, then every choice is equally probable. For example, choose 5 mice
at random from a group of 50. In making the last statement, we are assuming
that none of the 50 mice have distinguishing characteristics that would make
it either more or less likely to be chosen.

Example 2.2.5   Suppose that in a group of 10 persons, 4 are male. If 2 are
chosen at random, what is the probability that (1) both are male, (2) both are
female, (3) 1 is male and 1 is female?

Solution: Let A be the event that both are male, B the event that both are
female, and C the event of 1 male and 1 female. The sample space S consists
of pairs of people and contains $\binom{10}{2} = 45$ elementary events. There are
$\binom{4}{2} = 6$ ways to choose 2 males among the four present, so $P(A) = \frac{6}{45} = \frac{2}{15}$.
Similarly, there are $\binom{6}{2} = 15$ ways to choose 2 females, so $P(B) = \frac{15}{45} = \frac{1}{3}$.

Finally, there are four ways to choose 1 male and six ways to choose 1 female, yielding $6 \times 4 = 24$ ways to choose one of each. Thus $P(C) = \frac{24}{45} = \frac{8}{15}$. We note that $P(A) + P(B) + P(C) = 1$.

By means of the set complement, we can associate to any event E another event.

Definition 2.2.7   Complementary Event   *Given an event E in S, the complementary event $\overline{E}$ is the set of elementary events in S but not in E. Then, $\overline{E} = S \setminus E = $ the set complement of E in S.*

In an equiprobable space, the probability of $\overline{E}$ can be calculated in terms of the probability of E. If E is the union of n elementary events, then $\overline{E}$ is the union of $N - n$ elementary events. Thus

$$P(\overline{E}) = \frac{\text{number of events in } \overline{E}}{\text{number of events in } S} = \frac{N - n}{N} = 1 - \frac{n}{N} = 1 - P(E).$$

This yields the useful result $P(E) + P(\overline{E}) = 1$.

The set operations of union and intersection also have interpretations in terms of events. If E_1 and E_2 are arbitrary events, $E_1 \cup E_2$ is the event that occurs if and only if at least one of E_1 and E_2 occurs on a trial of the experiment. Also, $E_1 \cap E_2$ is the event that occurs if and only if both E_1 and E_2 occur. If $E_1 \cap E_2 = \emptyset$, E_1 and E_2 are said to be *mutually exclusive* events. The complement $\overline{E_1}$ of an event E_1 is the event that occurs if and only if E_1 does not occur on a trial of the experiment. We note that $E \cap \overline{E} = \emptyset$; that is, an event and its complementary event are mutually exclusive events.

Problems 2.2

1. What are the sample spaces of the following experiments? What is the probability of an elementary event in each of these spaces?
    (a) Draw a card at random from a standard 52 card deck.
    (b) Choose at random an integer from 1 to 10.
    (c) Choose 10 persons at random from a group of 30.
    (d) Dial at random a seven-digit number.
2. Suppose that S is a finite equiprobable space and that A and B are any two subsets of S.

  (a) Prove that $P(\varnothing) = 0$ from the definition of P.

  (b) Prove that $P(A \cup B) = P(A) + P(B) - P(A \cap B)$ and conclude that $P(A \cup B) \leq P(A) + P(B)$.

  (c) If A and B are mutually exclusive, prove that $P(A \cup B) = P(A) + P(B)$.

3. (a) Give examples of two events, A and B, in a finite equiprobable space S which have the property that $P(A \cap B) = P(A)P(B)$.

  (b) By means of examples, show that $P(A \cap B)$ is not equal to $P(A)P(B)$ in general. (*Hint:* What happens when $A = B$?)

4. Numbers from 1 to 100 are written on slips of paper and placed in a bowl. After the bowl is thoroughly shaken, one of the slips of paper is drawn at random.

  (a) What is the probability that the number appearing is divisible by 3?

  (b) What is the probability that the number appearing is divisible by 3 and by 5?

5. A professor assigns 20 different grades to the 20 students in his class. Because of a computer error, the grades are distributed at random on the transcripts of the students.

  (a) What is the probability that every student receives the correct grade?

  (b) What is the probability that exactly 19 students receive their correct grades?

6. A cage contains six white mice and four brown mice. Consider the experiment of drawing three mice at random from the cage.

  (a) Describe the sample space of this experiment.

  (b) Calculate the probabilities of the four possible distributions of color (three white, two white and a brown, and so on).

7. From a standard deck of 52 cards, 5 cards are chosen at random.

  (a) How many events are in the sample space of this experiment?

  (b) If the 5 cards chosen are observed and another 5 cards are drawn from the remaining 47 cards, how many events are in the sample space of this second experiment?

8. Of 20 persons who contract influenza at the same time, 15 have completely recovered within 3 days. Suppose that 5 persons were chosen at random from the 20. What is the probability that all 5 recover within 3 days? That exactly 4 recover? That none do?

9. The 250 people on a small island have been exposed to an infectious disease. Consider an experiment which consists of determining the number N of people on the island who have the disease.

  (a) What is the sample space of the experiment?

  (b) Will this space be an equiprobable space?

(c) If A is the event that $N \leq 50$, what is $\overline{A}$?

(d) If B is the event $N \geq 40$, what are $\overline{B}$, $A \cup B$, and $A \cap B$?

10. Ten persons are to be chosen at random from a group of 10 men and 10 women.

(a) What is the probability that 10 men are chosen?

(b) What is the probability that more men than women are chosen?

(c) What is the probability that at least 8 men are chosen?

11. A chimpanzee is placed at a toy typewriter with the letters A, B, C, D, and E. If the chimpanzee types four keys at random,

(a) What is the probability that the word "BEAD" is typed?

(b) What is the probability that all typed letters are the same?

2.3 Finite Probability Spaces

Equiprobable spaces are not the only sample spaces in which we are interested. For example, consider the experiment of tossing a fair coin two times in succession and observing the number of heads that appear. The sample space $S = \{0, 1, 2\}$ of this experiment is not an equiprobable space because the event {one head} is twice as probable as the event {no heads} or the event {two heads}. The event {one head} can occur in two ways (HT and TH), while the event {no heads} can occur in only one way (TT) and the event {two heads} can occur in only one way (HH).

To define a probability space in general, we must define the probability of all the events of a sample space S. We will require that the probability of the impossible event $\varnothing$ be 0 and the probability of the certain event S be 1. The probability of any event E will then be a number between 0 and 1. Finally, if two events E_1 and E_2 are mutually exclusive, they cannot happen together. Therefore, the probability that one or the other of these two events will occur on a trial of the experiment should be the sum of the probabilities of the separate events. This requirement is written $P(E_1 \cup E_2) = P(E_1) + P(E_2)$ when $E_1 \cap E_2 = \varnothing$. We summarize this motivation with the following definition.

Definition 2.3.1 Probability Space *A probability space S is a sample space, together with a function P defined on the events of S satisfying the axioms*

1. $P(\varnothing) = 0$, $P(S) = 1$.
2. $0 \leq P(E) \leq 1$ *for any event $E \subset S$.*
3. *If $E_1, E_2 \subset S$ and $E_1 \cap E_2 = \varnothing$, then $P(E_1 \cup E_2) = P(E_1) + P(E_2)$.*

The function P is called a *probability function*, and the number $P(E)$ is called the *probability* of the event E. A probability space is a sample space together with a probability function.

Example 2.3.1   Consider the experiment of tossing a fair coin three times in succession and observing the number of heads that appear.

1. What is the probability of each of the elementary events?
2. What is the probability that all three tosses give the same outcome?
3. If E_1 is the event {fewer than two heads} and E_2 is the event {two heads}, determine $P(E_1 \cup E_2)$.

Solution: The coin can be tossed three times in succession in eight ways: $HHH, HHT, HTH, THH, HTT, THT, TTH$, and TTT. Each of these ways is equally probable with probability $\frac{1}{8}$ (since the coin is fair). The experiment in which we are interested has the sample space $S = \{0, 1, 2, 3\}$. The event {0 heads} can occur on only one way (TTT) and therefore has probability $\frac{1}{8}$. The event {1 head} can occur in three ways and has probability $\frac{3}{8}$.

1. We conclude that $P(0) = \frac{1}{8}$, $P(1) = \frac{3}{8}$, $P(2) = \frac{3}{8}$, and $P(3) = \frac{1}{8}$.
2. All three tosses give the same outcome in two ways (HHH and TTT). Therefore, the probability of this event is $\frac{2}{8} = \frac{1}{4}$.
3. The event E_1 occurs if either no heads or one head appear. This can occur in four ways and the probability is $P(E_1) = \frac{4}{8} = \frac{1}{2}$. We have calculated $P(E_2) = \frac{3}{8}$. Since E_1 and E_2 are mutually exclusive ($E_1 \cap E_2 = \varnothing$), we have

$$P(E_1 \cup E_2) = P(E_1) + P(E_2) = \frac{1}{2} + \frac{3}{8} = \frac{7}{8}.$$

In a finite probability space, it is only necessary to know the probabilities of the elementary events to determine the probabilities of all events. The following theorem gives an elementary method for calculating the probability of any event.

Theorem 2.3.1   *In a finite probability space, the probability of an event E is the sum of the probabilities of the elementary events contained in E.*

Proof: The event E is a union of elementary events $E_1, E_2, \ldots, E_k$; $E = E_1 \cup E_2 \cup \cdots \cup E_k$. To prove that

$$P(E) = P(E_1 \cup E_2 \cup \cdots \cup E_k) = P(E_1) + P(E_2) + \cdots + P(E_k),$$

we will use property (3) of the probability function. If we define $F = E_2 \cup E_3 \cup \cdots \cup E_k$, we have $E = E_1 \cup F$ and $P(E) = P(E_1) + P(F)$, since $E_1 \cap F = \varnothing$. Similarly, if we define $G = E_3 \cup \cdots \cup E_k$, have $F = E_2 \cup G$ and $P(F) = P(E_2) + P(G)$. Therefore, $P(E) = P(E_1) + P(E_2) + P(G)$. We continue this procedure until the result is proved.

The following theorem contains a number of results that are very useful for the calculation of probabilities.

Theorem 2.3.2   *For any events A_1, A_2, and A_3 in a finite probability space,*

1. $P(A_1 \setminus A_2) + P(A_1 \cap A_2) = P(A_1)$.
2. $A_1 \subset A_2$ *implies that* $P(A_1) \leq P(A_2)$.
3. $P(A_1 \cup A_2) = P(A_1) + P(A_2) - P(A_1 \cap A_2)$.
4. $P(A_1 \cup A_2 \cup A_3) = P(A_1) + P(A_2) + P(A_3) - P(A_1 \cap A_2)$
    $\qquad\qquad - P(A_1 \cap A_3) - P(A_2 \cap A_3) + P(A_1 \cap A_2 \cap A_3)$.

Proof: Result 1 follows from the identity $A_1 = (A_1 \setminus A_2) \cup (A_1 \cap A_2)$ (see Figure 2.1). Since $A_1 \setminus A_2$ and $A_1 \cap A_2$ are disjoint sets, result 1 follows from probability function axiom 3. Since $A_2 = A_1 \cup (A_2 \setminus A_1)$ (see Figure 2.2), and since these two sets are disjoint, we have $P(A_2) = P(A_1) + P(A_2 \setminus A_1)$. Since $P(A_2 \setminus A_1) \geq 0$, result 2 follows. To prove 3 we note that $P(A_1 \cup A_2) = P(A_1 \setminus A_2) + P(A_2 \setminus A_1) + P(A_2 \cap A_1) = P(A_1) - P(A_1 \cap A_2) + P(A_2) - P(A_2 \cap A_1) + P(A_2 \cap A_1)$, which yields the desired result. To prove 4, we let $A_4 = A_2 \cup A_3$. Then $A_1 \cap A_4 = A_1 \cap (A_2 \cup A_3) = (A_1 \cap A_2) \cup (A_1 \cap A_3)$, which implies that $P(A_1 \cap A_4) = P(A_1 \cap A_2) +$

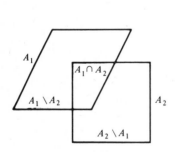

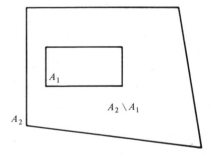

Figure 2.1

$A_1 = (A_1 \setminus A_2) \cup (A_1 \cap A_2)$

Figure 2.2

$A_1 \subset A_2, A_2 = A_1 \cup (A_2 \setminus A_1)$

$P(A_1 \cap A_3) - P(A_1 \cap A_2 \cap A_3)$.   Then $P(A_1 \cup A_4) = P(A_1) + P(A_4) - P(A_1 \cap A_4) = P(A_1) + P(A_2) + P(A_3) - P(A_2 \cap A_3) - [P(A_1 \cap A_2) + P(A_1 \cap A_3) - P(A_1 \cap A_2 \cap A_3)] = P(A_1) + P(A_2) + P(A_3) - P(A_2 \cap A_3) - P(A_1 \cap A_2) - P(A_1 \cap A_3) + P(A_1 \cap A_2 \cap A_3)$.

Theorem 2.3.2 contains results that will be used continually in the material which follows. The following examples are typical.

Example 2.3.2   In a large population of fruit flies, 25 per cent of the flies have an eye mutation, 50 per cent have a wing mutation, and 40 per cent of the flies with the eye mutation have the wing mutation.

1. What is the probability that a fly chosen at random from the population has at least one of the mutations?
2. What is the probability that the randomly chosen fly has the eye mutation but not the wing mutation?

Solution: Define E and W to be the events that a fly chosen at random has the eye mutation and the wing mutation, respectively.

1. The probability that the fly has one or both mutations is $P(E \cup W) = P(E) + P(W) - P(E \cap W)$. But $P(E) = .25$, $P(W) = .50$, and $P(E \cap W) = (.40)(.25) = .10$. Therefore, $P(E \cup W) = .25 + .50 - .10 = .65$.
2. The probability that the fly has the eye mutation but not the wing mutation is $P(E \setminus W) = P(E) - P(E \cap W) = .25 - .10 = .15$. We have used parts 1 and 3 of Theorem 2.3.2 in these calculations.

Example 2.3.3   In a tank containing three fish, A, B, and C, pellets of food are infrequently placed. Each time a pellet is dropped, the fish compete for it. Suppose that over a long period of time it is observed that either A or B is successful $\frac{1}{2}$ of the time and that either A or C is successful $\frac{3}{4}$ of the time.

1. What is the probability that A is successful?
2. Which fish is the best fed?

Solution: Let A be the event that fish A gets the food, and so on. Since only one of the fish can devour each particle of food, $P(A \cap B) = P(A \cap C) = P(B \cap C) = 0$. We are given $P(A \cup B) = \frac{1}{2}$ and $P(A \cup C) = \frac{3}{4}$. Applying part 3 of the theorem, we have $P(A \cup B) = P(A) + P(B) = \frac{1}{2}$ and $P(A \cup C) = P(A) + P(C) = \frac{3}{4}$. Hence $P(B) = \frac{1}{2} - P(A)$ and $P(C) = \frac{3}{4} - P(A)$. Since

$P(A) + P(B) + P(C) = 1$, we have $1 = P(A) + [\frac{1}{2} - P(A)] + [\frac{3}{4} - P(A)] = \frac{5}{4} - P(A)$, which gives $P(A) = \frac{1}{4}$. Similarly, $P(B) = \frac{1}{4}$ and $P(C) = \frac{1}{2}$. Fish C gets twice as much to eat as each of the other two.

Problems 2.3

1. Prove the following identities for events in the finite probability space S.
   (a) $P(\overline{A}) = 1 - P(A)$.
   (b) If $A_i \cap A_j = \varnothing$ for $i \neq j$, then $P(A_1 \cup A_2 \cup \cdots \cup A_r) = P(A_1) + P(A_2) + \cdots + P(A_r)$.

2. Let A and B be events of a finite probability space S such that $P(A \cap B) = \frac{1}{5}$, $P(\overline{A}) = \frac{1}{3}$, and $P(B) = \frac{1}{2}$. Determine (a) $P(A \cup B)$ and (b) $P(\overline{A} \cap \overline{B})$.

3. In a large population, 40 per cent have black hair, 40 per cent have brown hair, and 20 per cent have blonde hair. In this population, all the people with black hair have brown eyes, all the people with blonde hair have blue eyes, and half the brown-haired people have blue eyes, the other half having brown eyes. Let A_1, A_2, and A_3 be the events that a person has black, brown, and blonde hair, respectively, and let B_1 and B_2 be the events of brown and blue eyes, respectively.
   (a) Determine $P(A_1)$, $P(A_1 \cap B_1)$, and $P(A_1 \cap B_2)$.
   (b) Determine $P(B_1)$, $P(B_2)$, and $P(A_1 \cup B_2)$.
   (c) Describe the event $A_1 \cup A_2 \cup B_2$. Evaluate the probability of this event.

4. In a population of 2000 fruit flies, 250 exhibit a recessive wing characteristic W and 150 exhibit a recessive eye characteristic E. Suppose that 50 of the flies exhibit both characteristics. A fly is chosen at random from the population for a breeding experiment.
   (a) What is the probability that this fly exhibits W? E?
   (b) What is the probability that both W and E are present?
   (c) Calculate $P(E \cap \overline{W})$ and $P(\overline{E} \cap W)$.
   (d) Verify that $P(E \cup \overline{W}) = P(E) + P(\overline{W}) - P(E \cap \overline{W})$.

5. Nine polar bears have been tagged with numbers 1 to 9 in a study of their migration. Three bears are recaptured.
   (a) How many events are in the sample space of this experiment?
   (b) Let A_i be the event that the ith bear recaptured has an even number. What is $P(A_1)$? $P(A_2)$? $P(A_3)$?
   (c) Describe the events $A_1 \cup A_2$ and $A_1 \cup A_2 \cup A_3$.
   (d) Determine $P(A_1 \cap A_2)$, $P(A_1 \cap A_2 \cap A_3)$, and $P(A_1 \cup A_2 \cup A_3)$.

6. A certain plant population consists of individuals of three types labeled

AA, Aa, and aa. The numbers of each type are 200, 600, and 50, respectively. Suppose that a plant is chosen at random from this population.

(a) What is the probability that the plant is AA?

(b) What is the probability that the plant is either AA or Aa?

2.4 Conditional Probability

The probabilities that we assign to events depend on the information which is known about them. Between the extremes of no information and complete information, there are many levels of partial information which, if known, must be taken into account in the calculation of probabilities.

To illustrate this idea, consider the following three-part problem. Suppose a family has two children.

1. What is the probability that both children are boys?
2. If it is known that at least one of the children is a boy, what is the probability that both are boys?
3. If the oldest child is known to be a boy, what is the probability that both are boys?

   (In these problems, we will assume that either sex is equally likely at birth.)

All the parts of this problem look alike. After all, we are always asking for $P(BB)$, where BB denotes the event that the first child and the second child are boys. In part 1, the sample space consists of four equally likely events: $S_1 = \{BB, BG, GB, GG\}$. From the definition of probability in an equiprobable space, we have $P(BB) = \frac{1}{4}$. In part 2, the event GG has been eliminated from the same space since there must be at least one boy. Thus the sample space is now $S_2 = \{BB, BG, GB\}$ and $P(BB) = \frac{1}{3}$. Finally, the sample space for part 3 is $S_3 = \{BB, BG\}$ and $P(BB) = \frac{1}{2}$.

We now generalize the ideas in this example. Consider an experiment with probability space S. Suppose that A and E are two events contained in S. The conditional probability of an event A given E is defined as the probability that the event A has occurred when it is known that E has occurred. The notation $P(A \mid E)$ will be used to represent this conditional probability. In the example above (part 2) we are seeking $P(A \mid E)$, where E is the event that at least one of the children is a boy and A is the event of two boys. In part 1 we seek $P(A)$ and in part 3 we seek $P(A \mid F)$, where F is the event that the oldest child is a boy. As we have seen, all these probabilities are different.

To calculate $P(A \mid E)$ in general, it is necessary to know the probability that E occurs on a trial and the probability that A and E occur together on a trial. One procedure to determine whether $A \cap E$ has occurred is to first verify that E has occurred and, then, if E has indeed occurred, to verify that A has occurred. This suggests the formula

$$P(A \cap E) = P(E)P(A \mid E),$$

which will be used to define conditional probability.

Definition 2.4.1   Conditional Probability   *If E is an event in the probability space S with $P(E) > 0$, the conditional probability of an event A in S, given that E has occurred, is*

$$P(A \mid E) = \frac{P(A \cap E)}{P(E)}.$$

The conditional probability $P(A \mid E)$ can be interpreted as the "relative probability" of an event A relative to the event E. This is illustrated in Figure 2.3. $A \cap E$ is considered as an event in the sample space E.

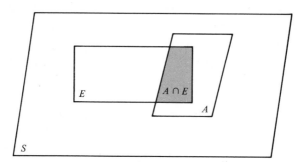

Figure 2.3. $A \cap E \subset E \subset S$

In an equiprobable space S, the following theorem allows us to calculate $P(A \mid E)$ by counting.

Theorem 2.4.1   *In an equiprobable space S with events A and E in S,*

$$P(A \mid E) = \frac{\text{number of elementary events in } A \cap E}{\text{number of elementary events in } E}.$$

Proof: We know that

$$P(A \cap E) = \frac{\text{number of elementary events in } A \cap E}{\text{number of elementary events in } S}$$

and

$$P(E) = \frac{\text{number of elementary events in } E}{\text{number of elementary events in } S}.$$

Since $P(A \mid E) = P(A \cap E)/P(E)$, the result follows.

Example 2.4.1   Suppose two fair dice are thrown.

1. What is the probability that at least one of the numbers showing is a 2?
2. If the sum is 6, what is the probability that one of the numbers is a 2?

Solution:

1. The sample space consists of all pairs (x, y), where x and y can assume any of the integer values 1 through 6. There are 36 such pairs. There are 11 pairs containing at least one 2. [The pair $(2, 2)$ contains two of them.] Thus P (at least one two) $= \frac{11}{36}$.
2. This is a conditional probability problem which we will solve two ways. Let A be the event of at least one 2 appearing and let E be the event of a sum of 6. There are five ways for the event E to occur: $(1, 5), (2, 4), (3, 3),$ $(4, 2),$ and $(5, 1)$. There are two ways for the event $A \cap E$ to occur: $(2, 4)$ and $(4, 2)$. Hence

$$P(A \mid E) = \frac{P(A \cap E)}{P(E)} = \frac{2/36}{5/36} = \frac{2}{5}.$$

Alternatively, we can consider the reduced sample space S_E consisting of all pairs that sum to 6. This is an equiprobable space containing five elements. Since the event $A \cap E$ contains two elementary events in S_E, we have $P(A \mid E) = \frac{2}{5}$.

Example 2.4.2   In a given fruit fly population, there are two mutations present: 25 per cent have a wing mutation, 15 per cent have an eye mutation, and 10 per cent have both. A fly is chosen at random.

1. If it has the wing mutation, what is the probability that it also has the eye mutation?
2. If it has the eye mutation, what is the probability that it also has the wing mutation?
3. What is the probability that it has at least one of the mutations?

Solution: Let W and E represent the events of a wing and eye mutation, respectively.

1. $P(E \mid W) = \dfrac{P(E \cap W)}{P(W)} = \dfrac{.10}{.25} = \dfrac{2}{5}$.

2. $P(W \mid E) = \dfrac{P(W \cap E)}{P(E)} = \dfrac{.10}{.15} = \dfrac{2}{3}$.

3. $P(W \cup E) = P(W) + P(E) - P(W \cap E) = .25 + .15 - .10 = \dfrac{3}{10}$.

Many of the applications of conditional probability are consequences of the following result, known as the *multiplication theorem of conditional probability.*

Theorem 2.4.2 *If A, B, and C are three events in the finite probability space S with $P(A) \neq 0$ and $P(A \cap B) \neq 0$, then*

$$P(A \cap B \cap C) = P(A)P(B \mid A)P(C \mid A \cap B).$$

Proof: Consider the two events $A \cap B$ and C. From the definition of conditional probability, $P(C \mid A \cap B) = P(A \cap B \cap C)/P(A \cap B)$ and $P(B \mid A) = P(A \cap B)/P(A)$. Therefore,

$$P(A)P(B \mid A)P(C \mid A \cap B) = P(A)\frac{P(A \cap B)}{P(A)}\frac{P(A \cap B \cap C)}{P(A \cap B)} = P(A \cap B \cap C)$$

(since the other terms cancel out).

Example 2.4.3 An aquarium contains three white fish, three red fish, and three blue fish. Three fish, chosen at random, are transferred to another aquarium. What is the probability that the three transferred fish are white?

Solution: There are two methods to solve this problem. The first method is simply to count all possibilities. There are

$$\binom{9}{3} = 84 \text{ ways}$$

to choose three fish from nine. Since there is only

$$\binom{3}{3} = 1 \text{ way}$$

to choose three white fish, the required probability is $\frac{1}{84}$. The second method uses the multiplication rule of conditional probability. Suppose that we choose the three fish one after the other. Define A, B, and C to be the events that the first, second, and third fish chosen are white. The probability that all selected fish are white is $P(A \cap B \cap C) = P(A)P(B \mid A)P(C \mid A \cap B)$. Clearly, $P(A) = \frac{3}{9} = \frac{1}{3}$, since there are three white fish among the nine. If the first fish chosen is white, there will be two white fish among the remaining eight fish. This means that $P(B \mid A) = \frac{2}{8} = \frac{1}{4}$. By similar reasoning, $P(C \mid A \cap B) = \frac{1}{7}$. Therefore, $P(A \cap B \cap C) = \frac{1}{3} \cdot \frac{1}{4} \cdot \frac{1}{7} = \frac{1}{84}$. With experience, this second method is much easier to apply than the counting method.

Example 2.4.4 On two ranches, A and B, each having 1000 cattle, there is an outbreak of hoof and mouth disease. The proportions of infected cattle are $\frac{1}{5}$ and $\frac{1}{4}$, respectively. One cow is chosen at random.

1. What is the probability that the chosen cow comes from ranch A and has the disease?
2. If 70 per cent of the infected cattle on each ranch are less than 1 year old, what is the probability that the chosen cow comes from ranch B, has the disease, and is more than 1 year old?

Solution:

1. Define $H = \{$cow has hoof and mouth disease$\}$. Then the required probability is

$$P(A \cap H) = P(A)P(H \mid A) = \frac{1000}{2000} \cdot \frac{1}{5} = \frac{1}{10}.$$

2. Define C to be the event that a cow chosen at random is more than 1 year old. The required probability is

$$P(B \cap H \cap C) = P(B)P(H \mid B)P(C \mid B \cap H) = \tfrac{1}{2} \cdot \tfrac{1}{4} \cdot \tfrac{3}{10} = \tfrac{3}{80}.$$

The multiplication theorem of conditional probability can be generalized easily to more than three events.

Theorem 2.4.3   *If $A_1, A_2, \ldots, A_n$ are events in the finite probability space S, and $P(A_1 \cap A_2 \cap \cdots \cap A_{n-1}) \neq 0$, then*

$$P(A_1 \cap A_2 \cap \cdots \cap A_n) = P(A_1)P(A_2 \mid A_1)P(A_3 \mid A_1 \cap A_2) \cdots$$
$$P(A_n \mid A_1 \cap A_2 \cap \cdots \cap A_{n-1}).$$

Proof: Define $B = A_1 \cap A_2 \cap \cdots \cap A_{n-1}$. Then $P(A_1 \cap A_2 \cap \cdots \cap A_n) = P(A_n \cap B) = P(B)P(A_n \mid B) = P(A_1 \cap A_2 \cap \cdots \cap A_{n-1})P(A_n \mid A_1 \cap A_2 \cdots \cap A_{n-1})$. This procedure is continued by defining $C = A_1 \cap A_2 \cap \cdots \cap A_{n-2}$ to find that $P(A_1 \cap A_2 \cap \cdots \cap A_{n-2} \cap A_{n-1}) = P(A_1 \cap A_2 \cap \cdots \cap A_{n-2}) \times P(A_{n-1} \mid A_1 \cap A_2 \cap \cdots \cap A_{n-2})$. After a finite number of steps, the required result is proved.

The conditional probability $P(A \mid B)$ gives the probability that A occurs when it is known that B occurs. It is natural to ask under what circumstances the probability of the event A is not affected by the occurrence of the event B.

Definition 2.4.2   Two Independent Events   *The events A and B in the finite probability space S are said to be independent if $P(A \mid B) = P(A)$. In other words, the probability that A occurs is not affected by the occurrence (or nonoccurrence) of B.*

Since $P(A \mid B) = P(A \cap B)/P(B)$, we conclude that $P(A) = P(A \cap B)/P(B)$ if A and B are independent events. This implies that $P(A \cap B) = P(A)P(B)$. This equation is an equivalent definition of independent events. If A and B are independent, we also have

$$P(B \mid A) = \frac{P(A \cap B)}{P(A)} = \frac{P(A)P(B)}{P(A)} = P(B).$$

Example 2.4.5 Suppose that a coin is tossed twice. Define A to be the event of a head on the first toss and B to be the event of a head on the second. The sample space $S = \{HH, HT, TH, TT\}$ of this experiment is an equiprobable space. The event $A \cap B$ is the event HH with probability $P(A \cap B) = \frac{1}{4}$. The events A and B have probabilities $P(A) = P(B) = \frac{1}{2}$. We note that $P(A \mid B) = P(A \cap B)/P(B) = \frac{1}{4}/\frac{1}{2} = \frac{1}{2} = P(A)$ and therefore that A and B are independent events.

Example 2.4.6 In a study of lung disease, 10,000 people over the age of 60 are examined. It is found that 4000 of this group have been steady smokers. Among the smokers 1800 have serious lung disorders. Among the non-smokers 1500 have serious lung disorders. Are smoking and lung disorder independent events?

Solution: Define A to be the event that a person chosen at random has been a steady smoker and B to be the event that the person has a serious lung disorder. Then $P(A) = 4{,}000/10{,}000 = .4$ and $P(B) = 3{,}300/10{,}000 = .33$. The conditional probability of smoking given a lung disorder is

$$P(A \mid B) = \frac{P(A \cap B)}{P(B)} = \frac{1{,}800}{10{,}000} \cdot \frac{10{,}000}{3{,}300} \approx .55 \neq P(A).$$

We conclude that A and B are not independent events.

Suppose that two events A and B are mutually exclusive ($A \cap B = \varnothing$). Then they are independent if and only if the probability of at least one of them is zero. To see this, we note that $0 = P(\varnothing) = P(A \cap B) = P(A)P(B)$ if A and B are independent. This serves to point out that independent events are not necessarily mutually exclusive events. On the other hand, there are mutually exclusive events which are not independent, so these two concepts are really very different.

We can extend the definition of independence to more than two events.

Definition 2.4.3 Three Independent Events *Three events A_1, A_2, and A_3 in the finite probability space are said to be independent if $P(A_1 \cap A_2 \cap A_3) = P(A_1)P(A_2)P(A_3)$ and if the three events are pairwise independent; that is, $P(A_1 \cap A_2) = P(A_1)P(A_2)$, $P(A_1 \cap A_3) = P(A_1)P(A_3)$, and $P(A_2 \cap A_3) = P(A_2)P(A_3)$.*

The following example shows that three events may be pairwise independent but not independent.

Example 2.4.7 Suppose that two cages each contain three white mice and three black mice. If one mouse is chosen from each cage, let A_1 be the event that the mouse taken from the first cage is white, A_2 the event that the mouse taken from the second cage is black, and A_3 the event that both mice are of the same color. Then, $P(A_1) = P(A_2) = \frac{1}{2}$ and $P(A_3) = P(BB \text{ or } WW) = \frac{1}{2}$. Also, $P(A_1 \cap A_2) = P(WB) = \frac{1}{4} = P(A_1)P(A_2)$. $P(A_2 \cap A_3) = P(BB) = \frac{1}{4} = P(A_2)P(A_3)$ and $P(A_1 \cap A_3) = P(WW) = \frac{1}{4} = P(A_1)P(A_3)$. Thus the three events are pairwise independent. But $A_1 \cap A_2 \cap A_3 = \emptyset$, so $P(A_1 \cap A_2 \cap A_3) = 0$, which is not equal to $P(A_1)P(A_2)P(A_3)$, and the three events are therefore not independent.

Problems 2.4

In Problems 1 to 6, we assume that a child of either sex is equally probable at birth.

1. A family has three children.
   (a) What is the probability that exactly two are girls?
   (b) If it is known that at least one is a girl, what is the probability that exactly two are girls?
   (c) If it is known that the oldest child is a girl, what is the probability that exactly two are girls?

2. What is the probability that a girl who is known to be from a family of four children has an older brother?

3. (a) How many children must a couple plan to have in order that the probability of at least one boy will be greater than 90 per cent?
   (b) How many children must a couple plan to have in order that the probability of at least one boy and one girl will be greater than 70 per cent?

4. (a) Determine the probability that a family of six children consists of three boys and three girls.
   (b) What is the probability that all children in a family of six children are of the same sex?

5. Suppose that a particular family has eight children, four boys and four girls. What is the probability that the oldest child is a boy? What is the probability that the four boys are all older than the four girls?

6. A couple has three children. Define the events A (first child is a girl), B (second child is a boy), C (third child is a boy), D (first two children are boys), and E (at least one child is a boy).
    (a) Calculate the probabilities of these five events.
    (b) Are A and D independent? A and E? B and E?
    (c) Prove that A, B, and C are independent.
    (d) Are the events B, C, and E independent?

7. A certain vaccine is 75 per cent effective in producing immunity. Two persons are vaccinated. Let A and B be the events that the first and second persons receive immunity, respectively. Are A and B independent? A and $\overline{B}$? $\overline{A}$ and B? $\overline{A}$ and $\overline{B}$?

8. Six persons have a disease that has a recovery rate of 98 per cent. What are the probabilities that (a) all will recover, (b) none will recover, and (c) exactly five will recover?

9. A student is studying chemistry, mathematics, and biology. He estimates that the probabilities of A's in these courses are $\frac{1}{2}$, $\frac{1}{3}$, and $\frac{1}{4}$, respectively. Assuming that his grades for the three courses are independent, what is the probability that he will receive no A's? An A in chemistry only?

10. Three rats learn to perform three different tasks (one rat to each task). The probabilities that the rats complete their tasks in 1 minute are $\frac{2}{3}$, $\frac{1}{2}$, and $\frac{1}{3}$. What is the probability that all three rats will complete their tasks in 1 minute? That exactly two will?

11. (a) Given that a throw of five dice produced at least one 3, what is the probability of two or more threes?
    (b) Two dice are thrown and the sum of the numbers showing is 6. What is the probability that one of the numbers is 4?
    (c) Given that a throw of seven dice produced at least one 5, what is the probability of two or more fives?

12. From a standard deck of 52 cards, one card is chosen at random. If the card is not a spade, it is returned to the deck, which is then shuffled and another card is chosen. This continues until a spade is chosen.
    (a) What is the probability that more than four cards will be chosen?
    (b) What is the probability that the third card will be the first spade if the first card is a diamond?
    (c) What is the probability that the first spade will appear on an even choice, that is, on the second or fourth or later choice? (*Hint:* Use the geometric series.)

13. In a certain midwestern city, during the month of August, the probability

of a thunderstorm is .25 and the probability of hail is .10 on any given day. The probability of hail during a thunderstorm is .3.

(a) Are the events "hail" and thunderstorm" independent?

(b) What is the probability of hail on a day when there is no thunderstorm?

14. On three ranches, *A*, *B*, and *C*, there is an outbreak of hoof and mouth disease. The proportions of infected cattle are $\frac{1}{6}$, $\frac{1}{4}$, and $\frac{1}{3}$, respectively. One cow is chosen at random from each ranch.

(a) What is the probability that exactly one has the disease?

(b) If exactly one is infected, what is the probability that the infected cow came from ranch *A*?

2.5 Bayes' Theorem

It is often useful to partition the sample space *S* into subsets $S_1, S_2, \ldots, S_n$. The subsets are disjoint and $S_1 \cup S_2 \cup \cdots \cup S_n = S$. For example, a group of people may be partitioned into subgroups corresponding to different blood types. The analysis of probabilities of events may be more meaningful or informative when it is done relative to the subsets rather than relative to the full sample space.

Figure 2.4 illustrates the following theorem.

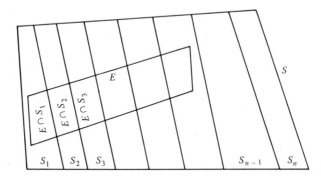

Figure 2.4. Partition of $E \subset S$ from a Partition of *S*

Theorem 2.5.1   *If the finite probability space S is partitioned into the disjoint subsets $S_1, S_2, \ldots, S_n$, then for any event E in S,*

$$P(E) = P(E \mid S_1)P(S_1) + P(E \mid S_2)P(S_2) + \cdots + P(E \mid S_n)P(S_n).$$

Proof: The set E can be expressed as the disjoint union

$$E = (E \cap S_1) \cup (E \cap S_2) \cup \cdots \cup (E \cap S_n)$$

(see Theorem 1.2.2). From the third axiom of probability,

$$P(E) = P(E \cap S_1) + P(E \cap S_2) + \cdots + P(E \cap S_n).$$

Using the formula for conditional probability, the result follows.

Example 2.5.1   In a mouse population $\frac{2}{3}$ of the colored mice are heterozygous colored (Cc), and $\frac{1}{3}$ are homozygous colored (CC). Suppose that an albino (cc) and a colored mouse chosen at random from the population are mated and a litter of four results. Given that the albino gene is recessive, what is the probability that all four offspring will be colored? (See Section 9.3 for an explanation of terminology.)

Solution: Clearly, the color of the offspring depends on whether the colored mouse is homozygous colored (CC) or heterozygous (Cc) for the albino gene. In the first case, all the offspring will be colored. In the second case, colored and albino offspring are equally probable. Thus

$$P(\text{all colored}) = P(\text{all colored} \mid CC)P(CC) + P(\text{all colored} \mid Cc)P(Cc)$$

$$= 1(\tfrac{1}{3}) + (\tfrac{1}{2})^4(\tfrac{2}{3}) = \tfrac{1}{3} + \tfrac{1}{24} = \tfrac{3}{8}.$$

Note that $P(\text{all colored} \mid CC)$ is 1, because all resulting offspring will be Cc, heterozygous colored.

In the above problem, we could have used a *tree diagram* to calculate the probabilities. There are two sources of uncertainty in this problem. First, we do not know whether the resulting offspring are all colored or not. The probabilities of the various possibilities are illustrated in the tree diagram of Figure 2.5. The first two "branches" correspond to the two possibilities CC and Cc with probabilities $\frac{1}{3}$ and $\frac{2}{3}$. From each of these possibilities there are two branches, corresponding to having all offspring colored or not all colored. If the colored mouse is CC, the offspring are all coloured. If the colored mouse is Cc, then the probability that all offspring are colored is $(\tfrac{1}{2})^4 = \tfrac{1}{16}$. In this case, the probability that at least one offspring is not colored is $1 - (\tfrac{1}{2})^4 = \tfrac{15}{16}$. To determine the probability that all four offspring

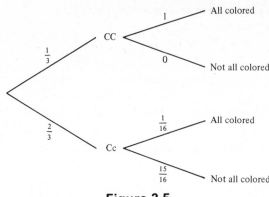

Figure 2.5

are colored, we add the products of the appropriate probabilities on the
tree. This gives $P(\text{all colored}) = \frac{1}{3} \cdot 1 + \frac{2}{3} \cdot \frac{1}{16} = \frac{3}{8}$.

Example 2.5.2   A laboratory contains three cages. Cage I contains two
brown and three white mice; cage II contains four brown and two white mice;
and cage III contains five brown and five white mice. A cage is chosen at
random, and a mouse is chosen at random from the cage. What is the probability that the mouse chosen is white?

Solution: This problem can be illustrated by the tree diagram of Figure 2.6.

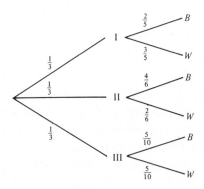

Figure 2.6

The probability that any cage will be chosen is $\frac{1}{3}$. If cage I is chosen, then
$P(B \mid I) = \frac{2}{5}$ and $P(W \mid I) = \frac{3}{5}$. Adding all three ways of choosing a white

mouse, we have $P(W) = (\frac{1}{3})(\frac{3}{5}) + (\frac{1}{3})(\frac{2}{6}) + (\frac{1}{3})(\frac{5}{10}) = \frac{43}{90}$. We can obtain the same result by quoting Theorem 2.5.1.

$$P(W) = P(W \mid \text{I})P(\text{I}) + P(W \mid \text{II})P(\text{II}) + P(W \mid \text{III})P(\text{III})$$
$$= (\tfrac{3}{5})(\tfrac{1}{3}) + (\tfrac{2}{6})(\tfrac{1}{3}) + (\tfrac{5}{10})(\tfrac{1}{3}) = \tfrac{43}{90}.$$

The tree diagram is useful, since it summarizes the given information and helps us to visualize the problem. Many of the problems at the end of this section can be solved very simply by the use of tree diagrams.

We may ask a different question in Example 2.5.2. Suppose that the mouse chosen turns out to be white. What is the probability that the mouse came from cage I? This probability is $P(\text{I} \mid W) = P(\text{I} \cap W)/P(W)$. We know that $P(\text{I} \cap W) = P(\text{I})P(W \mid \text{I}) = \frac{1}{3} \cdot \frac{3}{5} = \frac{1}{5}$, since $P(W \mid \text{I}) = P(\text{I} \cap W)/P(\text{I})$, and that $P(W) = P(W \mid \text{I})P(\text{I}) + P(W \mid \text{II})P(\text{II}) + P(W \mid \text{III})P(\text{III}) = \frac{43}{90}$. Therefore,

$$P(\text{I} \mid W) = \frac{P(\text{I})P(W \mid \text{I})}{P(\text{I})P(W \mid \text{I}) + P(\text{II})P(W \mid \text{II}) + P(\text{III})P(W \mid \text{III})}$$
$$= \frac{\frac{1}{5}}{\frac{43}{90}} = \frac{18}{43}.$$

The following theorem generalizes the above calculation. It looks imposing at first sight, but it can be proved very easily and is extremely useful in applications.

Theorem 2.5.2 Bayes' Theorem *If the finite probability space S is partitioned by disjoint subsets $S_1, S_2, \ldots, S_n$, then for any event E in S,*

$$P(S_k \mid E) = \frac{P(S_k)P(E \mid S_k)}{P(S_1)P(E \mid S_1) + P(S_2)P(E \mid S_2) + \cdots + P(S_n)P(E \mid S_n)},$$
$$k = 1, 2, \ldots, n.$$

Proof:

$$P(S_k \mid E) = \frac{P(S_k \cap E)}{P(E)} = \frac{P(S_k)P(S_k \cap E)}{P(S_k)P(E)} = \frac{P(S_k)}{P(E)}\left[\frac{P(E \cap S_k)}{P(S_k)}\right]$$
$$= \frac{P(S_k)P(E \mid S_k)}{P(E)}.$$

By using the expression for $P(E)$ given in Theorem 2.5.1, the result follows.

Example 2.5.3   Suppose that in a given large population there are equal numbers of males and females. In this population, 5 per cent of the men and .25 percent of the women are colorblind. A colorblind person is chosen at random. What is the probability that the person is male?

Solution: The population is divided into two disjoint subsets, males and females. We seek $P(M \mid CB)$. By Bayes' theorem,

$$P(M \mid CB) = \frac{P(M)P(CB \mid M)}{P(M)P(CB \mid M) + P(F)P(CB \mid F)} = \frac{(.05)(1/2)}{(.05)(1/2) + (.0025)(1/2)}$$

$$= \frac{2500}{2625} = \frac{20}{21}.$$

Example 2.5.4   Suppose that a women of blood type 0 and a man of blood type AB have a pair of boy twins of blood type B. If it is known that approximately one fourth of all sets of twins come from a single egg, what is the probability that these twins came from a single egg?

Solution: The set of boy twins is partitioned into those from a single egg and those from two eggs. From Bayes' theorem, the required probability can be written $P(\text{single egg} \mid \text{both type B}) =$

$$\frac{P(\text{single egg})P(\text{both type B} \mid \text{single egg})}{P(\text{single egg})P(\text{both type B} \mid \text{single egg}) + P(\text{2 eggs})P(\text{both type B} \mid \text{2 eggs})}.$$

To calculate these conditional probabilities, we must know that, when O and AB mate, 50 per cent of the offspring will be group A and 50 per cent will be group B. If the twins come from a single egg, they will have the same blood type. Since A and B are equally likely, $P(\text{both type B} \mid 1 \text{ egg}) = \frac{1}{2}$. If they come from two different eggs, each will have a probability of $\frac{1}{2}$ of being type B. Thus $P(\text{both type B} \mid 2 \text{ eggs}) = \frac{1}{4}$. Hence

$$P(\text{single egg} \mid \text{both type B}) = \frac{(\frac{1}{4})(\frac{1}{2})}{(\frac{1}{4})(\frac{1}{2}) + (\frac{3}{4})(\frac{1}{4})} = \frac{\frac{1}{8}}{\frac{5}{16}} = \frac{2}{5}.$$

Problems 2.5

1. A laboratory animal is either healthy (probability .9) or it is not. If it is healthy, it will be able to perform a certain task in 75 per cent of all attempts. If it is not healthy, it will be able to perform the task in 40 per cent of all attempts. Suppose that the animal is observed and it fails to complete the task. What is the probability that it is healthy?

2. A vaccine produces immunity against rubella in 95 per cent of cases. Suppose that, in a large population, 30 per cent have been vaccinated. Suppose also that a vaccinated person without immunity has the same probability of contracting rubella as an unvaccinated person. What is the probability that a person who contracts rubella has been vaccinated?

3. Between 1955 and 1966, the fraction of American men over 18 who smoked cigarettes declined from 57 to 51 per cent, while the fraction of women smokers increased from 28 to 33 per cent. Assuming that the population over 18 was 51 per cent women and 49 per cent men throughout this period, what were the probabilities in 1955 and 1966 that a smoker chosen at random was man?

4. A certain disease that occurs in 5 per cent of a population is very difficult to diagnose. One crude test for the disease gives a positive result (indicating the presence of the disease) in 60 per cent of cases when the patient has the disease and in 30 per cent of cases when the patient does not have the disease. Suppose that the test gives a positive result for a particular patient. What is the probability that he has the disease?

5. In a large population, equal numbers have black and brown hair. It is observed that 30 per cent of the people with black hair have blue eyes, as do 50 per cent of the people with brown hair. A person is chosen at random from those with black or brown hair and is observed to have blue eyes. What is the probability that this person has black hair?

6. In one industry, it is estimated that 3 per cent of the labor force are alcoholics with an absentee rate three times that of other workers. If a worker chosen at random is absent from work, what is the probability that this worker is an alcoholic?

7. Machines L, M, and N produce 25, 30, and 45 per cent of the output of a factory. The proportions of output that are defective are 1, 2, and 3 per cent, respectively. An item of output is chosen at random and is found to be defective. What are the probabilities that it came from L, M, and N?

8. An experimental procedure reassembles into a living amoeba the nucleus removed from one amoeba, the cytoplasm from a second and the outer membrane from a third. If all components come from the same strain, about 85 per cent reproduce normally. If components come from different strains of amoeba, about 1 per cent reproduce normally. Suppose that equal numbers of amoebae have been assembled from the same strain and from different strains. If one of these amoebae is chosen at random and is observed to reproduce normally, what is the probability that its components come from the same strain?

9. A laboratory rat is trained to perform four tasks, each within a time limit of 5 minutes. When a task is completed within the time limit, the rat pushes a lever and received a pellet of food. The probabilities of successful completion of the various tasks within the time limit are .8, .6, .4, and .2, respectively. Suppose that the rat begins a task chosen at random and is observed 5 minutes later with a pellet of food. What is the probability that the rat completed the first task? The second task?

10. Two cages contain experimental mice. One cage contains five brown and six white mice; the second cage contains two brown and five white mice. A cage is selected at random and a mouse chosen at random is removed from this cage. If the mouse is brown, what is the probability that it came from the second cage?

11. To test the contagiousness of several strains of bacteria, a large number of guinea pigs are kept in pairs in separate cages. One guinea pig of each pair is then infected with one of strain I, strain II, or strain III (equal numbers of guinea pigs being used for each strain). The proportions of healthy guinea pigs that become infected are found to be $\frac{1}{3}$, $\frac{1}{4}$, and $\frac{3}{4}$. A cage is chosen at random and it is observed that both animals have been infected. What are the probabilities that the infection was caused by strains I, II, and III?

12. The California Sunlovers (a football team) and the Wisconsin Icemen have had a long rivalry. The California team wins with probability .8 when the weather is warm and with probability .3 when it is cold. During the football season, it is warm 90 per cent of the time in California and 20 per cent of the time in Wisconsin. Suppose that the California team lost the last game with Wisconsin. (Assume that equal numbers of games are played in each state.)
   (a) What is the probability that the game was played in California?
   (b) What is the probability that the weather was cold?

13. In a large nursing home, it is estimated that 50 cent of the men and 30 per cent of the women have serious heart disorders. There are twice as many women as men in the institution. A patient chosen at random has a serious heart disorder. What is the probability that this patient is a man?

14. A large population is divided into two groups of equal size. One group receives a special diet high in unsaturated fats, and the control group eats a normal diet high in saturated fats. After 10 years on these diets, the incidence of cardiovascular disease in the two groups is 31 and 48 per cent, respectively. A person is chosen from the population at random and is found to have cardiovascular disease. What is the probability that this person belongs to the control group?

15. Rubella may result in major congenital malformations in children if it is contracted by the mother during the early stages of pregnancy. The chance of malformation is estimated to be 45, 20, and 5 per cent if rubella is contracted during the first, second, and third months of pregnancy, respectively. Suppose that the probability of contracting rubella is the same in any month of pregnancy and that a child is born with major congenital malformations resulting from rubella. What is the probability that the mother contracted rubella during the first month of pregnancy?

16. It is estimated that male cigarette smokers over the age of 40 are 10 times as likely to die of lung cancer as are male nonsmokers. Assuming that 60 per cent of this population are smokers, what is the probability that a male who dies of lung cancer had been a smoker?

17. (a) The probability that a moose will survive the winter is estimated to be 80 per cent if the moose is healthy and 30 per cent if the moose is not healthy. If 20 per cent of the moose population is not healthy, what proportion of the population will survive the winter?
    (b) If wolves kill 80 per cent of the healthy moose and 70 per cent of the sick moose that do not survive the winter, what proportion of the moose population is killed by wolves during the winter?

18. A rare disease occurs in .1 per cent of a population and is very difficult to diagnose. One crude test for the disease gives a positive result (indicating the presence of the disease) in 75 per cent of cases when the patient has the disease and in 25 per cent of cases when the patient does not have the disease. Suppose that the test gives a positive result for a person chosen at random. The test is then given again to this person and gives a negative result. Assuming that the tests are independent, what is the probability that this person has the disease?

19. It is estimated that 1 male child in 700 on average is born with an extra Y chromosome and that these children are 20 times as likely to show highly aggressive behavior. Accepting these figures, suppose that a boy is known to be highly aggressive. What is the probability that this child has an extra Y chromosome?

2.6 Repeated Trials: The Binomial and Multinomial Distributions

Up to this point, we have been concerned with the probabilities assigned to the outcomes of a trial of an experiment. In this section, we are interested in the probabilities of the possible outcomes when the same experiment is performed several times in succession.

As an example, consider an experiment with two possible outcomes. For convenience, let the two outcomes be referred to as "success" and "failure." Suppose that, on one trial of the experiment, the probability of success is p and the probability of failure is $q = 1 - p$. Now, suppose that the above experiment is repeated twice. This can be thought of as a new experiment with a sample space consisting of four elementary events—success followed by success; success, failure; failure, success; failure, failure. Assuming that the outcome on the second trial is not influenced by the outcome on the first trial, we assign the probabilities p^2, pq, qp, and q^2 to these four events. The total probability is

$$p^2 + pq + qp + q^2 = p^2 + 2pq + q^2 = (p + q)^2 = 1.$$

We define *n repeated trials* of the experiment with two outcomes to be the experiment consisting of the original experiment repeated n times. There are 2^n possible outcomes. Each of these outcomes has a probability $p^k q^{n-k}$, where k is the number of successes in the outcome of the n trials. Of course, we are assuming that the probability p of success does not change from trial to trial. The probability of success on the first trial is the same as the probability of success on the ninth trial, the eleventh trial, and so on.

In many important problems, we are more interested in the total number of successes in n trials than in the exact outcome of each trial. For example, if a coin is tossed 1000 times, we might be interested in how many heads actually occur. It is unlikely that we would be interested in the outcome of the 431st toss, say.

Example 2.6.1   Suppose that a fair coin is tossed three times in succession. What are the probabilities that 0 heads, 1 head, 2 heads, and 3 heads are thrown?

Solution: This experiment can be thought of as three repeated trials of an experiment with two outcomes. If throwing a head is defined to be success, then the probability of success on any trial is $p = \frac{1}{2}$. The sample space of the experiment is the equiprobable space $S = \{HHH, HHT, HTH, HTT, THH, THT, TTH, TTT\}$. Define $P(k)$ to be the probability of k successes (that is, the probability that k heads are thrown) for $k = 0, 1, 2, 3$. Then $P(0) = \frac{1}{8}$, $P(1) = \frac{3}{8}$, $P(2) = \frac{3}{8}$, and $P(3) = \frac{1}{8}$.

To generalize this example, define $f(n, k, p)$ to be the probability of exactly k successes in the n repeated trials.

Theorem 2.6.1   Binomial Distribution

$$f(n, k, p) = \binom{n}{k} p^k q^{n-k}.$$

Proof: If there are exactly k successes in the n trials, there must be $n - k$ failures. Since the trials are independent, the probability of having k successes followed by $n - k$ failures is $p^k q^{n-k}$. But k successes can occur in a number of different ways. From the n trials we "choose" k successes and $n - k$ failures. There are

$$\binom{n}{k} \text{ ways}$$

to choose k objects from n objects. Each of these ways has probability $p^k q^{n-k}$, and the theorem is proved.

Example 2.6.2   A coin is tossed eight times in succession. What is the probability that exactly five heads will occur?

Solution: In this case, $n = 8$, $k = 5$, $p = \frac{1}{2}$.

$$f(n, k, p) = f(8, 5, \tfrac{1}{2}) = \binom{8}{5}(\tfrac{1}{2})^5(\tfrac{1}{2})^3 = 56(\tfrac{1}{2})^8 = \tfrac{56}{256} = \tfrac{7}{32}.$$

Example 2.6.3   An average individual has the probability of $\frac{3}{5}$ of completing a certain task in 1 minute. Suppose that the task is attempted by 10 individuals. What is the probability of exactly seven successful completions of the task in 1 minute?

Solution: In this example, $n = 10$, $k = 7$, $p = \frac{3}{5}$.

$$f(n, k, p) = f(10, 7, \tfrac{3}{5}) = \binom{10}{7}(\tfrac{3}{5})^7(\tfrac{2}{5})^3 \approx .215.$$

Example 2.6.4   Suppose that an albino mouse and a homozygous colored mouse are mated. What is the probability that two of six mice in the second generation are albino? (For an explanation of terminology, see Section 9.3.)

Solution: In the first generation, all the mice will be colored since the albino gene is recessive. It is easy to see that in the second generation, $\frac{3}{4}$ of all the mice will be colored. Since all first generation mice are Cc, a cross of Cc with Cc will yield, with equal probabilities, CC, Cc, and cC, and cc, and only the cc offspring are albino. Thus $P(\text{albino}) = \frac{1}{4}$, and the problem reduces to the binomial distribution with $n = 6$, $k = 2$ and $p = \frac{1}{4}$. The required probability is

$$f(6, 2, \tfrac{1}{4}) = \binom{6}{2}(\tfrac{1}{4})^2(\tfrac{3}{4})^4 \approx .297.$$

The binomial distribution is applicable to repeated trials of an experiment with two outcomes. This can be generalized to an experiment with m outcomes. Let $E_1, E_2, \ldots, E_m$ be the outcomes of a trial of the experiment and $p_1, p_2, \ldots, p_m$ be the probabilities of these outcomes. Then $p_1 + p_2 + \cdots + p_m = 1$. Of n repeated trials of this experiment, there will be n_1 outcomes of type E_1, n_2 of $E_2, \ldots, n_m$ of E_m, where $n_1 + n_2 + \cdots + n_m = n$. Let $P(n_1, n_2, \ldots, n_m)$ denote the probability that this occurs.

Theorem 2.6.2   Multinomial Distribution

$$P(n_1, n_2, \ldots, n_m) = \binom{n}{n_1, n_2, \ldots, n_m} p_1^{n_1} p_2^{n_2} \cdots p_m^{n_m}.$$

Proof: Generalize the proof of the previous theorem. Each of the possible ways of having exactly $n_1, n_2, \ldots, n_m$ outcomes of the various types has probability $p_1^{n_1} p_2^{n_2} \cdots p_m^{n_m}$. Since there are

$$\binom{n}{n_1, n_2, \cdots, n_m} \text{ ways}$$

(see Theorem 1.5.2), the result follows.

The following examples illustrate the use of the multinomial distribution.

Example 2.6.5 In a certain large population, 70 per cent are right-handed, 20 per cent are left-handed, and 10 per cent are ambidextrous. If 10 persons are chosen at random from the population, what is the probability that all are right-handed? That 7 are right-handed, 2 are left-handed, and 1 is ambidextrous?

Solution: In this example, $n = 10$, $p_1 = .7$, $p_2 = .2$, and $p_3 = .1$.

$$P(10, 0, 0) = \binom{10}{10, 0, 0}(.7)^{10} = (.7)^{10} \approx .028,$$

$$P(7, 2, 1) = \binom{10}{7, 2, 1}(.7)^7(.2)^2(.1) \approx .119.$$

Example 2.6.6 Humans can be classified according to blood types into four mutually exclusive categories, O, A, B, and AB. In a large population, the proportions with the different blood types are .45, .40, .10, and .05, respectively. Suppose that six persons are chosen at random from the population. What are the probabilities that (1) three are of type O and three of type A and (2) none are of type AB?

Solution:

1. $P(3, 3, 0, 0) = \frac{6!}{3!3!}(.45)^3(.40)^3 \approx .117.$
2. $f(6, 0, .05) = \frac{6!}{6!0!}(.95)^6 \approx .735.$

It is apparent, for example, that if the above percentages are correct, a large number of people would have to be chosen to be reasonably certain that type AB would be found. To be 90 per cent certain, we would need to have the probability of "none of type AB" less than 10 per cent. That is, $(.95)^n < .1$, where n is the number of people chosen. Using logarithms to solve the equation $(.95)^n = .1$, we obtain $n \log (.95) = \log (.1)$ or $n \log 95 - n \log 100 = \log 1 - \log 10$ or $n (\log 95 - \log 100) = -\log 10$ or $n = \log 10(\log 100 - \log 95) = 1/(2 - 1.9777) \approx 44$.

Problems 2.6

1. An experiment has 95 per cent probability of success and 5 per cent probability of failure. The experiment is repeated five times. Determine the probabilities of the following events.
   (a) No successes.
   (b) No failures.
   (c) Four successes and one failure.

2. The weather in June can be good (probability 50 per cent), mediocre (probability 25 per cent), or bad (probability 25 per cent) on any given day. Assuming that the weather on one day does not influence the weather on any other day, what are the probabilities that during a week in June there are seven good days? Four good days, two mediocre days, and one bad day?

3. In a certain large population, 40 per cent of the people are black-haired, 40 per cent are brown-haired, and 20 per cent are blonde-haired. If 10 persons are chosen at random from the population, what are the probabilities of the following events?
   (a) Five black, 5 brown.
   (b) Four black, 4 brown, 2 blonde.
   (c) Three black, 3 brown, 4 blonde.

4. Colorblindness affects 1 per cent of a large population. Suppose that n people are chosen at random from this population. What is the probability that none of the n people are colorblind? How large must n be in order that this probability be less than 10 per cent?

5. A machine makes a product that must satisfy certain specifications. The probability that a given product item is acceptable is 95 per cent. A sample of 10 items is taken from the machine's production. What is the probability that all 10 items will be acceptable?

6. It is estimated that a single wolf attacking moose will be successful on 8 per cent of encounters. What is the probability that in five encounters no moose are successfully attacked?

7. Human brains have a well-defined auditory area in each hemisphere. From anatomical studies, it has been found that the auditory area of the left hemisphere is more developed in 65 per cent of cases, less developed in 10 per cent of cases, and equally developed in 25 per cent of cases. In a group of five persons chosen at random, what is the probability three, zero, and two people are in these respective categories?

8. In an introductory course in quantitative analysis, it is observed that an acceptable end point is achieved in 80 per cent of titrations. One student achieves an acceptable end point only once in six titrations. What is the probability of this occurring by chance? Do you think this student will become an experimental chemist?

9. A thyroid condition is treated by iodine therapy. It is found that 50 per cent of the patients improve quickly, 40 per cent are not noticeably affected, and 10 per cent actually become worse. Nine patients are treated. What are the probabilities that (a) all nine improve; (b) five improve, three stay the same and one becomes worse; and (c) three improve, three stay the same, and three become worse?

10. A treatment for a disease produces a cure in 75 per cent of cases. Six patients are treated. What are the probabilities that (a) all will be cured, (b) none will be cured, (c) four will be cured, and (d) at least four will be cured?

11. Consider n repeated trials of a binomial experiment with $p = q = \frac{1}{2}$.
    (a) If n is even, what is the probability of $n/2$ successes and $n/2$ failures?
    (b) Determine the probabilities of more successes than failures when n is odd and when n is even.

12. A wolf has a 50 per cent chance of obtaining food in each foraging period. What is the probability of success in more than half of all foraging periods if there are 31 periods? Fourteen periods?

13. (a) How many dice should be thrown in order that the probability of obtaining an odd number on at least one of the dice is greater than 90 per cent?
    (b) How many dice should be thrown in order that the probability of at least one five is greater than 50 per cent?

14. In a midwestern city, 50 per cent of the population favor stronger gun controls, 30 per cent favor weaker controls, and 20 per cent like things just as they are. Twelve people chosen at random are interviewed. What is the probability that (a) all favor stronger controls, (b) half favor stronger controls and half favor weaker controls, and (c) equal numbers favor the three alternatives?

15. A meteorologist applies for a research grant to travel to Spain to test the
    theory that the rain in Spain falls mainly on the plain. He plans to
    conduct his observations during a part of the year when the probability
    of rain on any given day on the plain is 20 per cent. (Assume that this
    probability is independent of the previous weather.)
    (a) How many days must he plan to spend in Spain to be 99 per cent
        certain that he will observe rain?
    (b) Suppose that the meteorologist has been awarded research funds for
        15 days in Spain and that, after 10 days, no rain has been observed.
        What is the probability that his trip will be a failure, that is, that no
        rain will be observed?

16. In a population of drosophila, 20 per cent have a wing mutation. If six
    flies are chosen at random from the population, what is the probability
    that two have the mutation? That at least one has the mutation? That
    fewer than five have the mutation?

17. In order to estimate the fish population of a lake,[3] samples are taken at
    periodic intervals (usually 24 hours) and the fishes are then tagged and
    returned to the lake. For the ith sample, define t_i to be the total number of
    fish, d_i the number of new (previously untagged) fish, and r_i the number of
    recaptured fish. Let N denote the total fish population of the lake and let
    M_i denote the number of tagged fish in the lake just before the ith sample
    is taken.
    (a) What is the probability that a fish chosen at random from the ith
        sample is tagged? Untagged?
    (b) Using the binomial distribution, calculate the probability of drawing
        r_i recaptures and d_i new fish in the t_i fish of the ith sample.

18. A general who has won five battles in a row is usually considered to be a
    great military leader. Assuming that the outcome of battles is determined
    by chance with 50 per cent probabilities of success and failure and that
    each general fights exactly five battles, what proportion of generals are
    great military leaders?

19. Caffeine and benzedrine are stimulants that seem to have some ability to
    counteract the depressing effects of alcohol. In a test of their relative
    effectiveness, 40 volunteers each consume 6 ounces of alcohol. The
    volunteers are then divided into 20 pairs, and one member of each pair
    receives benzedrine while the second member receives caffeine. As
    measured by certain tests, benzedrine brings about a more rapid recovery

[3] Z. Schnabel, "The Estimation of the Total Fish Population of a Lake," *American Mathematical Monthly*, 45:348–352 (1938).

in all 20 pairs. What is the probability of obtaining this result if there is no difference in the effects of caffeine and benzedrine?

2.7 Random Variables

It is often useful to assign numbers to the various outcomes of an experiment. For example, we could toss a coin three times and record the number of heads that occur; or the experiment may be to choose 10 persons from 1000 people and then test them for a certain disease. For the second experiment, we could associate to the outcome the number of persons from the 10 chosen who have the disease.

Definition 2.7.1   Random Variable   *A random variable X associated with a finite probability space S is a real-valued function defined on the sample space S.*

Definition 2.7.2   Range Space of a Random Variable   *The range space X(S) of a random variable X defined on the finite probability space S is the set of real numbers X(E), where E is an elementary event of S.*

Let us emphasize that a random variable is a function which assigns a real number to each outcome of an experiment. In the above example, the sample space consists of all possible choices of 10 persons from 1000 people. Let E represent one choice of 10 persons. Then a random variable X is defined by $X(E) =$ number of persons in E who have the disease. The range space $X(S)$ of this random variable is the set of numbers $0, 1, 2, \ldots, 10$.

Example 2.7.1   Consider the experiment of tossing a fair coin four times in succession. Define X to be the random variable equal to the total number of heads that occur. The range space of this random variable is $X(S) = \{0, 1, 2, 3, 4\}$. This is the set of values that X can take on. To each elementary event of the experiment, we associate a value of the random variable X. For example, the elementary event $HTTT$ corresponds to $X = 1$, since exactly one head comes up in the four trials.

To avoid confusion, a random variable will be denoted by a capital letter X and the values that the random variable takes on will be denoted by a lowercase letter x; X is a function and x is a real number. We will use the notation $P(X = x)$ to denote the probability that the random variable X takes on the value x. In Example 2.7.1, the probability that X takes on the value 2 is

$$P(X = 2) = \binom{4}{2}(\tfrac{1}{2})^4 = \tfrac{3}{8}.$$

(This is the binomial probability.) Similarly, $P(X = 0) = \tfrac{1}{16}$, $P(X = 1) = P(X = 3) = \tfrac{1}{4}$, and $P(X = 4) = \tfrac{1}{16}$. Note that these probabilities add up to 1. We can generalize the ideas of this example in the following definition.

Definition 2.7.3   Probability Function   *The probability function f associated to a random variable X is defined on the real line by*

$$f(x) = P(X = x) = \sum_i P(E_i),$$

where $x \in \mathbf{R}$ and the sum is over all elementary events E_i in S such that $X(E_i) = x$.

It should be emphasized that $f(x)$ is defined for every real number x. The domain of f is $D(f) = \mathbf{R}$ (see Section 1.3). In Example 2.7.1, we saw that $f(0) = \tfrac{1}{16}$, $f(1) = \tfrac{1}{4}$, $f(2) = \tfrac{3}{8}$, $f(3) = \tfrac{1}{4}$, and $f(4) = \tfrac{1}{16}$. In addition, we have $f(-2) = P(X = -2) = 0$, since it is impossible to throw -2 heads. Similarly, $f(\tfrac{1}{2}) = 0$. In fact, if x is not equal to 0, 1, 2, 3, or 4, then $f(x) = 0$.

Example 2.7.2   On any given day in July in a certain city, the probability of rain is $\tfrac{1}{6}$, the probability of cloudy weather (without rain) is $\tfrac{1}{3}$, and the probability of sunny weather is $\tfrac{1}{2}$. Consider the experiment of choosing a day in July at random and observing the weather in this city. Associate a random variable X to the outcomes of this experiment by defining $X = -3$ if it rains, $X = -1$ if it is cloudy, and $X = 4$ if it is sunny. The range space of X is $X(S) = \{-3, -1, 4\}$. The probability function $f(x) = P(X = x)$ is zero unless $x = -3, -1$, or 4. From the given information, we have $f(-3) = \tfrac{1}{6}$, $f(-1) = \tfrac{1}{3}$, and $f(4) = \tfrac{1}{2}$.

In many problems, we may be interested in knowing the probability that a random variable takes on values less than or equal to a certain number x.

For example, if we choose 10 persons from 1000 people and test them for a certain disease, we may be interested in the probability that no more than 5 of the 10 persons have the disease.

Example 2.7.3 In a large population, 50 per cent of the people have brown hair. Four persons are chosen at random from the population. What is the probability that no more than three of the four persons chosen have brown hair?

Solution: Define the random variable X to be the number of brown-haired people chosen. The probability that X takes on a value less than or equal to 3, written $P(X \le 3)$, is the sum of the probabilities that $X = 0$, $X = 1$, $X = 2$, and $X = 3$. These are binomial probabilities. Therefore, the required probability is

$$P(X \le 3) = P(X = 0) + P(X = 1) + P(X = 2) + P(X = 3)$$
$$= (\tfrac{1}{2})^4 + 4(\tfrac{1}{2})^4 + 6(\tfrac{1}{2})^4 + 4(\tfrac{1}{2})^4$$
$$= \tfrac{15}{16}.$$

Note that $P(X \le 3) = 1 - P(X = 4)$. (Why?)

The previous example suggests the following definition.

Definition 2.7.4 Distribution Function *The distribution function F associated to a random variable X is a function defined on the real numbers by*

$$F(x) = P(X \le x) = \sum_j P(E_j).$$

The sum is taken over all elementary events E_j such that $X(E_j) \le x$.

It turns out that for finite probability spaces $f(x)$ is the more useful function. The distribution function $F(x)$ plays a greater role when we study continuous probability spaces in Chapter 8.

Example 2.7.4 Consider the experiment of throwing two dice. The sample space S of this experiment consists of 36 outcomes $(1, 1), (1, 2), \ldots, (6, 6)$.

Suppose that we are interested in the sum of the outcomes on the two dice. This defines a random variable X with range space $X(S) = \{2, 3, 4, .., 12\}$. By a counting argument, the reader may calculate $f(x)$ and $F(x)$ for this random variable X. These functions are given in the following table:

X	2	3	4	5	6	7	8	9	10	11	12
$f(x)$	$\frac{1}{36}$	$\frac{2}{36}$	$\frac{3}{36}$	$\frac{4}{36}$	$\frac{5}{36}$	$\frac{6}{36}$	$\frac{5}{36}$	$\frac{4}{36}$	$\frac{3}{36}$	$\frac{2}{36}$	$\frac{1}{36}$
$F(x)$	$\frac{1}{36}$	$\frac{3}{36}$	$\frac{6}{36}$	$\frac{10}{36}$	$\frac{15}{36}$	$\frac{21}{36}$	$\frac{26}{36}$	$\frac{30}{36}$	$\frac{33}{36}$	$\frac{35}{36}$	$\frac{36}{36}$

The distribution function $F(x)$ is defined for all $x \in \mathbf{R}$. In the example, $F(\frac{5}{2}) = P(X \le \frac{5}{2}) = \frac{1}{36}$, since the probability that X is less than or equal to $\frac{5}{2}$ is equal to the probability that $X = 2$. Similarly, $F(6.75) = P(X \le 6.75) = P(X \le 6) = \frac{15}{36}$, and $F(1000) = P(X \le 1000) = 1$. In fact, it should be evident that as x increases, $F(x)$ increases in steps from 0 to 1. The distribution function for Example 2.7.4 is illustrated in Figure 2.7.

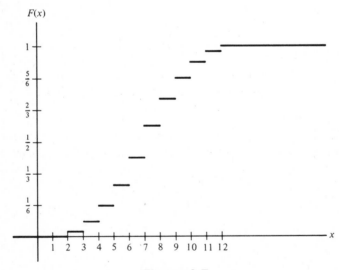

Figure 2.7

Example 2.7.5 Binomial Distribution   Consider n repeated trials of the binomial experiment with probability p of success and q of failure on one trial. The number of successes on the n trials is a random variable X whose probability function $f(x)$ is given by the binomial distribution $f(n, x, p)$.

Example 2.7.6   Suppose that a coin is tossed until the first head comes up. The sample space S of this experiment consists of all sequences of a number of tails followed by one head (we stop as soon as a head occurs). Thus $S = \{H, TH, TTH, TTTH, \dots\}$. The number of tosses is a random variable whose range space consists of all positive integers; $X(S) = \{1, 2, 3, \dots\}$. To calculate the probability function, we observe that the only way to have exactly x tosses is to have $x - 1$ tails followed by a head. Thus

$$f(x) = P(X = x) = (\tfrac{1}{2})^{x-1}(\tfrac{1}{2}) = (\tfrac{1}{2})^x.$$

The distribution function

$$F(x) = P(X \leq x) = P(X = 1) + P(X = 2) + \cdots + P(X = x) = \sum_{k=1}^{x} \frac{1}{2^k}.$$

This is a finite geometric series with sum $1 - 1/2^x$. (See Problem 1.6.14.)

Example 2.7.7   A certain type of skin graft is successful in 40 per cent of all cases. The skin graft is performed on a patient several times in succession until a graft takes. What is the probability that the graft will be successful (1) on the first attempt, (2) on the third attempt, and (3) on the nth attempt?

Solution:

1. The probability of success on any trial is $p = .40$. Therefore, the probability of success on the first trial is .40.
2. To have the first success on the third attempt, the first two grafts must be failures and the third must be successful. The probability of failure on any trial is $q = 1 - p = .60$. Therefore, the required probability is $(.6)^2(.4) = .144$.
3. To succeed for the first time on the nth trial, the operation must fail $n - 1$ times in succession and then succeed. The probability that this happens is $q^{n-1}p = (.6)^{n-1}(.4)$.

We can easily generalize the last two examples to obtain an important new distribution. Suppose that we perform a binomial experiment (with probability p of success on one trial) repeatedly, until the first success occurs. As in Example 2.7.6, we define a random variable X to be the number of trials required. The random variable X can take on any positive integer value, $X = 1, 2, 3, \dots$. The first success will occur on the nth trial if the first

$n - 1$ trials are failures and the nth trial is a success. The probability that this happens is $q^{n-1}p$. The probability function $f(x)$ of the random variable X is zero if x is not a positive integer and, if x is a positive integer, $f(x) = P(X = x) = q^{x-1}p$. The random variable X is said to have a *geometric distribution*. It is the first example that we have seen of a random variable whose range space is infinite, $X(S) = \{1, 2, 3, 4, \ldots\}$.

Example 2.7.8   Suppose that a predator has a .4 chance of success in catching a certain type of prey. How many prey must he stalk in order to be at least 90 per cent certain of catching at least one?

Solution: This is clearly an example of the geometric distribution with $p = .4$ and $q = .6$. Thus the probability that the first success occurs on the nth try is $(.6)^{n-1}(.4)$. We seek n such that $P(X \le n) \ge .9$. $X \le n$ means that the first success occurs on or before the nth try. But $P(X = 1) = .4$, $P(X = 2) = (.6)(.4) = .24$, $P(X = 3) = (.36)(.4) = .144$, $P(X = 4) = (.216)(.4) = .0864$, and $P(X = 5) = (.1296)(.4) = .05184$. Adding, we obtain $P(X \le 5) = \sum\limits_{k=1}^{5} P(X = k) = .92\ldots$, which is greater than 90 per cent, and so the predator must stalk at least five prey to be 90 per cent certain of at least one capture.

A much simpler solution can be obtained by noting that the predator must stalk n prey, where n is chosen so that $(.6)^n < .1$. Since $(.6)^5 \approx .078 < .1$, the predator must stalk five prey to be assured of a meal at least 90 per cent of the time.

Finally in this section, we emphasize the difference between the binomial and the geometric distributions. Both distributions are related to repeated trials of an experiment with two outcomes, success with probability p and failure with probability $q = 1 - p$. If we perform the experiment a fixed number n of times and ask for the probability of k successes in n trials, then we have the binomial distribution. For the geometric distribution, the number of times that we perform the experiment is not fixed. Instead, the experiment is repeated until the first success occurs. The random variable is the number of trials until the first success.

Problems 2.7

1. Consider an experiment with three elementary outcomes, E_1, E_2, and E_3, each with probability $\frac{1}{3}$. Define a random variable X by $X(E_1) = -1$, $X(E_2) = 0$, and $X(E_3) = 1$.

(a) What is the range space of the random variable X?

(b) What is the probability function for X?

(c) What is the distribution function for X?

2. Consider the experiment of choosing an integer from 1 to 10 at random. Define the random variable X by $X = 0$ if the number chosen is 3, 5, 6, or 9; $X = 1$ if the number chosen is 2, 4, 8, or 10; and $X = 2$ if the number chosen is 1 or 7.

(a) What is the sample space of this experiment? What are the elementary outcomes?

(b) What is the range space for X?

(c) Determine the probability function and the distribution function for X.

3. A fair coin is tossed four times. Let X be the number of heads obtained.

(a) What is the range space of this random variable?

(b) Determine the probability function and the distribution function of X.

4. A pair of dice are thrown five times in succession. Define the random variable X to be the number of times that a seven is thrown.

(a) What is the range space of X?

(b) What are the probability function and distribution function of X?

5. A hospital has 1000 patients at any one time. Define X to be the number of patients who will be hospitalized more than 5 days.

(a) What is the range space of X?

(b) Draw graphs of possible probability functions and distribution functions of X.

6. In a large drosophila population, 25 per cent of the flies have a wing mutation. Let X be the number of flies that have the mutation in a random sample of five chosen from the population. Determine the range space, the probability function, and the distribution function of the random variable X.

7. An experiment is completed successfully in 80 per cent of all attempts. Let X denote the number of successful experiments in a series of eight repetitions. Determine the range space, the probability function, and the distribution function of X.

8. The probability that a first-year student of English will know the meaning of the word "adumbrate" is estimated to be 10 per cent. In an English class of 20 first-year students, define X to be the number of students who can define this word correctly.

(a) What is the range space, the probability function, and the distribution function of X?

(b) What is the probability that $X \leq 4$?

(c) What is the probability that $X \geq 4$?

9. A rat is being trained to go through a maze. There are five possible paths, only one of which leads through the maze. Assume that the rat chooses a path at random until the correct path is chosen; also assume that an incorrect path will not be chosen a second time. Define a random variable X to be the number of incorrect paths chosen.

   (a) What are the range space, probability function, and distribution function of X?

   (b) What is the probability that $X \geq 3$? That $X > 0$?

10. In a gambling game, A has the first move and then B follows. The rules of the game are such that either competitor can win on any move with probability p. They continue playing until one of them wins. Define X to be the number of moves taken by A until the game ends and define Y to be the number of moves taken by B in the game.

    (a) What are the range spaces of X and Y?

    (b) Determine $P(X = 2)$, $P(Y = 1)$, and $P(Y > 2)$.

    (c) Prove that their respective probabilities of winning are $1/(2 - p)$ and $(1 - p)/(2 - p)$. [*Hint:* Use the geometric series (Example 1.6.8.)]

    (d) Is this a "fair" game for any value of p?

11. Two thirds of the children in a large school are absent because of an epidemic of influenza. In one class of 25 students, the teacher calls the roll. Define X to be the number of students called until one responds. What is the range space of X? What is the probability that $X = 10$, that is, that the tenth child called is the first one to be present? Determine $P(X \leq 2)$ and $P(X \geq 2)$.

12. A low-iodine diet produces enlarged thyroid glands in 60 per cent of the animals in a large population. Four enlarged glands are needed for an experiment.

    (a) What is the probability that four animals chosen at random will have enlarged thyroid glands?

    (b) Define X to be the number of animals selected at random until four of the selected animals have enlarged thyroids. What are the range space, probability function, and distribution function of X? Determine $P(X = 7)$ and $P(X > 5)$.

13. A storekeeper estimates that the probability of selling n loaves of bread on any given day is of the form $P(n) = kn$ for $n = 0, 1, 2, \ldots, 25$ and $P(n) = k(50 - n)$ for $n = 26, 27, \ldots, 50$.

(a) For what value of the constant k is $P(n)$ the probability function of a random variable? Define the random variable.

(b) What is the probability that the storekeeper will sell on a given day fewer than 26 loaves? Between 22 and 28 loaves? More than 45 loaves?

(c) If the storekeeper wishes to meet the demand for bread on at least 95 per cent of all days, how many loaves should he stock?

14. Suppose that there are 1000 animals in a certain region and that 10 traps are set out to capture them. Let p be the probability that a given animal will be caught on a given day. (The traps are inspected daily.) Assume that p is independent of the animal and the number of animals previously trapped. What is the probability that a particular animal will be caught on the third day? On the seventh day?

2.8 Expected Value and Variance

Consider a random variable X that can take on the values $x_1, x_2, \ldots, x_N$. The probabilities that X takes on these values are $f(x_1), f(x_2), \ldots, f(x_N)$, where $f(x) = P(X = x)$ is the probability function of the random variable. If the experiment that gives rise to X is repeated a large number of times, we expect the various values of the random variable to occur in proportions which are approximately equal to their probabilities. For example, if a fair coin is tossed 1000 times, we would expect approximately 500 heads to occur. (Of course, the probability of exactly 500 heads is extremely small.) If a fair die is thrown 600 times in succession, we expect to obtain approximately 100 ones, 100 twos, and so on. This intuitive idea leads us to define the expected value of a random variable.

Definition 2.8.1   Expected Value of a Random Variable   *The expected value $E(X)$ of a random variable X is defined by*

$$E(X) = x_1 f(x_1) + x_2 f(x_2) + \cdots + x_N f(x_N) = \sum_{i=1}^{N} x_i f(x_i),$$

where $x_1, x_2, \ldots, x_N$ are the values that X takes on and $f(x_1), f(x_2), \ldots, f(x_N)$ are the corresponding probabilities.

The expected value is the "average" outcome of the experiment. The value is sometimes called the expectation *or* mean *of X and is often denoted by the symbol μ.*

Example 2.8.1   Consider the experiment of throwing a single six-sided die. The outcomes are represented by a random variable X defined as the number of spots on the top face. The values that X takes on are 1, 2, 3, 4, 5, and 6, each with probability $\frac{1}{6}$. The expected value is

$$E(X) = 1(\tfrac{1}{6}) + 2(\tfrac{1}{6}) + 3(\tfrac{1}{6}) + 4(\tfrac{1}{6}) + 5(\tfrac{1}{6}) + 6(\tfrac{1}{6}) = 3.5.$$

Note that 3.5 is the average of the numbers 1 through 6.

Example 2.8.2   Consider the experiment of tossing a fair coin twice in succession. Define X to be the random variable equal to the number of heads that appear. Then, X takes on the values 0, 1, and 2 with probabilities $\frac{1}{4}, \frac{1}{2}$, and $\frac{1}{4}$, respectively. The expected value is $E(X) = 0(\tfrac{1}{4}) + 1(\tfrac{1}{2}) + 2(\tfrac{1}{4}) = 1$. This means that, on average, we expect one head in two tosses of a fair coin.

Example 2.8.3   Calculate the expected value of the random variable defined in Example 2.7.2.

Solution: The random variable takes on the values -3, -1, and 4. Therefore, $\mu = E(X) = -3f(-3) + (-1)f(-1) + 4f(4) = -3(\tfrac{1}{6}) + (-1)(\tfrac{1}{3}) + 4(\tfrac{1}{2}) = \tfrac{7}{6}$. In this example, we note that this random variable cannot take on its expected value, since $f(\tfrac{7}{6}) = P(X = \tfrac{7}{6}) = 0$. This should not be confusing if the expected value is viewed as an average.

The expected value can be defined in a similar way when the random variable takes on an infinite number of values. This is illustrated in the following example.

Example 2.8.4   Geometric Distribution   In the previous section, we defined the geometric distribution to be the probability distribution of a random variable X with range space the positive integers. The probability that X takes on the value x is $P(X = x) = q^{x-1}p$ for $x = 1, 2, 3, \ldots$. The expected value of X is

$$\mu = E(X) = \sum_{k=1}^{\infty} kP(X = k) = \sum_{k=1}^{\infty} kq^{k-1}p.$$

To simplify this expression, we compare $E(X)$ to $qE(X)$.

$$E(X) = p + 2qp + 3q^2p + \cdots + kq^{k-1}p + \cdots,$$
$$qE(X) = qp + 2q^2p + 3q^3p + \cdots + (k-1)q^{k-1}p + kq^kp + \cdots.$$

Subtracting,

$$(1 - q)E(X) = p + qp + q^2p + \cdots + q^kp + \cdots$$

$$= p(1 + q + q^2 + \cdots + q^k + \cdots) = \frac{p}{1 - q}.$$

We have used the fact that the sum of the geometric series $1 + q + q^2 + \cdots + q^k + \cdots$ is $1/(1 - q)$ if $0 < q < 1$ (see Example 1.6.8). But $1 - q = p$, and we have $pE(X) = p/p = 1$ or $E(X) = 1/p$. When the probability of success on one trial is p, the expected number of trials until the first success is $\mu = E(X) = 1/p$.

Example 2.8.5   In a large population, 25 per cent of the people have blue eyes. Volunteers are chosen at random from this population, one at a time, until a volunteer with blue eyes is chosen. What is the expected number of people that will be chosen?

Solution: The number of people that will be chosen is a random variable with geometric distribution. The probability of success on one trial is $p = .25$. Therefore, the expected number of trials is $\mu = 1/p = 1/.25 = 4$.

Example 2.8.6   A fair die is thrown until a 6 appears. What is the expected number of times that the die will be thrown?

Solution: The probability of success on one trial is $p = \frac{1}{6}$. The expected number of trials is $\mu = 1/p = 6$.

Example 2.8.7   Binomial Distribution   The random variable X which equals the number of successes in n repeated trials of the binomial experiment has probability function $f(n, x, p)$ with $x = 0, 1, 2, \ldots, n$. Therefore,

$$E(X) = \sum_{k=0}^{n} kf(n, k, p) = \sum_{k=1}^{n} k \frac{n!}{k!(n - k)!} p^k q^{n-k}$$

$$= np \sum_{k=1}^{n} \frac{(n - 1)!}{(k - 1)!(n - k)!} p^{k-1} q^{n-k}.$$

We now change the index of summation by letting $l = k - 1$. Then

$$E(X) = np \sum_{l=0}^{n-1} \frac{(n - 1)!}{l!(n - 1 - l)!} p^l q^{n-1-l} = np(p + q)^{n-1}$$

(from the binomial theorem). Since $p + q = 1$, $E(X) = np$. This is the very reasonable result that the expected number of successes in n trials is the number of trials multiplied by the probability of success on one trial.

Example 2.8.8 Suppose that a predator has a probability of .4 of capturing an individual prey on each encounter. What is the expected number of captures in 20 encounters?

Solution: This is an example of a binomial distribution with $n = 20$ and $p = .4$. The expected number is $E(X) = np = 20(.4) = 8$.

Example 2.8.9   It is estimated that 20 per cent of the adults in a large population are seriously overweight. Fifty adults are chosen at random from this population. What is the expected number that will be found to be overweight?

Solution: The number of overweight people in the sample of 50 is a random variable with binomial distribution with $p = .20$ and $n = 50$. The expected number of overweight people is $\mu = np = 50(.2) = 10$.

The expected value or mean of a random variable gives the average of the values that the random variables takes on if the experiment is performed many times. This gives us valuable but limited information. To illustrate what information is missing when we know only the mean, we consider two very different random variables, which take on the same values and have the same mean. In the first experiment, we toss a coin 100 times in succession. Define the random variable X_1 to be the number of heads that occur. Then, X_1 takes on the values $0, 1, 2, \ldots, 100$ with probabilities given by the binomial distribution. The expected value is $E(X_1) = np = 50$.

In the second experiment, we choose a piece of paper at random from a bowl containing 101 pieces of paper numbered 0 to 100. Define the random variable X_2 to be the number that appears on the piece of paper chosen. Then, X_2 takes on the values $0, 1, 2, \ldots, 100$, each with probability $\frac{1}{101}$. The expected value is $E(X_2) = 50 =$ the average of the numbers from 0 to 100.

There is a profound difference between these two random variables. If we toss a coin 100 times, it is very probable that we will obtain approximately 50 heads. We may obtain 42 or 53 or 57, say, but it is extremely improbable that we would obtain 5 or 92 heads. In the second experiment, all outcomes are equally likely to occur. It is just as probable that we draw the number 5 or 92 as it is that we draw the number 42 or 53. In the second experiment, the difference between the observed outcome and the expected value can be quite large. In the first experiment, it is unlikely that this difference will be very large. This leads us to ask by how much we can expect a random variable X to deviate from its mean. Since the deviation can be either positive or negative, we study the square of the deviation.

Definition 2.8.2   Variance   *The variance, var (X), of the random variable X is the expected value of the square of the difference between X and $\mu = E(X)$. This means that*

$$\text{var } (X) = E(X - \mu)^2 = \sum_{i=1}^{N} (x_i - \mu)^2 f(x_i).$$

Definition 2.8.3   Standard Deviation   *The standard deviation, σ, of the random variable X is the square root of the variance of X, $\sigma = \sqrt{\text{var } (X)}$. We use σ^2 as a convenient notation for var (X).*

The variance and standard deviation give measures of how far, on average, a random variable can be expected to differ from its mean. In this section, we will calculate the variances and standard deviations of a number of random variables. We will return to these concepts in Chapter 8. The following theorem simplifies the calculation of variances considerably.

Theorem 2.8.1   *The variance $\sigma^2 \equiv E(X - \mu)^2$ is given by the formula*

$$\sigma^2 = E(X^2) - (E(X))^2.$$

Proof:

$$E(X - \mu)^2 = \sum_{k=1}^{N} (x_k - \mu)^2 p_k$$

$$= \sum_{k=1}^{N} (x_k^2 - 2\mu x_k + \mu^2) p_k$$

$$= \sum_{k=1}^{N} x_k^2 p_k - 2\mu \sum_{k=1}^{N} x_k p_k + \mu^2 \sum_{k=1}^{N} p_k$$

$$= E(X^2) - 2\mu^2 + \mu^2$$

$$= E(X^2) - \mu^2 = E(X^2) - (E(X))^2.$$

We have used the fact that $\sum_{k=1}^{N} x_k p_k = \mu$ [by definition of $\mu = E(X)$].

Example 2.8.10 Consider the experiment of tossing a fair coin two times in succession. Define the random variable X to be the number of heads that appear. Then, X takes on the values 0, 1, and 2 with probabilities $\frac{1}{4}, \frac{1}{2}$, and $\frac{1}{4}$. The mean is $\mu = np = 2(\frac{1}{2}) = 1$. Therefore, the variance is

$$\text{var}(X) = E(X^2) - \mu^2 = 0^2(\tfrac{1}{4}) + 1^2(\tfrac{1}{2}) + 2^2(\tfrac{1}{4}) - 1^2 = \tfrac{1}{2} + 1 - 1 = \tfrac{1}{2}.$$

Therefore $\sigma^2 = \text{var}(X) = \frac{1}{2}$ and $\sigma = 1/\sqrt{2}$.

In the above example, we have calculated the variance and standard deviation of a binomial distribution. We generalize this calculation in the following example.

Example 2.8.11 Binomial Distribution Consider n repeated trials of the binomial experiment. The random variable X is the number of successes in the n trials. We have calculated $E(X) = np$. To determine σ^2, we must calculate $E(X^2)$. Since $k^2 = k(k-1) + k$, we have

$$E(X^2) = \sum_{k=0}^{n} k^2 \binom{n}{k} p^k q^{n-k} = \sum_{k=0}^{n} k(k-1) \binom{n}{k} p^k q^{n-k} + \sum_{k=0}^{\infty} k \binom{n}{k} p^k q^{n-k}.$$

We recognize the second sum as $E(X) = np$. It is easy to verify that

$$k(k-1)\binom{n}{k} = n(n-1)\binom{n-2}{k-2}.$$

Therefore, the first sum is equal to

$$n(n-1)p^2 \sum_{k=2}^{n} \binom{n-2}{k-2} p^{k-2} q^{(n-2)-(k-2)} = n(n-1)p^2 \sum_{j=0}^{n-2} \binom{n-2}{j} p^j q^{(n-2)-j}$$

$$= n(n-1)p^2(p+q)^{n-2}$$

$$= n(n-1)p^2.$$

Therefore, $E(X^2) = n(n-1)p^2 + np$,

$$\sigma^2 = E(X^2) - (E(X))^2 = n(n-1)p^2 + np - n^2p^2 = npq,$$

and $\sigma = \sqrt{npq}$. We have proved that the variance and standard deviation of the binomial distribution are npq and $\sqrt{npq}$.

Example 2.8.12   Define the random variable X_1 to be the number of heads obtained when a fair coin is tossed 100 times in succession. Calculate the variance and standard deviation of X_1.

Solution: The random variable X_1 has binomial distribution with $n = 100$ and $p = \frac{1}{2}$. The expected value is $\mu = E(X_1) = np = 50$. The variance is $\sigma^2 = \text{var}(X_1) = npq = 100(\frac{1}{2})(\frac{1}{2}) = 25$. The standard deviation is $\sigma = \sqrt{npq} = 5$.

Example 2.8.13   Consider the experiment of choosing at random a piece of paper from a bowl containing 101 pieces of paper numbered 0, 1, 2,..., 100. Define X_2 to be the random variable equal to the number on the chosen piece of paper. Calculate the variance and standard deviation of X_2.

Solution: In this example, $E(X_2) = 50$. To calculate the variance, we first calculate

$$E(X_2^2) = \sum_{i=0}^{100} i^2 f(i) = \tfrac{1}{101}(0^2 + 1^2 + 2^2 + 3^3 + \cdots + 100^2),$$

since $f(i) = \frac{1}{101}$ for $i = 0, 1, 2, .., 100$. In Appendix F, we prove that the sum of the squares of the first n integers is $n(n + 1)(2n + 1)/6$. This implies that

$$E(X_2^2) = \frac{1}{101}\frac{100(101)(201)}{6} = 3350.$$

Therefore, the variance of X_2 is $\sigma^2 = \text{var}(X) = E(X_2^2) - (E(X_2))^2 = 3350 - 2500 = 850$ and the standard deviation is $\sigma = \sqrt{850} \approx 29.15$. We notice that the variance of X_2 is very much greater than the variance of X_1 in the previous example.

Example 2.8.14   Geometric Distribution   For the geometric distribution, we have calculated $E(X) = \sum_{k=1}^{\infty} kq^{k-1}p = 1/p$. To calculate $E(X^2)$, we must sum the infinite series $E(X^2) = \sum_{k=1}^{\infty} k^2 q^{k-1}p$. Expanding this series, we find that

$$E(X^2) = 1^2 p + 2^2 qp + 3^2 q^2 p + \cdots + k^2 q^{k-1}p + (k + 1)^2 q^k p + \cdots,$$
$$qE(X^2) = \qquad\quad 1^2 qp + 2^2 q^2 p + \cdots + (k - 1)^2 q^{k-1}p + k^2 q^k p + \cdots.$$

Therefore,

$$(1 - q)E(X^2) = p + \sum_{k=1}^{\infty} [(k + 1)^2 - k^2]q^k p = p + \sum_{k=1}^{\infty} (2k + 1)q^k p.$$

This implies that $pE(X^2) = p + 2q \sum_{k=1}^{\infty} kq^{k-1}p + q \sum_{k=1}^{\infty} q^{k-1}p$. But $\sum_{k=1}^{\infty} kq^{k-1}p = E(X) = 1/p$ and $\sum_{k=1}^{\infty} q^{k-1}p = \sum_{k=1}^{\infty} f(k) = 1$, so that

$$pE(X^2) = p + \frac{2q}{p} + q$$

$$= \frac{p^2 + 2q + qp}{p} = \frac{p^2 + 2(1 - p) + (1 - p)p}{p} = \frac{2 - p}{p}.$$

Therefore,

$$E(X^2) = \frac{2 - p}{p^2} = \frac{1 + q}{p^2}$$

and

$$\sigma^2 = E(X^2) - (E(X))^2 = \frac{1 + q}{p^2} - \frac{1}{p^2} = \frac{q}{p^2}.$$

The standard deviation of the geometric distribution is $\sigma = \sqrt{q}/p$.

Example 2.8.15   A Learning Experiment   A child is given six objects, only one of which will pass through an opening in a frame. The child tries to pass the objects through the opening.

1. Assuming that the same object is not tried twice, what is the expected number of objects that will be tried?
2. Assuming that the six objects are equally likely to be chosen on each trial, what is the expected number of objects that will be tried?

Solution:

1. In this case, the number of objects tried is a random variable X which can take on values 1, 2, 3, 4, 5, and 6. Clearly, $P(X = 1) = \frac{1}{6}$. Also $P(X = 2) = \frac{5}{6}(\frac{1}{5}) = \frac{1}{6}$ and similarly, $P(X = 3) = P(X = 4) = P(X = 5) = P(X = 6) = \frac{1}{6}$. Therefore, $E(X) = \frac{1}{6}(1 + 2 + 3 + 4 + 5 + 6) = 3.5$.

2. The number of objects tried is a random variable which can take on the values 1, 2, 3, 4, ... (any positive integer). Again, we have $P(X = 1) = \frac{1}{6}$, but $P(X = 2) = \frac{5}{6}(\frac{1}{6})$, $P(X = 3) = (\frac{5}{6})^2(\frac{1}{6})$, ..., $P(X = n) = (\frac{5}{6})^{n-1}(\frac{1}{6})$, ... This is the geometric distribution with $p = \frac{1}{6}$. Therefore, $E(X) = 1/p = 6$.

This example can be thought of as a simple learning experiment. As the child learns to play this game systematically, the number of objects tried before one passes through should decrease from 6 or more to 3.5 or less. If the observed number of trials decreases steadily as the game is repeated, this can be taken as an indication of learning.

Problems 2.8

1. Consider the experiment of choosing an integer from 1 to 20 at random. Define the random variable X by $X = 0$ if the integer is divisible by 2, $X = 1$ if the integer is divisible by 3 but not by 2, and $X = 5$ otherwise.
   (a) Determine the range space, probability function, and probability distribution function of the random variable X.
   (b) Calculate the expected value, variance, and standard deviation of X.

2. (a) A fair coin is tossed five times. Define X to be the number of heads obtained. Determine the expected value, variance, and standard deviation of the random variable X.
   (b) Define Y to be the number of heads minus the number of tails obtained. What is the range space of Y? Calculate the expected value, variance, and standard deviation of Y.

3. A pair of dice are thrown four times in succession. Define the random variable X to be the number of times that a 7 is thrown. Calculate the expected value, variance, and standard deviation of X.

4. A test consists of five multiple-choice questions. Each question has three possible answers. A student guesses at all the answers.
   (a) What are the probabilities of the possible grades 0, 20, 40, 60, 80, and 100 per cent?
   (b) If a large number of students answer the test by guessing, what is the expected value of their average mark? What is the standard deviation.

5. In a large drosophila population, 25 per cent of the flies have a wing mutation. Three hundred flies are chosen at random from the population and examined for the wing mutation. Define X to be the number of flies in the sample that have the mutation. Determine the expected value, variance, and standard deviation of X.

6. Consider the rat and maze in Problem 2.7.9. Calculate the expected value, variance, and standard deviation of X, the number of incorrect paths chosen.

7. In Problem 2.7.11, calculate the expected value and standard deviation of X.

8. In Problem 2.7.12, what is the expected number of animals that will be chosen to have four animals with enlarged thyroids?

9. In Problem 2.7.13, what is the expected number of loaves that the storekeeper will sell on a given day (if he stocks at least 50 loaves)?

10. Calculate the variance of the random variable discussed in Example 2.8.3.

11. Test tubes are supplied from the storeroom of a biology department in units of 50. On any given day, the demand is equally likely to be for 0, 1, 2, ..., up to 10 units. What is the expected number of test tubes that will be supplied on a given day? What is the standard deviation?

12. In an experiment, *Drosophila pseudoobscura* were selected for positive and negative response to light. The flies were forced to pass through a maze in which 15 choices had to be made between light and dark passages. For each light passage chosen, a fly was given a score of 1 and the total score of each fly was recorded. A fly that chose all dark passages received a total score of 0; a fly that chose all light passages received a total score of 15. Assuming that the original population has no preference for light or dark passages, what is the probability distribution of the total score? What is the expected total score?

13. In constructing a mathematical model of a population, it is assumed that the probability p_n that a family has n children is given by $p_n = (.3)(.7)^n$.
   (a) What is the probability that a family has no children?
   (b) What is the probability that a family has fewer than four children?
   (c) What is the expected number of children in a family?

14. In the population model of Problem 13, it is assumed that the probability that a family with n children receives welfare assistance is given by $q_n = 1 - (.9)^{(n+1)/2}$.

   (a) What is the probability that a family with one child receives welfare assistance?
   (b) What is the probability that a family chosen at random has five children and receives welfare assistance?

15. A volunteer with a certain heart condition is needed for a medical study. Of 10 volunteers, only one has the heart condition. The volunteers are tested in turn until the volunteer with the heart condition is found.

Define X to be the number of volunteers tested. Determine the probability function of X and determine $E(X)$, the expected number of volunteers tested.

2.9 The Poisson Distribution

With the exception of the geometric distribution, we have considered only experiments with a finite number of outcomes. Our main example has been the binomial experiment with probabilities p and $q = 1 - p$ of success and failure. In this section, we will be concerned with the analysis of rare events. For rare events, the probability p of observing the event on one trial of the experiment is very small. We will consider n repeated trials of such an experiment, where n is very large. To define and study the Poisson distribution, we must use many of the properties of the exponential function. These are developed in Appendix D.

Example 2.9.1   Suppose that a certain rare disease occurs in .1 per cent of a given large population. If 5000 people are chosen at random from this population and tested for the disease, the expected number of people who have the disease is $E(X) = np = 5000(.001) = 5$. The probability that exactly k people have the disease is

$$P(k) = f(5000, k, .001) = \binom{5000}{k}(.001)^k(.999)^{n-k}$$

for $k = 0, 1, \ldots, 5000$. Calculating $P(10)$, say, using this formula can be extremely tedious. For values of k much larger than 5, $P(k)$ is extremely small. As an example, $P(5000) = (.001)^{5000}$. This number has 14,999 zeros after the decimal place. To avoid these calculations, we will develop a new distribution that is very useful as an approximation to the binomial distribution, which can be used whenever n is large and p is small.

The probability of k successes in n trials is $P(k)$, where $P(k) = f(n, k, p) = \binom{n}{k}p^k q^{n-k}$. Setting $\mu = np$ and $q = 1 - p$, we have

$$P(k) = \frac{n(n-1)\cdots(n-k+1)}{k!}\left(\frac{\mu}{n}\right)^k\left(1 - \frac{\mu}{n}\right)^{n-k}$$

$$= \frac{n}{n}\frac{n-1}{n}\frac{n-2}{n}\cdots\frac{n-k+1}{n}\frac{\mu^k}{k!}\left(1 - \frac{\mu}{n}\right)^n\left(1 - \frac{\mu}{n}\right)^{-k}.$$

Now, for n very large (relative to k), $(n - k)/n$ is very close to 1. Since $\mu/n = p$ is small, $(1 - \mu/n)^{-k}$ is also very close to 1. From Appendix D, $(1 - \mu/n)^n \to e^{-\mu}$ as $n \to \infty$. Thus $(1 - \mu/n)^n$ is close to $e^{-\mu}$ for n large. This leads to the approximate formula $P(k) \approx (\mu^k/k!)\, e^{-\mu}$, where k is any nonnegative integer. This formula gives an approximation to the binomial distribution with $\mu = np$, where n is large and p is small.

Definition 2.9.1   Poisson Distribution with Parameter μ   *The Poisson distribution with parameter μ is given by the probability function*

$$P(k) = \frac{\mu^k}{k!}\, e^{-\mu}$$

where k is any nonnegative integer.

In Example 2.9.1, we may approximate the binomial distribution by $f(5000, k, .001) \approx (5^k/k!)e^{-k}$. This is a much simpler formula for the probabilities.

In Definition 2.9.1, we have used the word "distribution" in reference to the Poisson approximation. This is justified only after we have verified that $P(k) = (\mu^k/k!)e^{-\mu}$ really is a probability function. It is clear that $P(k) > 0$ for $k = 0, 1, 2, \ldots$. We must show that the sum of the probabilities is 1. But

$$\sum_{k=0}^{\infty} P(k) = \sum_{k=0}^{\infty} \frac{\mu^k}{k!}\, e^{-\mu} = e^{-\mu} \sum_{k=0}^{\infty} \frac{\mu^k}{k!} = e^{-\mu}e^{\mu} = 1$$

(since $\sum_{k=0}^{\infty} \mu^k/k! = e^{\mu}$ by the definition of the exponential function in Appendix D). The mean or expected value of the Poisson distribution is

$$E(X) = \sum_{k=0}^{\infty} kP(k) = \sum_{k=1}^{\infty} k\frac{\mu^k}{k!}e^{-\mu} = e^{-\mu}\mu \sum_{k=1}^{\infty} \frac{\mu^{k-1}}{(k-1)!}.$$

Letting $l = k - 1$, we obtain

$$E(X) = \mu e^{-\mu} \sum_{l=0}^{\infty} \frac{\mu^l}{l!} = \mu e^{-\mu}e^{\mu} = \mu.$$

Thus the parameter in the Poisson distribution is also its mean.

The Poisson distribution is extremely important in many biological and physical problems. The potential applications are illustrated in the two examples that follow and in the problems at the end of the section.

Example 2.9.2   Radioactive Decay   Consider a sample of radioactive matter that produces, on an average, r impulses of radioactivity per second. In t seconds, the expected number of impulses is rt. This process can be described by a Poisson distribution. The sample consists of a very large number n of radioactive atoms, each atom having an extremely small probability p of decaying during any 1-second period. The expected number of decays in 1 second is $r = np$. The expected number of decays in t seconds is $rt = npt$. This is the mean of the Poisson distribution $P(k)$, which gives the probability of k decays in t seconds:

$$P(k) = \frac{(rt)^k}{k!}e^{-rt}.$$

For example, if there are three impulses of radioactivity per second on average, then the probability that there will be 10 impulses in a 5-second interval is

$$P(10) = \frac{[(3)(5)]^{10}}{10!}e^{-(3)(5)} = \frac{15^{10}}{10!}e^{-15}.$$

Example 2.9.3   Counting of Cells Under a Microscope   The problem of counting cells under a microscope is a familiar one to biologists. Under reasonable assumptions, there is an interesting interpretation of this problem as a problem involving the Poisson distribution. Suppose that n cells of a certain type are distributed at random over an area of a glass slide that is divided by a square grid into 900 (30 × 30) equal areas. Consider any one of these 900 areas. The probability that a particular cell is in this part of the grid is $p = \frac{1}{900}$. The process of placing the n cells on the slide can be considered as n repeated trials of a binomial experiment, with "success" defined as a cell being placed in the particular area of the grid.

If n is large, the Poisson approximation to the binomial distribution can be used to calculate the probability that a particular area of the grid contains k cells. The parameter $\mu = np = n/900$.

$$P(k) = \frac{(np)^k}{k!}e^{-np} = \left(\frac{n}{900}\right)^k \frac{e^{-n/900}}{k!}, \qquad k = 0, 1, 2, \ldots.$$

The proportion of the 900 areas that contain k cells is given by $P(k)$. The number of the areas that will contain k cells is $(900)P(k)$. For example, we expect, on average, $900e^{-n/900}$ of the areas to contain no cells.

This gives us a method of estimating the number of cells present by observing the number of areas of the square grid which contain no cells. If we observe, for example, that 75 of the areas of the square grid contain no cells, then $75 \approx 900e^{-n/900}$. Therefore, $n = 900 \log_e \frac{900}{75} \approx 2240$. (Values of the function $y = \log_e x$ are given in Table II.)

The assumption we have made is that the n cells are distributed at random throughout the square grid. If this assumption is justified, the Poisson distribution will give a very effective means of estimating the number of cells present on the slide.

We conclude this section by calculating the variance of the Poisson distribution. We have calculated $E(X) = \mu$, where X is the random variable such that $P(X = k) = (\mu^k/k!)e^{-\mu}$. To calculate $E(X^2)$, we again make use of the identity $k^2 = k(k - 1) + k$.

$$E(X^2) = e^{-\mu} \sum_{k=0}^{\infty} k^2 \frac{\mu^k}{k!} = e^{-\mu} \sum_{k=0}^{\infty} k(k-1)\frac{\mu^k}{k!} + e^{-\mu} \sum_{k=0}^{\infty} k\frac{\mu^k}{k!}$$

$$= e^{-\mu} \sum_{k=2}^{\infty} \frac{\mu^k}{(k-2)!} + E(X) = e^{-\mu} \mu^2 \sum_{k=2}^{\infty} \frac{\mu^{k-2}}{(k-2)!} + E(X).$$

Letting $j = k - 2$, we find that this last expression is equal to

$$\mu^2 e^{-\mu} \sum_{j=0}^{\infty} \frac{\mu^j}{j!} + E(X) = \mu^2 e^{-\mu} e^{\mu} + \mu$$

$$= \mu^2 + \mu.$$

Therefore, $\sigma^2 = E(X^2) - (E(X))^2 = \mu^2 + \mu - \mu^2 = \mu$. We conclude that the mean and variance of the Poisson distribution are both equal to μ.

We could have calculated the variance of the Poisson distribution by a much simpler argument. The Poisson distribution with mean μ is an approximation to the binomial distribution with mean $\mu = np$, where n is large and p is small. The variance of the binomial distribution is npq. If p is small, then $1 - p = q$ is approximately 1 and npq is approximately equal to np. But $\mu = np$, and we conclude that the variance of the Poisson distribution with mean μ is equal to μ.

Problems 2.9

1. Suppose that a rare disease occurs in 0.02 per cent of a large population. A sample of 20,000 people are chosen at random from this population and tested for the disease. What is the expected number of people in this sample who have the disease? What is the formula for the probability that exactly k people have the disease? (Give the binomial probability and the Poisson approximation.) What is the probability that the disease will not be observed in this sample?

2. For the Poisson distribution with parameter μ, prove that

   (a) $P(k + 1) = \dfrac{\mu}{k + 1} P(k).$

   (b) $P(k + 2) = \dfrac{\mu^2}{(k + 2)(k + 1)} P(k).$

   (c) If μ is a positive integer, $P(k)$ is a maximum when $k = \mu$.

3. About 1 child in 700 is born with Down's syndrome (mongolism). In a large hospital, there are 3500 births per year. What is the expected number of children born with Down's syndrome? What is the probability that more than 2 children are born with Down's syndrome?

4. Among 10,000 barley seedlings, 2 on average do not have the normal green color, as a result of spontaneous mutations which affect their chlorophyll. What is the probability that, among 10,000 barley seedlings chosen at random, exactly 2 do not have the normal green color?

5. A vaccine is believed to produce immunity against polio in 99.99 per cent of cases. Suppose that the vaccine is given to 10,000 people. What is the expected number of people who do not have immunity? What is the probability that exactly k people do not have immunity? What is the probability that fewer than 2 people do not have immunity?

6. A field is divided into 2500 squares of equal area. Dandelions are randomly distributed over the field, and it is observed that exactly 275 of the squares contain no dandelions. Assuming a Poisson distribution, give the formula for the number of squares containing exactly three dandelions. What is the formula for the number of squares containing three or more dandelions?

7. The catch from 1000 lobster traps was 1200 lobsters. Assuming a Poisson distribution, what is the number of traps that contain (a) no lobsters and (b) two or more lobsters?

8. A book has 400 pages and it is estimated that there are 400 misprints distributed at random through the book. Assuming a Poisson distribu-

tion, what is the number of pages that contain (a) no misprints, (b) exactly one misprint, and (c) more than two misprints?

9. In a study of the feeding habits of protozoa, it was assumed that the number of units of food eaten in 1 hour was a random variable with Poisson distribution with mean 10.
   (a) What is the probability that in a given hour a protozoan will eat nothing?
   (b) What is the probability that at least one unit of food will be consumed in a given hour by a particular protozoan?

10. In a study of detritus feeders in a lake, the lake bottom was divided into 1,000,000 squares of equal area. It was estimated that there was an average of two units of detritus food per square and the probability distribution of the number of units of food in a square was assumed to be Poisson with parameter 2. What is the number of squares which contain (a) no units of food, (b) exactly two units of food, and (c) four or more units of food?

11. A vaccine for an infectious disease produces an unfavorable reaction in 0.1 per cent of all cases and does not produce immunity in 0.2 per cent of all cases. Suppose that these effects are independent. If the vaccine is administered to 10,000 people, determine the probabilities (a) that no unfavorable reactions ensue and all have immunity and (b) that exactly one unfavorable reaction ensues and exactly two do not have immunity.

12. In a sample of radioactive matter, 10 impulses of radioactivity are produced per second on average. What is the probability distribution of the number of impulses emitted in a tenth of a second? In ten seconds?

13. The number of emergency cases admitted to a hospital in 1 hour is a random variable with Poisson distribution with mean 3. Determine the probabilities that in a given hour (a) no emergency cases are admitted and (b) more than three emergency cases are admitted.

14. A particular food produces an allergic reaction in 0.01 per cent of a large population. If 100,000 people eat this food on an average day, what is the expected number of people who will experience an allergic reaction? What is the probability distribution of the number of people (in this group of 100,000) who are allergic to this food?

15. It is estimated that there is on average one birth per hour in a certain city. What is the probability that, in a given hour, there will be no births? More than two births?

16. A supplier of microscopes receives an average of three orders for a particular microscope on a business day. Suppose that, on one day, there

are only four of these microscopes in stock. What is the probability that the supplier will be able to fill that day's orders for this microscope?

17. The frequency of tuberculosis in a large population is estimated to be 4 in 10,000. Suppose that 20,000 people chosen at random are tested for the disease, using a test that indicates the presence of the disease in 95 per cent of cases in which it is present and in 5 per cent of cases in which it is not.

   (a) What is the expected number of people who have the disease in this sample?

   (b) What is the expected number of people who will be found to have the disease by this test (and will be required to take further tests)?

   (c) What is the expected number of people who have tuberculosis which is not detected by this test?

   (d) What is the probability that no case of tuberculosis in this sample will escape detection?

18. The death rates from sudden coronary attacks are believed to be affected by the hardness of the local water supply. In a study, the annual death rates per 1,000,000 people from sudden coronary attacks were 2000, where the water supply was soft, and 1200 where the water supply was hard. Suppose that two similar towns each contain 1000 people, but that town I has soft water and town II has hard water. What is the probability that, in a given year, there are no sudden coronary attacks in the two towns? That there are two attacks in each town?

19. (a) It is estimated that 60 per cent of the population in the United States receive diagnostic X-rays annually. Assuming a Poisson distribution what proportion of the population in a given year receive exactly two diagnostic X-rays? More than two diagnostic X-rays?

   (b) What is the expected number of diagnostic X-rays per capita in a given year?

   (c) What factors could affect the validity of the assumption of a Poisson distribution?

20. During the migration period, the number of ducks arriving at a lake in a wildlife preserve averages 200 per hour. Assuming that ducks fly in flocks of 40 during migration and that the arrivals of the flocks at the lake are independent, estimate the probability that no ducks arrive in a given 1-hour period and the probability that more than 100 ducks arrive in the given 1-hour period.

21. To study the distribution of ant nests in a large open field, the field is divided into 1600 squares of equal area. It is observed that exactly 400

of the squares contain no ant nests. Assuming that the ant nests are distributed at random over the field, estimate the total number of nests in this field. How would territorial behavior (competition between nests) tend to affect this estimate?

Vectors and Matrices

3.1 Vectors

The description of biological processes often involves studying the inter-relations of a very large number of variables. For example, to study a natural ecosystem it may be necessary to record or model the growth and interaction of 100 or more plant and animal species. To keep track of all the relevant variables and to make possible the mathematical description of such complex systems, a simple mathematical framework must be developed. A natural way to do this is by means of vectors and matrices.

Definition 3.1.1   Row and Column Vectors   *An n-component row vector* **x** *is a set of n real numbers written in a row in a definite order,* $\mathbf{x} = (x_1, x_2, \ldots, x_n)$. *An n-component column vector* **y** *is a set of n real numbers written in a column in a definite order,*

$$\mathbf{y} = \begin{pmatrix} y_1 \\ y_2 \\ \vdots \\ y_n \end{pmatrix}.$$

The numbers x_i and y_i are the ith components of **x** *and* **y**, *respectively. Two n-component row (column) vectors are said to be* equal *if their corresponding components are equal. The* zero *vector (row or column) has all components zero.*

$$0 = (0, 0, \ldots, 0).$$

Example 3.1.1   The following are examples of vectors.

1. $\begin{pmatrix} \frac{1}{2} \\ \frac{1}{4} \\ \frac{1}{4} \end{pmatrix}$ (three-component column vector).

2. $(3, 4, 6)$ (three-component row vector).
3. $(1, -1, 0, 2)$ (four-component row vector).

Example 3.1.2   Vector Geometry   Two-component row (or column) vectors can be thought of as points in a plane with the first component being the coordinate on the horizontal axis and the second component the coordinate on the vertical axis (Figure 3.1). The vector $\mathbf{x} = (x_1, x_2)$ is also represented by an arrow from the origin to the point (x_1, x_2) in the plane.

We now introduce two operations with vectors, which are suggested by the vectors shown in Figure 3.1.

Definition 3.1.2   Vector Addition   *If* $\mathbf{x} = (x_1, x_2, \ldots, x_n)$ *and* $\mathbf{y} = (y_1, y_2, \ldots, y_n)$ *are two n-component row vectors, their vector sum is*

$$\mathbf{z} = \mathbf{x} + \mathbf{y} = (x_1 + y_1, x_2 + y_2, \ldots, x_n + y_n).$$

This is an n-component row vector whose components are the sums of the corresponding components of the vectors **x** *and* **y**.

The same definition holds for the addition of two n-component column vectors. We emphasize that two vectors can be added if and only if they are both either row vectors or column vectors and they have the same number of components.

Definition 3.1.3   Scalar Multiplication   *Let* $\mathbf{x} = (x_1, x_2, \ldots, x_n)$ *be an n-component row vector and let c be a real number (or scalar). Then, c***x** *is the*

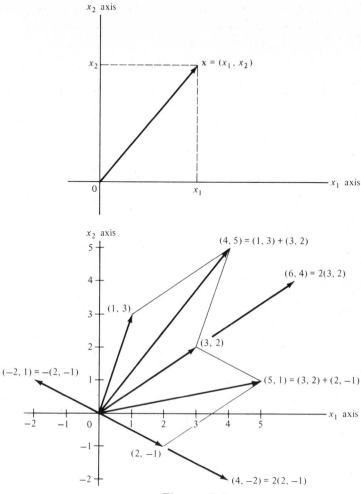

Figure 3.1

vector obtained from **x** *by multiplying each component of* **x** *by c.*

$$c\mathbf{x} = (cx_1, cx_2, \ldots, cx_n).$$

Example 3.1.3 If $\mathbf{x} = (1, 2, 3, 4)$ and $\mathbf{y} = (3, 6, 9, 12)$, verify that $3\mathbf{x} - \mathbf{y} = \mathbf{0}$.

Solution: We must verify that the vectors $3\mathbf{x}$ and $-\mathbf{y}$ add to $\mathbf{0}$. But $3\mathbf{x} = (3, 6, 9, 12)$ and $-\mathbf{y} = (-1)\mathbf{y} = (-3, -6, -9, -12)$. Therefore

$$3\mathbf{x} - \mathbf{y} = (3 - 3, 6 - 6, 9 - 9, 12 - 12) = (0, 0, 0, 0) = \mathbf{0}.$$

Vector addition and scalar multiplication have an obvious geometrical interpretation, which is illustrated in Figure 3.2 for two-component vectors.

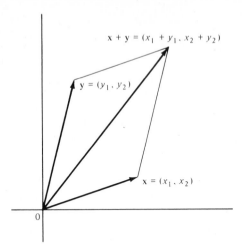

Figure 3.2. The Addition of Vectors

From the definition of the sum of two vectors, we can define the sum of any number of vectors. For example, if $\mathbf{x} = (x_1, x_2, \ldots, x_n), \mathbf{y} = (y_1, y_2, \ldots, y_n)$, and $\mathbf{z} = (z_1, z_2, \ldots, z_n)$, then $\mathbf{x} + \mathbf{y} + \mathbf{z} = (x_1 + y_1 + z_1, x_2 + y_2 + z_2, \ldots, x_n + y_n + z_n)$. If $\mathbf{x}_1, \mathbf{x}_2, \ldots, \mathbf{x}_m$ are m n-component row vectors and $c_1, c_2, \ldots, c_m$ are scalars, then the vector $\mathbf{x} = c_1 \mathbf{x}_1 + c_2 \mathbf{x}_2 + \cdots + c_m \mathbf{x}_m$ is called a *linear combination* of $\mathbf{x}_1, \mathbf{x}_2, \ldots, \mathbf{x}_m$. We will often be interested in determining whether a given vector can be written as a linear combination of other given vectors.

Definition 3.1.4   Linear Dependence and Independence   *The n-component row vectors $\mathbf{x}_1, \mathbf{x}_2, \ldots, \mathbf{x}_m$ are said to be* linearly dependent *if there exist scalars $c_1, c_2, \ldots, c_m$ not all equal to zero such that $c_1 \mathbf{x}_1 + c_2 \mathbf{x}_2 + \cdots + c_m \mathbf{x}_m = \mathbf{0}$. Otherwise, the vectors are said to be* linearly independent.

Example 3.1.4   The vectors $\mathbf{x}_1 = (1, 0, 0, \ldots, 0)$, $\mathbf{x}_2 = (0, 1, 0, \ldots, 0), \ldots$, $\mathbf{x}_n = (0, 0, 0, \ldots, 1)$ are linearly independent. If $c_1 \mathbf{x}_1 + c_2 \mathbf{x}_2 + \cdots + c_n \mathbf{x}_n = \mathbf{0}$, this implies that $(c_1, c_2, \ldots, c_n) = (0, 0, \ldots, 0)$ and therefore $c_1 = c_2 = \cdots = c_n = 0$. Every n-component row vector can be written as a linear combination of these vectors. If $\mathbf{y} = (y_1, y_2, \ldots, y_n)$, then $\mathbf{y} = y_1 \mathbf{x}_1 + y_2 \mathbf{x}_2 + \cdots + y_n \mathbf{x}_n$.

Example 3.1.5   Are the vectors $x_1 = (1, 2, 3)$, $x_2 = (4, -3, 7)$, and $x_3 = (-2, 7, -1)$ linearly dependent or independent?

Solution: To prove linear dependence, we must find real numbers c_1, c_2, and c_3 (not all zero) such that $c_1 x_1 + c_2 x_2 + c_3 x_3 = 0$. Then, we must have $(c_1, 2c_1, 3c_1) + (4c_2, -3c_2, 7c_2) + (-2c_3, 7c_3, -c_3) = (0, 0, 0)$. This leads to three equations:

$$c_1 + 4c_2 - 2c_3 = 0,$$
$$2c_1 - 3c_2 + 7c_3 = 0,$$
$$3c_1 + 7c_2 - c_3 = 0.$$

In later sections, we study such systems of equations in general. For the moment, note that one solution is $c_1 = 2, c_2 = -1, c_3 = -1$. Therefore, $2x_1 - x_2 - x_3 = 0$, and the three vectors are linearly dependent.

It is possible to define a product of two n-component (row or column) vectors. To motivate the definition, consider an aquarium that contains n_1 fish of one species, n_2 fish of a second species, and n_3 fish of a third species. A natural vector to define would be the population vector $\mathbf{n} = (n_1, n_2, n_3)$. An average fish of the first species may require q_1 units of food per day. For the second and third species, the corresponding consumptions are q_2 and q_3. The consumption vector is $\mathbf{q} = (q_1, q_2, q_3)$. The total daily requirements are $n_1 q_1 + n_2 q_2 + n_3 q_3$. This suggests the following definition.

Definition 3.1.5   Inner Product   *Let* $\mathbf{x} = (x_1, x_2, \ldots, x_n)$ *and* $\mathbf{y} = (y_1, y_2, \ldots, y_n)$ *be two n-component vectors. The inner product, written* $\mathbf{x} \cdot \mathbf{y}$, *of* $\mathbf{x}$ *and* $\mathbf{y}$ *is*

$$\mathbf{x} \cdot \mathbf{y} = x_1 y_1 + x_2 y_2 + \cdots + x_n y_n = \sum_{k=1}^{n} x_k y_k.$$

The inner product is also called the *dot product* or the *scalar product* (since $\mathbf{x} \cdot \mathbf{y}$ is a scalar or real number). It is defined only when $\mathbf{x}$ and $\mathbf{y}$ have the same number of components. Note that $\mathbf{x} \cdot \mathbf{y} \cdot \mathbf{z}$ is, in general, not defined since $\mathbf{x} \cdot \mathbf{y}$ is a scalar and $\mathbf{z}$ is a vector.

Example 3.1.6   Calculate the inner product of $\mathbf{x} = (1, 1, 2, 3)$ and $\mathbf{y} = (1, 0, -1, 2)$.

Solution: $\mathbf{x} \cdot \mathbf{y} = 1(1) + 1(0) + 2(-1) + 3(2) = 1 + 0 - 2 + 6 = 5.$

Example 3.1.7   Expected Value as an Inner Product   The expected value E of the outcome of an experiment is $E = x_1 p_1 + x_2 p_2 + \cdots + x_n p_n$ (Definition 2.8.1). This can be interpreted as the inner product of the "outcome vector" $\mathbf{x} = (x_1, x_2, \ldots, x_n)$ and the "probability vector" $\mathbf{p} = (p_1, p_2, \ldots, p_n)$. In Example 2.8.1, $\mathbf{x} = (1, 2, 3, 4, 5, 6)$, $\mathbf{p} = (\frac{1}{6}, \frac{1}{6}, \frac{1}{6}, \frac{1}{6}, \frac{1}{6}, \frac{1}{6})$, and $E = \mathbf{x} \cdot \mathbf{p} = 3.5$. Probability vectors will be studied in Chapter 5 in connection with Markov chains.

The inner product can also be used to define the length or *norm* of a vector. If $\mathbf{x} = (x_1, x_2)$, then, referring again to Figure 3.1, we see that the length of the vector from $(0, 0)$ to (x_1, x_2) is given, using the Pythagorean theorem, by $\sqrt{x_1^2 + x_2^2}$. But $x_1^2 + x_2^2 = x_1 x_1 + x_2 x_2 = \mathbf{x} \cdot \mathbf{x}$. This leads to the following definition.

Definition 3.1.6   Norm of a Vector   *Let* $\mathbf{x} = (x_1, x_2 \ldots, x_n)$ *be an n-component vector. Then the* norm *of* $\mathbf{x}$, *denoted by* $\|\mathbf{x}\|$, *is given by*

$$\|\mathbf{x}\| = \sqrt{\mathbf{x} \cdot \mathbf{x}} = \sqrt{x_1^2 + x_2^2 + \cdots x_n^2}.$$

Example 3.1.8   Calculate $\|\mathbf{x}\|$ and $\|\mathbf{y}\|$, where $\mathbf{x} = (1, 2, 3)$ and

$$\mathbf{y} = \begin{pmatrix} 1 \\ -3 \\ 0 \\ 2 \\ 4 \end{pmatrix}.$$

Solution: $\|\mathbf{x}\| = \sqrt{1^2 + 2^2 + 3^2} = \sqrt{14}$ and

$$\|\mathbf{y}\| = \sqrt{1^2 + (-3)^2 + 0^2 + 2^2 + 4^2} = \sqrt{30}.$$

Problems 3.1

1. Write the vector $(1, 1, 2)$ as a linear combination of the vectors $(1, 1, 1)$, $(0, 1, 1)$, and $(0, 0, 1)$. Are these four vectors linearly independent?

2. Determine constants c_1 and c_2 such that $c_1(1, 3) + c_2(-1, 2) = (1, 8)$. Draw a diagram that illustrates this solution geometrically.

3. Verify the following properties of the inner product.
   (a) $(\mathbf{x} + \mathbf{y}) \cdot \mathbf{z} = \mathbf{x} \cdot \mathbf{z} + \mathbf{y} \cdot \mathbf{z}$.
   (b) $\mathbf{x} \cdot (\mathbf{y} + \mathbf{z}) = \mathbf{x} \cdot \mathbf{y} + \mathbf{x} \cdot \mathbf{z}$.

4. Define the vectors $\mathbf{u} = (2, -3, 7)$ and $\mathbf{v} = (3, 1, -4)$. Calculate
   (a) $\mathbf{u} + 2\mathbf{v}$.   (b) $3\mathbf{u} - 2\mathbf{v}$.   (c) $\|\mathbf{u}\|$.
   (d) $\|\mathbf{v}\|$.   (e) $\|\mathbf{u} + \mathbf{v}\|$.   (f) $\mathbf{u} \cdot \mathbf{v}$.

5. Verify that the following sets of vectors are linearly dependent.
   (a) $(2, 1), (3, 2), (1, -1)$.
   (b) $(1, 1, 1), (2, -1, -1), (0, -3, -3)$.
   (c) $(1, 0, 0, 1), (0, 1, -1, 0), (3, 2, -2, 3)$.

6. If $\mathbf{x}_1, \mathbf{x}_2, \ldots, \mathbf{x}_m$ are linearly dependent n-component row vectors, prove that at least one of these vectors can be written as a linear combination of the others.

7. If d_1, d_2, and d_3 are the lengths of the three sides of a triangle, the *law of cosines* from trigonometry states that $d_3^2 = d_1^2 + d_2^2 - 2d_1 d_2 \cos \theta$, where θ is the angle between the sides with lengths d_1 and d_2. If $\mathbf{x}$ and $\mathbf{y}$ are any two-component vectors, the law of cosines can be written $\|\mathbf{x} - \mathbf{y}\|^2 = \|\mathbf{x}\|^2 + \|\mathbf{y}\|^2 - 2\|\mathbf{x}\| \|\mathbf{y}\| \cos \theta$. By writing $\|\mathbf{x} - \mathbf{y}\|^2 = (\mathbf{x} - \mathbf{y}) \cdot (\mathbf{x} - \mathbf{y})$, conclude that $\mathbf{x} \cdot \mathbf{y} = \|\mathbf{x}\| \|\mathbf{y}\| \cos \theta$. (This identity is sometimes used to define the inner product in two dimensions.)

8. Using Problem 7, determine the cosine of the angle between the following pairs of vectors.
   (a) $(1, 2), (3, 4)$.   (b) $(1, 1), (1, 3)$.
   (c) $(0, 1), (5, 12)$.   (d) $(-1, -1), (2, 3)$.

9. From Problem 7, prove that $|\mathbf{x} \cdot \mathbf{y}| \leq \|\mathbf{x}\| \|\mathbf{y}\|$, where $\mathbf{x}$ and $\mathbf{y}$ are any two-component vectors. (This relation is called the *Schwarz inequality*. It is valid for any two n-component vectors $\mathbf{x}$ and $\mathbf{y}$).

10. Use Problem 9 to prove that $\|\mathbf{x} + \mathbf{y}\| \leq \|\mathbf{x}\| + \|\mathbf{y}\|$. (This relation between the norms of two vectors and the norm of their sum is known as the *triangle inequality*. In two dimensions, it can be interpreted as the statement that the length of one side of a triangle is less than or equal to the sum of the lengths of the other two sides.)

11. Verify the Schwarz inequality and the triangle inequality for the following pairs of vectors.
    (a) $\mathbf{x} = (0, 1), \quad \mathbf{y} = (3, 4)$.
    (b) $\mathbf{x} = (1, 2, 2), \quad \mathbf{y} = (0, 3, 4)$.

(c) $x = (1, 1, 1),\quad y = (1, 0, -1)$.

(d) $x = (1, 3, -2, 2),\quad y = (2, 6, -4, 4)$.

12. Two n-component vectors x and y are said to be *orthogonal* if their inner product is zero. Prove that, in two dimensions, the vectors x and y are orthogonal if and only if they are at right angles to each other.

13. Which of the following pairs of vectors are orthogonal?

(a) $x = (1, 2),\quad y = (-2, 1)$.

(b) $x = (1, 2),\quad y = (1, -2)$.

(c) $x = (3, -7, 2),\quad y = (7, 3, 0)$.

(d) $x = (1, 1, -1, -3),\quad y = (3, 1, 1, 1)$.

14. The vectors $(1, 0, 0)$, $(0, 1, 0)$, and $(0, 0, 1)$ are said to be a *basis* for $\mathbf{R}^3$. Verify (a) that any vector $x = (x_1, x_2, x_3)$ can be written as a linear combination of these basis vectors and (b) that this is an orthogonal basis, that is, the basis vectors are orthogonal in pairs.

15. Determine the constants c such that the following pairs of vectors are orthogonal.

(a) $x = (2, 1),\quad y = (1, c)$.

(b) $x = (1, 1, 1),\quad y = (0, c^2, -c)$.

(c) $x = (-1, 2, 2),\quad y = (3, 4, c)$.

16. The *population vector* of an ecosystem composed of m coexisting species is defined to be an m-component row vector $\mathbf{n} = (n_1, n_2, \ldots, n_m)$, where the ith component n_i is the population of the ith species.

(a) If the population of each species doubles, what is the new population vector?

(b) If all species except the first become extinct, what is the population vector?

(c) If two individuals of each species are added to the ecosystem, what is the new population vector?

(d) If two isolated ecosystems with population vectors $\mathbf{n}^{(1)}$ and $\mathbf{n}^{(2)}$ become connected, prove that the population vector of the new ecosystem is $\mathbf{n}^{(1)} + \mathbf{n}^{(2)}$.

3.2 Matrices

Definition 3.2.1   Matrix   *An $m \times n$ matrix $A = (a_{ij})$ is a rectangular array of mn numbers arranged in a definite order in m rows and n columns.*

$$A = \begin{pmatrix} a_{11} & a_{12} & \cdots & a_{1j} & \cdots & a_{1n} \\ a_{21} & a_{22} & \cdots & a_{2j} & \cdots & a_{2n} \\ \vdots & \vdots & & \vdots & & \vdots \\ a_{i1} & a_{i2} & \cdots & a_{ij} & \cdots & a_{in} \\ \vdots & \vdots & & \vdots & & \vdots \\ a_{m1} & a_{m2} & \cdots & a_{mj} & \cdots & a_{mn} \end{pmatrix}.$$

The number a_{ij} appearing in the ith row and jth column is the ijth component of A. For convenience, the matrix A is often denoted $A = (a_{ij})$.

Example 3.2.1   The following are $m \times n$ matrices for various values of m and n.

1. $A = \begin{pmatrix} 1 & 6 \\ 3 & 5 \\ 4 & 7 \end{pmatrix}$, $m = 3, n = 2$.

2. $B = \begin{pmatrix} 1 & 2 & 4 \\ -1 & 0 & 2 \end{pmatrix}$, $m = 2, n = 3$.

3. $C = \begin{pmatrix} 1 & 1 & 0 \\ 0 & 1 & 1 \\ -1 & 2 & 2 \end{pmatrix}$, $m = n = 3$.

4. $I = \begin{pmatrix} 1 & 0 & 0 \\ 0 & 1 & 0 \\ 0 & 0 & 1 \end{pmatrix}$, $m = n = 3$.

Vectors can be thought of as special cases of matrices. An n-component row vector is a $1 \times n$ matrix and an n-component column vector is an $n \times 1$ matrix. Matrices that are not vectors will be denoted by capital letters.

Example 3.2.2   Five laboratory animals are fed on three different foods. If c_{ij} is defined to be the daily consumption of the ith food by the jth animal, then

$$C = (c_{ij}) = \begin{pmatrix} c_{11} & c_{12} & c_{13} & c_{14} & c_{15} \\ c_{21} & c_{22} & c_{23} & c_{24} & c_{25} \\ c_{31} & c_{32} & c_{33} & c_{34} & c_{35} \end{pmatrix}$$

is a 3×5 matrix that records all daily consumptions. This is a convenient way to keep records.

A *square matrix* has the same number of rows as columns. The components $a_{11}, a_{12}, \ldots, a_{nn}$ of an $n \times n$ square matrix $A = (a_{ij})$ are called the *diagonal components* of A. If the nondiagonal components of a square matrix A are all zero, then A is said to be *diagonal*. The $n \times n$ diagonal matrix I whose diagonal components are all equal to 1 is called the $n \times n$ *identity matrix*.

Definition 3.2.2   Addition and Scalar Multiplication of Matrices
If $A = (a_{ij})$ and $B = (b_{ij})$ are two $m \times n$ matrices, the sum of A and B is $A + B = (a_{ij} + b_{ij})$. If c is a scalar, we define the matrix $cA = (ca_{ij})$. The ijth components of $A + B$ and cA are $a_{ij} + b_{ij}$ and ca_{ij}, respectively.

Example 3.2.3   If

$$A = \begin{pmatrix} 1 & 2 & 3 \\ -1 & 0 & 4 \end{pmatrix} \quad \text{and} \quad B = \begin{pmatrix} 5 & 7 & 0 \\ -2 & 1 & 3 \end{pmatrix},$$

calculate $2A$, $A + B$, and $2A + B$.

Solution:

$$2A = \begin{pmatrix} 2 & 4 & 6 \\ -2 & 0 & 8 \end{pmatrix}, \quad A + B = \begin{pmatrix} 6 & 9 & 3 \\ -3 & 1 & 7 \end{pmatrix}, \quad 2A + B = \begin{pmatrix} 7 & 11 & 6 \\ -4 & 1 & 11 \end{pmatrix}.$$

It is possible to define a product of two matrices using a definition suggested by the inner product of n-component vectors.

Definition 3.2.3   Product of Two Matrices   *If $A = (a_{ij})$ is an $m \times n$ matrix and $B = (b_{ij})$ is an $n \times p$ matrix, the product of A and B is the $m \times p$*

matrix $C = AB = (c_{ij})$, whose ijth component is $c_{ij} = \sum_{k=1}^{n} a_{ik}b_{kj}$. In other words, c_{ij} is the inner product of the ith row of A and the jth column of B.

It is important to note that the product AB of two matrices is defined only when the number of columns of A is equal to the number of rows of B. If $\mathbf{a}_i$ is the n-component row vector $\mathbf{a}_i = (a_{i1}, a_{i2}, \ldots, a_{in})$ and $\mathbf{b}_j$ is the n-component column vector

$$\mathbf{b}_j = \begin{pmatrix} b_{1j} \\ b_{2j} \\ \vdots \\ b_{nj} \end{pmatrix},$$

$$c_{ij} = \mathbf{a}_i \cdot \mathbf{b}_j = a_{i1}b_{1j} + a_{i2}b_{2j} + \cdots + a_{in}b_{nj} = \sum_{k=1}^{n} a_{ik}b_{kj}.$$

Example 3.2.4 If

$$A = \begin{pmatrix} 1 & 2 & 3 \\ 4 & 5 & 6 \end{pmatrix}, \quad B = \begin{pmatrix} 1 & 1 \\ 1 & 2 \\ 1 & 3 \end{pmatrix}, \quad C = \begin{pmatrix} 1 & -1 & 0 \\ 0 & 1 & 1 \\ 2 & 1 & 1 \end{pmatrix},$$

calculate AB, BA, AC, BC, and C^2.

Solution:

$$AB = \begin{pmatrix} 1 & 2 & 3 \\ 4 & 5 & 6 \end{pmatrix} \begin{pmatrix} 1 & 1 \\ 1 & 2 \\ 1 & 3 \end{pmatrix} = \begin{pmatrix} 1+2+3 & 1+4+9 \\ 4+5+6 & 4+10+18 \end{pmatrix} = \begin{pmatrix} 6 & 14 \\ 15 & 32 \end{pmatrix}.$$

Similarly,

$$BA = \begin{pmatrix} 1 & 1 \\ 1 & 2 \\ 1 & 3 \end{pmatrix} \begin{pmatrix} 1 & 2 & 3 \\ 4 & 5 & 6 \end{pmatrix} = \begin{pmatrix} 5 & 7 & 9 \\ 9 & 12 & 15 \\ 13 & 17 & 21 \end{pmatrix}.$$

Also,

$$AC = \begin{pmatrix} 7 & 4 & 5 \\ 16 & 7 & 11 \end{pmatrix} \quad \text{and} \quad C^2 = \begin{pmatrix} 1 & -2 & -1 \\ 2 & 2 & 2 \\ 4 & 0 & 2 \end{pmatrix}$$

The product BC is not defined, since the number of columns of B (two) is not equal to the number of rows of C (three).

Example 3.2.5   If

$$A = \begin{pmatrix} 1 & 2 \\ -1 & 0 \end{pmatrix} \quad \text{and} \quad B = \begin{pmatrix} 2 & 0 \\ 1 & 3 \end{pmatrix},$$

calculate AB and BA.

Solution:

$$AB = \begin{pmatrix} 1 & 2 \\ -1 & 0 \end{pmatrix}\begin{pmatrix} 2 & 0 \\ 1 & 3 \end{pmatrix} = \begin{pmatrix} 1(2) + 2(1) & 1(0) + 2(3) \\ -1(2) + 0(1) & -1(0) + 0(3) \end{pmatrix} = \begin{pmatrix} 4 & 6 \\ -2 & 0 \end{pmatrix},$$

$$BA = \begin{pmatrix} 2 & 0 \\ 1 & 3 \end{pmatrix}\begin{pmatrix} 1 & 2 \\ -1 & 0 \end{pmatrix} = \begin{pmatrix} 2(1) + 0(-1) & 2(2) + 0(0) \\ 1(1) + 3(-1) & 1(2) + 3(0) \end{pmatrix} = \begin{pmatrix} 2 & 4 \\ -2 & 2 \end{pmatrix}.$$

Examples 3.2.4 and 3.2.5 illustrate that, even when AB and BA are both defined, they are not necessarily equal. The multiplication of matrices is very different from the ordinary multiplication of real numbers.

Example 3.2.6   If

$$A = \begin{pmatrix} a_{11} & a_{12} & a_{13} \\ a_{21} & a_{22} & a_{23} \\ a_{31} & a_{32} & a_{33} \end{pmatrix} \quad \text{and} \quad I = \begin{pmatrix} 1 & 0 & 0 \\ 0 & 1 & 0 \\ 0 & 0 & 1 \end{pmatrix},$$

then $A = IA = AI$. The matrix I is the 3×3 identity matrix. It plays the role in multiplication of 3×3 matrices that is played by the number 1 in ordinary multiplication.

Example 3.2.7   **First and Second Order Contact to a Contagious Disease**
Suppose that three persons have contracted a contagious disease. A second group of six persons is questioned to determine who has been in contact with the three infected persons. A third group of seven persons is then questioned to determine contacts with any of the six persons in the second group. Define the 3×6 matrix $A = (a_{ij})$ by defining $a_{ij} = 1$ if the jth person in the second group has had contact with the ith person in the first group and $a_{ij} = 0$

otherwise. Similarly, define the 6 × 7 matrix $B = (b_{ij})$ by defining $b_{ij} = 1$ if the jth person in the third group has had contact with the ith person in the second group and $b_{ij} = 0$ otherwise. These two matrices describe the direct or first order contacts between the groups.

For example, we could have

$$A = \begin{pmatrix} 0 & 0 & 1 & 0 & 1 & 0 \\ 1 & 0 & 0 & 1 & 0 & 0 \\ 0 & 0 & 1 & 1 & 0 & 1 \end{pmatrix} \quad \text{and} \quad B = \begin{pmatrix} 0 & 0 & 1 & 0 & 0 & 1 & 0 \\ 0 & 0 & 1 & 1 & 0 & 0 & 0 \\ 1 & 0 & 0 & 0 & 0 & 1 & 1 \\ 0 & 0 & 1 & 1 & 0 & 0 & 0 \\ 0 & 1 & 0 & 1 & 0 & 0 & 0 \\ 1 & 0 & 0 & 0 & 0 & 1 & 0 \end{pmatrix}$$

In this case, we have $a_{24} = 1$, which means that the fourth person in the second group has had contact with the second infected person. Also, $b_{33} = 0$, which means that the third person in the third group has not had contact with the third person in the second group.

We may be interested in studying the indirect or second order contacts between the seven persons in the third group and the three infected people. The matrix product $C = AB$ describes these second order contacts. The ijth component $c_{ij} = \sum_{k=1}^{6} a_{ik}b_{kj}$ gives the number of second order contacts between the jth person in the third group and the ith person in the infected group. With the given matrices A and B, we have

$$C = AB = \begin{pmatrix} 1 & 1 & 0 & 1 & 0 & 1 & 1 \\ 0 & 0 & 2 & 1 & 0 & 1 & 0 \\ 2 & 0 & 1 & 1 & 0 & 2 & 1 \end{pmatrix}.$$

The component $c_{23} = 2$ implies that there are two second order contacts between the third person in the third group and the second contagious person. Note that the sixth person in the third group has had $1 + 1 + 2 = 4$ indirect contacts with the infected group. Only the fifth person has had no contacts.

Theorem 3.2.1   Associative Law for Matrix Multiplication   *Suppose that $A = (a_{ij})$, $B = (b_{ij})$, and $C = (c_{ij})$ are $m \times n$, $n \times p$, and $p \times q$ matrices, respectively. Then the matrix product of AB and C is equal to the matrix product of A and BC, that is, $(AB)C = A(BC)$.*

Proof: Define $D = (d_{ij}) = AB$. Then $d_{ij} = \sum_{k=1}^{n} a_{ik}b_{kj}$. The ijth component of $(AB)C = DC$ is $\sum_{l=1}^{p} d_{il}c_{lj} = \sum_{k=1}^{n} \sum_{l=1}^{p} a_{ik}b_{kl}c_{lj}$. Now define $E = (e_{ij}) = BC$. Then, $e_{ij} = \sum_{l=1}^{p} b_{il}c_{lj}$ and the ijth component of $A(BC) = AE$ is $\sum_{k=1}^{n} a_{ik}e_{kj} = \sum_{k=1}^{n} \sum_{l=1}^{p} a_{ik}b_{kl}c_{lj}$. But this is the ijth component of $(AB)C$, and the theorem is proved.

Theorem 3.2.1 can be extended to more complicated matrix products. For example, if AB, BC, and CD are defined, then

$$ABCD = (AB)(CD) = A(BC)D.$$

The following definition is fundamental for the study of square matrices.

Definition 3.2.4   Inverse of a Matrix   *The $n \times n$ square matrix A is said to be invertible if there exists an $n \times n$ matrix, written A^{-1}, which satisfies $AA^{-1} = A^{-1}A = I$. The matrix A^{-1} is the inverse of the matrix A.*

The inverse of nonsquare matrices is not defined and not all square matrices have inverses. [For example, consider A to be the zero matrix, $a_{ij} = 0$, for all i and j. It is clear that there is no matrix A^{-1} such that $A^{-1}(0) = I$]. In the next two sections of this chapter, methods will be developed to determine whether a given square matrix has an inverse and, when it does, to calculate it. We conclude this section by proving two important results about matrix inverses.

Theorem 3.2.2   *If the $n \times n$ matrix A is invertible, the inverse is unique.*

Proof: If B and C are two inverses of A, then $AB = BA = AC = CA = I$. Therefore, $B = BI = B(AC) = (BA)C = IC = C$. We conclude that $B = C$ and that a square matrix has at most one inverse.

Theorem 3.2.3   *If A and B are two $n \times n$ matrices with inverses A^{-1} and B^{-1}, then AB is invertible with inverse $(AB)^{-1} = B^{-1}A^{-1}$.*

Proof: Since the inverse is unique if it exists, it is only necessary to verify that $B^{-1}A^{-1}$ is an inverse of AB. But $(B^{-1}A^{-1})(AB) = B^{-1}(A^{-1}A)B = B^{-1}IB = B^{-1}B = I$ and $(AB)(B^{-1}A^{-1}) = A(BB^{-1})A^{-1} = AIA^{-1} = AA^{-1} = I$. Therefore $(AB)^{-1} = B^{-1}A^{-1}$.

Problems 3.2

1. If

$$A = \begin{pmatrix} 1 & 2 & 3 \\ 0 & 1 & 1 \\ 0 & 0 & -1 \end{pmatrix} \quad \text{and} \quad B = \begin{pmatrix} 1 & -1 & -1 \\ 0 & 1 & -1 \\ 0 & 0 & 1 \end{pmatrix},$$

calculate AB and BA.

2. For

$$A = \begin{pmatrix} 0 & 5 & 3 \\ 0 & 0 & 3 \\ 0 & 0 & 0 \end{pmatrix},$$

calculate A^2, A^3, and A^4. What is A^{100}?

3. For

$$A = \begin{pmatrix} 1 & 1 \\ -1 & 1 \end{pmatrix},$$

calculate A^2, A^3, and A^4.

4. For

$$A = \begin{pmatrix} 2 & 1 & -1 \\ 1 & 1 & 0 \\ 1 & -1 & 1 \end{pmatrix},$$

calculate A^2 and A^3.

5. (a) Prove that a diagonal matrix has an inverse if and only if all the diagonal components are not zero.

(b) What are the inverses of

$$A = \begin{pmatrix} 2 & 0 & 0 \\ 0 & 3 & 0 \\ 0 & 0 & 1 \end{pmatrix} \quad \text{and} \quad B = \begin{pmatrix} 5 & 0 & 0 & 0 \\ 0 & 4 & 0 & 0 \\ 0 & 0 & -3 & 0 \\ 0 & 0 & 0 & 4 \end{pmatrix}?$$

6. (a) Suppose that A and B are two $n \times n$ diagonal matrices. Calculate AB and prove that $AB = BA$.

(b) Calculate a matrix A such that

$$A^3 = \begin{pmatrix} 27 & 0 & 0 \\ 0 & -8 & 0 \\ 0 & 0 & 1 \end{pmatrix}.$$

7. A square matrix with all elements below the main diagonal equal to zero is called an *upper triangular matrix*. This means that the square matrix $A = (a_{ij})$ is upper triangular if $a_{ij} = 0$ when $i > j$. Prove that the product of two upper triangular matrices is an upper triangular matrix.

8. If

$$A = \begin{pmatrix} 0 & 1 & 2 \\ 0 & 0 & 1 \\ 0 & 0 & 0 \end{pmatrix},$$

determine the smallest integer k such that the matrix A^k is the zero matrix.

9. Suppose that $A = (a_{ij})$ is an $n \times n$ matrix such that $a_{ij} = 0$ when $i \geq j$. Prove that there is an integer k such that A^k is the zero matrix.

10. If A is a 3×3 matrix such that

$$A\begin{pmatrix} 1 \\ 0 \\ 0 \end{pmatrix} = \begin{pmatrix} 1 \\ 3 \\ 5 \end{pmatrix}, \quad A\begin{pmatrix} 0 \\ 1 \\ 0 \end{pmatrix} = \begin{pmatrix} 2 \\ 2 \\ 0 \end{pmatrix}, \quad A\begin{pmatrix} 0 \\ 0 \\ 1 \end{pmatrix} = \begin{pmatrix} 1 \\ -1 \\ 2 \end{pmatrix},$$

what are the components of A?

11. Suppose that $A_1, A_2, \ldots, A_m$ are $n \times n$ matrices with inverses. Prove that the inverse matrix of the product $A_1 A_2 \cdots A_m$ is $A_m^{-1} \cdots A_2^{-1} A_1^{-1}$.

12. The *transpose* of the $m \times n$ matrix A is the $n \times m$ matrix A^t obtained by interchanging the rows and columns of A. For example,

$$\begin{pmatrix} 1 & 2 & 3 \\ 4 & 5 & 6 \end{pmatrix}^t = \begin{pmatrix} 1 & 4 \\ 2 & 5 \\ 3 & 6 \end{pmatrix}.$$

The *ij*th component of A^t is a_{ji}, the *ji*th component of A. Calculate the transposes of the following matrices.

(a) $A = \begin{pmatrix} 1 & 2 \\ 5 & 7 \end{pmatrix}$.

(b) $A = \begin{pmatrix} 1 & -1 & 0 \\ -1 & 2 & 1 \\ 2 & 2 & -1 \end{pmatrix}$.

(c) $A = \begin{pmatrix} 1 & 2 & 3 & 4 \\ -1 & -2 & -3 & -4 \end{pmatrix}$.

(d) $A = \begin{pmatrix} 1 \\ 4 \\ -1 \end{pmatrix}$.

13. (a) Prove that the transpose of the transpose of a matrix is equal to the
original matrix; that is, $(A^t)^t = A$.
(b) If the product AB is defined, prove that $(AB)^t = B^t A^t$.

14. A square matrix A is *symmetric* if it is equal to its transpose; that is,
$A^t = A$ or $a_{ij} = a_{ji}$ for all i and j. A square matrix is *skew-symmetric* if
$A^t = -A$ or $a_{ij} = -a_{ji}$ for all i and j. Which of the following matrices
are symmetric or skew-symmetric?

(a) $\begin{pmatrix} 1 & 4 \\ 4 & 1 \end{pmatrix}$.   (b) $\begin{pmatrix} 1 & 4 \\ -4 & 1 \end{pmatrix}$.   (c) $\begin{pmatrix} 0 & 4 \\ -4 & 0 \end{pmatrix}$.

(d) $\begin{pmatrix} 1 & 2 & 3 \\ 2 & 1 & 0 \\ 3 & 0 & 1 \end{pmatrix}$.   (e) $\begin{pmatrix} 1 & 1 & 1 \\ -1 & 1 & 1 \\ -1 & -1 & 1 \end{pmatrix}$.   (f) $\begin{pmatrix} 0 & 0 & 1 \\ 0 & 0 & 0 \\ -1 & 0 & 0 \end{pmatrix}$.

15. Suppose that A and B are two $n \times n$ symmetric matrices.
(a) Prove that $A + B$ is a symmetric matrix.
(b) Prove that AB is a symmetric matrix if and only if $AB = BA$.

16. Suppose that A and B are two $n \times n$ skew-symmetric matrices.
(a) Prove that $A + B$ is a skew-symmetric matrix.
(b) Prove that AB is a symmetric matrix if and only if $AB = BA$.

17. (a) Given the $n \times n$ matrix A, prove that $A + A^t$ and AA^t are symmetric
matrices.
(b) Verify that the components of AA^t and $A^t A$ are all nonnegative.

18. For

$$A = \begin{pmatrix} 1 & -2 & 0 \\ 3 & 0 & 1 \end{pmatrix},$$

calculate AA^t and $A^t A$.

19. An $n \times n$ matrix A has the property that its matrix product with any
$n \times n$ matrix is the zero matrix. Prove that A is the zero matrix.

20. Consider an ecosystem containing n competing species. Define the
consumption matrix $A = (a_{ij})$ to be the $n \times n$ matrix in which a_{ij} is the
average amount of the jth species consumed per day by an average
individual of the ith species. What types of behavior are described by the
following consumption matrices?

(a) $A = \begin{pmatrix} 0 & \frac{1}{2} & \frac{1}{2} \\ \frac{1}{2} & 0 & \frac{1}{2} \\ \frac{1}{2} & \frac{1}{2} & 0 \end{pmatrix}$.   (b) $A = \begin{pmatrix} 0 & \frac{1}{2} & \frac{1}{2} \\ 1 & 0 & 0 \\ 0 & 1 & 0 \end{pmatrix}$.

21. Suppose that consumption of one unit of the ith species in Problem 20 produces a return of r_i calories to a predator. Define **r** to be the n-component column vector whose ith component is r_i. Give a biological interpretation of the components of the vector $A\mathbf{r}$.

22. As in Example 3.2.7, suppose that two persons have contracted a contagious disease. A second group of five persons has been in possible contact with the two infected persons and a third group of four persons has been in possible contact with the second group. Describe the second order contacts between the third group and the two infected persons when the first order (or direct) contacts are given by the following matrices.

(a) $A = \begin{pmatrix} 1 & 1 & 0 & 0 & 1 \\ 0 & 1 & 1 & 1 & 0 \end{pmatrix}$, $B = \begin{pmatrix} 1 & 1 & 0 & 1 \\ 0 & 1 & 0 & 1 \\ 0 & 0 & 1 & 0 \\ 0 & 0 & 1 & 1 \\ 0 & 1 & 1 & 0 \end{pmatrix}$.

(b) $A = \begin{pmatrix} 1 & 1 & 1 & 1 & 1 \\ 0 & 0 & 1 & 1 & 0 \end{pmatrix}$, $B = \begin{pmatrix} 0 & 1 & 1 & 1 \\ 1 & 0 & 1 & 0 \\ 0 & 1 & 0 & 0 \\ 0 & 0 & 0 & 0 \\ 1 & 1 & 1 & 1 \end{pmatrix}$.

3.3 Systems of Linear Equations

We begin this section by considering the following system of two linear equations in two unknowns, x_1 and x_2.

$$a_{11}x_1 + a_{12}x_2 = b_1$$
$$a_{21}x_1 + a_{22}x_2 = b_2. \tag{3.1}$$

The coefficients $a_{11}, a_{12}, a_{21}, a_{22}, b_1$, and b_2 are given constants and the problem is to determine all solutions of system (3.1). There are three possibilities. There may be no solutions, a unique solution, or an infinite number of solutions. This can be seen geometrically, since the two equations $a_{11}x_1 +$

$a_{12}x_2 = b_1$ and $a_{21}x_1 + a_{22}x_2 = b_2$ represent two straight lines in the plane (Figure 3.3). If the two lines intersect in one point, there is a unique solution given by the coordinates of the point of intersection. If the lines are coincident, every point on the line gives a solution. Finally, if the lines are parallel, then there is no point of intersection and system (3.1) has no solution.

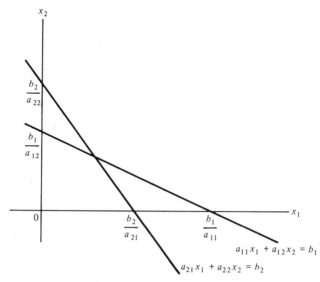

(a) Lines intersecting at one point

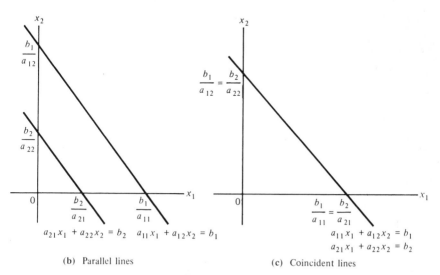

(b) Parallel lines

(c) Coincident lines

Figure 3.3

System (3.1) can be solved algebraically. Multiply the first equation by a_{22} and the second equation by a_{12} and subtract. Then $(a_{11}a_{22} - a_{21}a_{12})x_1 = a_{22}b_1 - a_{12}b_2$. Similarly, we obtain $(a_{11}a_{22} - a_{21}a_{12})x_2 = a_{11}b_2 - a_{21}b_1$ by eliminating x_1 from the equations. The solution of (3.1) is

$$x_1 = \frac{a_{22}b_1 - a_{12}b_2}{a_{11}a_{22} - a_{21}a_{12}}, \qquad x_2 = \frac{a_{11}b_2 - a_{21}b_1}{a_{11}a_{22} - a_{21}a_{12}}. \qquad (3.2)$$

Equations (3.2) are valid if $a_{11}a_{22} - a_{21}a_{12} \neq 0$. In this case, (3.2) gives the unique solution to system (3.1). If $a_{11}a_{22} - a_{21}a_{12} = 0$, then the equations leading to (3.2) yield $0 = a_{22}b_1 - a_{12}b_2$ and $0 = a_{11}b_2 - a_{21}b_1$. Thus, in order for there to be a solution, we must have $a_{22}b_1 = a_{12}b_2$ and $a_{11}b_2 = a_{21}b_1$. But then, $a_{11}/a_{21} = b_1/b_2 = a_{12}/a_{22}$, and the two equations in (3.1) are proportional. That is, any numbers x_1 and x_2 that satisfy the first equation automatically satisfy the second. Since there are infinitely many pairs of numbers x_1 and x_2 that satisfy this first equation, there will be an infinite number of solutions to system (3.1). The three possibilities correspond to the geometric possibilities in Figure 3.3.

System (3.1) can be written in another way. Define the matrix

$$A = \begin{pmatrix} a_{11} & a_{12} \\ a_{21} & a_{22} \end{pmatrix}$$

and the vectors

$$\mathbf{b} = \begin{pmatrix} b_1 \\ b_2 \end{pmatrix} \qquad \text{and} \qquad \mathbf{x} = \begin{pmatrix} x_1 \\ x_2 \end{pmatrix}.$$

Then, system (3.1) can be rewritten as $A\mathbf{x} = \mathbf{b}$. The two equations of the linear system are replaced by a single vector equation. The number $a_{11}a_{22} - a_{21}a_{12}$ plays such an important role in the solution of (3.1) that we introduce a name for it.

Definition 3.3.1   Determinant of a 2 × 2 Matrix   *The determinant of the matrix*

$$A = \begin{pmatrix} a_{11} & a_{12} \\ a_{21} & a_{22} \end{pmatrix},$$

written

$$\det A = \begin{vmatrix} a_{11} & a_{12} \\ a_{21} & a_{22} \end{vmatrix},$$

is defined as $\det A = a_{11}a_{22} - a_{21}a_{12}$.

In terms of the determinant of A, we can summarize our results about the solutions of $Ax = b$ in the following theorem.

Theorem 3.3.1 *The equation* $Ax = b$, *where* A, x, *and* b *are defined as above, has a unique solution vector* x *if and only if* $\det A \neq 0$. *If* $\det A = 0$, *there is either no solution or an infinite number of solutions.*

Example 3.3.1 Solve the linear systems

1. $x_1 + x_2 = 1$
    $4x_1 - 2x_2 = 1$.

2. $x_1 + x_2 = 1$
    $2x_1 + 2x_2 = 2$.

3. $3x_1 - x_2 = 2$
    $6x_1 - 2x_2 = 3$.

Solution:

1.
$$A = \begin{pmatrix} 1 & 1 \\ 4 & -2 \end{pmatrix}$$

and $\det A = -2 - 4 = -6$. The unique solution is $x_1 = x_2 = \frac{1}{2}$.

2.
$$A = \begin{pmatrix} 1 & 1 \\ 2 & 2 \end{pmatrix}$$

and $\det A = 0$. We notice that the second equation is just twice the first equation. The value of x_1 can be chosen arbitrarily and then $x_2 = 1 - x_1$. There are an infinite number of solutions.

3.
$$A = \begin{pmatrix} 3 & -1 \\ 6 & -2 \end{pmatrix}$$

and det $A = 0$. This system has no solutions. The first equation implies that $6x_1 - 2x_2 = 4$, which cannot be satisfied if the second equation is satisfied.

System (3.1) can be studied in yet another way. Define the vectors

$$\mathbf{a}_1 = \begin{pmatrix} a_{11} \\ a_{21} \end{pmatrix} \quad \text{and} \quad \mathbf{a}_2 = \begin{pmatrix} a_{12} \\ a_{22} \end{pmatrix}.$$

Then system (3.1) becomes

$$x_1 \mathbf{a}_1 + x_2 \mathbf{a}_2 = \mathbf{b}. \tag{3.3}$$

To solve this vector equation, the given vector $\mathbf{b}$ must be expressed as a linear combination of the given vectors $\mathbf{a}_1$ and $\mathbf{a}_2$. Let us consider the special case $\mathbf{b} = \mathbf{0}$. Then system (3.3) becomes

$$x_1 \mathbf{a}_1 + x_2 \mathbf{a}_2 = \mathbf{0}.$$

There is always one solution to this equation: $x_1 = x_2 = 0$. There will be another solution if and only if the vectors $\mathbf{a}_1$ and $\mathbf{a}_2$ are linearly dependent. But (3.3) can also be written as

$$a_{11}x_1 + a_{12}x_2 = 0$$
$$a_{21}x_1 + a_{22}x_2 = 0.$$

Since there always is the solution $x_1 = x_2 = 0$, the system will have another solution if and only if det $A = 0$. This is summarized in the next theorem.

Theorem 3.3.2   *If*

$$A = \begin{pmatrix} a_{11} & a_{12} \\ a_{21} & a_{22} \end{pmatrix}, \quad \mathbf{a}_1 = \begin{pmatrix} a_{11} \\ a_{21} \end{pmatrix}, \quad \mathbf{a}_2 = \begin{pmatrix} a_{12} \\ a_{22} \end{pmatrix},$$

then det $A = 0$ *if and only if* $\mathbf{a}_1$ *and* $\mathbf{a}_2$, *the columns of A, are linearly dependent.*

Theorems 3.3.1 and 3.3.2 can be generalized to $n \times n$ matrices. In this section we will study the following generalization of (3.1).

Definition 3.3.2   General Linear System   *The general linear system of m equations in n unknowns is the system*

$$
\begin{aligned}
a_{11}x_1 + a_{12}x_2 + \cdots + a_{1n}x_n &= b_1 \\
a_{21}x_1 + a_{22}x_2 + \cdots + a_{2n}x_n &= b_2 \\
&\ \ \vdots \\
a_{m1}x_1 + a_{m2}x_2 + \cdots + a_{mn}x_n &= b_m.
\end{aligned}
\tag{3.4}
$$

The system is said to be homogeneous *if* $b_1 = b_2 = \cdots = b_m = 0$. *Otherwise, the system is* inhomogeneous.

To avoid trivial cases, we will assume that at least one of the coefficients a_{ij} is not zero. By relabeling the variables and interchanging rows if necessary, we may assume that $a_{11} \neq 0$. We now divide the first equation by a_{11} and use the resulting equation to eliminate the terms involving x_1 from the other equations. This is accomplished first by multiplying the first equation by $-a_{21}$ and adding it to the second equation (remember, a_{11} is replaced by 1). Then the first equation is multiplied by $-a_{31}$ and added to the third. This is continued until the term in each equation (except the first) that multiplies x_1 has become zero. System (3.4) is now replaced by

$$
\begin{aligned}
x_1 + a'_{12}x_2 + a'_{13}x_2 + \cdots + a'_{1n}x_n &= b'_1 \\
a'_{22}x_2 + a'_{23}x_3 + \cdots + a'_{2n}x_n &= b'_2 \\
&\ \ \vdots \\
a'_{m2}x_2 + a'_{m3}x_3 + \cdots + a'_{mn}x_n &= b'_m,
\end{aligned}
\tag{3.5}
$$

where $a'_{12} = a_{12}/a_{11}, a'_{13} = a_{13}/a_{11}, \cdots$, and $a'_{22} = a_{22} - a_{21}a_{12}/a_{11}, a'_{ij} = a_{ij} - a_{i1}a_{1j}/a_{11}$ for $i = 2, 3, \ldots, m$ and $j = 2, 3, \ldots, n$. Also $b'_1 = b_1/a_{11}$ and $b'_i = b_i - a_{i1}b_1/a_{11}$ for $i = 2, 3, \ldots, m$.

We now repeat this process with the last $m - 1$ equations in system (3.5). This is a linear system of $m - 1$ equations in $n - 1$ unknowns. If not all coefficients are zero, we may assume that $a'_{22} \neq 0$ by relabeling variables and rearranging equations if necessary. We then divide the second row by a_{22} and use the second equation, exactly as above, to eliminate all terms involving x_2 from equations $3, 4, \ldots, m$. The system is then

$$x_1 + a'_{12}x_2 + a'_{13}x_3 + \cdots + a'_{1n}x_n = b'_1$$
$$x_2 + a''_{23}x_3 + \cdots + a''_{2n}x_n = b''_2$$
$$a''_{33}x_3 + \cdots + a''_{3n}x_n = b''_3$$
$$\vdots \qquad\qquad \vdots$$
$$a''_{m3}x_3 + \cdots + a''_{mn}x_n = b''_m.$$

This process is continued, until we finally obtain the system

$$x_1 + c_{12}x_2 + c_{13}x_3 + \cdots + c_{1l}x_l + \cdots + c_{1n}x_n = d_1$$
$$x_2 + c_{23}x_3 + \cdots + c_{2l}x_l + \cdots + c_{2n}x_n = d_2$$
$$x_3 + \cdots + c_{3l}x_l + \cdots + c_{3n}x_n = d_3$$
$$\vdots \qquad\qquad \vdots \qquad \vdots$$
$$x_l + \cdots + c_{ln}x_n = d_l \qquad (3.6)$$
$$0 = d_{l+1}$$
$$0 = d_{l+2}$$
$$\vdots$$
$$0 = d_m$$

There are three possibilities that must be examined separately.

Case I: $l < n$ and $d_{l+1} = d_{l+2} = \cdots = d_m = 0$. In this case, $x_{l+1}, x_{l+2}, \ldots, x_n$ can be chosen arbitrarily. From the lth equation, $x_l = d_l - c_{l,l+1}x_{l+1} - \cdots - c_{ln}x_n$. The other variables, $x_{l-1}, x_{l-2}, \ldots, x_1$, can be determined successively in terms of $x_{l+1}, x_{l+2}, \ldots, x_{l+n}$. There are an infinite number of solutions.

Case II: Here, $l < n$ and at least one of $d_{l+1}, d_{l+2}, \ldots, d_m$, say d_{l+i}, is not zero. In this case, there is no solution, since we have $0 \neq d_{l+i} = 0$. The original system (3.4) is then said to be *inconsistent*.

Case III: $l = n$. The system (3.6) in this case is

$$x_1 + c_{12}x_2 + \cdots + c_{1,n-1}x_{n-1} + c_{1n}x_n = d_1$$
$$x_2 + \cdots + c_{2,n-1}x_{n-1} + c_{2n}x_n = d_2$$
$$\vdots \qquad \vdots \qquad \vdots \qquad\qquad (3.7)$$
$$x_{n-1} + c_{n-1,n}x_n = d_{n-1}$$
$$x_n = d_n.$$

This system has a unique solution $x_n = d_n, x_{n-1} = d_{n-1} - c_{n-1,n}d_n$, and so on.

The procedure for going from system (3.4) to system (3.6) is called the *method of row reduction*. This method transforms any system of m equations in n unknowns to a much simpler system. It is important to note that if $m < n$ (there are more unknowns than equations), then $l < n$ and system (3.4) has either an infinite number of solutions or no solutions (but never a unique solution).

Example 3.3.2   Solve the following systems of linear equations.

1. $2x_1 + 8x_2 + 6x_3 = 20$
   $4x_1 + 2x_2 - 2x_3 = -2$
   $3x_1 - x_2 + x_3 = 11.$

2. $x_1 + x_2 + x_3 = 3$
   $2x_1 - x_2 + 3x_3 = 7$
   $4x_1 + x_2 + 5x_3 = 20.$

3. $x_1 + x_2 + x_3 = 3$
   $2x_1 - x_2 + 3x_3 = 7$
   $4x_1 + x_2 + 5x_3 = 13.$

Solution:
1. Divide the first row by 2 to obtain

$$x_1 + 4x_2 + 3x_3 = 10$$
$$4x_1 + 2x_2 - 2x_3 = -2$$
$$3x_1 - x_2 + x_3 = 11.$$

Now eliminate the x_1 terms from the second and third equations by subtracting four times the first row from the second row and by subtracting three times the first row from the third row. This gives

$$x_1 + 4x_2 + 3x_3 = 10$$
$$-14x_2 - 14x_3 = -42$$
$$-13x_2 - 8x_3 = -19.$$

Divide the second and third rows by -14 and -1, respectively.

$$x_1 + 4x_2 + 3x_3 = 10$$
$$x_2 + x_3 = 3$$
$$13x_2 + 8x_3 = 19.$$

Multiply the second row by 13, and subtract from the third row,

$$x_1 + 4x_2 + 3x_3 = 10$$
$$x_2 + x_3 = 3$$
$$-5x_3 = -20.$$

Finally, divide the third row by -5.

$$x_1 + 4x_2 + 3x_3 = 10$$
$$x_2 + x_3 = 3$$
$$x_3 = 4.$$

The solution is then $x_3 = 4$, $x_2 = 3 - x_3 = -1$, $x_1 = 10 - 4x_2 - 3x_3 = 2$. This is an example of Case III ($l = n$) and the solution is unique.

2. In this example, $a_{11} = 1$. Multiply the first row by 2 and 4 and subtract from the second and third rows, respectively. This yields

$$x_1 + x_2 + x_3 = 3$$
$$-3x_2 + x_3 = 1$$
$$-3x_2 + x_3 = 8.$$

Continuing the reduction, we subtract the second row from the third row.

$$x_1 + x_2 + x_3 = 3$$
$$x_2 - \tfrac{1}{3}x_3 = -\tfrac{1}{3}$$
$$0 = 7.$$

Clearly, this system of equations is inconsistent and has no solution. This is an example of Case II.

3. Proceeding as before, we obtain

$$x_1 + x_2 + x_3 = 3$$
$$-3x_1 + x_3 = 1$$
$$-3x_1 + x_3 = 1.$$

This reduces to

$$x_1 + x_2 + x_3 = 3$$
$$x_2 - \tfrac{1}{3}x_3 = -\tfrac{1}{3}$$
$$0 = 0.$$

This is an example of Case I. There are an infinite number of solutions, all satisfying $x_2 = (x_3 - 1)/3$ and $x_1 = (10 - 4x_3)/3$, where x_3 can have any value.

To conclude this section, we now consider homogeneous systems of linear equations as a special case of the gneral linear system (3.4).

$$a_{11}x_1 + a_{12}x_2 + \cdots + a_{1n}x_n = 0$$
$$a_{21}x_1 + a_{22}x_2 + \cdots + a_{2n}x_n = 0$$
$$\vdots \qquad \vdots \qquad\quad \vdots \quad \vdots \qquad\qquad (3.8)$$
$$a_{m1}x_1 + a_{m2}x_2 + \cdots + a_{mn}x_n = 0.$$

This system can be studied by the method of row reduction. It can be reduced to the form

$$x_1 + c_{12}x_2 + \cdots + c_{1l}x_l + \cdots + c_{1n}x_n = 0$$
$$x_2 + \cdots + c_{2l}x_l + \cdots + c_{2n}x_n = 0$$
$$\vdots \qquad\qquad \vdots$$
$$x_l + \cdots + c_{ln}x_n = 0 \qquad\qquad (3.9)$$
$$0 = 0$$
$$\vdots$$
$$0 = 0.$$

System (3.8) [and (3.9)] always has the solution $x_1 = x_2 = \cdots = x_n = 0$. It is clear from (3.9) that there will be other solutions if and only if $l < n$. A case in which this always happens is given by the following theorem.

Theorem 3.3.3   *A homogeneous linear system with more unknowns than equations $(n > m)$ always has an infinite number of solutions.*

System (3.8) can be studied as a single vector equation. Define

$$\mathbf{a}_1 = \begin{pmatrix} a_{11} \\ a_{21} \\ \vdots \\ a_{m1} \end{pmatrix}, \quad \mathbf{a}_2 = \begin{pmatrix} a_{12} \\ a_{22} \\ \vdots \\ a_{m2} \end{pmatrix}, \dots, \quad \mathbf{a}_n = \begin{pmatrix} a_{1n} \\ a_{2n} \\ \vdots \\ a_{mn} \end{pmatrix}.$$

Then, (3.8) is equivalent to the vector equation

$$x_1\mathbf{a}_1 + x_2\mathbf{a}_2 + \cdots + x_n\mathbf{a}_n = \mathbf{0}. \qquad (3.10)$$

This equation has a nonzero solution if and only if the vectors $\mathbf{a}_1, \mathbf{a}_2, \dots, \mathbf{a}_n$ are linearly dependent.

Example 3.3.3   Coexistence of Bacteria   Three species of bacteria coexist in a test tube and feed on three resources. Suppose that a bacterium of the ith species consumes on average an amount c_{ij} of the jth resource per day. Define $c_i = (c_{i1}, c_{i2}, c_{i3})$ to be the consumption vector for the ith species. Suppose that $c_1 = (1, 1, 1)$, $c_2 = (1, 2, 3)$, and $c_3 = (1, 3, 5)$ and suppose that there are 15,000 units of the first resource supplied each day to the test tube,

30,000 units of the second resource, and 45,000 units of the third resource. Assuming that all the resources are consumed, what are the populations of the three species that can coexist in this environment?

Solution: Let x_1, x_2, and x_3 be the populations of the three species that can be supported by the resources. The x_1 individuals of the first species consume x_1 units of each resource. The x_2 bacteria of the second species consume x_2, $2x_2$, and $3x_2$ units of the first, second, and third resources, respectively. The corresponding consumptions of the third species are x_3, $3x_3$, and $5x_3$ units. Equating the total consumption of each resource to the amount available, we obtain

$$x_1 + x_2 + x_3 = 15{,}000$$
$$x_1 + 2x_2 + 3x_3 = 30{,}000$$
$$x_1 + 3x_2 + 5x_3 = 45{,}000.$$

Simplifying by row reduction, we have

$$x_1 + x_2 + x_3 = 15{,}000 \qquad\qquad x_1 + x_2 + x_3 = 15{,}000$$
$$x_2 + 2x_3 = 15{,}000 \quad \text{or} \quad x_2 + 2x_3 = 15{,}000$$
$$2x_2 + 4x_3 = 30{,}000 \qquad\qquad\qquad 0 = 0.$$

This system does not have a unique solution but $x_1 = x_3 = (15{,}000 - x_2)/2$. The bacteria populations must be nonnegative. Therefore, $0 \le x_2 \le 15{,}000$ and $0 \le x_1 = x_3 \le 7500$. The total population that can coexist is 15,000, and the populations of the three species satisfy $x_1 = x_3$ and $x_2 = 15{,}000 - 2x_1$ if all resources are consumed.

Problems 3.3

1. By the method of row reduction, solve the following linear systems.

   (a) $x_1 + x_2 + x_3 = 8$

   $5x_1 + 2x_2 + 4x_3 = 14$

   $3x_1 - 2x_2 + x_3 = 12.$

   (b) $x_1 + 2x_2 - x_3 = 2$

   $2x_1 - 6x_2 + x_3 = -11$

   $x_1 - x_2 - 2x_3 = -12.$

2. Three species of birds are introduced to a new environment with a total initial population of 10,000. The populations of the three species are

observed to be growing at annual rates of 3, 4 and 5 per cent for species I, II and III, respectively. It is estimated that the total population increase in the first year will be 380 and that the increase in numbers of the first species will equal the increase of the third species. Determine the initial populations of the three species.

3. For which value of k is the following system of linear equations inconsistent?

$$x_1 + x_2 + x_3 = 5$$
$$5x_1 + 2x_2 + 2x_3 = 16$$
$$x_1 \qquad + kx_3 = 10.$$

4. For which values of λ is the following system of equations inconsistent?

$$2x_1 + x_2 + x_3 = 0$$
$$3x_1 + 2x_2 + 2x_3 = 0$$
$$x_1 - x_2 + \lambda x_3 = 0.$$

5. By row reduction, determine all solutions of the following linear systems.

(a) $\quad y_1 + y_2 - y_3 = 3$ $\qquad$ (b) $\quad z_1 + z_2 + 4z_3 = 0$
$\quad\ 2y_1 - y_2 + 3y_3 = 2$ $\qquad\qquad\ 3z_1 - z_2 + z_3 = 4$
$\quad\ y_1 + 4y_2 + 7y_3 = 20.$ $\qquad\qquad\ z_1 - 3z_2 + 7z_3 = 4.$

6. Three species of bacteria coexist in a test tube and feed on three resources. For the ith species, define the consumption vector $\mathbf{c}_i = (c_{i1}, c_{i2}, c_{i3})$, where c_{ij} is the average consumption of the jth food in one day by an individual of the ith species. Suppose that $\mathbf{c}_1 = (1, 1, 1)$, $\mathbf{c}_2 = (1, 2, 3)$, and $\mathbf{c}_3 = (1, 3, 5)$. Suppose that 20,000 units of the first food, 30,000 units of the second food, and 40,000 units of the third food are supplied each day to the test tube. Assuming that all three resources are consumed, what populations of the three species can coexist in this environment? Are these populations unique?

7. Evaluate the following 2×2 determinants.

(a) $\det \begin{pmatrix} 1 & 6 \\ 2 & 7 \end{pmatrix}.$ $\qquad$ (b) $\det \begin{pmatrix} 5 & 2 \\ 10 & 4 \end{pmatrix}.$ $\qquad$ (c) $\begin{vmatrix} 1 & 3 \\ 2 & 7 \end{vmatrix}.$

(d) $\begin{vmatrix} 4 & 2 \\ 2 & 2 \end{vmatrix}.$ $\qquad$ (e) $\det \begin{pmatrix} 6 & 2 \\ 1 & 1 \end{pmatrix}.$

8. Which of the following systems of equations has a nontrivial solution?
   (a) $3x_1 - 7x_2 = 0$ (b) $x_1 - x_2 = 0$

   $\quad 7x_1 - 4x_2 = 0.$ $-3x_1 + 3x_2 = 0.$

   (c) $x_1 + x_2 + x_3 = 0$ (d) $2x_1 + x_2 - x_3 = 0$

   $\quad x_1 - x_2 + x_3 = 0$ $x_1 + 4x_2 + 3x_3 = 0.$

   $\quad x_1 + x_2 - x_3 = 0.$

9. Solve the following linear systems by the method of row reduction.
   (a) $2x_1 + x_2 - 2x_3 = 1$ (b) $3x - y + 2z = \quad 7$

   $\quad x_1 - 3x_2 + 4x_3 = 2$ $5x + 3y - 4z = -1$

   $\quad 3x_1 + x_2 - x_3 = 3.$ $x + y + 2z = \quad 9.$

10. Consider the system $Ax = b$ of n equations in n unknowns.
    (a) Assume that there is a nonzero vector w such that $Aw = 0$. Prove that, if x is a solution of $Ax = b$, then $x + w$ is also a solution. Are there any other solutions?
    (b) Assume that there is no nonzero vector w such that $Aw = 0$. Prove that, if x is a solution of $Ax = b$, then this solution is unique.

11. Determine all solutions of the following linear systems.
    (a) $x_1 + x_2 - x_3 = 6$ (b) $x_1 + x_2 + x_3 + x_4 = 5$

    $\quad 3x_1 + 6x_2 - 3x_3 = 16.$ $x_1 + x_2 - x_3 - x_4 = 0.$

12. A protozoan consumes three foods: euglena, tetrahymena, and chlamydomonas. It is observed that, for each unit of E consumed, two units of T and three units of C are consumed on average. If the net energy return from the consumption of one unit of each food is one unit of energy, what consumptions provide a net return of 12 energy units to the protozoan?

13. The activities of a grazing animal can be classified roughly into three categories: (1) grazing, (2) moving (to new grazing areas or to avoid predators), and (3) resting. The net energy gain (above maintenance requirements) from grazing is 200 calories per hour. The net energy losses in moving and resting are 150 and 50 calories per hour, respectively.
    (a) How should the day be divided among the three activities so that the energy gains during grazing exactly compensate energy losses during moving and resting?
    (b) Is this division of the day unique?

14. Suppose that the grazing animal of Problem 13 must rest for at least 6 hours every day. How should the day be divided?

15. (a) Suppose that to avoid overgrazing the grazing animal of Problem 13 must spend equal times moving and grazing. How should the day be divided?

    (b) Suppose that the net energy gain from grazing is $150 + Q$ calories per hour, where Q is a parameter which indicates the quality of the grazing land. (The larger Q is, the higher is the net energy gain.) Prove that, for Q negative, the animal cannot support itself in the long run if equal times are spent grazing and moving. For Q positive, determine the number of hours that must be spent on grazing as a function of Q.

16. By means of a simple model, prove that competition between two species for two resources is most severe when they require the resources in exactly the same proportions. (This is related to the *competitive exclusion principle* discussed in Section 9.2.)

17. Suppose that an experiment has five elementary outcomes with probabilities p_1, p_2, p_3, p_4, and p_5. If $p_1 = p_2 + p_3, p_3 + p_4 = 2p_2, p_2 + p_3 + p_4 = p_5$, and $p_1 + p_2 = p_5$, determine the probabilities of the five elementary outcomes.

18. Suppose that $c_1 = (1, 1, 1)$, $c_2 = (1, 2, 3)$, and $c_3 = (1, 2, 5)$ in Example 3.3.3. If 10,000, 20,000, and 50,000 units of the three foods are supplied each day to the test tube and if all resources are consumed, what populations of the three species can coexist in this environment?

19. The result in Problem 18 suggests that species that can adapt their consumptions to the foods available have a selective advantage. Generalize Example 3.3.3 to give a model of this phenomenon.

3.4 The Inverse of a Matrix

The general linear system (3.4) can be written

$$Ax = b, \tag{3.11}$$

where

$$x = \begin{pmatrix} x_1 \\ x_2 \\ \vdots \\ x_n \end{pmatrix}, \quad b = \begin{pmatrix} b_1 \\ b_2 \\ \vdots \\ b_m \end{pmatrix}, \quad A = \begin{pmatrix} a_{11} & a_{12} \cdots a_{1n} \\ a_{21} & a_{22} \cdots a_{2n} \\ \vdots & \vdots \\ a_{m1} & a_{m2} \cdots a_{mn} \end{pmatrix}.$$

In this section, we consider only the case $m = n$. Therefore, A is a square matrix and the system (3.4) has the same number of equations as unknowns.

If A is invertible, there exists a matrix A^{-1} such that $A^{-1}A = AA^{-1} = I$. Assuming that A is invertible, we multiply equation (3.11) on both sides by A^{-1}. This yields $A^{-1}A\mathbf{x} = A^{-1}\mathbf{b}$. But $A^{-1}A\mathbf{x} = I\mathbf{x} = \mathbf{x}$, and we conclude that $\mathbf{x} = A^{-1}\mathbf{b}$. To check that this is a solution, we note that $A(A^{-1}\mathbf{b}) = (AA^{-1})\mathbf{b} = I\mathbf{b} = \mathbf{b}$.

The problem of solving (3.11) is thus reduced to the problem of calculating the inverse of the matrix A, if it exists. In the next section, a method will be given to determine whether a given square matrix is invertible. In this section, we will develop a simple method for calculating the inverse.

Consider again Example 3.3.2(1). After row reduction, the system of linear equations becomes

$$x_1 + 4x_2 + 3x_3 = 10$$
$$x_2 + x_3 = 3$$
$$x_3 = 4.$$

To reduce this system to an even simpler form, add -3 and -1 times the third row to the first and second rows, respectively. This gives

$$x_1 + 4x_2 = -2$$
$$x_2 = -1$$
$$x_3 = 4.$$

Finally, add -4 times the second equation to the first equation to obtain

$$x_1 = 2$$
$$x_2 = -1$$
$$x_3 = 4.$$

This can be written $I\mathbf{x} = \mathbf{c}$, where

$$I = \begin{pmatrix} 1 & 0 & 0 \\ 0 & 1 & 0 \\ 0 & 0 & 1 \end{pmatrix} \quad \text{and} \quad \mathbf{c} = \begin{pmatrix} 2 \\ -1 \\ 4 \end{pmatrix}.$$

The vector $\mathbf{c}$ is the solution of the original system $Ax = \mathbf{b}$, where

$$A = \begin{pmatrix} 2 & 8 & 6 \\ 4 & 2 & -2 \\ 3 & -1 & 1 \end{pmatrix} \quad \text{and} \quad \mathbf{b} = \begin{pmatrix} 20 \\ -2 \\ 11 \end{pmatrix}.$$

Since the solution is unique, we must have $\mathbf{c} = A^{-1}\mathbf{b}$.

The original system of equations $Ax = \mathbf{b} = I\mathbf{b}$ is changed, by row operations, to the system $Ix = A^{-1}\mathbf{b}$. The operations that transform A to I also transform I to A^{-1}. To calculate the inverse of A, it is only necessary to perform the same operations on the rows of I that were performed on the rows of A.

To illustrate this method of calculation of the inverse, let us once more consider Example 3.3.2(1). The matrix A of the system is augmented by adjoining the identity matrix I and the column vector $\mathbf{b}$ to A.

$$\left(\begin{array}{ccc|ccc|c} 2 & 8 & 6 & 1 & 0 & 0 & 20 \\ 4 & 2 & -2 & 0 & 1 & 0 & -2 \\ 3 & -1 & 1 & 0 & 0 & 1 & 11 \end{array} \right).$$

The matrix A in this *augmented matrix* is then reduced to I by row operations. The same operations are applied to the whole augmented matrix. Proceeding as before, divide the first row by 2.

$$\left(\begin{array}{ccc|ccc|c} 1 & 4 & 3 & \frac{1}{2} & 0 & 0 & 10 \\ 4 & 2 & -2 & 0 & 1 & 0 & -2 \\ 3 & -1 & 1 & 0 & 0 & 1 & 11 \end{array} \right).$$

Subtract 4 and 3 times the first row from the second and third rows, respectively.

$$\left(\begin{array}{ccc|ccc|c} 1 & 4 & 3 & \frac{1}{2} & 0 & 0 & 10 \\ 0 & -14 & -14 & -2 & 1 & 0 & -42 \\ 0 & -13 & -8 & -\frac{3}{2} & 0 & 1 & -19 \end{array} \right).$$

Now, divide the second row by -14.

$$\left(\begin{array}{ccc|ccc|c} 1 & 4 & 3 & \frac{1}{2} & 0 & 0 & 10 \\ 0 & 1 & 1 & \frac{2}{14} & -\frac{1}{14} & 0 & 3 \\ 0 & -13 & -8 & -\frac{3}{2} & 0 & 1 & -19 \end{array} \right).$$

Add -4 and $+13$ times the second row to the first and third rows, respectively.

$$\left(\begin{array}{ccc|ccc|c}
1 & 0 & -1 & -\frac{1}{14} & \frac{4}{14} & 0 & -2 \\
0 & 1 & 1 & \frac{2}{14} & -\frac{1}{14} & 0 & 3 \\
0 & 0 & 5 & \frac{5}{14} & -\frac{13}{14} & 1 & 20
\end{array}\right).$$

Divide the third row by 5.

$$\left(\begin{array}{ccc|ccc|c}
1 & 0 & -1 & -\frac{1}{14} & \frac{4}{14} & 0 & -2 \\
0 & 1 & 1 & \frac{2}{14} & -\frac{1}{14} & 0 & 3 \\
0 & 0 & 1 & \frac{5}{70} & -\frac{13}{30} & \frac{1}{5} & 4
\end{array}\right).$$

Finally, add the third row to the first row and subtract it from the second row.

$$\left(\begin{array}{ccc|ccc|c}
1 & 0 & 0 & 0 & \frac{7}{70} & \frac{14}{70} & 2 \\
0 & 1 & 0 & \frac{5}{70} & \frac{8}{70} & -\frac{14}{70} & -1 \\
0 & 0 & 1 & \frac{5}{70} & -\frac{13}{70} & \frac{14}{70} & 4
\end{array}\right).$$

Therefore

$$A^{-1} = \tfrac{1}{70}\begin{pmatrix} 0 & 7 & 14 \\ 5 & 8 & -14 \\ 5 & -13 & 14 \end{pmatrix} \quad \text{and} \quad \mathbf{x} = \begin{pmatrix} 2 \\ -1 \\ 4 \end{pmatrix}$$

is the solution of $A\mathbf{x} = \mathbf{b}$. The solution can be verified by multiplication.

$$A^{-1}A = \tfrac{1}{70}\begin{pmatrix} 0 & 7 & 14 \\ 5 & 8 & -14 \\ 5 & -13 & 14 \end{pmatrix}\begin{pmatrix} 2 & 8 & 6 \\ 4 & 2 & -2 \\ 3 & -1 & 1 \end{pmatrix}$$

$$= \tfrac{1}{70}\begin{pmatrix} 70 & 0 & 0 \\ 0 & 70 & 0 \\ 0 & 0 & 70 \end{pmatrix} = \begin{pmatrix} 1 & 0 & 0 \\ 0 & 1 & 0 \\ 0 & 0 & 1 \end{pmatrix} = I$$

and

$$A^{-1}\mathbf{b} = \tfrac{1}{70}\begin{pmatrix} 0 & 7 & 14 \\ 5 & 8 & -14 \\ 5 & -13 & 14 \end{pmatrix}\begin{pmatrix} 20 \\ -2 \\ 11 \end{pmatrix} = \tfrac{1}{70}\begin{pmatrix} 140 \\ -70 \\ 280 \end{pmatrix} = \begin{pmatrix} 2 \\ -1 \\ 4 \end{pmatrix}.$$

This *augmented matrix method* will always work whenever A^{-1} exists. When A is not invertible, a row of zeros will appear in the reduced form of A. In Example 3.3.2(2), we arrive at the system

$$\begin{aligned} x_1 + x_2 + x_3 &= 3 \\ -3x_2 + x_3 &= 1 \\ 0 &= 7. \end{aligned}$$

The last row of the matrix corresponding to this system is a row of zeros. When this occurs, there will be either no solution or an infinite number of solutions. Since there is a unique solution when A is invertible, the inverse does not exist when there is a row of zeros in the reduced form of A.

Example 3.4.1   Calculate the inverses of the following matrices.

1. $A = \begin{pmatrix} 1 & 1 & 1 \\ 1 & 2 & 1 \\ 2 & 3 & 4 \end{pmatrix}$.

2. $B = \begin{pmatrix} 1 & 2 & 4 & 6 \\ 0 & 1 & 2 & 0 \\ 0 & 0 & 1 & 2 \\ 0 & 0 & 0 & 2 \end{pmatrix}$.

Solution:
1. Write the augmented matrix.

$$\left(\begin{array}{ccc|ccc} 1 & 1 & 1 & 1 & 0 & 0 \\ 1 & 2 & 1 & 0 & 1 & 0 \\ 2 & 3 & 4 & 0 & 0 & 1 \end{array}\right).$$

Subtract the first row from the second row and subtract the first two rows from the third row.

$$\begin{pmatrix} 1 & 1 & 1 & \vdots & 1 & 0 & 0 \\ 0 & 1 & 0 & \vdots & -1 & 1 & 0 \\ 0 & 0 & 2 & \vdots & -1 & -1 & 1 \end{pmatrix}.$$

Finally, substract the second row and one half the third row from the first row and divide the third row by 2.

$$\begin{pmatrix} 1 & 0 & 0 & \vdots & \frac{5}{2} & -\frac{1}{2} & -\frac{1}{2} \\ 0 & 1 & 0 & \vdots & -1 & 1 & 0 \\ 0 & 0 & 1 & \vdots & -\frac{1}{2} & -\frac{1}{2} & \frac{1}{2} \end{pmatrix}.$$

Therefore

$$A^{-1} = \begin{pmatrix} \frac{5}{2} & -\frac{1}{2} & -\frac{1}{2} \\ -1 & 1 & 0 \\ -\frac{1}{2} & -\frac{1}{2} & \frac{1}{2} \end{pmatrix}.$$

2. Form the augmented matrix.

$$\begin{pmatrix} 1 & 2 & 4 & 6 & \vdots & 1 & 0 & 0 & 0 \\ 0 & 1 & 2 & 0 & \vdots & 0 & 1 & 0 & 0 \\ 0 & 0 & 1 & 2 & \vdots & 0 & 0 & 1 & 0 \\ 0 & 0 & 0 & 2 & \vdots & 0 & 0 & 0 & 1 \end{pmatrix}.$$

Subtract the fourth row from the third row and divide the fourth row by 2.

$$\begin{pmatrix} 1 & 2 & 4 & 6 & \vdots & 1 & 0 & 0 & 0 \\ 0 & 1 & 2 & 0 & \vdots & 0 & 1 & 0 & 0 \\ 0 & 0 & 1 & 0 & \vdots & 0 & 0 & 1 & -1 \\ 0 & 0 & 0 & 1 & \vdots & 0 & 0 & 0 & \frac{1}{2} \end{pmatrix}.$$

Subtract twice the second row and six times the fourth row from the first row, and subtract twice the third row from the second row.

$$\left(\begin{array}{cccc|cccc} 1 & 0 & 0 & 0 & 1 & -2 & 0 & -3 \\ 0 & 1 & 0 & 0 & 0 & 1 & -2 & 2 \\ 0 & 0 & 1 & 0 & 0 & 0 & 1 & -1 \\ 0 & 0 & 0 & 1 & 0 & 0 & 0 & \frac{1}{2} \end{array}\right).$$

Therefore,

$$B^{-1} = \begin{pmatrix} 1 & -2 & 0 & -3 \\ 0 & 1 & -2 & 2 \\ 0 & 0 & 1 & -1 \\ 0 & 0 & 0 & \frac{1}{2} \end{pmatrix}.$$

This calculation should be verified by calculating BB^{-1} and $B^{-1}B$ and showing that they are both equal to I.

Problems 3.4

1. Calculate the inverses of the following matrices by the augmented matrix method.

(a) $A = \begin{pmatrix} 1 & 2 & 3 & 4 \\ 0 & 1 & 2 & 3 \\ 0 & 0 & 1 & 2 \\ 0 & 0 & 0 & 1 \end{pmatrix}$.   (b) $B = \begin{pmatrix} 1 & 3 & -1 \\ 3 & 5 & -2 \\ 2 & 1 & -1 \end{pmatrix}$.

(c) $C = \begin{pmatrix} 1 & 1 & 5 \\ 0 & 9 & -7 \\ 1 & 6 & 1 \end{pmatrix}$.   (d) $D = \begin{pmatrix} 1 & 1 & 1 \\ 3 & -1 & 11 \\ 2 & 1 & 4 \end{pmatrix}$.

2. By calculating the inverses, solve the following systems of linear equations.

(a) $\quad x_1 - \frac{1}{2}x_2 \qquad\qquad = 1$ (b) $x_1 + x_2 + x_3 = 6$

$\quad -\frac{1}{2}x_1 + x_2 - \frac{1}{2}x_3 = 0$ $x_1 + x_2 - x_3 = 0$

$\quad\qquad -\frac{1}{2}x_2 + x_3 = 2.$ $x_1 + 2x_2 - x_3 = -1.$

3. Suppose that A is an $n \times n$ matrix and that $\mathbf{a}, \mathbf{b}, \mathbf{y}$, and $\mathbf{z}$ are n-component column vectors such that $A\mathbf{y} = \mathbf{a}$ and $A\mathbf{z} = \mathbf{b}$.
   (a) Solve the vector equations $A\mathbf{x} = \mathbf{b} + 3\mathbf{a}$ and $A\mathbf{w} = 2\mathbf{a} + 5\mathbf{b}$ for the vectors $\mathbf{x}$ and $\mathbf{w}$.
   (b) When are the vectors $\mathbf{x}$ and $\mathbf{w}$ unique?
   (c) If the matrix A has an inverse, evaluate $A^{-1}(\mathbf{a} + \mathbf{b})$ and $A^{-1}(5\mathbf{a} + 6\mathbf{b})$.

4. Use the method of row reduction to determine the inverses of the following 2×2 matrices.

   (a) $\begin{pmatrix} 4 & 2 \\ 1 & 6 \end{pmatrix}$.   (b) $\begin{pmatrix} 1 & -1 \\ -1 & 1 \end{pmatrix}$.   (c) $\begin{pmatrix} 1 & 2 \\ 3 & 7 \end{pmatrix}$.

5. (a) Prove that the matrix

$$A = \begin{pmatrix} a_{11} & a_{12} \\ a_{21} & a_{22} \end{pmatrix}$$

has the inverse

$$A^{-1} = \frac{1}{\det A}\begin{pmatrix} a_{22} & -a_{12} \\ -a_{21} & a_{11} \end{pmatrix}$$

if $\det A \neq 0$.

   (b) When is the matrix A equal to its own inverse?

6. Determine the inverses of the following matrices.

   (a) $\begin{pmatrix} 1 & 1 \\ 0 & 1 \end{pmatrix}$.   (b) $\begin{pmatrix} 1 & 1 & 1 \\ 0 & 1 & 1 \\ 0 & 0 & 1 \end{pmatrix}$.   (c) $\begin{pmatrix} 1 & 1 & 1 & 1 \\ 0 & 1 & 1 & 1 \\ 0 & 0 & 1 & 1 \\ 0 & 0 & 0 & 1 \end{pmatrix}$.

7. Suppose that the $n \times n$ matrix A is upper-triangular (Problem 3.2.7). Prove that A has an inverse if and only if all the components of A on the main diagonal are nonzero.

8. Calculate the inverses of the following matrices (if they exist).

   (a) $\begin{pmatrix} 4 & 6 & -3 \\ 0 & 0 & 7 \\ 0 & 0 & 5 \end{pmatrix}$.   (b) $\begin{pmatrix} 1 & 0 & 0 \\ 0 & 5 & 6 \\ 0 & 0 & 2 \end{pmatrix}$.

(c) $\begin{pmatrix} 5 & 1 & 2 & 1 \\ 0 & 1 & 1 & 1 \\ 0 & 0 & 1 & 0 \\ 0 & 0 & 0 & 4 \end{pmatrix}$.   (d) $\begin{pmatrix} 1 & 2 & 0 & 3 \\ 0 & 2 & 0 & 0 \\ 0 & 0 & 1 & 1 \\ 0 & 0 & 0 & 3 \end{pmatrix}$.

9. If

$$A = \begin{pmatrix} 2 & 2 & 4 \\ 1 & 0 & 1 \\ 0 & 1 & 0 \end{pmatrix} \quad \text{and} \quad B = \begin{pmatrix} 1 & 1 & 1 \\ -1 & 1 & 0 \\ 2 & 0 & 0 \end{pmatrix},$$

calculate A^{-1}, B^{-1}, AB, and $(AB)^{-1}$. Verify that $(AB)^{-1} = B^{-1}A^{-1}$.

10. An $n \times n$ matrix A with inverse A^{-1} is said to be an *orthogonal matrix* if its inverse is equal to its transpose, $A^{-1} = A^t$. (A^t is defined in Problem 3.2.12.) If A and B are orthogonal $n \times n$ matrices, prove that AB is orthogonal.

11. Verify that

$$A = \begin{pmatrix} \sin t & -\cos t \\ \cos t & \sin t \end{pmatrix}$$

is an orthogonal matrix for any real number t.

12. If

$$A = \begin{pmatrix} \sin t & -\cos t \\ \cos t & \sin t \end{pmatrix}$$

for some real number t and if

$$\mathbf{x} = \begin{pmatrix} x_1 \\ x_2 \end{pmatrix}$$

is any two-component column vector, verify that $\|A\mathbf{x}\| = \|\mathbf{x}\|$.

13. Which of the following matrices are orthogonal?

(a) $\begin{pmatrix} \dfrac{1}{\sqrt{2}} & -\dfrac{1}{\sqrt{2}} \\ \dfrac{1}{\sqrt{2}} & \dfrac{1}{\sqrt{2}} \end{pmatrix}$.   (b) $\begin{pmatrix} 0 & -1 \\ 1 & 0 \end{pmatrix}$.

(c) $\begin{vmatrix} \dfrac{1}{\sqrt{3}} & \dfrac{1}{\sqrt{2}} & \dfrac{1}{\sqrt{6}} \\[2mm] -\dfrac{1}{\sqrt{3}} & \dfrac{1}{\sqrt{2}} & -\dfrac{1}{\sqrt{6}} \\[2mm] \dfrac{1}{\sqrt{3}} & 0 & -\dfrac{2}{\sqrt{6}} \end{vmatrix}.$ (d) $\begin{vmatrix} \dfrac{1}{\sqrt{3}} & \dfrac{1}{\sqrt{2}} & -\dfrac{1}{\sqrt{6}} \\[2mm] -\dfrac{1}{\sqrt{3}} & \dfrac{1}{\sqrt{2}} & \dfrac{1}{\sqrt{6}} \\[2mm] \dfrac{1}{\sqrt{3}} & 0 & -\dfrac{2}{\sqrt{6}} \end{vmatrix}.$

14. (a) If A is an $n \times n$ matrix with an inverse, prove that the homogeneous system of equations $A\mathbf{x} = \mathbf{0}$ for the unknown n-component column vector $\mathbf{x}$ has only the trivial solution $\mathbf{x} = \mathbf{0}$.

(b) Use part (a) to prove that the following matrices are not invertible.

(i) $\begin{pmatrix} 3 & 4 \\ 6 & 8 \end{pmatrix}.$ (ii) $\begin{pmatrix} 1 & -1 \\ -1 & 1 \end{pmatrix}.$ (iii) $\begin{pmatrix} 1 & 2 & 3 \\ 4 & 5 & 6 \\ 5 & 7 & 9 \end{pmatrix}.$

3.5 Determinants and Cramer's Rule

The determinant of a 2×2 matrix was defined in Section 3.3,

$$\det \begin{pmatrix} a_{11} & a_{12} \\ a_{21} & a_{22} \end{pmatrix} = a_{11}a_{22} - a_{12}a_{21}.$$

We make use of this definition to define the determinant of a 3×3 matrix.

$$\det \begin{pmatrix} a_{11} & a_{12} & a_{13} \\ a_{21} & a_{22} & a_{23} \\ a_{31} & a_{32} & a_{33} \end{pmatrix} = a_{11} \det \begin{pmatrix} a_{22} & a_{23} \\ a_{32} & a_{33} \end{pmatrix} - a_{12} \det \begin{pmatrix} a_{21} & a_{23} \\ a_{31} & a_{33} \end{pmatrix}$$

$$+ a_{13} \det \begin{pmatrix} a_{21} & a_{22} \\ a_{31} & a_{32} \end{pmatrix}.$$

To calculate a 3×3 determinant, it is necessary to calculate three 2×2 determinants.

This suggests the *general definition* of the *determinant of an* $n \times n$ *matrix* as a linear combination of $(n-1) \times (n-1)$ determinants.

$$
\det \begin{pmatrix} a_{11} & a_{12} & \cdots & a_{1n} \\ a_{21} & a_{22} & \cdots & a_{2n} \\ \vdots & \vdots & & \vdots \\ a_{n1} & a_{n2} & \cdots & a_{nn} \end{pmatrix}
$$

(3.12)

$$
= a_{11} \det A_{11} - a_{12} \det A_{12} + \cdots + (-1)^{n+1} a_{1n} \det A_{1n}
$$

$$
= \sum_{j=1}^{n} (-1)^{j+1} a_{1j} \det A_{1j},
$$

where A_{1j} is the $(n-1) \times (n-1)$ matrix obtained from A by omitting the first row and jth column. This is an inductive definition of the determinant of an $n \times n$ matrix. Note that $(-1)^{j+1} = -1$ when j is even, and $= +1$ when j is odd. A common notation for det A is $|A|$.

Example 3.5.1   Calculate the determinants of the following matrices.

1.
$$
A = \begin{pmatrix} 1 & 1 & 2 \\ 1 & 2 & 1 \\ 2 & 3 & 4 \end{pmatrix}.
$$

2.
$$
B = \begin{pmatrix} 2 & 3 & -2 \\ 1 & -2 & 3 \\ 4 & -1 & 4 \end{pmatrix}.
$$

3.
$$
C = \begin{pmatrix} 1 & 2 & 3 & 4 \\ 0 & 1 & 2 & 0 \\ 0 & 1 & 1 & 1 \\ 0 & 0 & 1 & 2 \end{pmatrix}.
$$

Solution:

1. $\det A = 1 \det \begin{pmatrix} 2 & 1 \\ 3 & 4 \end{pmatrix} - 1 \det \begin{pmatrix} 1 & 1 \\ 2 & 4 \end{pmatrix} + 2 \det \begin{pmatrix} 1 & 2 \\ 2 & 3 \end{pmatrix}$

$= [2(4) - 3(1)] - [1(4) - 2(1)] + 2[1(3) - 2(2)]$

$= 5 - 2 - 2 = 1.$

2. $\det B = 2 \det \begin{pmatrix} -2 & 3 \\ -1 & 4 \end{pmatrix} - 3 \det \begin{pmatrix} 1 & 3 \\ 4 & 4 \end{pmatrix} - 2 \det \begin{pmatrix} 1 & -2 \\ 4 & -1 \end{pmatrix}$

$= 2(-8 + 3) - 3(4 - 12) - 2(-1 + 8)$

$= -10 + 24 - 14 = 0.$

3. $\det C = 1 \det \begin{pmatrix} 1 & 2 & 0 \\ 1 & 1 & 1 \\ 0 & 1 & 2 \end{pmatrix} - 2 \det \begin{pmatrix} 0 & 2 & 0 \\ 0 & 1 & 1 \\ 0 & 1 & 2 \end{pmatrix} + 3 \det \begin{pmatrix} 0 & 1 & 0 \\ 0 & 1 & 1 \\ 0 & 0 & 2 \end{pmatrix}$

$- 4 \det \begin{pmatrix} 0 & 1 & 2 \\ 0 & 1 & 1 \\ 0 & 0 & 1 \end{pmatrix}$

$= \left[\det \begin{pmatrix} 1 & 1 \\ 1 & 2 \end{pmatrix} - 2 \det \begin{pmatrix} 1 & 1 \\ 0 & 2 \end{pmatrix} \right] - 2 \left[-2 \det \begin{pmatrix} 0 & 1 \\ 0 & 2 \end{pmatrix} \right]$

$+ 3 \left[-\det \begin{pmatrix} 0 & 1 \\ 0 & 2 \end{pmatrix} \right] - 4 \left[-\det \begin{pmatrix} 0 & 1 \\ 0 & 1 \end{pmatrix} + 2 \det \begin{pmatrix} 0 & 1 \\ 0 & 0 \end{pmatrix} \right]$

$= [2 - 1 - 4] + 0 + 0 + 0$

$= -3.$

Example 3.5.2 For 3×3 matrices, there is a simple way to calculate the determinant. First, if

$$A = \begin{pmatrix} a_{11} & a_{12} & a_{13} \\ a_{21} & a_{22} & a_{23} \\ a_{31} & a_{32} & a_{33} \end{pmatrix},$$

define

by writing the first and second columns of A as the fourth and fifth columns of B. The determinant is the sum of all the products of components along the diagonals shown. Diagonals pointing down have $+$ signs and diagonals pointing up have $-$ signs.

$$\det A = a_{11}a_{22}a_{33} + a_{12}a_{23}a_{31} + a_{13}a_{21}a_{32} - a_{31}a_{22}a_{13}$$
$$- a_{32}a_{23}a_{11} - a_{33}a_{21}a_{12}.$$

The reader should verify that this agrees with the earlier definition.

Example 3.5.3   For

$$A = \begin{pmatrix} 2 & 3 & -2 \\ 1 & -2 & 3 \\ 4 & -1 & 4 \end{pmatrix},$$

calculate the determinant.

Solution:

$$\begin{aligned} \det A &= 2(-2)(4) + 3(3)(4) + (-2)(1)(-1) - 4(-2)(-2) \\ &\quad -(-1)(3)(2) - 4(1)(3) \\ &= -16 + 36 + 2 - 16 + 6 - 12 \\ &= 0. \end{aligned}$$

The following theorem generalizes the results concerning 2×2 linear systems of Section 3.3.

Theorem 3.5.1   *Given an $n \times n$ matrix A and the n-component column vector $\mathbf{b}$, then the system of equations $A\mathbf{x} = \mathbf{b}$ has a unique solution if and only if any one of the following equivalent statements is true.*
1. *A is invertible.*
2. *$\det A \neq 0$.*
3. *The columns of A are linearly independent vectors.*
4. *A can be reduced by row operations to the identity matrix.*

This theorem indicates why the determinant of the matrix A plays such an important role in the study of linear systems. If $\det A \neq 0$, A^{-1} exists and the system $A\mathbf{x} = \mathbf{b}$ has the unique solution $\mathbf{x} = A^{-1}\mathbf{b}$. If $\det A = 0$, then A is not invertible and the system $A\mathbf{x} = \mathbf{b}$ has either no solution or an infinite number of solutions.

Determinants can also be used to give an explicit formula for the solution of the system $A\mathbf{x} = \mathbf{b}$ of n linear equations in n unknowns. As before,

$$A = \begin{pmatrix} a_{11} & a_{12} & \cdots & a_{1n} \\ a_{21} & a_{22} & \cdots & a_{2n} \\ \vdots & & & \\ a_{n1} & a_{n2} & \cdots & a_{nn} \end{pmatrix}.$$

Define $D = \det A$. We define the following n determinants.

$$D_1 = \det \begin{pmatrix} b_1 & a_{12} & \cdots & a_{1n} \\ b_2 & a_{22} & \cdots & a_{2n} \\ \vdots & \vdots & & \vdots \\ b_n & a_{n2} & \cdots & a_{nn} \end{pmatrix}, \quad D_2 = \det \begin{pmatrix} a_{11} & b_1 & \cdots & a_{1n} \\ a_{21} & b_2 & \cdots & a_{2n} \\ \vdots & \vdots & & \vdots \\ a_{n1} & b_n & \cdots & a_{nn} \end{pmatrix}, \ldots,$$

$$D_n = \det \begin{pmatrix} a_{11} & a_{12} & \cdots & b_1 \\ a_{21} & a_{22} & \cdots & b_2 \\ \vdots & \vdots & & \vdots \\ a_{n1} & a_{n2} & \cdots & b_n \end{pmatrix}.$$

For example, D_j is the determinant of the matrix obtained from A by replacing the jth column of A by the column vector $\mathbf{b}$. The importance of these determinants is seen in the following theorem.

Theorem 3.5.2   Cramer's Rule   *If A is an $n \times n$ matrix with $D = \det A \neq 0$, then the unique solution of the system $A\mathbf{x} = \mathbf{b}$ is given by*

$$x_1 = \frac{D_1}{D}, \quad x_2 = \frac{D_2}{D}, \quad \cdots, \quad x_n = \frac{D_n}{D}. \tag{3.13}$$

To prove Theorems 3.5.1 and 3.5.2 would require a more extended discussion of the properties of the determinant than will be given here. The determinant method for solving linear systems with more than three unknowns can be extremely tedious; it is often simpler to use the elementary method of row reduction.

Example 3.5.4   Solve

$$2x + 8y + 6z = 20$$
$$4x + 2y - 2z = -2$$
$$3x - y + z = 11.$$

Solution:

$$D = \begin{vmatrix} 2 & 8 & 6 \\ 4 & 2 & -2 \\ 3 & -1 & 1 \end{vmatrix} = (2)(2)(1) + (8)(-2)(3) + 6(-1)(4) - (6)(2)(3)$$
$$- (8)(4)(1) - (2)(-1)(-2)$$
$$= 4 - 48 - 24 - 36 - 32 - 4 = -140 \neq 0.$$

$$D_1 = \begin{vmatrix} 20 & 8 & 6 \\ -2 & 2 & -2 \\ 11 & -1 & 1 \end{vmatrix} = 40 - 176 + 12 - 132 + 16 - 40 = -280.$$

Similarly,

$$D_2 = \begin{vmatrix} 2 & 20 & 6 \\ 4 & -2 & -2 \\ 3 & 11 & 1 \end{vmatrix} = 140 \quad \text{and} \quad D_3 = \begin{vmatrix} 2 & 8 & 20 \\ 4 & 2 & -2 \\ 3 & -1 & 11 \end{vmatrix} = -560.$$

Thus,

$$x = \frac{D_1}{D} = \frac{-280}{-140} = 2, \quad y = \frac{D_2}{D} = \frac{140}{-140} = -1, \quad z = \frac{D_3}{D} = \frac{-560}{-140} = 4.$$

In this example, we have used x, y, and z as the variables instead of x_1, x_2, and x_3.

Problems 3.5

1. Evaluate the determinants of the following matrices.

(a) $\begin{pmatrix} 1 & 4 \\ 4 & 2 \end{pmatrix}$.

(b) $\begin{pmatrix} 2 & 5 \\ 3 & \frac{15}{2} \end{pmatrix}$.

(c) $\begin{pmatrix} 0 & 4 & 7 \\ -1 & 3 & 1 \\ 2 & -2 & 5 \end{pmatrix}$.

(d) $\begin{pmatrix} 1 & 1 & 1 \\ 2 & 3 & 4 \\ 4 & 9 & 16 \end{pmatrix}$.

2. By Cramer's rule, determine the solutions of the following systems of linear equations.

(a) $x_1 + 4x_2 = 13$
    $4x_1 + 2x_2 = 10.$

(b) $x_1 + x_2 + x_3 = 1$
    $2x_1 + 3x_2 + 4x_3 = 3$
    $4x_1 + 9x_2 + 16x_3 = 11.$

3. Solve the following linear system by the augmented matrix method and by Cramer's rule.

$$x_1 + x_2 + x_3 = 6$$
$$2x_1 - x_2 \qquad = 0$$
$$2x_1 \qquad + x_3 = 3.$$

4. Using Theorem 3.5.1, determine whether the following sets of vectors are linearly dependent or independent.
   (a) $(2, 1, 0), (4, 1, 1), (0, 1, 3)$.
   (b) $(1, 5, 1), (2, 0, 0), (1, 4, 2)$.
   (c) $(1, 2, 3), (3, 4, 5), (4, 6, 8)$.
   (d) $(1, -1, 0, 0), (4, 0, 2, 0), (0, 1, 1, 1), (-2, 1, 0, -1)$.

5. Without solving, determine which of the following linear systems have nontrivial solutions.

   (a) $x_1 - x_2 + x_3 = 0$ (b) $x_1 + 2x_2 - x_3 - x_4 = 0$
        $2x_1 + 2x_2 + x_3 = 0$ $x_1 - x_2 + x_3 + x_4 = 0$
        $5x_1 + 7x_2 + 5x_3 = 0.$ $2x_1 + 3x_3 + x_3 - 2x_4 = 0$
                                               $2x_1 + 2x_2 + \frac{1}{2}x_3 - \frac{3}{2}x_4 = 0.$

   (The following problems give some additional properties of determinants which are frequently useful in calculating their values.)

6. Assuming Theorem 3.5.1, prove that if an $n \times n$ matrix A has two rows (or columns) that are identical, then det $A = 0$.

7. From Equation (3.12), prove that

   $$\det A = \sum_{j=1}^{n} (-1)^{i+j} a_{ij} \det A_{ij} = \sum_{i=1}^{n} (-1)^{i+j} a_{ij} \det A_{ij}.$$

   This means that the determinant can be calculated by expanding along any row or column. (Before attempting a general proof, the reader should verify that the result is correct for 2×2 and 3×3 matrices.)

8. Verify that, if any two rows (or columns) of an $n \times n$ matrix are interchanged, the determinant of the new matrix is the negative of the determinant of the original matrix.

9. Verify that, if any row (or column) of an $n \times n$ matrix is multiplied by a constant, the determinant is multiplied by the same constant.

10. Using Problems 6 and 9, prove that, if one row (or column) of an $n \times n$ matrix A is a constant multiple of another row (or column), then det $A = 0$.

11. Given three $n \times n$ matrices $A = (a_{ij})$, $B = (b_{ij})$, and $C = (c_{ij})$ that are identical except in the kth row, where $c_{kj} = a_{kj} + b_{kj}$, prove that det $C =$ det $A +$ det B. (A similar result holds for columns.)

12. Using Problems 10 and 11, prove that, if a multiple of one row (or column) of an $n \times n$ matrix A is added to another row (or column), then the determinant of the new matrix is equal to the determinant of A.

13. Prove that, if an $n \times n$ matrix A is upper triangular, then its determinant is equal to the product of its diagonal elements.

14. Using the results of the previous problems, calculate the determinant of

$$B = \begin{pmatrix} 2 & 3 & -2 \\ 1 & -2 & 3 \\ 4 & -1 & 4 \end{pmatrix}$$

by carrying out the following steps
(a) Show that

$$\det B = -\det \begin{pmatrix} 1 & -2 & 3 \\ 2 & 3 & -2 \\ 4 & -1 & 4 \end{pmatrix} = -\det B_1 .$$

(b) Show that

$$-\det B_1 = -\det \begin{pmatrix} 1 & -2 & 3 \\ 0 & 7 & -8 \\ 0 & 7 & -8 \end{pmatrix} = -\det B_2 .$$

(c) Show that $\det B_2 = 0$.

15. Calculate the determinant of

$$B = \begin{pmatrix} 1 & 1 & 2 \\ 1 & 2 & 1 \\ 2 & 3 & 4 \end{pmatrix}$$

by showing that

(a) $\det B = \det \begin{pmatrix} 1 & 1 & 2 \\ 0 & 1 & -1 \\ 0 & 1 & 0 \end{pmatrix} = \det B_1 .$

(b) $\det B_1 = \det \begin{pmatrix} 1 & 1 & 2 \\ 0 & 1 & -1 \\ 0 & 0 & 1 \end{pmatrix} = \det B_2 .$

(c) $\det B_2 = 1$.

16. Use the methods outlined in Problems 14 and 15 to calculate the determinants of the following matrices.

(a) $\begin{pmatrix} 3 & 1 & 2 \\ 2 & 4 & 4 \\ -2 & 1 & 3 \end{pmatrix}$.

(b) $\begin{pmatrix} 2 & 1 & 1 \\ -1 & 1 & 3 \\ 4 & 2 & 2 \end{pmatrix}$.

(c) $\begin{pmatrix} 1 & 2 & 3 & 1 \\ 1 & -1 & 1 & 4 \\ 2 & 1 & 0 & 1 \\ 1 & 1 & 2 & 4 \end{pmatrix}$.

(d) $\begin{pmatrix} 2 & 0 & 2 & 1 \\ 1 & 1 & 0 & 2 \\ 0 & 1 & 1 & 2 \\ -1 & -1 & 0 & 1 \end{pmatrix}$.

Which of these matrices has an inverse?

17. If

$$A = \begin{pmatrix} 2 & 2 & 4 \\ 4 & 3 & 6 \\ 1 & -1 & 2 \end{pmatrix} \quad \text{and} \quad \mathbf{b} = \begin{pmatrix} -2 \\ 2 \\ 1 \end{pmatrix},$$

use Cramer's rule to solve the vector equation $A\mathbf{x} = \mathbf{b}$.

18. Solve the following systems by Cramer's rule.

(a) $3x_1 - 3x_2 - 2x_3 = 10$
    $x_1 + 2x_2 - 4x_3 = 3$
    $5x_1 - 9x_2 - 4x_3 = 12.$

(b) $3x_1 + x_2 - 2x_3 = 3$
    $x_1 - 2x_2 - 3x_3 = 1$
    $2x_1 + 3x_2 + x_3 = 2.$

19. It can be proved that, if A and B are $n \times n$ matrices, then det $(AB) = $ det (A) det (B). Using this result, prove that AB has an inverse if and only if both A and B have inverses. Prove also that det $(A^{-1}) = 1/\det (A)$. (These results can be very useful in calculations.)

3.6 Eigenvalues and Eigenvectors

The $n \times n$ matrix A acts on a given n-component column vector $\mathbf{x}$ to produce another n-component column vector $\mathbf{y} = A\mathbf{x}$. Are there any vectors $\mathbf{x}$ such that $A\mathbf{x}$ is just a constant multiple of $\mathbf{x}$? This question leads to the concepts of the eigenvalues and eigenvectors of a matrix. The importance of these concepts is suggested in the study of Markov chains in Chapter 5. This is only one of the many applications of eigenvalues and eigenvectors of a matrix.

Definition 3.6.1 **Eigenvalues and Eigenvectors** *The $n \times n$ matrix A is said to have the eigenvalue λ corresponding to an eigenvector* $\mathbf{x}$ *if the (real or complex) number λ and the (nonzero) vector* $\mathbf{x}$ *satisfy*

$$A\mathbf{x} = \lambda\mathbf{x}. \tag{3.14}$$

Equation (3.14) is the eigenvalue equation or the eigenvector equation of the matrix A.

To solve the eigenvalue equation, there must exist a nontrivial solution of the homogeneous linear system $(A - \lambda I)\mathbf{x} = \mathbf{0}$. From Theorem 3.5.1, the necessary and sufficient condition for a nontrivial solution is det $(A - \lambda I) = 0$.

$$\det(A - \lambda I) = \begin{vmatrix} a_{11} - \lambda & a_{12} & a_{13} & \cdots & a_{1n} \\ a_{21} & a_{22} - \lambda & a_{23} & \cdots & a_{2n} \\ a_{31} & a_{32} & a_{33} - \lambda & \cdots & a_{3n} \\ \vdots & \vdots & \vdots & & \vdots \\ a_{n1} & a_{n2} & a_{n3} & \cdots & a_{nn} - \lambda \end{vmatrix}.$$

If this determinant were explicitly written out according to the definition of the previous section, it could be written as a polynomial in λ by grouping terms with similar powers of λ. Since, for λ very large, det $(A - \lambda I)$ is approximately det $(-\lambda I) = (-\lambda)^n$, we can see that det $(A - \lambda I)$ is a polynomial of degree n. The results are summarized in the following definition and theorem.

Definition 3.6.2 **Characteristic Equation** *The characteristic polynomial of the $n \times n$ matrix A is the nth-degree polynomial* det $(A - \lambda I)$ *in the variable λ. The characteristic equation of A is the polynomial equation* det $(A - \lambda I) = 0$.

Theorem 3.6.1 *If A is an $n \times n$ matrix, then λ is an eigenvalue of A if and only if λ satisfies the characteristic equation of A,* det $(A - \lambda I) = 0$.

Corollary 3.6.2 *The $n \times n$ matrix A has eigenvalue 0 if and only if* det $A = 0$.

If

$$A = \begin{pmatrix} a_{11} & a_{12} \\ a_{21} & a_{22} \end{pmatrix},$$

then the characteristic equation of A is

$$0 = \det (A - \lambda I) = \begin{vmatrix} a_{11} - \lambda & a_{12} \\ a_{21} & a_{22} - \lambda \end{vmatrix} = (a_{11} - \lambda)(a_{22} - \lambda) - a_{12}a_{21}$$

$$= \lambda^2 - (a_{11} + a_{22})\lambda + (a_{11}a_{22} - a_{12}a_{21}).$$

The theory of eigenvalues and eigenvectors of matrices is extremely extensive and fundamental to any further study of linear algebra. It is important to be able to calculate eigenvectors and eigenvalues of given matrices as in the following examples.

Example 3.6.1 Calculate the eigenvalues and eigenvectors of

$$A = \begin{pmatrix} 3 & -6 \\ 2 & -5 \end{pmatrix}.$$

Solution: The characteristic polynomial is

$$\det (A - \lambda I) = \det \begin{pmatrix} 3 - \lambda & -6 \\ 2 & -5 - \lambda \end{pmatrix}$$

$$= (3 - \lambda)(-5 - \lambda) + 12 = \lambda^2 + 2\lambda - 3.$$

The characteristic equation is $\lambda^2 + 2\lambda - 3 = (\lambda + 3)(\lambda - 1) = 0$. The eigenvalues of A are $\lambda_1 = -3$ and $\lambda_2 = 1$. The eigenvector corresponding to $\lambda_1 = -3$ satisfies $A\mathbf{x} = -3\mathbf{x}$ or

$$\begin{pmatrix} 3 & -6 \\ 2 & -5 \end{pmatrix}\begin{pmatrix} x_1 \\ x_2 \end{pmatrix} = \begin{pmatrix} -3x_1 \\ -3x_2 \end{pmatrix}, \qquad \begin{aligned} 3x_1 - 6x_2 &= -3x_1 \\ 2x_1 - 5x_2 &= -3x_2. \end{aligned}$$

Therefore, the components of the eigenvector corresponding to $\lambda_1 = -3$ are equal. Any vector

$$\begin{pmatrix} x_1 \\ x_1 \end{pmatrix}$$

is an eigenvector of A if $x_1 \neq 0$. In particular, the vector

$$\begin{pmatrix} 1 \\ 1 \end{pmatrix}$$

is an eigenvector.

For $\lambda_2 = 1$, the eigenvector equation is $A\mathbf{x} = \mathbf{x}$ or

$$\begin{pmatrix} 3 & -6 \\ 2 & -5 \end{pmatrix}\begin{pmatrix} x_1 \\ x_2 \end{pmatrix} = \begin{pmatrix} x_1 \\ x_2 \end{pmatrix} \qquad \begin{array}{l} 3x_1 - 6x_2 = x_1 \\ 2x_1 - 5x_2 = x_2. \end{array}$$

The components of the eigenvector satisfy $x_1 = 3x_2$. If $x_2 \neq 0$, any vector of the form

$$\begin{pmatrix} 3x_2 \\ x_2 \end{pmatrix}$$

is an eigenvector of A corresponding to the eigenvalue 1. In particular,

$$\begin{pmatrix} 3 \\ 1 \end{pmatrix}$$

is an eigenvector.

Example 3.6.2   Calculate the eigenvalues and eigenvectors of

$$A = \begin{pmatrix} 1 & 3 & 0 \\ 7 & 1 & 5 \\ 0 & -4 & 1 \end{pmatrix}.$$

Solution:

$$\det (A - \lambda I) = \det \begin{pmatrix} 1 - \lambda & 3 & 0 \\ 7 & 1 - \lambda & 5 \\ 0 & -4 & 1 - \lambda \end{pmatrix}$$

$$= (1 - \lambda)^3 - 21(1 - \lambda) + 20(1 - \lambda)$$
$$= (1 - \lambda)((1 - \lambda)^2 - 1) = (1 - \lambda)(1 - 2\lambda + \lambda^2 - 1)$$
$$= (1 - \lambda)\lambda(\lambda - 2) = -\lambda(\lambda - 1)(\lambda - 2).$$

The eigenvalues of A are $\lambda_1 = 0$, $\lambda_2 = 1$, and $\lambda_3 = 2$. The corresponding eigenvector equations are (1) $A\mathbf{x} = 0$, (2) $A\mathbf{x} = \mathbf{x}$, (3) $A\mathbf{x} = 2\mathbf{x}$.

1. $\begin{pmatrix} 1 & 3 & 0 \\ 7 & 1 & 5 \\ 0 & -4 & 1 \end{pmatrix}\begin{pmatrix} x_1 \\ x_2 \\ x_3 \end{pmatrix} = \begin{pmatrix} 0 \\ 0 \\ 0 \end{pmatrix},$ $\begin{array}{l} x_1 + 3x_2 = 0 \\ 7x_1 + x_2 + 5x_3 = 0 \\ -4x_2 + x_3 = 0. \end{array}$

From the first equation, $x_1 = -3x_2$. From the third equation, $x_3 = 4x_2$. If $x_1 = -3x_2$ and $x_3 = 4x_2$, the second equation is satisfied. Therefore, if $x_2 \neq 0$, the vector

$$\begin{pmatrix} -3x_2 \\ x_2 \\ 4x_2 \end{pmatrix}$$

is an eigenvector. In particular,

$$\begin{pmatrix} -3 \\ 1 \\ 4 \end{pmatrix}$$

is an eigenvector.

2.
$$\begin{pmatrix} 1 & 3 & 0 \\ 7 & 1 & 5 \\ 0 & -4 & 1 \end{pmatrix}\begin{pmatrix} x_1 \\ x_2 \\ x_3 \end{pmatrix} = \begin{pmatrix} x_1 \\ x_2 \\ x_3 \end{pmatrix}, \qquad \begin{aligned} x_1 + 3x_2 &= x_1 \\ 7x_1 + x_2 + 5x_3 &= x_2 \\ -4x_2 + x_3 &= x_3. \end{aligned}$$

Therefore, $x_2 = 0$ and $7x_1 + 5x_3 = 0$. An eigenvector corresponding to $\lambda_2 = 1$ is

$$\mathbf{x} = \begin{pmatrix} 5 \\ 0 \\ -7 \end{pmatrix}.$$

3.
$$\begin{pmatrix} 1 & 3 & 0 \\ 7 & 1 & 5 \\ 0 & -4 & 1 \end{pmatrix}\begin{pmatrix} x_1 \\ x_2 \\ x_3 \end{pmatrix} = \begin{pmatrix} 2x_1 \\ 2x_2 \\ 2x_3 \end{pmatrix}, \qquad \begin{aligned} x_1 + 3x_2 &= 2x_1 \\ 7x_1 + x_2 + 5x_3 &= 2x_2 \\ -4x_2 + x_3 &= 2x_3. \end{aligned}$$

Therefore, $x_1 = 3x_2$ and $x_3 = -4x_2$, and an eigenvector corresponding to $\lambda_3 = 2$ is

$$\begin{pmatrix} 3 \\ 1 \\ -4 \end{pmatrix}.$$

In Chapter 5 we will study an eigenvalue problem for $n \times n$ probability matrices in a form slightly different from the eigenvalue problems of this section. We will prove that if P is an $n \times n$ probability matrix (see Definition

5.1.2), then there exists an n-component row vector $\mathbf{t}$ such that $\mathbf{t}P = \mathbf{t}$. This equation can be written

$$(t_1, t_2, \ldots, t_n)\begin{pmatrix} p_{11} & p_{12} & \cdots & p_{1n} \\ p_{21} & p_{22} & \cdots & p_{2n} \\ \vdots & \vdots & & \vdots \\ p_{n1} & p_{n2} & \cdots & p_{nn} \end{pmatrix} = (t_1, t_2, \ldots, t_n).$$

The theory of eigenvectors and eigenvalues could have been developed in exactly the same way by studying the equation $\mathbf{x}A = \lambda\mathbf{x}$ instead of $A\mathbf{x} = \lambda\mathbf{x}$. In particular, there is a nontrivial eigenvector $\mathbf{t}$ satisfying $\mathbf{t}P = \mathbf{t}$ (in other words, 1 is an eigenvalue) if and only if $\mathbf{t}(P - I) = \mathbf{0}$ has a nontrivial solution. This happens if and only if det $(P - I) = 0$.

Problems 3.6

1. If A is an $n \times n$ diagonal matrix, prove that the eigenvalues of A are the components along the diagonal.

2. If A is an $n \times n$ matrix with inverse A^{-1}, prove that the eigenvalues of A^{-1} are the reciprocals of the eigenvalues of A. In other words, prove that if λ is an eigenvalue of A, then $1/\lambda$ is an eigenvalue of A^{-1}.

3. If $\mathbf{x}$ is an eigenvector of the $n \times n$ matrix A corresponding to the eigenvalue 0 and if $A\mathbf{y} = \mathbf{b}$, where $\mathbf{b}$ is a nonzero vector, determine an infinite number of solutions $\mathbf{z}$ of the vector equation $A\mathbf{z} = \mathbf{b}$.

4. Determine all eigenvalues and eigenvectors of the following matrices.
   (a) $\begin{pmatrix} 2 & 1 \\ 2 & 3 \end{pmatrix}$.   (b) $\begin{pmatrix} 5 & 1 \\ -1 & 3 \end{pmatrix}$.   (c) $\begin{pmatrix} 4 & 3 \\ 2 & 3 \end{pmatrix}$.

5. Determine all eigenvalues and eigenvectors of the following matrices.
   (a) $\begin{pmatrix} 0 & -1 \\ 1 & 0 \end{pmatrix}$.   (b) $\begin{pmatrix} 1 & 2 \\ -1 & 3 \end{pmatrix}$.   (c) $\begin{pmatrix} 1 & -1 \\ 1 & 1 \end{pmatrix}$.

   (These examples illustrate that eigenvalues can be complex numbers. See Appendix E.)

6. Determine all eigenvalues and eigenvectors of

$$A = \begin{pmatrix} 2 & 0 & 1 \\ -3 & 5 & 0 \\ -2 & 4 & -1 \end{pmatrix}.$$

7. Determine the eigenvalues and eigenvectors of the following matrices.

(a) $\begin{pmatrix} 3 & 2 & 4 \\ 2 & 0 & 2 \\ 4 & 2 & 3 \end{pmatrix}$.

(b) $\begin{pmatrix} \frac{2}{3} & \frac{2}{3} & \frac{1}{3} \\ -\frac{2}{3} & -\frac{1}{3} & \frac{2}{3} \\ \frac{1}{3} & -\frac{2}{3} & \frac{2}{3} \end{pmatrix}$.

8. Prove that the eigenvalues of an upper-triangular matrix are simply the diagonal components of the matrix. What are the corresponding eigenvectors?

9. Determine all eigenvalues and eigenvectors of the following matrices.

(a) $\begin{pmatrix} 1 & 1 \\ 0 & 1 \end{pmatrix}$.

(b) $\begin{pmatrix} 3 & 0 \\ 0 & 3 \end{pmatrix}$.

(c) $\begin{pmatrix} -1 & 0 \\ 1 & 2 \end{pmatrix}$.

10. Determine all eigenvalues and eigenvectors of the following matrices.

(a) $\begin{pmatrix} 2 & 1 & 3 \\ 0 & 2 & -1 \\ 0 & 0 & 1 \end{pmatrix}$.

(b) $\begin{pmatrix} 1 & -1 & -3 \\ 0 & 0 & 1 \\ 0 & 0 & -4 \end{pmatrix}$.

(c) $\begin{pmatrix} 2 & 0 & 0 \\ 0 & 3 & 0 \\ 1 & 1 & 4 \end{pmatrix}$.

11. Determine all eigenvalues and eigenvectors of the following matrices.

(a) $\begin{pmatrix} 9 & -3 & 2 & 1 \\ 0 & 4 & -4 & 1 \\ 0 & 0 & 1 & 0 \\ 0 & 0 & 0 & -3 \end{pmatrix}$.

(b) $\begin{pmatrix} 6 & 1 & 0 & -5 \\ 0 & 8 & 3 & 4 \\ 0 & 0 & -6 & 2 \\ 0 & 0 & 0 & -8 \end{pmatrix}$.

12. The characteristic polynomial of the $n \times n$ matrix A is the nth-degree polynomial $\det (A - \lambda I)$.

(a) Prove that the leading term (the highest power of λ) is $(-1)^n \lambda^n$.

(b) Prove that the constant term (the term independent of λ) is $\det A$ by setting $\lambda = 0$ in the characteristic polynomial.

(c) Conclude that the product of the eigenvalues of A is equal to $\det A$.

13. Consider an ecosystem which contains m coexisting species. Define the population vector $\mathbf{n}(t)$ of the ecosystem at time t to be the m-component column vector whose ith component $n_i(t)$ is the population of the ith species at time t. Define $A(t)$ to be the transition matrix of the ecosystem from time t to time $t + 1$. This means that $\mathbf{n}(t + 1) = A(t)\mathbf{n}(t)$. Consider the special case

$$\mathbf{n}(0) = \begin{pmatrix} 100 \\ 100 \\ 200 \end{pmatrix} \quad \text{and} \quad A(t) = \begin{pmatrix} 1 + \dfrac{t}{20} & 0 & 0 \\ 0 & 1 & \dfrac{t}{20} \\ 0 & 0 & 1 - \dfrac{t}{10} \end{pmatrix}.$$

(a) Calculate $\mathbf{n}(1)$, $\mathbf{n}(2)$, $\mathbf{n}(3)$, and $\mathbf{n}(4)$.

(b) Describe in biological terms the evolution of this ecosystem containing three competing species through the first four time periods.

(c) When does the third species become extinct?

14. Referring to Problem 13, prove that in general $\mathbf{n}(t) = A(t - 1)A(t) \cdots A(1)A(0)\mathbf{n}(0)$. If $\mathbf{n}(21) = \mathbf{n}(20)$, prove that the matrix $A(20)$ has eigenvalue 1. What is the biological interpretation of the equation $\mathbf{n}(21) = \mathbf{n}(20)$?

Linear Programming 4

4.1 Introduction

The problem of determining the maximum or minimum of a given function occurs in many applications of mathematics to biology. It is not difficult to think of examples of such problems. What is the maximum rate at which oxygen can be supplied to the tissues of an animal? What is the minimum rate of reproduction necessary for the survival of a given species? What is the maximum biomass that can be supported by a given ecosystem? How can the food requirements of a predator be satisfied with minimum expenditure of energy? When an animal moves from one point to another in its domain, it may attempt to minimize the time required or to minimize the energy expended. A study of migrating salmon[1] indicates that the salmon swim at a speed that minimizes the total energy expended to reach their spawning sites.

Maximization and minimization problems in biology are usually subject to constraints or limits on the biological variables. These constraints may be a result of physiological limits or of limits of availability of resources. For

[1] J. R. Brett, "The Swimming Energetics of Salmon," *Scientific American*, 213:80–85 (Aug. 1965).

161

example, the reproduction rate of a given species may be limited by a lengthy gestation period or by availability of space or food. Some constraints on biological variables are obvious consequences of their definitions. For example, the speed of a migrating salmon cannot be negative.

In this chapter, we consider the special problem of maximizing or minimizing linear functions of several variables subject to linear constraints. Instead of giving a general definition of this problem at this point, let us consider the following example, which will then be generalized in the later sections.

Example 4.1.1   A mountain lake in a national park is stocked each spring with two species of fish, S_1 and S_2. The average weight of the fish stocked is 4 pounds for S_1 and 2 pounds for S_2. Two foods, F_1 and F_2, are available in the lake. The average requirement of a fish of species S_1 is 1 unit of F_1 and 3 units of F_2 each day. The corresponding requirement of S_2 is 2 units of F_1 and 1 unit of F_2. If 500 units of F_1 and 900 units of F_2 are available daily, how should the lake be stocked to maximize the weight of fish supported by the lake?

Solution: Let x_1 and x_2 denote the numbers of fish of species S_1 and S_2 stocked in the lake. The total weight W of fish stocked is given by

$$W = 4x_1 + 2x_2, \tag{4.1}$$

the total consumption of food F_1 is $x_1 + 2x_2$, and the total consumption of food F_2 is $3x_1 + x_2$. Since 500 units of F_1 and 900 units of F_2 are available, we have

$$x_1 + 2x_2 \leq 500 \quad \text{and} \quad 3x_1 + x_2 \leq 900. \tag{4.2}$$

Finally, we have the obvious constraints

$$x_1 \geq 0 \quad \text{and} \quad x_2 \geq 0, \tag{4.3}$$

since the lake cannot be stocked with a negative number of fish of either species.

This is a typical problem of linear programming: determine the maximum of W subject to the constraints (4.2) and (4.3). This particular problem can

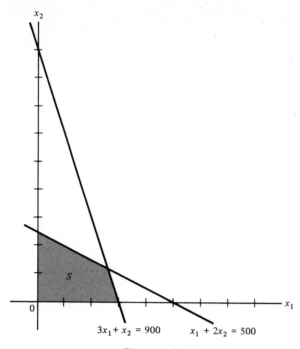

$$3x_1 + x_2 = 900 \qquad x_1 + 2x_2 = 500$$

Figure 4.1

be solved by a geometrical method. In Figure 4.1, the straight lines $x_1 + 2x_2 = 500$ and $3x_1 + x_2 = 900$ are plotted in the $x_1 x_2$ plane.

The solution of this problem satisfies the constraints (4.2) and (4.3). Therefore, only points in the first quadrant of the $x_1 x_2$ plane that lie below the line $x_1 + 2x_2 = 500$ and to the left of the line $3x_1 + x_2 = 900$ need be considered. These points lie in the shaded region denoted by S in Figure 4.1. To solve the problem, we must determine the maximum of $W = 4x_1 + 2x_2$ over the region S. To do this, we draw the lines $4x_1 + 2x_2 = c$ for different values of the constant c (Figure 4.2). On each of the parallel lines, the value of W is equal to c at every point on the line. Therefore, to determine the maximum of W in S, we determine the line $4x_1 + 2x_2 = c$ with the largest value of c which intersects the set S. The value of c increases as the parallel lines move to the right. From Figure 4.2, the maximum value of W is the value of c for the line $4x_1 + 2x_2 = c$ that passes through P, the intersection of $x_1 + 2x_2 = 500$ and $3x_1 + x_2 = 900$. The coordinates of P are $x_1 = 260$ and $x_2 = 120$, giving a value of $W = 4(260) + 2(120) = 1280$. We conclude that the lake can support a maximum weight of 1280 pounds if 260 fish of species S_1 and 120 fish of species S_2 are stocked.

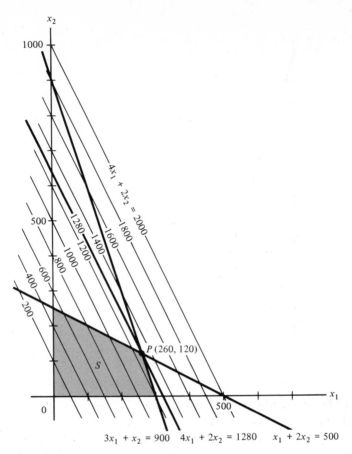

Figure 4.2

The reader should verify that other points in the set S have coordinates x_1 and x_2, which give smaller values for W. In the rest of this chapter, we generalize the ideas contained in this example.

Problems 4.1

1. Determine the number of fish of species S_1 and of species S_2 that can coexist in the lake (Example 4.1.1) with total weight 1200 pounds. Plot the corresponding points in the $x_1 x_2$ plane (Figure 4.1).

2. Suppose that 1000 units of F_1 and 1800 units of F_2 are available daily in Example 4.1.1. How should the lake be stocked to maximize the weight of fish supported by the lake?

3. As in Problem 2, how should the lake be stocked if 1000 units of F_1 and 1000 units of F_2 are available daily?

4. Suppose that m types of food are available in a lake in fixed daily amounts and that the daily requirements for these foods of average fish of n species are known. Formulate a general problem of stocking the lake in order to maximize (a) the number of fish supported by the lake and (b) the weight of fish supported by the lake. (These are maximum problems of linear programming.)

5. In Example 4.1.1, how should the lake be stocked in order to maximize the total number of fish supported by the lake?

6. In Problem 4.1.2, how should the lake be stocked to maximize the total number of fish supported by the lake?

7. In Problem 4.1.3, how should the lake be stocked to maximize the total number of fish supported by the lake? In this case, what is the total weight of fish stocked in the lake?

8. In the production of fertilizers, three chemicals are combined in different mixtures or grades and are sold in 100-pound units. Suppose that the three chemicals cost 20, 15, and 5 cents a pound, respectively. In all mixtures, there must be at least 20 pounds of the first chemical, and the amount of the third chemical in a mixture must not be greater than the amount of the second chemical. What is the mixture that minimizes the cost of a unit of fertilizer? (This is a minimum problem of linear programming.)

9. Find the maximum of

$$w = 3x_1 + 8x_2$$

subject to the constraints

$$x_1 + 2x_2 \leq 4$$
$$3x_1 + 2x_2 \leq 6.$$

4.2 Convex Sets and Linear Inequalities

In this section, we will study subsets of $\mathbf{R}^n$, the space of n-component row vectors. First, we consider the special case $\mathbf{R}^2$ of two-component row vectors. We have a convenient representation of $\mathbf{R}^2$ as the two-dimensional space of

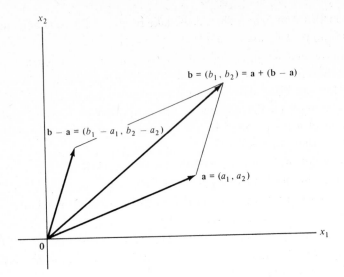

Figure 4.3

analytic geometry. Let $\mathbf{a} = (a_1, a_2)$ and $\mathbf{b} = (b_1, b_2)$ be two points (or vectors) in $\mathbf{R}^2$. The vectors $\mathbf{a}, \mathbf{b}$ and $\mathbf{b} - \mathbf{a}$ are drawn in Figure 4.3.

If t is a real number between 0 and 1, then $t(\mathbf{b} - \mathbf{a})$ is a vector in the direction of the vector $\mathbf{b} - \mathbf{a}$. This is drawn in Figure 4.4. By addition of vectors, the vector $\mathbf{a} + t(\mathbf{b} - \mathbf{a})$ is the vector drawn in Figure 4.4. We conclude that

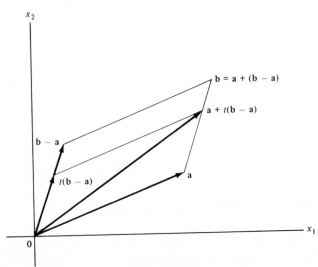

Figure 4.4

the vector $\mathbf{a} + t(\mathbf{b} - \mathbf{a})$ for $0 \le t \le 1$ represents a point on the straight line segment joining the points $\mathbf{a}$ and $\mathbf{b}$. As t increases from 0 to 1, the point $\mathbf{a} + t(\mathbf{b} - \mathbf{a})$ goes from the point $\mathbf{a}$ to the point $\mathbf{b}$. Note that we may write $\mathbf{a} + t(\mathbf{b} - \mathbf{a}) = (1 - t)\mathbf{a} + t\mathbf{b}$.

Example 4.2.1 $\mathbf{a} = (1, 0)$ and $\mathbf{b} = (0, 1)$ (Figure 4.5). The line joining the points $\mathbf{a}$ and $\mathbf{b}$ in this example has the equation $x_1 + x_2 = 1$. The line segment joining $\mathbf{a}$ and $\mathbf{b}$ is the set of points

$$\{(1 - t)\mathbf{a} + t\mathbf{b} : 0 \le t \le 1\} = \{(1 - t, t) : 0 \le t \le 1\}.$$

For example, $t = \frac{1}{3}$ corresponds to the point $(\frac{2}{3}, \frac{1}{3})$ on the line segment.

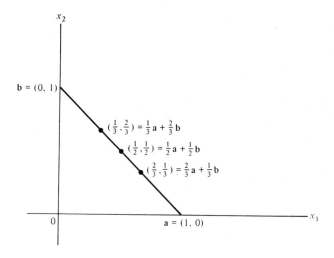

Figure 4.5

Definition 4.2.1 Convex Set in $\mathbf{R}^2$ *Let S be a subset of $\mathbf{R}^2$, the set of two component row vectors. The subset S is a convex set if every point on the straight line segment joining two points of S is a point of S. Equivalently, a set S in $\mathbf{R}^2$ is convex if, whenever $\mathbf{a}, \mathbf{b} \in S$, then $(1 - t)\mathbf{a} + t\mathbf{b} \in S$ for $0 \le t \le 1$.*

Example 4.2.2 In Figure 4.6, sets (a) and (b) are convex, while sets (c) and (d) are not convex.

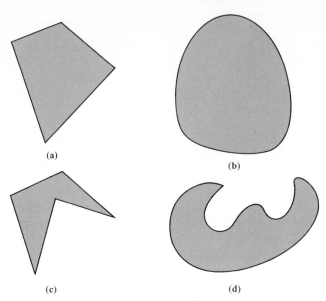

(a)

(b)

(c)   (d)

Figure 4.6

Example 4.2.3   Prove that the set $S = \{(x_1, x_2) : x_1^2 + x_2^2 \le 1\}$ is convex. The set S is the set of points on and inside the circle with center the origin and radius 1 in $\mathbf{R}^2$. (Figure 4.7).

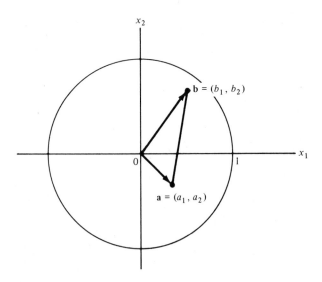

Figure 4.7

Solution: We must prove that, if **a** and **b** belong to S, then $(1 - t)\mathbf{a} + t\mathbf{b}$ belongs to S for $0 \leq t \leq 1$. Recall the triangle inequality (Problem 3.1.10) for the length of the sum of two vectors. If **a** and **b** belong to S, then $\|\mathbf{a}\| \leq 1$ and $\|\mathbf{b}\| \leq 1$. For definiteness, suppose that $\|\mathbf{a}\| \leq \|\mathbf{b}\|$. Then

$$\|(1 - t)\mathbf{a} + t\mathbf{b}\| \leq \|(1 - t)\mathbf{a}\| + \|t\mathbf{b}\|$$
$$= (1 - t)\|\mathbf{a}\| + t\|\mathbf{b}\| \qquad \text{(since } 0 \leq t \leq 1)$$
$$\leq (1 - t)\|\mathbf{b}\| + t\|\mathbf{b}\| = \|\mathbf{b}\| \leq 1.$$

We conclude that, for $0 \leq t \leq 1, (1 - t)\mathbf{a} + t\mathbf{b}$ belongs to S. In other words, the set S is convex.

We now generalize these concepts from $\mathbf{R}^2$ to $\mathbf{R}^n$.

Definition 4.2.2   Straight Line Segment in $\mathbf{R}^n$ *Let* $\mathbf{a} = (a_1, a_2, \ldots, a_n)$ *and* $\mathbf{b} = (b_1, b_2, \ldots, b_n)$ *be two points (or vectors) in* $\mathbf{R}^n$. *The straight line segment joining* **a** *and* **b** *is the set of points* $\{(1 - t)\mathbf{a} + t\mathbf{b} : 0 \leq t \leq 1\}$.

Definition 4.2.3   Convex Set in $\mathbf{R}^n$ *A set S in $\mathbf{R}^n$ is convex if, whenever* $\mathbf{a}, \mathbf{b} \in S$, *then* $(1 - t)\mathbf{a} + t\mathbf{b} \in S$ *for* $0 \leq t \leq 1$.

Theorem 4.2.1   *The intersection of any two convex sets in $\mathbf{R}^n$ is a convex set.*

Proof: Let S_1 and S_2 be convex sets in $\mathbf{R}^n$. Suppose that **a** and **b** belong to the intersection $S_1 \cap S_2$. Then $\mathbf{a}, \mathbf{b} \in S_1$ and $\mathbf{a}, \mathbf{b} \in S_2$. Therefore, $(1 - t)\mathbf{a} + t\mathbf{b}$ belongs to S_1 and to S_2 for $0 \leq t \leq 1$. We conclude that $(1 - t)\mathbf{a} + t\mathbf{b}$ belongs to $S_1 \cap S_2$ for $0 \leq t \leq 1$ and that $S_1 \cap S_2$ is convex. (The same argument proves that the intersection of any number of convex sets is convex.)

Example 4.2.4   The sets $S_1 = \{(x_1, x_2) : x_1^2 + x_2^2 \leq 1\}$ and $S_2 = \{(x_1, x_2) : x_1 \geq \frac{1}{2}\}$ are convex sets in $\mathbf{R}^2$. Their intersection $S_1 \cap S_2 = \{(x_1, x_2) : x_1^2 + x_2^2 \leq 1, x_1 \geq \frac{1}{2}\}$ is therefore a convex set. This is illustrated in Figure 4.8.

The next definition and theorem are central to the theory of linear programming.

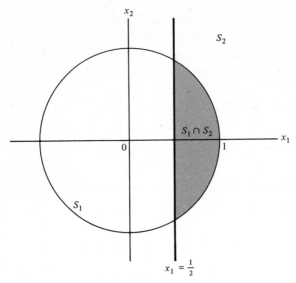

Figure 4.8

Definition 4.2.4   Linear Function on $\mathbf{R}^n$ *A function f defined for* $\mathbf{x} = (x_1, x_2, \ldots, x_n) \in \mathbf{R}^n$ *is linear if*

$$f(\mathbf{x}) = c_1 x_1 + c_2 x_2 + \cdots + c_n x_n \tag{4.4}$$

for some constants $c_1, c_2, \ldots, c_n$.

Theorem 4.2.2   *Suppose that f is a linear function on $\mathbf{R}^n$ and that $\mathbf{a}$ and $\mathbf{b}$ are two vectors of $\mathbf{R}^n$. Then the linear function f takes on values between $f(\mathbf{a})$ and $f(\mathbf{b})$ on the line segment joining $\mathbf{a}$ and $\mathbf{b}$.*

Proof: We may assume that $f(\mathbf{a}) \leq f(\mathbf{b})$. The same proof works if $f(\mathbf{b}) \leq f(\mathbf{a})$. If $0 \leq t \leq 1$, then $(1 - t)\mathbf{a} + t\mathbf{b}$ is a vector on the line segment joining $\mathbf{a}$ and $\mathbf{b}$. Evaluating f at $(1 - t)\mathbf{a} + t\mathbf{b}$, we have

$$
\begin{aligned}
f((1 - t)\mathbf{a} + t\mathbf{b}) &= c_1[(1 - t)a_1 + t b_1] + c_2[(1 - t)a_2 + t b_2] \\
&\quad + \cdots + c_n[(1 - t)a_n + t b_n] \\
&= (1 - t)[c_1 a_1 + c_2 a_2 + \cdots + c_n a_n] \\
&\quad + t[c_1 b_1 + c_2 b_2 + \cdots + c_n b_n] \\
&= (1 - t)f(\mathbf{a}) + t f(\mathbf{b}).
\end{aligned}
$$

But $f(\mathbf{a}) \le f(\mathbf{b})$. Therefore,

$$(1 - t)f(\mathbf{a}) + tf(\mathbf{a}) \le f((1 - t)\mathbf{a} + t\mathbf{b}) \le (1 - t)f(\mathbf{b}) + tf(\mathbf{b})$$

or

$$f(\mathbf{a}) \le f((1 - t)\mathbf{a} + t\mathbf{b}) \le f(\mathbf{b}).$$

This completes the proof.

Corollary 4.2.3 *If f is a linear function on $\mathbf{R}^n$ and if $\mathbf{a}$ and $\mathbf{b}$ are two vectors of $\mathbf{R}^n$ such that $f(\mathbf{a}) = f(\mathbf{b})$, then f is constant on the straight line segment joining $\mathbf{a}$ and $\mathbf{b}$.*

Example 4.2.5 What are the values taken on by $f(\mathbf{x}) = x_1 - x_2 + 5x_3$ on the line segment joining $(1, 0, 2)$ and $(3, 2, 0)$ in $\mathbf{R}^3$?

Solution: We have $f(1, 0, 2) = 11$ and $f(3, 2, 0) = 1$. Therefore, the linear function decreases from 11 to 1 along the line segment from $(1, 0, 2)$ to $(3, 2, 0)$.

The equation of a straight line in two dimensions $(\mathbf{R}^2)$ can be written

$$a_1 x_1 + a_2 x_2 = b, \tag{4.5}$$

where a_1, a_2, and b are real numbers. This straight line divides $\mathbf{R}^2$ into three disjoint subsets, $\mathbf{R}^2 = S_1 \cup S_2 \cup S_3$, where

$$S_1 = \{(x_1, x_2) : a_1 x_1 + a_2 x_2 = b\},$$
$$S_2 = \{(x_1, x_2) : a_1 x_1 + a_2 x_2 > b\}, \tag{4.6}$$
$$S_3 = \{(x_1, x_2) : a_1 x_1 + a_2 x_2 < b\}.$$

The points of S_1 are the points on the straight line.

Definition 4.2.5 **Open and Closed Half-Planes** *For real numbers a_1, a_2 and b, the sets S_2 and S_3 defined above are called open half-planes in $\mathbf{R}^2$. The sets $S_1 \cup S_2 = \{(x_1, x_2) : a_1 x_1 + a_2 x_2 \ge b\}$ and $S_1 \cup S_3 = \{(x_1, x_2) : a_1 x_1 + a_2 x_2 \le b\}$ are called closed half-planes in $\mathbf{R}^2$.*

Example 4.2.6 The three sets $S_1 = \{(x_1, x_2) : 3x_1 + 2x_2 = 6\}$, $S_2 = \{(x_1, x_2) : 3x_1 + 2x_2 > 6\}$, and $S_3 = \{(x_1, x_2) : 3x_1 + 2x_2 < 6\}$ are shown in Figure 4.9. The set $S_1 \cup S_2$ is the closed half-plane whose points are the points of $\mathbf{R}^2$ not in S_3. These are the points on and above the straight line $3x_1 + 2x_2 = 6$.

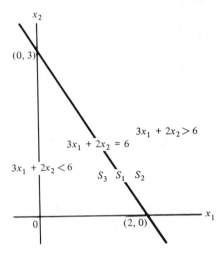

Figure 4.9

These concepts can be generalized to n dimensions ($\mathbf{R}^n$).

Definition 4.2.6　Hyperplane in $\mathbf{R}^n$　*Given real numbers $a_1, a_2, \ldots, a_n, b$, the set $H = \{\mathbf{x} = (x_1, x_2, \ldots, x_n) : a_1 x_1 + a_2 x_2 + \cdots + a_n x_n = b\}$ is a hyperplane in $\mathbf{R}^n$.*

Definition 4.2.7　Open and Closed Half-Spaces in $\mathbf{R}^n$　*The sets*

$$\{\mathbf{x} = (x_1, x_2, \ldots, x_n) : a_1 x_1 + a_2 x_2 + \cdots + a_n x_n < b\}$$

and

$$\{\mathbf{x} = (x_1, x_2, \cdots, x_n) : a_1 x_1 + a_2 x_2 + \cdots + a_n x_n > b\}$$

are called open half-spaces. The corresponding closed half-spaces are the sets

$$\{\mathbf{x} : a_1 x_1 + a_2 x_2 + \cdots + a_n x_n \leq b\}$$

and

$$\{\mathbf{x} : a_1 x_1 + a_2 x_2 + \cdots + a_n x_n \geq b\}.$$

The hyperplane H (previous definition) is called the bounding hyperplane *of these half-spaces.*

Theorem 4.2.4 *Half-spaces in* $\mathbf{R}^n$ *are convex sets.*

Proof: If $a_1 x_1 + a_2 x_2 + \cdots + a_n x_n = b$ is the bounding hyperplane of the half-space, define the linear function $f(\mathbf{x}) = a_1 x_1 + a_2 x_2 + \cdots + a_n x_n$. Suppose that $\mathbf{x}$ and $\mathbf{y}$ belong to one of the half-spaces with this bounding hyperplane. For definiteness, suppose that $\mathbf{x}$ and $\mathbf{y}$ belong to $\{\mathbf{x} : a_1 x_1 + a_2 x_2 + \cdots + a_n x_n \leq b\}$. Then $f(\mathbf{x}) \leq b$ and $f(\mathbf{y}) \leq b$. From Theorem 4.2.2, we know that $f((1 - t)\mathbf{x} + t\mathbf{y})$ is a number between $f(\mathbf{x})$ and $f(\mathbf{y})$ if $0 \leq t \leq 1$. Therefore, $f((1 - t)\mathbf{x} + t\mathbf{y}) \leq b$, and the result is proved. The proof is the same for all open and closed half-spaces with this bounding hyperplane.

In linear programming, we will be interested in regions of $\mathbf{R}^n$ defined as the intersection of half-spaces.

Definition 4.2.8 Polyhedral Convex Set *The intersection C of a finite number of closed half-spaces is called a polyhedral convex set.*

Definition 4.2.9 Corner Point of a Polyhedral Convex Set *A point* $\mathbf{x}$ *in* $\mathbf{R}^n$ *is called a corner point of a polyhedral convex set C if* $\mathbf{x} \in C$ *and if* $\mathbf{x}$ *is the point of intersection of n of the bounding hyperplanes that determine C. A corner point is also called an* extreme point *or a* vertex *of the polyhedral convex set.*

Definition 4.2.10 Solution Set of a System of Linear Inequalities *The set of vectors* $\mathbf{x} = (x_1, x_2, \ldots, x_n)$ *in* $\mathbf{R}^n$ *whose components satisfy the m linear inequalities*

$$
\begin{aligned}
a_{11} x_1 + a_{12} x_2 + \cdots + a_{1n} x_n &\leq b_1 \\
a_{21} x_1 + a_{22} x_2 + \cdots + a_{2n} x_n &\leq b_2 \\
\vdots \qquad\qquad \vdots \qquad\qquad \vdots \qquad\qquad \vdots & \\
a_{m1} x_1 + a_{m2} x_2 + \cdots + a_{mn} x_n &\leq b_m
\end{aligned}
\tag{4.7}
$$

is called the solution set of the system of linear inequalities.

In Definition 4.2.10, the inequalities have all been written with the "$\leq$" sign. By multiplying any of the inequalities by -1, the direction of the inequality is reversed. The solution set is defined for systems of linear inequalities with both "$\leq$" and "$\geq$" signs. System (4.7) can be written in a more compact form. Define $A = (a_{ij})$ to be the $m \times n$ matrix whose ijth component is a_{ij} and define the column vectors

$$\mathbf{x} = \begin{pmatrix} x_1 \\ x_2 \\ \vdots \\ x_n \end{pmatrix} \quad \text{and} \quad \mathbf{b} = \begin{pmatrix} b_1 \\ b_2 \\ \vdots \\ b_m \end{pmatrix},$$

where $\mathbf{x}$ has n components and $\mathbf{b}$ has m components. Then $A\mathbf{x}$ is an m-component column vector each of whose components is less than or equal to the corresponding component of $\mathbf{b}$. This is written

$$A\mathbf{x} \leq \mathbf{b}. \tag{4.8}$$

The vector inequality (4.8) is equivalent to the m inequalities of (4.7). The solution set of (4.7) has a relatively simple geometrical structure as shown by the following theorem.

Theorem 4.2.5   *The solution set of a system of m linear inequalities (4.7) is a polyhedral convex set in $\mathbf{R}^n$.*

Proof: The vectors of $\mathbf{R}^n$ satisfying the linear inequalities of (4.7) must lie in each of the m half-spaces of $\mathbf{R}^n$ defined by the inequalities. The solution set is the intersection of a finite number of closed half-spaces and is therefore a polyhedral convex set.

Example 4.2.7   Graph the solution set of the following system of linear inequalities in $\mathbf{R}^2$.

$$\begin{aligned} x_1 + 3x_2 &\leq 6 \\ x_1 - x_2 &\leq 2 \\ x_1 &\geq 0. \end{aligned}$$

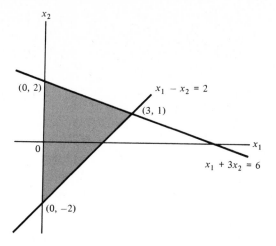

Figure 4.10

Solution: The shaded area of Figure 4.10 is the required solution set. The corner points of the solution set are $(0, 2)$, $(0, -2)$, and $(3, 1)$. These are the intersections of the lines bounding the solution set. For example, $(3, 1)$ is the intersection of the lines $x_1 + 3x_2 = 6$ and $x_1 - x_2 = 2$.

The polyhedral convex set or solution set of Example 4.2.7 is bounded. (Every point of the set can be enclosed in one circle with center the origin.) This is not always the case. A half-space is an example of a polyhedral convex set, which is clearly not bounded.

Example 4.2.8 Determine the corner points of the solution set of the following system of linear inequalities in $\mathbf{R}^2$.

$$
\begin{aligned}
-\ x_1 + 2x_2 &\le 4 \\
3x_1 + 2x_2 &\le 6 \\
x_1 &\ge 0.
\end{aligned}
$$

Solution: The solution set is the shaded area of Figure 4.11. The corner points are $(0, 2)$ and $(\frac{1}{2}, \frac{9}{4})$. Note that $(0, 3)$ is not a corner point, since it does not satisfy the first inequality. This solution set is not bounded.

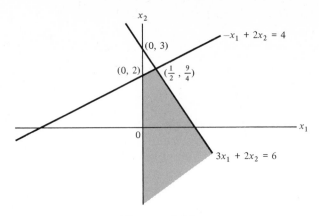

Figure 4.11

Example 4.2.9 Determine the solution set in $\mathbf{R}^2$ of the following system of linear inequalities.

$$x_1 - x_2 \geq 0$$
$$x_1 + x_2 \leq 1$$
$$x_2 \geq 2.$$

Solution: Suppose that $\mathbf{x} = (x_1, x_2)$ satisfies the first two inequalities. Then $x_1 \geq x_2$ and $x_1 + x_2 \leq 1$ or $1 \geq x_1 + x_2 \geq 2x_2$, which implies that $x_2 \leq \frac{1}{2}$. But this contradicts the third inequality. Since there can be no points in $\mathbf{R}^2$ that satisfy all three inequalities, the solution set is empty. This is illustrated in Figure 4.12.

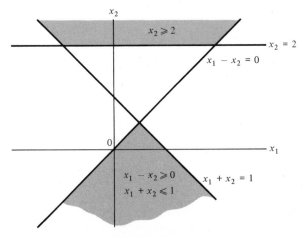

Figure 4.12

Problems 4.2

1. Given the vectors $\mathbf{a} = (2, 3)$ and $\mathbf{b} = (-1, 2)$ in $\mathbf{R}^2$, draw the vectors $(1 - t)\mathbf{a} + t\mathbf{b}$ for $t = -1, 0, \frac{1}{3}, \frac{2}{3}, 1$, and 2.

2. Which of the following sets of vectors $\mathbf{x}$ in $\mathbf{R}^2$ are nonempty convex sets? Draw a diagram to illustrate each set which is nonempty.

   (a) $S_1 = \{\mathbf{x} = (x_1, x_2) : x_1 \geq 1, x_2 \leq 1\}$.

   (b) $S_2 = \{\mathbf{x} = (x_1, x_2) : \dfrac{x_1^2}{4} + \dfrac{x_2^2}{9} \leq 1\}$.

   (c) $S_3 = \{\mathbf{x} = (x_1, x_2) : x_1 + x_2 \leq 1, x_1 \geq 0, x_2 \geq 2\}$.

   (d) $S_4 = \{\mathbf{x} = (x_1, x_2) : x_1^2 + x_2^2 \leq 4 \text{ or } (x_1 - 3)^2 + x_2^2 \leq 4\}$.

   (e) $S_5 = \{\mathbf{x} = (x_1, x_2) : 1 \leq x_1^2 + x_2^2 \leq 4\}$.

3. (a) Determine all planes in $\mathbf{R}^3$ that pass through $(1, 0, 0)$.

   (b) Which of the planes in (a) also pass through $(0, 1, 0)$?

   (c) Which of the planes in (b) also pass through $(1, 1, 1)$?

4. (a) Determine all hyperplanes in $\mathbf{R}^4$ that pass through $(1, 1, 1, 0)$ and $(0, 1, 1, 1)$.

   (b) Which of the hyperplanes in (a) also pass through $(0, 0, 0, 0)$?

   (c) Which of the hyperplanes in (b) also pass through $(-1, 0, 0, 0)$?

5. Describe the set X of points in $\mathbf{R}^2$ that satisfy $2x_1 + x_2 \geq 1, x_1 + 2x_2 \geq 1$, $x_1 + x_2 \leq 3, x_1 \geq 0$, and $x_2 \geq 0$. Draw a diagram of this set.

6. Determine the maximum and minimum values of the following linear functions over the set X of the previous problem. Illustrate the solutions by diagrams.

   (a) $f(x_1, x_2) = 3x_1 + 5x_2$

   (b) $g(x_1, x_2) = 2x_1 + x_2$

   (c) $h(x_1, x_2) = 2x_1 + 2x_2$.

7. Prove that the solution set in $\mathbf{R}^3$ of the following system of linear inequalities is empty.

$$x_1 + x_2 + x_3 \leq 5$$
$$x_1 - 2x_2 + x_3 \leq 2$$
$$x_3 \geq 0$$
$$x_1 + x_3 \geq 6.$$

8. Determine the values taken on by the following linear functions $f(\mathbf{x})$ on the line segment between the given points $\mathbf{a}$ and $\mathbf{b}$ in $\mathbf{R}^3$.

   (a) $f(\mathbf{x}) = x_1 - x_2 + 3x_3, \mathbf{a} = (1, 0, 0), \mathbf{b} = (0, 0, 1)$.

(b) $f(\mathbf{x}) = x_1 + 2x_2 - 3x_3, \mathbf{a} = (1, 1, 0), \mathbf{b} = (0, 0, -1)$.

(c) $f(\mathbf{x}) = 2x_1 - 2x_2 + x_3, \mathbf{a} = (1, 1, 1), \mathbf{b} = (2, 0, 2)$.

9. By a geometrical method, determine the maximum and minimum values of the nonlinear function $f(x_1, x_2) = (x_1 - 1)^2 + (x_2 + 1)^2$ on the convex set in $\mathbf{R}^2$ determined by $x_1 + 2x_2 \leq 5$, $x_1 \geq 0$, and $x_2 \geq 0$. [*Hint:* Draw the curves $f(x_1, x_2) = $ constant for different values of the constant.]

10. By a geometrical method, determine the maximum and minimum of the linear function $f(x_1, x_2) = 3x_1 + 4x_2$ over the convex set in $\mathbf{R}^2$ determined by $x_1^2 + x_2^2 \leq 25$, $x_1 \geq 0$, and $x_2 \geq 0$.

11. The storeroom of a biology department decides to stock at least 1000 beakers of one size and at least 1500 beakers of a second size. If the total number of beakers stocked is not to exceed 4000 because of space limitations, draw a diagram showing all possible combinations of the two beakers that can be stocked.

12. If in Problem 11 a beaker of the first type costs 10 cents and a beaker of the second type costs 8 cents, what are the maximum and minimum possible total values of the beakers in stock? Illustrate the solution of this problem by means of a diagram.

13. Determine the matrices A and the column vectors $\mathbf{b}$ such that the following systems of linear inequalities can be written in the form $A\mathbf{x} \leq \mathbf{b}$.

(a) $x_1 + 2x_2 \leq 5$
$\qquad x_1 \geq 0$
$\qquad x_2 \geq 0$.

(b) $x_1 - 2x_2 \leq 4$
$\qquad x_1 + x_2 \leq 2$
$\qquad x_1 + 3x_2 \leq 3$.

(c) $x_1 + x_2 + x_3 \leq 1$
$\qquad x_1 + 3x_2 + 4x_3 \leq 3$
$\qquad\qquad x_1 \geq 0$
$\qquad\qquad x_2 \geq 0$

(d) $x_1 - x_2 + 2x_3 \leq 5$
$\qquad x_1 + 3x_2 - x_3 \leq 5$
$\qquad -x_1 + x_2 + x_3 \leq 5$.

14. Draw diagrams in $\mathbf{R}^2$ of the solution sets of the vector inequalities $A\mathbf{x} \leq \mathbf{b}$ for the given matrices A and vectors $\mathbf{b}$.

(a) $A = \begin{pmatrix} 1 & 1 \\ 0 & -1 \end{pmatrix}$, $\mathbf{b} = \begin{pmatrix} 1 \\ 0 \end{pmatrix}$.

(b) $A = \begin{pmatrix} 2 & 4 \\ 4 & 2 \end{pmatrix}$, $\mathbf{b} = \begin{pmatrix} 3 \\ 3 \end{pmatrix}$.

(c) $A = \begin{pmatrix} 1 & 2 \\ 0 & -1 \\ -1 & 0 \end{pmatrix}$, $\mathbf{b} = \begin{pmatrix} 5 \\ 0 \\ 0 \end{pmatrix}$.

(d) $A = \begin{pmatrix} 2 & 2 \\ 1 & 3 \\ -1 & 0 \\ 0 & -1 \end{pmatrix}$, $\mathbf{b} = \begin{pmatrix} 3 \\ 5 \\ 0 \\ 0 \end{pmatrix}$.

15. The fitness sets of Levins[2] provide an interesting application of the idea of convex sets in $\mathbf{R}^n$. A genotype of a particular species may have different "fitness" W_1 and W_2 in two different environments. Suppose that all genotypes of the species correspond to the points of the convex set $W_1^2 + W_2^2 \leq 10$, $3W_2 - W_1 \geq 0$, and $3W_1 - W_2 \geq 0$. Which genotype (point of this set) maximizes W_1? W_2? $W_1 + W_2$? (These are the fittest genotypes in the first environment, the second environment, and a mixed environment, respectively.)

16. Graph the fitness set $W_1^2 + W_2^2 \leq 32$, $2W_2 - W_1 \geq 0$, $2W_1 - W_2 \geq 0$, and $(W_1 - 5)^2 + (W_2 - 5)^2 \leq 25$. Which point of this set maximizes W_1? W_2? $W_1 + W_2$? Which points minimize these functions?

4.3 Linear Programming: The Corner Point Method

In the first section of this chapter, we solved a problem of maximizing the weight of fish supported by a lake subject to constraints of food availability. This is an example of the general problem that will be studied in the remaining sections of this chapter.

Definition 4.3.1   Linear Programming Problem   *Determine the vector* $\mathbf{x} = (x_1, x_2, \ldots, x_n)$ *in* $\mathbf{R}^n$ *that maximizes or minimizes the linear function*

$$f(\mathbf{x}) = c_1 x_1 + c_2 x_2 + \cdots + c_n x_n \tag{4.9}$$

and satisfies the $m + n$ linear inequalities

$$
\begin{aligned}
a_{11}x_1 + a_{12}x_2 + \cdots + a_{1n}x_n &\leq b_1 \\
a_{21}x_1 + a_{22}x_2 + \cdots + a_{2n}x_n &\leq b_2 \\
\vdots \qquad \vdots \qquad\qquad \vdots \qquad &\ \vdots \\
a_{m1}x_1 + a_{m2}x_2 + \cdots + a_{mn}x_n &\leq b_m \\
x_1 \geq 0, \quad x_2 \geq 0, \quad \ldots, \quad x_n &\geq 0.
\end{aligned}
\tag{4.10}
$$

To introduce some terminology, the linear function $f(\mathbf{x})$ is called the *objective function* of the problem. The solution set of system (4.10) of linear inequalities is called the *constraint set* of the problem. Any vector $\mathbf{x}$ in the

[2] R. Levins, *Evolution in Changing Environments*, Princeton University Press, Princeton, N.J., 1968.

constraint set is a *feasible solution*. An *optimal solution* is a vector **x*** for which the objective function takes on its maximum (minimum) value over the constraint set.

Example 4.3.1   Formulate the problem of Example 4.1.1 as a linear programming maximum problem.

Solution: We must determine the vector $\mathbf{x} = (x_1, x_2)$ in $\mathbf{R}^2$ that maximizes the total weight $W = 4x_1 + 2x_2$. This is the objective function $f(\mathbf{x})$. The constraints on **x** are

$$x_1 + 2x_2 \le 500 \qquad 3x_1 + x_2 \le 900$$
$$x_1 \ge 0, \qquad\qquad x_2 \ge 0.$$

The constraint set is the shaded area of Figure 4.2. The optimal solution was found to be $\mathbf{x}^* = (260, 120)$.

The general maximum problem of linear programming can be written in a convenient matrix and vector notation. Define the vectors $\mathbf{c} = (c_1, c_2, \dots, c_n)$,

$$\mathbf{x} = \begin{pmatrix} x_1 \\ x_2 \\ \vdots \\ x_n \end{pmatrix}, \quad \mathbf{b} = \begin{pmatrix} b_1 \\ b_2 \\ \vdots \\ b_m \end{pmatrix}$$

and the matrix

$$A = \begin{pmatrix} a_{11} & a_{12} & \cdots & a_{1n} \\ a_{21} & a_{22} & \cdots & a_{2n} \\ \vdots & \vdots & & \vdots \\ a_{m1} & a_{m2} & \cdots & a_{mn} \end{pmatrix}.$$

The general problem is then to maximize or minimize $f(\mathbf{x}) = \mathbf{c} \cdot \mathbf{x}$ subject to the constraints $A\mathbf{x} \le \mathbf{b}$ and $\mathbf{x} \ge \mathbf{0}$. (The vector inequality $\mathbf{u} \le \mathbf{v}$ means that the vectors **u** and **v** have the same number of components and each component of **u** is less than or equal to the corresponding component of **v**.)

There is an elementary method for solving the maximum or minimum problem of linear programming when the constraint set is bounded, that is, when the constraint set occupies a finite region of $\mathbf{R}^n$. Unfortunately, the method is not practical if the number of variables is large. The method is based on the following theorem.

Theorem 4.3.1   *The linear function* $f(\mathbf{x}) = \mathbf{c} \cdot \mathbf{x}$ *defined on a bounded convex polyhedral set S takes on its maximum and minimum values at corner points of S.*

Proof: The proof will be given in $\mathbf{R}^2$ (two dimensions) by a simple geometrical argument, although the idea of the proof is the same in $\mathbf{R}^n$. The bounded polyhedral convex set S is represented in Figure 4.13. The set S has

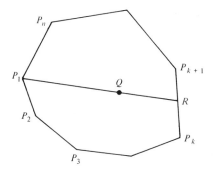

Figure 4.13

n corner points $P_1, P_2, P_3, \ldots, P_k, P_{k+1}, \ldots, P_n$. Evaluating $f(\mathbf{x}) = \mathbf{c} \cdot \mathbf{x}$ at these n corner points, we assume that the maximum of $f(\mathbf{x})$ over the corner points occurs at P_1. Therefore, $f(P_k) \leq f(P_1)$ for $k = 2, 3, \ldots, n$. Suppose that Q is any point of the set S. We must prove that $f(Q) \leq f(P_1)$. Draw the line segment P_1QR, where R is the point of intersection of the line joining P_1 and Q with the boundary of S. Suppose that R is on the line segment joining P_k and P_{k+1}. Then (Theorem 4.2.2), $f(R)$ lies between $f(P_k)$ and $f(P_{k+1})$. Therefore, $f(R) \leq f(P_1)$. But Q is on the line segment joining P_1 and R; therefore, $f(R) \leq f(Q) \leq f(P_1)$. But Q is any point of S, and we conclude that the maximum of $f(\mathbf{x}) = \mathbf{c} \cdot \mathbf{x}$ over S occurs at a corner point. The proof for the minimum of $f(\mathbf{x})$ is exactly the same.

A second proof of this result can be obtained by considering the parallel straight lines in $\mathbf{R}^2$ given by $f(\mathbf{x}) = c_1 x_1 + c_2 x_2 = l$ for different values of the constant l. As l increases, the lines $\mathbf{c} \cdot \mathbf{x} = l$ move across the plane (Figure 4.14). Each corner point of the polyhedral convex set S corresponds to a

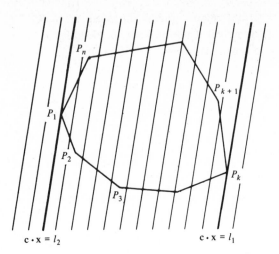

Figure 4.14. Parallel Lines $c \cdot x = l$

definite value of l. Suppose that l_1 and l_2 are the minimum and maximum values of l over the corner points of S. Then, from elementary geometry, $l_1 \leq f(\mathbf{x}) \leq l_2$ for any $\mathbf{x}$ in S. This is the result. We note that the theorem is not necessarily true if S is unbounded, since the minimum and maximum values of l over the set S do not exist in general.

Theorem 4.3.1 suggests the following three-step method for solving the maximum and minimum problems of linear programming:

1. Determine all corner points of the polyhedral convex set defined by the constraints (4.10).
2. Evaluate $f(\mathbf{x})$ at each corner point.
3. Determine the corner points where $f(\mathbf{x})$ takes on its maximum and minimum values. [If $f(\mathbf{x})$ takes on its maximum value at two corner points, then it takes on this value at every point on the line joining the corner points.]

The corner points determined by these steps will be optimal solutions of the maximum and minimum problems. This method will always produce the solutions if they exist. The practical difficulty with this method is that the number of constraints $(m + n)$ may be large in many applications. If this is the case, step (1) may require the solution of

$$\binom{m + n}{n} = \frac{(m + n)!}{m!n!}$$

systems of m linear equations. If $m = n = 5$, for example, there are 252 corner points to be determined. Some simplifications may be possible but, in general, the problem is an extremely tedious one when solved by this method. In the last two sections of this chapter, we develop a more systematic method which avoids the necessity of determining all the corner points.

In Example 4.1.1, we found that the maximum value 1280 of $W = 4x_1 + 2x_2$ over the constraint set occurs at the corner point $(260, 120)$. The minimum value of W is 0 and occurs at the corner point $(0, 0)$. The reader should determine all other corner points and verify that the values of W at these corner points are between 0 and 1280.

The following example illustrates that a problem in three dimensions with six constraints can be quite complex.

Example 4.3.2   Determine the optimal solutions of the maximum and minimum problems for the objective function $f(\mathbf{x}) = x_1 - x_2 + x_3$ over the constraint set defined by
(a) $x_1 + x_2 + x_3 \leq 15.$
(b) $2x_1 + x_2 + 2x_3 \leq 26.$
(c) $5x_1 + 2x_2 + 3x_3 \leq 43.$
(d) $x_1 \geq 0.$
(e) $x_2 \geq 0.$
(f) $x_3 \geq 0.$

Solution: Since we are in $\mathbf{R}^3$, to determine the corner points we determine the

$$\binom{6}{3} = 20$$

intersections of the six bounding planes taken three at a time. If the intersection satisfies all the constraints, it is a corner point and we evaluate $f(\mathbf{x})$ there. The results of these calculations are given in Table 4.1. The constraints (a), (b), and (c) determine the system of equations

$$x_1 + x_2 + x_3 = 15$$
$$2x_1 + x_2 + 2x_3 = 26$$
$$5x_1 + 2x_2 + 3x_3 = 43.$$

By the methods of Chapter 3, this has the unique solution $x_1 = 1, x_2 = 4,$

$x_3 = 10$. The constraints (d), (e), and (f) are obviously satisfied by this solution; therefore $(1, 4, 10)$ is a corner point. Evaluating $f(\mathbf{x})$, we find that $f(1, 4, 10) = 1 - 4 + 10 = 7$. Similarly, constraints (a), (c), and (e) give the system

$$x_1 + x_3 = 15$$
$$5x_1 + 3x_3 = 43,$$

(since $x_2 = 0$) with the solution $(-1, 0, 16)$, which is not a corner point, since constraint (d) is violated.

Table 4.1

Constraints	Intersection	Corner Point?	$f(\mathbf{x}) = x_1 - x_2 + x_3$
(a), (b) (c)	$(1, 4, 10)$	Yes	7
(a), (b), (d)	$(0, 4, 11)$	Yes	7
(a), (b), (e)	None (system is inconsistent)		
(a), (b), (f)	$(11, 4, 0)$	No	(Constraint (c) is not satisfied)
(a), (c), (d)	$(0, 2, 13)$	No	(Constraint (b) is not satisfied)
(a), (c), (e)	$(-1, 0, 16)$	No	(Constraint (d) is not satisfied)
(a), (c), (f)	$(\frac{13}{3}, \frac{32}{3}, 0)$	Yes	$-\frac{19}{3}$
(a), (d), (e)	$(0, 0, 15)$	No	(Constraints (b) and (c) are not satisfied)
(a), (d), (f)	$(0, 15, 0)$	Yes	-15
(a), (e), (f)	$(15, 0, 0)$	No	
(b), (c), (d)	$(0, 8, 9)$	No	
(b), (c), (e)	$(2, 0, 11)$	Yes	13
(b), (c), (f)	$(-9, 44, 0)$	No	
(b), (d), (e)	$(0, 0, 13)$	Yes	13
(b), (d), (f)	$(0, 26, 0)$	No	
(b), (e), (f)	$(13, 0, 0)$	No	
(c), (d), (e)	$(0, 0, \frac{43}{3})$	No	
(c), (d), (f)	$(0, \frac{43}{2}, 0)$	No	
(c), (e), (f)	$(\frac{43}{5}, 0, 0)$	Yes	$\frac{43}{5}$
(d), (e), (f)	$(0, 0, 0)$	Yes	0

The polyhedral convex set has eight corner points. The optimal solution of the maximum problem is $f(\mathbf{x}^*) = 13$ at either $\mathbf{x}^* = (2, 0, 11)$ or $\mathbf{x}^* = (0, 0, 13)$. The optimal solution of the minimum problem is $f(\mathbf{y}^*) = -15$ at $\mathbf{y}^* = (0, 15, 0)$. [Note that the maximum value of $f(\mathbf{x})$ occurs at every point on the edge joining the corner points $(2, 0, 11)$ and $(0, 0, 13)$.]

Example 4.3.3   A predator requires 10 units of food A, 12 units of food B, and 12 units of food C as its average daily consumption. These requirements are satisfied by feeding on two prey species. One prey of species I provides 5, 2, and 1 units of foods A, B, and C, respectively. An individual prey of species II provides 1, 2, and 4 units of A, B, and C, respectively. To capture and digest a prey of species I requires 3 units of energy on average. The corresponding energy requirement for species II is 2 units of energy. How many prey of each species should the predator capture to meet its food requirements with minimum expenditure of energy?

Solution: Let x_1 and x_2 be the average daily consumptions of prey of species I and II, respectively. The energy expenditure E is a linear function g of these consumptions, $E = g(\mathbf{x}) = 3x_1 + 2x_2$, measured in units of energy. The daily consumptions of foods A, B, and C are $5x_1 + x_2, 2x_1 + 2x_2$, and $x_1 + 4x_2$, respectively. Since these consumptions must satisfy the daily requirements of the predator, we have the following minimum problem of linear programming. Minimize $g(\mathbf{x}) = 3x_1 + 2x_2$ subject to the constraints $5x_1 + x_2 \geq 10, 2x_1 + 2x_2 \geq 12, x_1 + 4x_2 \geq 12, x_1 \geq 0$, and $x_2 \geq 0$. The constraint set for this problem is unbounded (Figure 4.15). The corner

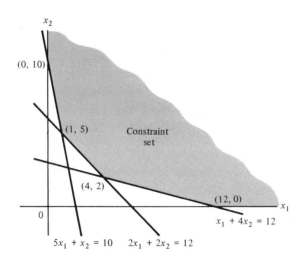

Figure 4.15

points of the constraint set are $(0, 10), (1, 5), (4, 2)$, and $(12, 0)$. The corresponding values of $g(\mathbf{x})$ are 20, 13, 16, and 36. Therefore, the diet that satisfies the food requirements of the predator with least expenditure of energy is one unit of the first prey and five units of the second prey as average daily consumptions. By specializing on one prey, the predator would be able to satisfy its requirements, but only with greatly increased consumptions and with much higher expenditures of energy.

Problems 4.3

1. Determine, by a graphical method, the maximum and minimum values of the following linear functions over the given polyhedral convex sets in $\mathbf{R}^2$.

(a) $f(\mathbf{x}) = 2x_1 + x_2$, $S_1 = \{\mathbf{x} = (x_1, x_2) : x_1 \geq 0, x_2 \geq 0, x_1 + x_2 \leq 1\}$.

(b) $f(\mathbf{x}) = x_1 - x_2$, $S_2 = \{\mathbf{x} = (x_1, x_2) : x_1 \geq 1, x_2 \leq 1, 5x_1 - 2x_2 \leq 20\}$.

(c) $f(\mathbf{x}) = 4x_1 + 5x_2$, $S_3 = \{\mathbf{x} = (x_1, x_2) : x_1 \geq -1, x_2 \geq x_1, x_1 + x_2 \leq 10\}$.

2. Determine the corner points of the solution sets of the following systems of linear inequalities in $\mathbf{R}^2$.

(a) $x_1 + 2x_2 \leq 5$
    $3x_1 + 2x_2 \leq 7$
    $x_1 + x_2 \geq 1$.

(b) $x_1 - x_2 \leq 5$
    $3x_1 - 2x_2 \leq 6$
    $x_1 \geq 1$.

(c) $x_1 + x_2 \leq 4$
    $-x_1 - 2x_2 \leq 2$
    $x_1 \leq 2$.

(d) $x_1 - 3x_2 \leq -1$
    $x_1 + x_2 \geq 2$
    $x_1 \geq 0$.

3. Determine the corner points of the solution sets of the following systems of linear inequalities in $\mathbf{R}^3$.

(a) $x_1 + x_2 + x_3 \leq 5$
    $2x_1 + x_2 + 3x_3 \leq 6$
    $x_1 \geq 0$
    $x_2 \geq 0$
    $x_3 \geq 0$.

(b) $x_1 - x_2 - x_3 \leq 5$
    $-x_1 + x_2 + 2x_3 \leq 6$
    $2x_1 - x_2 + x_3 \leq 7$
    $x_1 \geq 0$
    $x_2 \geq 0$
    $x_3 \geq 0$.

4. Determine the maximum and minimum values of the linear function $f(x_1, x_2, x_3) = -x_1 + x_2 + 2x_3$ over the polyhedral convex sets of Problem 3.

5. Suppose that the corner points of a polyhedral convex set in $\mathbf{R}^2$ are $(1, 1)$, $(1, 4)$, $(3, 7)$, and $(5, 6)$. Determine a system of linear inequalities whose solution set has these corner points (and no other corner points).

6. Determine the maximum and minimum values of the following linear functions over the solution set of Problem 5.
   (a) $f(x_1, x_2) = x_1 + 2x_2$.   (b) $f(x_1, x_2) = -x_1 - 3x_2$.
   (c) $f(x_1, x_2) = -2x_1 - 2x_2$.   (d) $f(x_1, x_2) = 3x_1 + 9x_2$.

7. In a large hospital, surgical operations are classified into three categories, according to their average times of 30 minutes, 1 hour, and 2 hours. The hospital receives a fee of $100, $150, or $200 for an operation in categories I, II, or III, respectively. If the hospital has eight operating rooms which are in use an average of 10 hours per day, how many operations of each type should the hospital schedule in order (a) to maximize its revenue and (b) to maximize the total number of operations?

8. If the total number of operations in Problem 7 cannot exceed 120, how many operations of each type should the hospital schedule in order (a) to maximize its revenue and (b) to maximize the total number of operations?

9. The activities of a grazing animal can be classified roughly into (a) grazing, (b) moving (to avoid predators or to new grazing areas), and (c) resting. The net energy gain above maintenance requirements during grazing is 200 calories per hour. The net energy losses in moving and resting are 150 calories and 50 calories per hour, respectively. If the animal must rest at least 6 hours per day and if the grazing time cannot exceed the moving time (to avoid predators and to prevent overgrazing), how should the day be divided to maximize the net energy gain?

10. Two foods contain carbohydrates and proteins only. Food I costs 50 cents per pound and is 90 per cent carbohydrates (by weight). Food II costs $1 per pound and is 60 per cent carbohydrates. What diet of these two foods provides 2 pounds of carbohydrates and 1 pound of proteins at minimum cost? What is the cost per pound of this diet?

11. Continuing Problem 10, suppose that a third food is available which costs $2 per pound and is 30 per cent carbohydrates and 70 per cent protein. What diet of these three foods provides 2 pounds of carbohydrates and 1 pound of proteins at minimum cost? What is the cost per pound of this diet?

12. A company producing canned mixed fruit has a stock of 10,000 pounds of pears, 12,000 pounds of peaches, and 8000 pounds of cherries. The company produces three fruit mixtures, which it sells in 1-pound cans. The first mixture is half pears and half peaches and sells for 30 cents. The second mixture has equal amounts of the three fruits and sells for 40 cents. The third mixture is half peaches and half cherries and sells

for 50 cents. How many cans of each mixture should be produced to maximize the return?

4.4 The Dual Problem

There is a remarkable connection between the maximum and minimum problems of linear programming. To every maximum problem is associated a minimum problem, called the dual of the maximum problem. Conversely, the maximum problem is the dual of the associated minimum problem. This association is useful, since the solution of one problem is closely related to the solution of the dual problem.

Definition 4.4.1   Dual Problems of Linear Programming   *The following maximum and minimum problems are called dual problems.*
    1. *Maximize*

$$f(\mathbf{x}) = c_1 x_1 + c_2 x_2 + \cdots + c_n x_n$$

subject to

$$a_{11} x_1 + a_{12} x_2 + \cdots + a_{1n} x_n \leq b_1$$
$$a_{21} x_1 + a_{22} x_2 + \cdots + a_{2n} x_n \leq b_2$$
$$\vdots \qquad \vdots \qquad\qquad \vdots \qquad \vdots \tag{4.11}$$
$$a_{m1} x_1 + a_{m2} x_2 + \cdots + a_{mn} x_n \leq b_m$$
$$x_1 \geq 0, \quad x_2 \geq 0, \ldots, x_n \geq 0.$$

    2. *Minimize*

$$g(\mathbf{y}) = b_1 y_1 + b_2 y_2 + \cdots + b_m y_m$$

subject to

$$a_{11} y_1 + a_{21} y_2 + \cdots + a_{m1} y_m \geq c_1$$
$$a_{12} y_1 + a_{22} y_2 + \cdots + a_{m2} y_m \geq c_2$$
$$\vdots \qquad \vdots \qquad\qquad \vdots \qquad \vdots \tag{4.12}$$
$$a_{1n} y_1 + a_{2n} y_2 + \cdots + a_{mn} y_m \geq c_n$$
$$y_1 \geq 0, \quad y_2 \geq 0, \quad \ldots, \quad y_m \geq 0.$$

To write the dual problems in a matrix and vector notation, define the $n \times m$ matrix A^t as the matrix whose ijth component is equal to the jith component of A. The matrix A^t is called the *transpose* of the matrix A. The dual problems are then

1. Maximize $f(\mathbf{x}) = \mathbf{c} \cdot \mathbf{x}$ subject to $A\mathbf{x} \leq \mathbf{b}$ and $\mathbf{x} \geq \mathbf{0}$.
2. Minimize $g(\mathbf{y}) = \mathbf{b} \cdot \mathbf{y}$ subject to $A^t\mathbf{y} \geq \mathbf{c}$ and $\mathbf{y} \geq \mathbf{0}$.

Example 4.4.1   Determine the dual problem of the following linear programming problem, and write both problems in a matrix and vector notation. Minimize $g(\mathbf{y}) = 3y_1 + 4y_2 + 6y_3$ subject to $y_1 + 3y_2 + 5y_3 \geq 7$, $2y_1 + y_2 + 4y_3 \geq 10$, $y_1 \geq 0$, $y_2 \geq 0$, and $y_3 \geq 0$.

Solution: The dual problem is to maximize $f(\mathbf{x}) = 7x_1 + 10x_2$ subject to $x_1 + 2x_2 \leq 3$, $3x_1 + x_2 \leq 4$, $5x_1 + 4x_2 \leq 6$, $x_1 \geq 0$, and $x_2 \geq 0$. Define the 3×2 matrix

$$A = \begin{pmatrix} 1 & 2 \\ 3 & 1 \\ 5 & 4 \end{pmatrix}$$

and the vectors

$$\mathbf{b} = \begin{pmatrix} 3 \\ 4 \\ 6 \end{pmatrix}, \qquad \mathbf{c} = (7 \quad 10).$$

Then, A^t is the 2×3 matrix

$$A^t = \begin{pmatrix} 1 & 3 & 5 \\ 2 & 1 & 4 \end{pmatrix}.$$

The maximum problem is to maximize $f(\mathbf{x}) = \mathbf{c} \cdot \mathbf{x}$ subject to $A\mathbf{x} \leq \mathbf{b}$ and $\mathbf{x} \geq \mathbf{0}$. The dual minimum problem is to minimize $g(\mathbf{y}) = \mathbf{b} \cdot \mathbf{y}$ subject to $A^t\mathbf{y} \geq \mathbf{c}$ and $\mathbf{y} \geq \mathbf{0}$. Note that $\mathbf{x} \in \mathbf{R}^2$ and $\mathbf{y} \in \mathbf{R}^3$.

The usefulness of defining the dual problem is indicated by the following theorem.

Theorem 4.4.1 Fundamental Theorem of Linear Programming
*Let f be the objective function of a linear programming maximum problem and
let g be the objective function of the corresponding dual minimum problem.
Then the maximum problem for f has a solution if and only if the minimum
problem for g has a solution. Furthermore,* $\mathbf{x}^*$ *and* $\mathbf{y}^*$ *are optimal solutions of the
two problems if and only if* $f(\mathbf{x}^*) = g(\mathbf{y}^*)$.

We will not prove this theorem here, except to note that the corner point
method of the previous section allows us to determine the solutions of the
dual problems when they exist. Therefore, in particular cases, the statements
of the theorem can be verified. In the next section, a more systematic procedure will be developed for determining the optimal solutions $\mathbf{x}^*$ and $\mathbf{y}^*$
when they exist. Optimal solutions will not exist if the constraint set is empty
(as in Example 4.2.8). In this case, the linear programming problem is said
to be *infeasible*. If the constraint set is not empty, the problem is *feasible*.
Finally, if the constraint set is unbounded, the problem may fail to have a
solution. In this case the problem is *feasible* and *unbounded*.

Example 4.4.2 The linear programming problem of minimizing $g(\mathbf{y}) =$
$y_1 - y_2$ subject to the constraints $y_1 + y_2 \geq 1$, $y_1 \geq 0$, and $y_2 \geq 0$ is an
example of a feasible, unbounded problem. The objective function g does not
have a minimum over the constraint set. (The minimum is unbounded.)

An application of the dual problem and the fundamental theorem is
illustrated in the following example.

Example 4.4.3 A dog food manufacturer advertises that one can of its
all-meat product provides the minimum daily requirements for carbohydrates
and proteins of the average 20-pound dog. The meats available for use are
beef, horsemeat, and liver. An ounce of beef costs 1.5 cents and yields .5
ounce of carbohydrate and .2 ounce of protein. An ounce of horsemeat costs
1 cent and yields .6 ounce of carbohydrate and .1 ounce of protein. An ounce
of liver costs 2 cents and yields .4 ounce of carbohydrate and .3 ounce of
protein. The minimum daily requirements of the average 20-pound dog are
estimated to be 6 ounces of carbohydrates and 3.1 ounces of protein. What
combination of the three meats should the manufacturer choose to satisfy
these requirements at minimum cost?

Solution: Let y_1, y_2, and y_3 denote the number of ounces of beef, horsemeat, and liver, respectively, to be used in the product. Then, the problem is

to minimize

$$g(\mathbf{y}) = 1.5y_1 + y_2 + 2y_3$$

subject to

$$.5y_1 + .6y_2 + .4y_3 \geq 6$$
$$.2y_1 + .1y_2 + .3y_3 \geq 3.1$$
$$y_1 \geq 0, \quad y_2 \geq 0, \quad y_3 \geq 0.$$

This problem can be solved by determining all corner points of the constraint set. Instead, we will consider the dual problem. Maximize

$$f(\mathbf{x}) = 6x_1 + 3.1x_2$$

subject to

$$.5x_1 + .2x_2 \leq 1.5$$
$$.6x_1 + .1x_2 \leq 1$$
$$.4x_1 + .3x_2 \leq 2$$
$$x_1 \geq 0, \quad x_2 \geq 0.$$

The constraint set for the dual problem is drawn in Figure 4.16. The corner points of the constraint set are $(0, 0)$, $(\frac{10}{6}, 0)$, $(\frac{5}{7}, \frac{40}{7})$, and $(0, \frac{20}{3})$. The corresponding values of f are 0, 10, 22, and $\frac{62}{3}$. The optimal solution of the dual maximum problem is $x_1 = \frac{5}{7}, x_2 = \frac{40}{7}$ and the maximum of f is $f(\mathbf{x}^*) = 22$ cents. By the fundamental theorem of linear programming, the minimum cost to the manufacturer is 22 cents. The optimal solution $\mathbf{y}^* = (y_1, y_2, y_3)$ of the minimum problem satisfies

$$g(\mathbf{y}^*) = 1.5y_1 + y_2 + 2y_3 = 22 \text{ cents.}$$

To determine a corner point of the constraint set that satisfies this equation, we assume that the diet constraints are just satisfied; that is,

$$.5y_1 + .6y_2 + .4y_3 = 6$$
$$.2y_1 + .1y_2 + .3y_3 = 3.1$$

One solution of this system of three equations is $y_1 = 0, y_2 = 4$, and $y_3 = 9$.

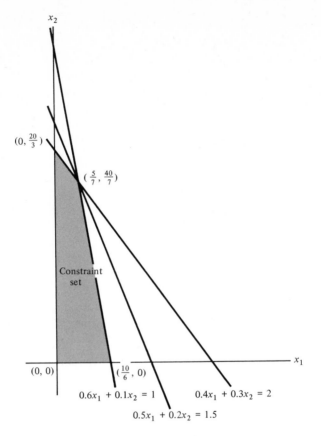

Figure 4.16

Since this solution satisfies the other constraints, it is a corner point, and we conclude that it is an optimal solution. This means that the manufacturer should use no beef, 4 ounces of horsemeat, and 9 ounces of liver to satisfy the requirements at minimum cost. A second solution is 8 ounces of beef, no horsemeat, and 5 ounces of liver. The reader should verify these solutions by solving the original problem by the corner point method.

Problems 4.4

1. A lake is to be stocked with three species of fish. The average weights of the fish stocked for species I, II, and III are 4, 2, and 1 pound. Species I requires 2 units of food A and 3 units of food B per day per average individual. The corresponding requirements for species II are 1 unit of

food A and 2 units of food B and, for species III, 1 unit of food A and $\frac{1}{2}$ unit of food B. Suppose that 1000 units of food A and 1000 units of food B are available per day. How should the lake be stocked to maximize the weight of fish supported by the lake? (Solve by the corner point method.)

2. Determine the dual problem for Problem 1 and solve by the corner point method. Use this solution to determine the solution of Problem 1.

3. What are the dual problems of the following linear programming problems?

(a) Maximize

$$f = x_1 + x_2 - 3x_3$$

subject to

$$x_1 + x_2 + x_3 \leq 5$$
$$x_1 - 2x_2 + 2x_3 \leq 6$$
$$2x_1 - x_2 + x_3 \leq 4$$
$$x_1 \geq 0, \quad x_2 \geq 0, \quad x_3 \geq 0.$$

(b) Maximize

$$f = x_1 - x_2 + x_3 - x_4$$

subject to

$$x_1 + x_2 + x_3 + x_4 \leq 1$$
$$x_1 \geq 0, x_2 \geq 0, x_3 \geq 0, x_4 \geq 0.$$

(c) Minimize

$$g = y_1 + \tfrac{1}{2}y_2 + y_3$$

subject to

$$y_1 + y_2 + 2y_3 \geq 1$$
$$2y_1 - y_2 - y_3 \geq 1$$
$$y_1 \geq 0, y_2 \geq 0, y_3 \geq 0.$$

4. What are the dual problems of the linear programming maximum problems to maximize $f(\mathbf{x}) = \mathbf{c} \cdot \mathbf{x}$ subject to $A\mathbf{x} \le \mathbf{b}$ and $\mathbf{x} \ge \mathbf{0}$ for the following matrices A and vectors $\mathbf{b}$ and $\mathbf{c}$.

(a) $A = \begin{pmatrix} 1 & 1 \\ 2 & 3 \end{pmatrix}, \mathbf{b} = \begin{pmatrix} 5 \\ 7 \end{pmatrix}, \mathbf{c} = \begin{pmatrix} 0 \\ 1 \end{pmatrix}.$

(b) $A = \begin{pmatrix} 1 & 0 \\ 4 & -1 \\ -2 & 1 \end{pmatrix}, \mathbf{b} = \begin{pmatrix} 1 \\ 3 \\ 5 \end{pmatrix}, \mathbf{c} = \begin{pmatrix} 1 \\ -1 \end{pmatrix}.$

(c) $A = \begin{pmatrix} 1 & -1 \\ 0 & 2 \\ 2 & 3 \end{pmatrix}, \mathbf{b} = \begin{pmatrix} 0 \\ 5 \\ 13 \end{pmatrix}, \mathbf{c} = \begin{pmatrix} 4 \\ 2 \end{pmatrix}.$

5. What are the duals of the maximum problems in Problem 4.3.7? Solve these dual problems and verify the fundamental theorem of linear programming.

6. What is the dual of Problem 4.3.9? Solve this dual problem and verify the fundamental theorem of linear programming.

7. (a) What is the major difficulty in applying the corner point method?
   (b) When is it reasonable to expect that the dual problem will be easier to solve than the original problem?
   (c) How does the solution of the dual problem assist in the solution of the original problem?

8. What is the dual of Problem 4.3.10? Solve this dual problem and use the solution to solve the original problem.

9. What is the dual of Problem 4.3.12? Solve this dual problem and use the solution to solve the original problem.

10. How does the fundamental theorem of linear programming give a method of verifying the solution of a maximum or minimum problem of linear programming?

4.5 The Simplex Method

The corner point method of Section 4.3 can become extremely tedious if the number of variables and constraints is large. As we observed, if n is the number of variables and $m + n$ the number of constraints, then

$$\binom{n + m}{n}$$

systems of n equations in n unknowns must be solved. If $n = m = 6$, for example, we would have to evaluate the solutions of 924 6×6 systems. The *simplex method* is a more systematic method of solving such problems in that it avoids the necessity of calculating all the corner points of the constraint set.

The idea behind this method is very simple. We start at any corner point and choose an edge of the constraint set along which the objective function increases (or decreases if we have a minimum problem). If there is no such edge, then the objective function already attains its maximum at the initial corner point (by Theorem 4.3.1) and the problem is solved. On the other hand, if there is an edge along which f increases, we proceed along that edge to the next corner point and then repeat this process. (The "next" corner point will always exist if the constraint set is bounded.) Since there are a finite number of corner points, an optimal solution will be reached after a finite number of steps. We will never return to the same corner point twice, since the objective function increases at each step. Beside the fact that this method avoids the necessity of determining all corner points, a great advantage of the simplex method is that its steps can be efficiently performed by a computer.

To illustrate the steps of this method, we consider the following example (which could, of course, be solved easily by the corner point method).

Example 4.5.1   Maximize

$$f = x_1 + x_2$$

subject to

$$x_1 + 2x_2 \leq 1$$
$$3x_1 + x_2 \leq 2$$
$$x_1 \geq 0, \quad x_2 \geq 0.$$

Solution: We define the *slack variables* $s_1 = 1 - x_1 - 2x_2$ and $s_2 = 2 - 3x_1 - x_2$ and use them to replace the first two inequalities by equalities. The following problem is equivalent to the original one. Maximize

$$f = x_1 + x_2 + 0 \cdot s_1 + 0 \cdot s_2$$

subject to

$$x_1 + 2x_2 + s_1 \qquad = 1$$
$$3x_1 + x_2 \qquad + s_2 = 2$$
$$x_1 \geq 0, \quad x_2 \geq 0, \quad s_1 \geq 0, \quad s_2 \geq 0.$$

At this point, we introduce some terminology which will be useful in the discussion that follows. The variables s_1 and s_2 are called *basic variables*; x_1 and x_2, *nonbasic variables*. In this problem, we can obtain a feasible solution (corner point) by setting each of the nonbasic variables equal to zero. The first corner point is $x_1 = 0, x_2 = 0, s_1 = 1, s_2 = 2$, and f takes the value 0 there. Note that, in the expression for f, the variables with the nonzero coefficients (x_1 and x_2) are the nonbasic variables. The three equations relating f and the variables are written in the form

I
$$x_1 + 2x_2 + s_1 \qquad = 1$$
$$3x_1 + x_2 \qquad + s_2 = 2$$
$$x_1 + x_2 \qquad = f.$$

By forming linear combinations of these three equations, we will rewrite f in terms of other variables in such a way that the coefficients of the variables that appear (always the nonbasic variables) will be negative. Since all variables are greater than or equal to zero, f will then attain its maximum when these nonbasic variables are set to zero. To simplify, divide the second equation by 3.

II
$$x_1 + 2x_2 + s_1 \qquad = 1$$
$$x_1 + \tfrac{1}{3}x_2 \qquad + \tfrac{1}{3}s_2 = \tfrac{2}{3}$$
$$x_1 + x_2 \qquad = f.$$

We now use the second equation to eliminate x_1 from the first and third equations by multiplying it by -1 and adding the result to the other equations. We obtain

III
$$\tfrac{5}{3}x_2 + s_1 - \tfrac{1}{3}s_2 = \tfrac{1}{3}$$
$$x_1 + \tfrac{1}{3}x_2 \qquad + \tfrac{1}{3}s_2 = \tfrac{2}{3}$$
$$\tfrac{2}{3}x_2 \qquad - \tfrac{1}{3}s_2 = f - \tfrac{2}{3}.$$

Since f is now expressed in terms of x_2 and s_2, these are the new nonbasic variables and setting them to zero yields $f = \tfrac{2}{3}$ at the corner point $x_1 = \tfrac{2}{3}$, $x_2 = 0, s_1 = \tfrac{1}{3}, s_2 = 0$. The value of f has increased by going to this new corner point.

Since f still contains the variable x_2 with positive coefficient, we multiply the first equation by $\tfrac{3}{5}$ and then use this equation to eliminate x_2 from the second and third equations.

IV
$$x_2 + \tfrac{3}{5}s_1 - \tfrac{1}{5}s_2 = \tfrac{1}{5}$$
$$x_1 + \tfrac{1}{3}x_2 \qquad + \tfrac{1}{3}s_2 = \tfrac{2}{3}$$
$$\tfrac{2}{3}x_2 \qquad - \tfrac{1}{3}s_2 = f - \tfrac{2}{3}.$$

V
$$x_2 + \tfrac{3}{5}s_1 - \tfrac{1}{5}s_2 = \tfrac{1}{5}$$
$$x_1 \qquad - \tfrac{1}{5}s_1 + \tfrac{2}{5}s_2 = \tfrac{3}{5}$$
$$\qquad - \tfrac{2}{5}s_1 - \tfrac{1}{5}s_2 = f - \tfrac{4}{5}.$$

Since all the coefficients in the expression for f are negative, the problem is solved. Setting the final nonbasic variables s_1 and s_2 to zero, we obtain $f = \tfrac{4}{5}$ at the corner point $x_1 = \tfrac{3}{5}, x_2 = \tfrac{1}{5}, s_1 = s_2 = 0$.

The reader should note that the method used here is nothing other than the row reduction used in Chapter 3.

There is another advantage of the simplex method. By solving the maximum problem by this technique, we obtain, with no extra calculation, the solution of the dual minimum problem. The dual problem for Example 4.5.1 is: minimize

$$g(y_1, y_2) = y_1 + 2y_2$$

subject to

$$y_1 + 3y_2 \geq 1$$
$$2y_1 + y_2 \geq 1$$
$$y_1 \geq 0, \quad y_2 \geq 0.$$

The solution of this problem can be read from the equation $f = \tfrac{4}{5} - \tfrac{2}{5}s_1 - \tfrac{1}{5}s_2$. The values of y_1 and y_2 are the negatives of the coefficients of s_1 and s_2 in the equation for f. The optimal solution of the dual minimum problem is $y_1 = \tfrac{2}{5}, y_2 = \tfrac{1}{5}$. We verify that $g(\tfrac{2}{5}, \tfrac{1}{5}) = \tfrac{2}{5} + 2(\tfrac{1}{5}) = \tfrac{4}{5}$, which is the maximum value of f. Therefore, by Theorem 4.4.1, we have a solution of the dual minimum problem.

Returning to the general maximum problem of linear programming, we introduce the *slack variables* $s_1, s_2, \ldots, s_m$. Maximize

$$f = c_1 x_1 + c_2 x_2 + \cdots + c_n x_n + 0 \cdot s_1 + 0 \cdot s_2 + \cdots + 0 \cdot s_m$$

subject to

$$a_{11}x_1 + a_{12}x_2 + \cdots + a_{1n}x_n + s_1 \qquad\qquad = b_1$$
$$a_{21}x_1 + a_{22}x_2 + \cdots + a_{2n}x_n \qquad + s_2 \qquad = b_2$$
$$\vdots \qquad \vdots \qquad\qquad\qquad\qquad \vdots \qquad (4.13)$$
$$a_{m1}x_1 + a_{m2}x_2 + \cdots + a_{mn}x_n \qquad\qquad + s_m = b_m$$
$$x_1 \geq 0, \quad x_2 \geq 0,\ldots, x_n \geq 0, \quad s_1 \geq 0, \quad s_2 \geq 0,\ldots, s_m \geq 0.$$

We will, at this point, make the assumption that $b_1, b_2,\ldots, b_m$ are non-negative so that when the nonbasic variables $x_1, x_2,\ldots, x_n$ are set equal to zero, yielding $s_1 = b_1, s_2 = b_2,\ldots, s_m = b_m$, we are at a corner point. (*Important point:* The only requirement for a corner point if the equality constraints are satisfied is that all variables have nonnegative values.) There are many important problems where some of the b_i are negative. In this case, $x_1 = x_2 = \cdots = x_n = 0, s_1 = b_1, s_2 = b_2,\ldots, s_m = b_m$ is *not* a corner point. There is a modification of the simplex method if this situation should arise; this will be discussed in the next section.

A convenient way to write the information contained in this problem is to define the *initial simplex tableau*.

x_1	x_2	$\cdots$	x_n	s_1	s_2	$\cdots$	s_m		
a_{11}	a_{12}	$\cdots$	a_{1n}	1	0	$\cdots$	0	b_1	s_1
a_{21}	a_{22}	$\cdots$	a_{2n}	0	1	$\cdots$	0	b_2	s_2
$\vdots$	$\vdots$			$\vdots$	$\vdots$		$\vdots$	$\vdots$	$\vdots$
a_{m1}	a_{m2}	$\cdots$	a_{mn}	0	0	$\cdots$	1	b_m	s_m
c_1	c_2	$\cdots$	c_n	0	0	$\cdots$	0	f	

$$(4.14)$$

The variables on the right side indicate that $s_1, s_2,\ldots, s_n$ are the basic variables. All other entries conform exactly to the equations they replace. The first line, for example, reads

$$a_{11}x_1 + a_{12}x_2 + \cdots + a_{1n}x_n + s_1 + 0 \cdot s_2 + \cdots + 0 \cdot s_m = b_1.$$

The entries in the bottom row (except for the entry in the lower right-hand corner) are called the *indicators* of the tableau. The object of the simplex method is to form linear combinations of the rows of (4.13) so that f will be written as a linear combination of the variables $x_1, x_2,\ldots, x_n, s_1, s_2,\ldots, s_m$ with all coefficients negative or zero. To do this, we work with the tableau (4.14) and successively eliminate all positive indicators. The procedure is as follows:

(a) First, choose any column in (4.14) with a positive indicator (if there is more than one positive indicator, select any one) and consider all *positive* components in that column. Suppose the jth column is chosen.

(4.15)

(b) Then define a *pivot* of this column to be a positive component a_{ij} in the jth column such that b_i/a_{ij} is a minimum for all such a_{ij}. If this minimum is not unique, choose any element in the jth column giving the minimum quotient.

Example 4.5.2 The circled components are the pivots of the following initial simplex tableaux.

1.

x_1	x_2	x_3	s_1	s_2		
②	③	2	1	0	2	s_1
-1	0	4	0	1	2	s_2
1	2	0	0	0	f	

2.

x_1	x_2	x_3	x_4	s_1	s_2		
②	-1	④	6	1	0	1	s_1
3	③	2	7	0	1	2	s_2
1	1	$\frac{1}{2}$	-2	0	0	f	

3.

x_1	x_2	x_3	x_4	x_5	s_1	s_2	s_3	s_4		
①	1	3	4	②	1	0	0	0	1	s_1
0	2	1	2	1	0	1	0	0	2	s_2
-1	⑤	0	1	②	0	0	1	0	1	s_3
1	-4	1	3	2	0	0	0	1	3	s_4
1	1	-1	0	2	0	0	0	0	f	

For example, in (3), 5 is the pivot in the second column, since the indicator 1 is positive and $\frac{1}{5}$ is the minimum of $\frac{1}{1}$, $\frac{2}{2}$, and $\frac{1}{5}$. Note that the pivot in the fifth column is not unique.

(c) Suppose that a_{ij} is a pivot in the ith row of the jth column. The next step is to divide the ith row by a_{ij} and then use the new ith row to eliminate all other entries of the jth column (including, of course, the positive indicator).

(4.16)

(d) The final step in the pivoting operation is to replace the basic variable on the right-hand side of the ith row by the nonbasic variable in the jth column on the top. For example, in Example 4.5.2(3), if we pivoted on $a_{32} = 5$, the basic variable s_3 (third row) would be removed (and become nonbasic) and the nonbasic variable x_2 (second column) would become basic.

From the way the pivot was chosen, it can be shown that two important things happen. First, all the nonnegative entries in the last column remain nonnegative. This is important because the entries in the last column are the values of the basic variables that must be nonnegative in order that we be at a corner point of the constraint set. (Remember, the nonbasic variables are always set to zero.) Second, after each pivot the number in the lower right-hand corner decreases. This means that f, the objective function, increases. To see this, suppose that, after one pivot, $f - 1$ becomes $f - 3$, say. Then the equation reads $f - 3 =$ a linear combination of nonbasic variables, each of which is set to zero to determine a corner point. Thus $f - 3 = 0$ or $f = 3$ at a corner point. Since, originally, we had $f = 1$, the objective function has increased.

The pivoting process outlined above is repeated until all positive indicators have been eliminated. We finally reach a *terminal tableau* when the indicators are all negative or zero.

Example 4.5.3   The following is a terminal tableau.

x_1	x_2	x_3	x_4	s_1	s_2	s_3	s_4		
0	1	0	1	$\frac{1}{3}$	1	2	0	2	x_4
1	2	0	0	$\frac{1}{2}$	$\frac{1}{2}$	1	0	3	x_1
0	0	1	0	0	1	3	0	1	x_3
0	1	0	0	2	0	4	2	0	x_2
0	-1	0	0	$-\frac{1}{3}$	$-\frac{1}{2}$	-1	0	$f-2$	

The solutions of the maximum and dual minimum problems can be read from this tableau. We have $f = 2$ by setting the nonbasic variables equal to zero, yielding the corner point $x_1 = 3, x_2 = 0, x_3 = 1, x_4 = 2, s_1 = 0, s_2 = 0,$ $s_3 = 0, s_4 = 0$. An optimal solution for the dual problem is $g = 2$, which occurs for $y_1 = \frac{1}{3}, y_2 = \frac{1}{2}, y_3 = 1$, and $y_4 = 0$. (The reader should verify that this corner point satisfies the constraints of the dual minimum problem.) Note that it is possible for a basic variable to have the value zero (in this case $x_2 = 0$).

Example 4.5.4 Maximize

$$f = x_1 + x_2 + x_3$$

subject to

$$x_1 + 2x_2 + 3x_3 \le 1$$
$$2x_1 + x_2 + x_3 \le 2$$
$$x_1 \ge 0, \quad x_2 \ge 0, \quad x_3 \ge 0.$$

Solution: The initial simplex tableau is

I

x_1	x_2	x_3	s_1	s_2		
①	2	3	1	0	1	s_1
2	1	1	0	1	2	s_2
1	1	1	0	0	f	

If we start with the first column (it has a positive indicator), we can either pivot on $a_{11} = 1$ or $a_{21} = 2$, since $\frac{1}{1} = \frac{2}{2} = 1$. Choosing a_{11} (since it already has the value 1), we pivot to obtain

II

x_1	x_2	x_3	s_1	s_2		
1	2	3	1	0	1	x_1
0	-3	-5	-2	1	0	s_2
0	-1	-2	-1	0	$f - 1$	

Here we subtracted twice the first row from the second row and subtracted the first row from the third row. We observe that all indicators are negative or zero; therefore, this is a terminal tableau. Setting the nonbasic variables to zero, we have the solution $f = 1$ at the corner point $x_1 = 1$, $x_2 = x_3 = s_1 = s_2 = 0$.

The dual problem is to minimize $g = y_1 + 2y_2$ subject to $y_1 + 2y_2 \geq 1$, $2y_1 + y_2 \geq 1$, $3y_1 + y_2 \geq 1$, $y_1 \geq 0$, and $y_2 \geq 0$. The solution is $y_1 = 1$, $y_2 = 0$ (the negatives of the coefficients of s_1 and s_2 in the terminal equation for f). The minimum value of g is, of course, 1, which is equal to the maximum of f. (*Problem:* Prove that this is not the only optimal solution of the dual problem by solving with the corner point method or by choosing a different initial pivot.)

Example 4.5.5 Determine terminal tableaux for the following initial simplex tableaux.

1.

x_1	x_2	x_3	s_1	s_2	
②	3	2	1	0	2	s_1
-1	0	4	0	1	2	s_2
1	2	0	0	0	f

2.

x_1	x_2	x_3	x_4	s_1	s_2	
2	-1	4	6	1	0	1	s_1
3	③	2	7	0	1	2	s_2
1	1	$\frac{1}{2}$	-2	0	0	f

Solution: 1. Using the circled component as pivot, divide the first row by 2.

II

x_1	x_2	x_3	s_1	s_2	
1	$\frac{3}{2}$	1	$\frac{1}{2}$	0	1	s_1
-1	0	4	0	1	2	s_2
1	2	0	0	0	f

III

x_1	x_2	x_3	s_1	s_2		
1	$\left(\frac{3}{2}\right)$	1	$\frac{1}{2}$	0	1	x_1
0	$\frac{3}{2}$	5	$\frac{1}{2}$	1	3	s_2
0	$\frac{1}{2}$	-1	$-\frac{1}{2}$	0	$f-1$	

We have used the pivot row to eliminate the components in the first column. Pivoting again about the circled component, we multiply the first row by $\frac{2}{3}$.

IV

x_1	x_2	x_3	s_1	s_2		
$\frac{2}{3}$	1	$\frac{2}{3}$	$\frac{1}{3}$	0	$\frac{2}{3}$	x_1
0	$\frac{3}{2}$	5	$\frac{1}{2}$	1	3	s_2
0	$\frac{1}{2}$	-1	$-\frac{1}{2}$	0	$f-1$	

V

x_1	x_2	x_3	s_1	s_2		
$\frac{2}{3}$	1	$\frac{2}{3}$	$\frac{1}{3}$	0	$\frac{2}{3}$	x_1
-1	0	4	0	1	2	s_2
$-\frac{1}{3}$	0	$-\frac{4}{3}$	$-\frac{2}{3}$	0	$f-\frac{4}{3}$	

There are no positive indicators in tableau V and, therefore, this is a terminal tableau. The reader should verify that this terminal tableau can be reached more easily if the initial pivot is the component 3 in the first row and second column.

2. Again using the circled component as initial pivot, divide the second row by 3. Then add the second row to the first row and subtract it from the third row.

II

x_1	x_2	x_3	x_4	s_1	s_2		
2	-1	4	6	1	0	1	s_1
1	1	$\frac{2}{3}$	$\frac{7}{3}$	0	$\frac{1}{3}$	$\frac{2}{3}$	s_2
1	1	$\frac{1}{2}$	-2	0	0	f	

204   Linear Programming   [Ch. 4]

III

	x_1	x_2	x_3	x_4	s_1	s_2		
	3	0	$\frac{14}{3}$	$\frac{25}{3}$	1	$\frac{1}{3}$	$\frac{5}{3}$	s_1
	1	1	$\frac{2}{3}$	$\frac{7}{3}$	0	$\frac{1}{3}$	$\frac{2}{3}$	x_2
	0	0	$-\frac{1}{6}$	$-\frac{13}{3}$	0	$-\frac{1}{3}$	$f - \frac{2}{3}$	

We observe that III is a terminal tableau.

Example 4.5.6   Determine the maximum of

$$f(x_1, x_2, x_3) = x_1 - x_2 + x_3$$

subject to the constraints

$$x_1 + x_2 + x_3 \leq 15$$
$$2x_1 + x_2 + 2x_3 \leq 26$$
$$5x_1 + 2x_2 + 3x_3 \leq 43$$
$$x_1 \geq 0, \quad x_2 \geq 0, \quad x_3 \geq 0.$$

Solution: We have solved this problem by the corner point method in Example 4.3.2. To use the simplex method, we write the initial simplex tableau.

I

	x_1	x_2	x_3	s_1	s_2	s_3		
	1	1	1	1	0	0	15	s_1
	2	1	②	0	1	0	26	s_2
	5	2	3	0	0	1	43	s_3
	1	−1	1	0	0	0	f	

Choosing as pivot the circled component. divide the second row by 2, and then use the new second row to eliminate the components in the third column of the other rows.

II

	x_1	x_2	x_3	s_1	s_2	s_3		
	1	1	1	1	0	0	15	s_1
	1	$\frac{1}{2}$	1	0	$\frac{1}{2}$	0	13	s_2
	5	2	3	0	0	1	43	s_3
	1	-1	1	0	0	0	f	

III

	x_1	x_2	x_3	s_1	s_2	s_3		
	0	$\frac{1}{2}$	0	1	$-\frac{1}{2}$	0	2	s_1
	1	$\frac{1}{2}$	1	0	$\frac{1}{2}$	0	13	x_3
	2	$\frac{1}{2}$	0	0	$-\frac{3}{2}$	1	4	s_3
	0	$-\frac{3}{2}$	0	0	$-\frac{1}{2}$	0	$f-13$	

There are positive indicators in III and, therefore, it is a terminal tableau. Since $f = 13 - \frac{3}{2}x_2 - \frac{1}{2}s_2$, the maximum value of f is 13, and it occurs when $x_2 = s_2 = 0$. The first three rows of the terminal tableau are equivalent to the equations

$$\frac{1}{2}x_2 \qquad\qquad + s_1 - \tfrac{1}{2}s_2 \qquad\quad = 2$$
$$x_1 + \tfrac{1}{2}x_2 + x_3 \qquad\qquad + \tfrac{1}{2}s_2 \qquad = 13$$
$$2x_1 + \tfrac{1}{2}x_2 \qquad\qquad\qquad - \tfrac{3}{2}s_2 + s_3 = 4.$$

If $x_2 = s_2 = 0$, we have $s_1 = 2$, $x_1 + x_3 = 13$, and $2x_1 + s_3 = 4$. Since $s_3 \geq 0$, we conclude that $0 \leq 2x_1 \leq 4$ or $0 \leq x_1 \leq 2$. Therefore, any point in $\mathbf{R}^3$ of the form $(x_1, 0, 13 - x_1)$ with $0 \leq x_1 \leq 2$ is an optimal solution. In particular, the corner points $(0, 0, 13)$ and $(2, 0, 11)$ are optimal solutions as found in Example 4.3.2. In this example, it is evident that the simplex method is more efficient than the corner point method.

Example 4.5.7 Determine, by the simplex method, the minimum of the linear function $g(y_1, y_2) = 3y_1 + 2y_2$ subject to

$$5y_1 + y_2 \geq 10$$
$$2y_1 + 2y_2 \geq 12$$
$$y_1 + 4y_2 \geq 12$$
$$y_1 \geq 0, \quad y_2 \geq 0.$$

Solution: This is Example 4.3.3. The dual problem is to maximize $f(x_1, x_2, x_3)$
$= 10x_1 + 12x_2 + 12x_3$ subject to $5x_1 + 2x_2 + x_3 \le 3$, $x_1 + 2x_2 + 4x_3 \le 2$,
$x_1 \ge 0$, $x_2 \ge 0$, and $x_3 \ge 0$. The initial simplex tableau for this problem is

I

x_1	x_2	x_3	s_1	s_2		
5	2	1	1	0	3	s_1
1	②	4	0	1	2	s_2
10	12	12	0	0	f	

Choose as pivot the circled component; divide the second row by 2; and use
the new second row to clear the second column.

II

x_1	x_2	x_3	s_1	s_2		
5	2	1	1	0	3	s_1
$\frac{1}{2}$	1	2	0	$\frac{1}{2}$	1	s_2
10	12	12	0	0	f	

III

x_1	x_2	x_3	s_1	s_2		
④	0	-3	1	-1	1	s_1
$\frac{1}{2}$	1	2	0	$\frac{1}{2}$	1	x_2
4	0	-12	0	-6	$f - 12$	

Pivoting again, we divide the first row by 4 and then clear the first column.

IV

x_1	x_2	x_3	s_1	s_2		
1	0	$-\frac{3}{4}$	$\frac{1}{4}$	$-\frac{1}{4}$	$\frac{1}{4}$	s_1
$\frac{1}{2}$	1	2	0	$\frac{1}{2}$	1	x_2
4	0	-12	0	-6	$f - 12$	

V

	x_1	x_2	x_3	s_1	s_2		
	1	0	$-\frac{3}{4}$	$\frac{1}{4}$	$-\frac{1}{4}$	$\frac{1}{4}$	x_1
	0	1	$\frac{19}{8}$	$-\frac{1}{8}$	$\frac{5}{8}$	$\frac{7}{8}$	x_2
	0	0	-9	-1	-5	$f-13$	

The terminal tableau V implies that $f = 13 - 9x_3 - s_1 - 5s_2$. We conclude that the maximum of f is 13 when $x_3 = s_1 = s_2 = 0$. Therefore, the minimum value of g is also 13, and it occurs when $y_1 = 1$, $y_2 = 5$ (the negatives of the coefficients of s_1 and s_2 in the terminal equation for f). This is the result found by a simpler geometrical method in Example 4.3.3.

Before concluding this section, we discuss a possible difficulty that may arise. Suppose that we reach an intermediate tableau, where there is a positive indicator, say in the jth column, but all other entries in that column are negative. Since a pivot must be positive, it is therefore impossible to continue. But we have not obtained a solution, since there is a positive indicator. If this situation should arise, we must have an unbounded solution. This is best illustrated by a simple example.

Example 4.5.8 Maximize

$$f = x_1 + 2x_2$$

subject to

$$x_1 - 4x_2 \leq 1$$
$$-3x_1 + 2x_2 \leq 6$$
$$x_1 \geq 0, \quad x_2 \geq 0.$$

Solution: The initial simplex tableau is

I

	x_1	x_2	s_1	s_2		
	①	-4	1	0	1	s_1
	-3	2	0	1	6	s_2
	1	2	0	0	f	

Pivoting on the circled entry, we obtain

II

	x_1	x_2	s_1	s_2		
	1	-4	1	0	1	x_1
	0	-10	3	1	9	s_2
	0	6	-1	0	$f-1$	

It is now impossible to continue, even though one of the indicators is still positive. Rewriting the equations, we have (after setting $s_1 = 0$)

$$x_1 - 4x_2 = 1$$
$$-10x_2 + s_2 = 9$$
$$f = 6x_2 + 1.$$

Let x_2 be any positive number (say c). Then, to satisfy the first two equations, we choose $x_1 = 4c + 1$ and $s_2 = 9 + 10c$. With these choices for x_1, x_2, s_1, and s_2, all the constraints are satisfied and $f = 6c + 1$. Since c was arbitrary, the objective function can be made as large as desired and the solution is unbounded.

Problems 4.5

1. Following the steps of Example 4.5.1, determine the maximum of $f(x_1, x_2)$ $= x_1 + 2x_2$ subject to $x_1 + x_2 \leq 3$, $2x_1 + x_2 \leq 3$ and $x_1 \geq 0, x_2 \geq 0$. What are the slack variables? What is the initial corner point?

2. Continuing Problem 1, what is the dual problem? What is the solution of the dual problem?

3. Illustrate Problems 1 and 2 by means of diagrams. By means of arrows, indicate the progress from an initial corner point to a final corner point in the maximum problem.

4. Following the steps of Example 4.5.1, determine the maximum of $f(x_1, x_2, x_3) = x_1 + x_2 + x_3$ subject to $x_1 + 2x_2 + x_3 \leq 4$, $x_1 + x_2 + 2x_3 \leq 5$ and $x_1 \geq 0, x_2 \geq 0, x_3 \geq 0$. What are the slack variables? What is the initial corner point?

5. What is the dual problem of Problem 4? Without additional calculations, determine the solution of the dual problem. Verify the solutions

of the dual problem and the original problem by means of the funda-mental theorem of linear programming.

6. What are the initial simplex tableaux for the following maximum problems?

(a) Maximize

$$f = x_1 + 2x_2 + 3x_3$$

subject to

$$x_1 + x_2 - x_3 \leq 1$$
$$x_1 - x_2 + x_3 \leq 2$$
$$-x_1 + x_2 + x_3 \leq 3$$
$$x_1 \geq 0, \quad x_2 \geq 0, \quad x_3 \geq 0.$$

(b) Maximize

$$f = x_1 - x_2 + x_3$$

subject to

$$x_1 + x_2 + 2x_3 \leq 5$$
$$2x_1 + x_2 + x_3 \leq 7$$
$$2x_1 - x_2 + 3x_3 \leq 8$$
$$x_1 + 2x_2 + 5x_3 \leq 9$$
$$x_1 \geq 0, \quad x_2 \geq 0, \quad x_3 \geq 0.$$

7. What are the initial simplex tableaux for the following minimum problems?

(a) Minimize

$$g = y_1 + y_2 + y_3$$

subject to

$$y_1 - y_2 + y_3 \geq 0$$
$$2y_1 + y_2 + y_3 \geq 2$$
$$4y_1 + y_2 - 4y_3 \geq 3$$
$$y_1 \geq 0, \quad y_2 \geq 0, \quad y_3 \geq 0.$$

(b) Minimize

$$g = y_1 + 2y_2 + y_3$$

subject to

$$2y_1 + y_2 + y_3 \geq 5$$
$$y_1 + y_2 - y_3 \geq 4$$
$$y_1 \geq 0, \quad y_2 \geq 0, \quad y_3 \geq 0.$$

8. Determine the pivots of the following initial simplex tableaux.

(a)

2	−1	2	1	0	1
−2	0	3	0	1	2
1	1	1	0	0	f

(b)

1	1	1	1	0	1
−1	0	1	0	1	2
2	1	3	0	0	f

(c)

1	2	3	1	0	0	5
2	3	1	0	1	0	3
3	1	2	0	0	1	1
2	−1	3	0	0	0	f

9. Determine terminal tableaux for the initial tableaux of Problem 8. Write out the corresponding dual maximum and minimum problems and determine their solutions.

10. By the simplex method, determine the maximum of $f(x_1, x_2, x_3) = 2x_1 + x_2 + x_3$ subject to the inequality constraints $x_1 + x_2 + x_3 \leq 5$, $x_1 - 2x_2 + 3x_3 \leq 7$ and $x_1 \geq 0, x_2 \geq 0, x_3 \geq 0$.

11. Determine the solution of the dual of Problem 10. Verify both solutions by the fundamental theorem of linear programming.

12. By the simplex method, determine the maximum of $f(x_1, x_2, x_3) = 2x_1 - 2x_2 + x_3$ subject to the inequality constraints $x_1 + x_2 \leq 10$, $x_2 + x_3 \leq 12$, $x_1 + x_3 \leq 8$ and $x_1 \geq 0, x_2 \geq 0, x_3 \geq 0$.

13. Determine the solution of the dual of Problem 12. Verify both solutions by the fundamental theorem of linear programming.

14. By the simplex method, determine the minimum of $g(y_1, y_2) = y_1 + 3y_2$ subject to the inequality constraints $y_1 + 5y_2 \geq 9$, $y_1 + 4y_2 \geq 8$, $y_1 + 2y_2 \geq 5, y_1 \geq 0$ and $y_2 \geq 0$. Verify the solution by the fundamental theorem of linear programming.

4.6 The Simplex Method (Continued)

In the last section, we made the assumption that, in the initial simplex tableau, all the elements in the right-hand column were nonnegative. This ensured us that initially, and after each pivot, we had a feasible solution (corner point). There are many important problems, however, in which initial negative values of the b_i appear. This can happen in one of two ways. First, a constraint such as $-2x_1 + 3x_2 \leq -1$ may be natural for a given problem. More often, we have a maximization problem with a constraint such as $x_1 + 2x_2 \geq 1$. In order to put this constraint into the standard form, we multiply both sides of the inequality by -1 to obtain $-x_1 - 2x_2 \leq -1$, and this value of b_i is negative. This situation often occurs in biological problems. Suppose, for example, that an organism acts so as to maximize food consumption subject to bounds on expendable energy. This is a typical maximization problem, except that one more constraint must be added: in order to satisfy maintenance requirements, a certain minimum amount of energy must be used. The problem could then become, for example: maximize

$$f = c_1 x_1 + c_2 x_2$$

subject to

$$a_{11}x_1 + a_{12}x_2 \leq b_1$$
$$a_{21}x_1 + a_{22}x_2 \leq b_2$$
$$a_{31}x_1 + a_{32}x_2 \geq b_3$$

where

$$b_1, b_2, b_3 \geq 0.$$

After multiplication by -1, the last constraint reads

$$-a_{31}x_1 - a_{32}x_2 \leq -b_3 < 0.$$

When this situation arises, it is necessary to perform pivoting operations so as to arrive at a tableau in which all the elements in the right-hand column are nonnegative. Then we can proceed with the simplex method as before. Sometimes this preliminary phase of the simplex method is called *phase I* and the method described in the previous section is called *phase II*. We emphasize that the idea behind phase I is to obtain a feasible solution. Phase II cannot be started until phase I has been completed.

We now set down the rules for choosing a pivot in phase I.

(a) Let b_i be the lowest negative element in the right-hand column and let a_{ij} be any *negative* element in the ith row. Then we pivot in the jth column.

(4.17)

(b) For a_{ij} and for each positive element a_{lj} *below* a_{ij} in the jth column (except for the bottom row), form the quotients b_l/a_{lj}. Let b_k/a_{kj} be the smallest such quotient. Then a_{kj} is the pivot. In case of ties, choose any element in the jth column giving the minimum quotient.

(*Note:* In the rules above, the words lowest and below refer to position, not numerical value. In other words, a_{lj} is below a_{ij} if $l > i$, that is, if a_{lj} is farther "south.")

Example 4.6.1 Maximize

$$f = 5x_1 + x_2$$

subject to

$$4x_1 + 3x_2 \leq 12$$
$$-2x_1 + 3x_2 \geq 6$$
$$x_1 \geq 0, \quad x_2 \geq 0.$$

Solution: The initial simplex tableau is (after multiplying the second constraint by -1)

I

x_1	x_2	s_1	s_2		
4	3	1	0	12	s_1
2	⊝3	0	1	-6	s_2
5	1	0	0	f	

Here, -6 is the lowest (in fact, the only) negative element in the right-hand column. Since -3 is the only negative element in this row, it is the pivot. We divide the second row by -3 and use it to eliminate the entries in the second column obtaining, successively,

II

x_1	x_2	s_1	s_2		
4	3	1	0	12	s_1
$-\frac{2}{3}$	1	0	$-\frac{1}{3}$	2	s_2
-5	1	0	0	f	

III

x_1	x_2	s_1	s_2		
⑥	0	1	1	6	s_1
$-\frac{2}{3}$	1	0	$-\frac{1}{3}$	2	x_2
$\frac{17}{3}$	0	0	$\frac{1}{3}$	$f-2$	

We now have a feasible solution: $x_1 = 0$, $x_2 = 2$, $s_1 = 6$, $s_2 = 0$. We can proceed to eliminate the positive indicators to find the maximum. Pivoting on the circled entry, we obtain

IV

x_1	x_2	s_1	s_2		
①	0	$\frac{1}{6}$	$\frac{1}{6}$	1	s_1
$-\frac{2}{3}$	1	0	$-\frac{1}{3}$	2	x_2
$\frac{17}{3}$	0	0	$\frac{1}{3}$	$f-2$	

V

x_1	x_2	s_1	s_2		
1	0	$\frac{1}{6}$	$\frac{1}{6}$	1	x_1
0	1	$\frac{1}{9}$	$-\frac{2}{9}$	$\frac{8}{3}$	x_2
0	0	$-\frac{17}{18}$	$-\frac{11}{18}$	$f-\frac{23}{3}$	

Since all the indicators are negative, the problem is completed, and we find that $f = \frac{23}{3}$ at the optimal corner point $x_1 = 1$, $x_2 = \frac{8}{3}$, $s_1 = s_2 = 0$. The dual problem is: minimize

$$g = 12y_1 - 6y_2$$

subject to

$$4y_1 + 2y_2 \geq 5$$
$$3y_1 - 3y_2 \geq 1$$
$$y_1 \geq 0, \quad y_2 \geq 0.$$

The solution to this problem is $y_1 = \frac{17}{18}, y_2 = \frac{11}{18}$ with a minimum value of $\frac{23}{3}$.

Example 4.6.2   Maximize

$$f = x_1 + 3x_2 + 5x_3$$

subject to

$$x_1 - x_2 + 2x_3 \geq 1$$
$$x_1 - x_2 + 2x_3 \leq 1$$
$$x_1 + 2x_2 - 3x_3 \geq 2$$
$$x_1 \geq 0, \quad x_2 \geq 0, \quad x_3 \geq 0.$$

Solution: We first multiply the first and third constraints by -1 and then add the slack variables to arrive at the initial simplex tableau.

I

x_1	x_2	x_3	s_1	s_2	s_3		
-2	-1	-1	1	0	0	-1	s_1
1	-1	2	0	1	0	1	s_2
$-①$	-2	3	0	0	1	-2	s_3
1	3	5	0	0	0		f

The lowest negative entry is in the third row. Our pivot could be either -1

or -2. Choosing -1, we obtain

II

	x_1	x_2	x_3	s_1	s_2	s_3		
	0	3	-7	1	0	-2	3	s_1
	0	$-\textcircled{3}$	5	0	1	1	-1	s_2
	1	2	-3	0	0	-1	2	x_1
	0	1	8	0	0	1	$f-2$	

Our next pivot must be in the second row. The only negative entry is -3, which has the positive entry $+2$ below it. Forming the quotients $-1/-3$ and $2/2$, we find that -3 must be the pivot. Performing this next operation, we obtain

III

	x_1	x_2	x_3	s_1	s_2	s_3	1	
	0	0	-2	1	1	-1	2	s_1
	0	1	$-\frac{5}{3}$	0	$-\frac{1}{3}$	$-\frac{1}{3}$	$\frac{1}{3}$	x_2
	1	0	$\frac{1}{3}$	0	$\frac{2}{3}$	$-\frac{1}{3}$	$\frac{4}{3}$	x_1
	0	0	$\frac{29}{3}$	0	$\frac{1}{3}$	$\frac{4}{3}$	$f-\frac{7}{3}$	

At this point, phase I is completed, and we have found a feasible solution, $x_1 = \frac{4}{3}$, $x_2 = \frac{1}{3}$, $x_3 = 0$, $s_1 = 2$, $s_2 = 0$, $s_3 = 0$. The value of the objective function at this point is $\frac{7}{3}$. This is not the maximum, since some of the indicators are positive. The remaining steps of this example are left as an exercise.

There is still one question remaining. What does it mean when there is a negative entry in the right-hand column but it is impossible to pivot? This occurs when all the entries in a row with a negative right-hand entry are positive. It should be clear that this indicates an infeasible solution, since such a row would represent the equation

$$a_{i1}x_1 + a_{i2}x_2 + \cdots + a_{in}x_n + a_{i,n+1}s_1 + a_{i,n+2}s_2 + \cdots + a_{i,n+m}s_m = b_i.$$

If all the a_{ij}'s are positive and b_i is negative, this is an obvious impossibility, and we conclude that there is no feasible solution.

Example 4.6.3   Maximize

$$f = 3x_1 + 5x_2$$

subject to

$$x_1 + x_2 \leq 1$$
$$-3x_1 + 2x_2 \geq 6$$
$$x_1 \geq 0, \quad x_2 \geq 0.$$

Solution: Multiplying the second constraint by -1, we obtain the initial simplex tableau

I

x_1	x_2	s_1	s_2		
1	1	1	0	1	s_1
3	$-②$	0	1	-6	s_2
3	5	0	0	f	

Pivoting on the circled entry, we obtain

II

x_1	x_2	s_1	s_2		
$\frac{5}{2}$	0	1	$\frac{1}{2}$	-2	s_1
$-\frac{3}{2}$	1	0	$-\frac{1}{2}$	3	x_2
$\frac{21}{2}$	0	0	$\frac{5}{2}$	$f - 15$	

It is now impossible to continue, even though there is a negative entry in the first row. We conclude that this problem is infeasible. This is easy to see if we draw the constraint set (Figure 4.17). There are no positive values of x_1 and x_2 that satisfy the constraints.

Problems 4.6

1. Determine the maximum of $f(x_1, x_2) = x_1 + 3x_2$ subject to $2x_1 + x_2 \leq 10$; $3x_1 - x_2 \geq 9$; and $x_1 \geq 0$, $x_2 \geq 0$. Solve this problem by (a) a geometrical method and (b) the simplex method.

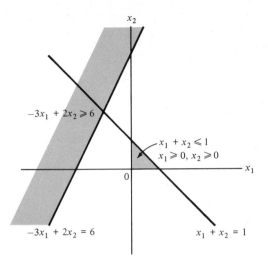

Figure 4.17

2. Determine the maximum of $f(x_1, x_2) = 2x_1 + 5x_2$ subject to $x_1 - x_2 \leq 12$; $x_1 + x_2 \leq 12$; $2x_1 - x_2 \geq 12$; and $x_1 \geq 0$, $x_2 \geq 0$. Solve this problem by (a) a geometrical method and (b) the simplex method.

3. Determine the maximum of $f(x_1, x_2, x_3) = x_1 + x_2 + x_3$ subject to $x_1 + x_2 + 2x_3 \leq 5$; $x_1 - x_2 + x_3 \leq 3$; $x_1 - x_3 \geq 1$; and $x_1 \geq 0$, $x_2 \geq \geq 0$, $x_3 \geq 0$.

4. Determine the minimum of $g(y_1, y_2) = 10y_1 - 9y_2$ subject to $2y_1 - 3y_2 \geq 1$; $y_1 + y_2 \geq 3$; and $y_1 \geq 0$, $y_2 \geq 0$.

5. Determine the minimum of $g(y_1, y_2, y_3) = 12y_1 + 12y_2 - 12y_3$ subject to $y_1 + y_2 - 2y_3 \geq 2$; $-y_1 + y_2 + y_3 \geq 5$; and $y_1 \geq 0$, $y_2 \geq 0$.

6. Determine the maximum of $f(x_1, x_2, x_3) = x_1 + 3x_2 + 4x_3$ subject to $2x_1 - 5x_2 + x_3 \geq 4$; $4x_1 + 3x_2 - 2x_3 \leq 2$; $x_1 + 5x_2 + x_3 \leq 3$; and $x_1 \geq 0$, $x_2 \geq 0$, $x_3 \geq 0$.

Markov Chains and Game Theory | 5

5.1 The Transition Matrix

To introduce the ideas of Markov chains, consider the following example. A mouse is placed in a box divided into three compartments, as shown in Figure 5.1. In the absence of other information, it is reasonable to assume that

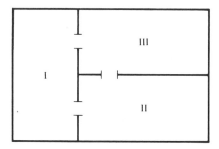

Figure 5.1

the mouse will choose a door at random to move from one compartment to another. If this assumption is valid, the mouse can be expected to be, in the long run, in each of the three compartments equally often. Wecker[1] describes a series of experiments with prairie deer mice that can be analyzed in these terms. By allowing the mice to travel between 10 compartments, half in an open field and half in a wooded area, Wecker is able to study the strength of the preference of prairie deer mice for the field habitat over the wooded habitat.

The theory of Markov chains provides an appropriate framework for the study of the mouse experiment of Figure 5.1. In the language of Markov chains, we have an experimental system with three possible states. The movement of the mouse from compartment I to compartment III, for example, is referred to as a transition of the system from the first state to the third state. The study of Markov chains is the study of the probabilities associated with the possible transitions between the states of an experimental system.

The following definitions and theorems form the basis of the theory of Markov chains.

Definition 5.1.1   Probability Vector   *The n-component row vector* $\mathbf{p} = (p_1, p_2, \ldots, p_n)$ *is a probability vector if all its components are nonnegative and if the sum of the components is* $1; p_1 + p_2 + \cdots + p_n = 1.$

Definition 5.1.2   Probability Matrix   *The* $n \times n$ *matrix* $P = (p_{ij})$ *is a probability matrix if each of its rows is a probability vector. This means that all components* p_{ij} *are nonnegative and the sum of the components in each row of* P *is* 1.

Example 5.1.1   The following are examples of probability matrices.

1. $\begin{pmatrix} 1 & 0 \\ \frac{1}{2} & \frac{1}{2} \end{pmatrix}.$
2. $\begin{pmatrix} \frac{1}{2} & \frac{1}{2} \\ \frac{1}{4} & \frac{3}{4} \end{pmatrix}.$
3. $\begin{pmatrix} \frac{1}{4} & \frac{1}{4} & \frac{1}{2} \\ 0 & 1 & 0 \\ \frac{1}{3} & \frac{1}{3} & \frac{1}{3} \end{pmatrix}.$
4. $\begin{pmatrix} 1 & 0 & 0 \\ \frac{1}{4} & \frac{1}{3} & \frac{5}{12} \\ 0 & 0 & 1 \end{pmatrix}.$

Probability vectors and matrices have a number of interesting and useful properties, as shown in the following theorems.

Theorem 5.1.1   *If* $\mathbf{p}$ *is an n-component probability vector and* $P = (p_{ij})$ *is an* $n \times n$ *probability matrix, then* $\mathbf{p}P$ *is an n-component probability vector.*

[1] S. C. Wecker. "Habitat Selection," *Scientific American*, 211:109–116 (Oct. 1964).

Proof: The given probability vector is $\mathbf{p} = (p_1, p_2, \ldots, p_n)$ and the given $n \times n$ probability matrix is $P = (p_{ij})$. Define the vector $\mathbf{r} = \mathbf{p}P$. Then,

$$\mathbf{r} = \mathbf{p}P = (p_1, p_2, \ldots, p_n) \begin{pmatrix} p_{11} & p_{12} & \cdots & p_{1n} \\ p_{21} & p_{22} & \cdots & p_{2n} \\ \vdots & \vdots & & \vdots \\ p_{n1} & p_{n2} & & p_{nn} \end{pmatrix}$$

$$= (p_1 p_{11} + p_2 p_{21} + \cdots + p_n p_{n1}, \ldots, p_1 p_{1n} + p_2 p_{2n} + \cdots + p_n p_{nn}).$$

In other words, $\mathbf{r} = \mathbf{p}P$ is an n-component row vector, $\mathbf{r} = (r_1, r_2, \ldots, r_n)$, with ith component $r_i = p_1 p_{1i} + p_2 p_{2i} + \cdots + p_n p_{ni}$. Each r_i is nonnegative, since it is the sum of nonnegative numbers. To prove that $\mathbf{r}$ is a probability vector, we must prove that the sum of the components is 1.

$$\sum_{i=1}^{n} r_i = \sum_{i=1}^{n} (p_1 p_{1i} + p_2 p_{2i} + \cdots + p_n p_{ni})$$

$$= \sum_{i=1}^{n} p_1 p_{1i} + \sum_{i=1}^{n} p_2 p_{2i} + \cdots + \sum_{i=1}^{n} p_n p_{ni}$$

$$= p_1 \sum_{i=1}^{n} p_{1i} + p_2 \sum_{i=1}^{n} p_{2i} + \cdots + p_n \sum_{i=1}^{n} p_{ni}$$

$$= p_1 + p_2 + \cdots + p_n = 1.$$

We have used the fact that

$$\sum_{i=1}^{n} p_{1i} = \sum_{i=1}^{n} p_{2i} = \cdots = \sum_{i=1}^{n} p_{ni} = 1,$$

since the sum of the components in each row of P is 1. We conclude that $\mathbf{p}P$ is a probability vector.

Theorem 5.1.2   *If* $P = (p_{ij})$ *and* $Q = (q_{ij})$ *are* $n \times n$ *probability matrices, the matrix product* PQ *is an* $n \times n$ *probability matrix.*

Proof: Define the matrix product $R = PQ = (r_{ij})$. This is an $n \times n$ matrix whose ijth component is $r_{ij} = \sum_{k=1}^{n} p_{ik} q_{kj}$. Since the components of P and Q are nonnegative, the components of $R = PQ$ are nonnegative. The sum of the components in the ith row of R is

$$\sum_{j=1}^{n} r_{ij} = \sum_{j=1}^{n} \sum_{k=1}^{n} p_{ik} q_{kj} = \sum_{k=1}^{n} \left(\sum_{j=1}^{n} p_{ik} q_{kj} \right)$$

$$= \sum_{k=1}^{n} \left(p_{ik} \sum_{j=1}^{n} q_{kj} \right) = \sum_{k=1}^{n} p_{ik} = 1.$$

In this calculation, the order of the double sum has been reversed and we have used the facts that $\sum_{k=1}^{n} p_{ik} = \sum_{j=1}^{n} q_{kj} = 1$, since P and Q are probability matrices. We have proved that PQ is a probability matrix.

Corollary 5.1.3   *If P is an $n \times n$ probability matrix, then $P^2, P^3, \ldots, P^m$, $\ldots$ are probability matrices.*

Proof: This result follows immediately from Theorem 5.1.2. Setting $Q = P$, we have that P^2 is a probability matrix. Similarly, if $Q = P^2$, we conclude that $PQ = P^3$ is a probability matrix; and so on.

Example 5.1.2   If

$$P = \begin{pmatrix} \frac{1}{4} & \frac{1}{4} & \frac{1}{2} \\ 0 & 1 & 0 \\ \frac{1}{3} & \frac{1}{3} & \frac{1}{3} \end{pmatrix} \quad \text{and} \quad Q = \begin{pmatrix} 1 & 0 & 0 \\ \frac{1}{4} & \frac{1}{3} & \frac{5}{12} \\ 0 & 0 & 1 \end{pmatrix},$$

verify that PQ is a probability matrix.

Solution:

$$PQ = \begin{pmatrix} \frac{1}{4} & \frac{1}{4} & \frac{1}{2} \\ 0 & 1 & 0 \\ \frac{1}{3} & \frac{1}{3} & \frac{1}{3} \end{pmatrix} \begin{pmatrix} 1 & 0 & 0 \\ \frac{1}{4} & \frac{1}{3} & \frac{5}{12} \\ 0 & 0 & 1 \end{pmatrix} = \begin{pmatrix} \frac{5}{16} & \frac{1}{12} & \frac{29}{48} \\ \frac{1}{4} & \frac{1}{3} & \frac{5}{12} \\ \frac{5}{12} & \frac{1}{9} & \frac{17}{36} \end{pmatrix}.$$

The components of PQ are nonnegative. Since $\frac{5}{16} + \frac{1}{12} + \frac{29}{48} = \frac{1}{4} + \frac{1}{3} + \frac{5}{12} = \frac{5}{12} + \frac{1}{9} + \frac{17}{36} = 1$, we conclude that PQ is a probability matrix. This is a special case of the general result, Theorem 5.1.2.

We now prove a result concerning the eigenvalue problem for an $n \times n$ probability matrix.

Theorem 5.1.4   *If $P = (p_{ij})$ is an $n \times n$ probability matrix, then there exists a (nonzero) n-component row vector $\mathbf{t}$ such that $\mathbf{t}P = \mathbf{t}$.*

Proof: From the theory of eigenvalues and eigenvectors (3.6), a nontrivial vector $\mathbf{t}$ exists satisfying $\mathbf{t}P = \mathbf{t}$ if and only if $\det(P - I) = 0$. But $\det(P - I) = 0$ if and only if the columns of the matrix $P - I$ are linearly dependent (Theorem 3.5.1). Since P is a probability matrix, we know that

$$p_{11} + p_{12} + \cdots + p_{1n} = p_{21} + p_{22} + \cdots + p_{2n} = \cdots$$
$$= p_{n1} + p_{n2} + \cdots + p_{nn} = 1.$$

These equations can be written in the following equivalent vector form.

$$\begin{pmatrix} p_{11} - 1 \\ p_{21} \\ \vdots \\ p_{n1} \end{pmatrix} + \begin{pmatrix} p_{12} \\ p_{22} - 1 \\ \vdots \\ p_{n2} \end{pmatrix} + \cdots + \begin{pmatrix} p_{1n} \\ p_{2n} \\ \vdots \\ p_{nn} - 1 \end{pmatrix} = \begin{pmatrix} 0 \\ 0 \\ \vdots \\ 0 \end{pmatrix}.$$

The vectors on the left of this equation are exactly the columns of $P - I$. Therefore, the columns of $P - I$ are linearly dependent, and the result is proved.

A vector $\mathbf{t}$ satisfying $\mathbf{t}P = \mathbf{t}$ is called a *fixed vector* of the matrix P. If $\mathbf{t}$ is a probability vector and P is a probability matrix with $\mathbf{t}P = \mathbf{t}$, then $\mathbf{t}$ is called a *fixed probability vector* of P.

To define Markov chains, consider an experiment with a finite sample space $S = \{E_1, E_2, \ldots, E_n\}$, where the E_i $(i = 1, 2, \ldots, n)$ are the elementary events (Definition 2.2.3). Consider a sequence (or *chain*) of trials of this experiment. The experimental system is said to be in the *state* E_i on the mth trial if E_i is the outcome of the mth trial of the experiment.

Definition 5.1.3   Markov Chain   *A sequence of trials of the experiment is a Markov chain if the outcome of the mth trial depends only on the outcome of the $(m - 1)$st trial and not on the outcomes of earlier trials.*

A Markov chain is characterized by the probabilities that the system goes from any state to any state on successive trials. This leads to the following definition.

Definition 5.1.4   Transition Matrix of a Markov Chain   *The transition matrix of a Markov chain is the $n \times n$ matrix $P = (p_{ij})$ whose ij*th *component p_{ij} is the probability that the experimental system goes from state E_i to state E_j on successive trials of the experiment.*

The transition matrix of a Markov chain is a probability matrix. The probabilities p_{ij} are nonnegative and the sum $p_{i1} + p_{i2} + \cdots + p_{in} = 1$, since it represents the probability that the system goes from state E_i to either state E_1 or state $E_2 \ldots$ or state E_n on a trial of the experiment. Of course, this always happens.

Example 5.1.3   Consider the mouse experiment (Figure 5.1). The three states E_1, E_2, and E_3 of the experimental system correspond to the mouse being in compartments I, II, and III. Assuming that the mouse is equally likely to choose either door to leave a compartment, we obtain the transition matrix

$$P = (p_{ij}) = \begin{pmatrix} p_{11} & p_{12} & p_{13} \\ p_{21} & p_{22} & p_{23} \\ p_{31} & p_{32} & p_{33} \end{pmatrix} = \begin{pmatrix} 0 & \frac{1}{2} & \frac{1}{2} \\ \frac{1}{2} & 0 & \frac{1}{2} \\ \frac{1}{2} & \frac{1}{2} & 0 \end{pmatrix}.$$

For example, the probability that the mouse goes from compartment II to compartment I on successive trials is $p_{21} = \frac{1}{2}$.

Example 5.1.4   The weather in Montreal is good, indifferent, or bad on any given day. If the weather is good today, it will be good tomorrow with probability .60, indifferent with probability .20, and bad with probability .20. If the weather is indifferent today, tomorrow it will be good, indifferent, or bad with probabilities .25, .50, and .25, respectively. Finally, if the weather is bad today, the probabilities are .25, .25, and .50 of good, indifferent, or bad weather tomorrow. This can be described as a Markov chain of trials of an experiment with three outcomes E_1, E_2, and E_3, corresponding to good, indifferent, and bad weather on any given day. The transition matrix for this Markov chain is

$$P = \begin{pmatrix} \frac{3}{5} & \frac{1}{5} & \frac{1}{5} \\ \frac{1}{4} & \frac{1}{2} & \frac{1}{4} \\ \frac{1}{4} & \frac{1}{4} & \frac{1}{2} \end{pmatrix}.$$

The transition probability p_{ij} is also called the *one-step transition probability* from the state E_i to the state E_j, since it refers to successive trials of the experiment. The *two-step transition probability* $p_{ij}^{(2)}$ is defined as the probability that the experimental system goes from state E_i on one trial to state E_j two trials later. To go from state E_i to the state E_j in two steps, the system may pass through any intermediate state E_k ($k = 1, 2, \ldots, n$). For example, the system may go from E_i to E_1 to E_j. The probability of this is $p_{i1}p_{1j}$, since, by assumption, the successive trials are independent. We conclude that the two-step transition probability $p_{ij}^{(2)}$ is the sum of n terms, one term for each of the possible intermediate states.

$$p_{ij}^{(2)} = p_{i1}p_{1j} + p_{i2}p_{2j} + \cdots + p_{in}p_{nj} = \sum_{k=1}^{n} p_{ik}p_{kj}. \qquad (5.1)$$

This should be recognized as the ijth component of the matrix P^2. For this reason, the matrix P^2 is called the *two-step transition matrix* of the Markov chain.

The *m-step transition probability* $p_{ij}^{(m)}$ is defined as the probability that the system goes from state E_i on one trial to state E_j m trials later. The following theorem generalizes the result for the two-step transition probabilities.

Theorem 5.1.5   *For the Markov chain with $n \times n$ transition matrix $P = (p_{ij})$, the m-step transition probability $p_{ij}^{(m)}$ is the ij*th *component of the matrix P^m.*

Proof: This result will be proved by induction. (Many of the proofs in this chapter use mathematical induction. The reader is referred to Appendix F for a discussion of this important principle.) It is obviously true for $m = 1$ by the definition of P. Suppose that the theorem is true for $m = k$. Define $Q = P^k = (q_{ij})$. The system can go from state E_i to state E_j in $k + 1$ trials by going from state E_i to state E_l on one trial and then from state E_l to state E_j on k trials. The probability of doing this is $p_{il}q_{lj}$. The intermediate state E_l can be $E_1, E_2, \ldots, E_n$. The probability that the system goes from state E_i to state E_j in $k + 1$ trials is therefore the sum of n probabilities.

$$p_{ij}^{(k+1)} = p_{i1}q_{1j} + p_{i2}q_{2j} + \cdots + p_{in}q_{nj}.$$

But this is the ijth component of $PQ = PP^k = P^{k+1}$. This completes the proof.

Example 5.1.5   Calculate the two-step transition matrices for the Markov chains of Examples 5.1.3 and 5.1.4.

Solution: In Example 5.1.3, the (one-step) transition matrix is

$$P = \begin{pmatrix} 0 & \frac{1}{2} & \frac{1}{2} \\ \frac{1}{2} & 0 & \frac{1}{2} \\ \frac{1}{2} & \frac{1}{2} & 0 \end{pmatrix}$$

and, therefore,

$$P^2 = \begin{pmatrix} \frac{1}{2} & \frac{1}{4} & \frac{1}{4} \\ \frac{1}{4} & \frac{1}{2} & \frac{1}{4} \\ \frac{1}{4} & \frac{1}{4} & \frac{1}{2} \end{pmatrix}.$$

This implies that, for example, the probability of going from compartment I on one trial to compartment II two trials later is $p_{12}^{(2)} = \frac{1}{4}$. In Example 5.1.4

$$P = \begin{pmatrix} \frac{3}{5} & \frac{1}{5} & \frac{1}{5} \\ \frac{1}{4} & \frac{1}{2} & \frac{1}{4} \\ \frac{1}{4} & \frac{1}{4} & \frac{1}{2} \end{pmatrix}$$

and, therefore,

$$P^2 = \begin{pmatrix} \frac{46}{100} & \frac{27}{100} & \frac{27}{100} \\ \frac{27}{80} & \frac{29}{80} & \frac{24}{80} \\ \frac{27}{80} & \frac{24}{80} & \frac{29}{80} \end{pmatrix}.$$

For example, this tells us that, if the weather is good today, the probability of good weather two days from today is $p_{11}^{(2)} = \frac{46}{100}$. By calculating P^7, the probabilities of good, indifferent, and bad weather one week from today can be found. This would be a good exercise in matrix multiplication.

In the study of Markov chains, we are often interested in the long-run probabilities that the system is in the various states. Over many trials, on what proportion of trials will the system be found to be in state E_i for $i = 1, 2, \ldots, n$? In the weather model of Example 5.1.4, it is of interest to know the proportions of days over the next few years that will have good, indifferent, or bad weather. This problem is studied in the next section.

Problems 5.1

1. Which of the following matrices are probability matrices?

(a) $\begin{pmatrix} .99 & .02 & -.01 \\ 0 & 1.0 & 0 \\ .98 & .01 & .01 \end{pmatrix}$.

(b) $\begin{pmatrix} \frac{1}{2} & \frac{1}{4} & \frac{1}{8} & \frac{1}{8} \\ 0 & \frac{1}{2} & \frac{1}{2} & 0 \\ \frac{1}{2} & \frac{1}{4} & \frac{1}{8} & \frac{1}{8} \end{pmatrix}$.

(c) $\begin{pmatrix} 1 & 0 & 0 \\ 0 & 1 & 0 \\ 0 & 0 & 1 \end{pmatrix}$.

2. Calculate the two-step transition matrix for the following transition matrix

$$P = \begin{pmatrix} \frac{1}{2} & \frac{1}{4} & \frac{1}{4} \\ \frac{1}{4} & \frac{1}{2} & \frac{1}{4} \\ \frac{1}{4} & \frac{1}{4} & \frac{1}{2} \end{pmatrix}.$$

   In particular, evaluate $p_{11}^{(2)}$, $p_{12}^{(2)}$, and $p_{13}^{(2)}$.

3. Consider the experiment of placing a mouse in the box drawn in Figure 5.2. Assuming that the mouse is equally likely to choose any door to leave a compartment, describe this experiment as a Markov chain and determine the transition matrix.

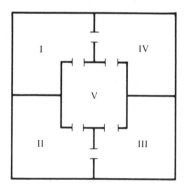

Figure 5.2

4. Determine the two-step transition matrix for the mouse experiment of Problem 3.

5. Calculate a fixed vector and a fixed probability vector for the probability matrix

$$P = \begin{pmatrix} \frac{3}{5} & \frac{1}{5} & \frac{1}{5} \\ \frac{1}{4} & \frac{1}{2} & \frac{1}{4} \\ \frac{1}{4} & \frac{1}{4} & \frac{1}{2} \end{pmatrix}.$$

6. Calculate the two-step transition matrices for the following transition matrices.

(a) $\begin{pmatrix} 1 & 0 \\ \frac{1}{2} & \frac{1}{2} \end{pmatrix}$.

(b) $\begin{pmatrix} \frac{1}{2} & \frac{1}{2} \\ \frac{1}{2} & \frac{1}{2} \end{pmatrix}$.

(c) $\begin{pmatrix} \frac{2}{3} & \frac{1}{3} \\ \frac{1}{3} & \frac{2}{3} \end{pmatrix}$.

(d) $\begin{pmatrix} \frac{1}{3} & \frac{1}{3} & \frac{1}{3} \\ \frac{1}{3} & \frac{1}{3} & \frac{1}{3} \\ \frac{1}{3} & \frac{1}{3} & \frac{1}{3} \end{pmatrix}$.

(e) $\begin{pmatrix} \frac{1}{2} & \frac{1}{3} & \frac{1}{6} \\ \frac{1}{6} & \frac{1}{2} & \frac{1}{3} \\ \frac{1}{3} & \frac{1}{6} & \frac{1}{2} \end{pmatrix}$.

(f) $\begin{pmatrix} \frac{2}{3} & \frac{1}{6} & \frac{1}{6} \\ \frac{1}{6} & \frac{2}{3} & \frac{1}{6} \\ \frac{1}{6} & \frac{1}{6} & \frac{2}{3} \end{pmatrix}$.

7. Calculate a fixed probability vector for the transition matrix

$$P = \begin{pmatrix} \frac{2}{3} & \frac{1}{6} & \frac{1}{6} \\ \frac{1}{4} & \frac{1}{2} & \frac{1}{4} \\ 0 & 0 & 1 \end{pmatrix}.$$

Is this fixed probability vector unique?

8. Calculate a fixed probability vector for the transition matrix

$$P = \begin{pmatrix} \frac{1}{2} & \frac{1}{4} & \frac{1}{4} \\ 0 & 1 & 0 \\ 0 & 0 & 1 \end{pmatrix}.$$

Is this fixed probability vector unique?

9. Prove by mathematical induction that the nth power of the transition matrix

$$P = \begin{pmatrix} \frac{1}{2} & \frac{1}{4} & \frac{1}{4} \\ 0 & 1 & 0 \\ 0 & 0 & 1 \end{pmatrix}$$

is .

$$P^n = \begin{pmatrix} \dfrac{1}{2^n} & \dfrac{2^n - 1}{2^{n+1}} & \dfrac{2^n - 1}{2^{n+1}} \\ 0 & 1 & 0 \\ 0 & 0 & 1 \end{pmatrix}.$$

What is $\lim_{n \to \infty} P^n$? What is the probability of going from the first state to the third state in exactly n steps if n is very large? (By the notation $\lim_{n \to \infty} P^n$, we mean the matrix that P^n approaches as n becomes very large. The symbol $\lim_{n \to \infty} P^n$ is read "the limit of P^n as n approaches infinity.")

10. By mathematical induction, prove that the nth power of the transition matrix

$$P = \begin{pmatrix} \frac{1}{2} & 0 & \frac{1}{2} \\ 0 & \frac{1}{2} & \frac{1}{2} \\ 0 & 0 & 1 \end{pmatrix}$$

is

$$P^n = \begin{pmatrix} \dfrac{1}{2^n} & 0 & 1 - \dfrac{1}{2^n} \\ 0 & \dfrac{1}{2^n} & 1 - \dfrac{1}{2^n} \\ 0 & 0 & 1 \end{pmatrix}.$$

What is $\lim_{n \to \infty} P^n$? Give an interpretation of this limiting matrix.

11. On any given day, a person is either healthy or ill. If the person is healthy today, the probability that he will be healthy tomorrow is estimated to be 98 per cent. If the person is ill today, the probability that he will be healthy tomorrow is 30 per cent. Describe the sequence of states of health as a Markov chain. What is the transition matrix?

12. (a) If the person of Problem 11 is ill today, what are the probabilities that he will recover tomorrow, 2 days from now, and 3 days from now?

   (b) What is the expected number of days that a person ill today will remain ill?

13. (a) If the person of Problem 11 is healthy today, what is the probability that he was healthy yesterday?

   (b) If the person is ill today, what is the probability that he was ill yesterday?

14. It is estimated that, if a critically ill patient lives for one more day, then the probability is .8 that the patient will survive another day. Describe this process as a Markov chain. What is the transition matrix?

15. (a) What is the probability that the patient of Problem 14 will live for at least 5 days?

   (b) What is the expected number of days that this patient will live?

5.2 Regular Markov Chains

The transition matrix $P = (p_{ij})$ of a Markov chain gives the probabilities of the n^2 possible transitions of the experimental system on a trial of the experiment. In some applications of Markov chains, we may have information concerning the initial state of the system. For example, we may know that the system is initially in state E_1 or we may know the probabilities that the system begins in the n possible states $E_1, E_2, \ldots, E_n$. A natural problem is then to determine the probability that the system is in state E_j after m steps. To solve this problem, we first introduce the following definitions.

Definition 5.2.1   Initial Probability Distribution   *The initial probability distribution of a Markov chain is the n-component row vector* $\mathbf{p}^{(0)} = (p_1^{(0)}, p_2^{(0)}, \ldots, p_n^{(0)})$, *where* $p_i^{(0)}$ *is the probability that the system begins in state* E_i *for* $i = 1, 2, \ldots, n$.

Example 5.2.1   If the experimental system is initially in state E_1, the initial probability distribution is $\mathbf{p}^{(0)} = (1, 0, \ldots, 0)$. If the experimental system is equally likely to be in any of the n states initially, then the initial probability distribution is $\mathbf{p}^{(0)} = (1/n, 1/n, \ldots, 1/n)$. It is important to note that the initial probability distribution is a probability vector.

Definition 5.2.2   The *m*th-step Probability Distribution   *The mth-step probability distribution of a Markov chain is the n-component row vector* $\mathbf{p}^{(m)} = (p_1^{(m)}, p_2^{(m)}, \ldots, p_n^{(m)})$, *where* $p_i^{(m)}$ *is the probability that the system is in state* E_i *on the mth trial for* $i = 1, 2, \ldots, n$.

The following theorem says that, given the initial probability distribution and the transition matrix of a Markov chain, the probability distributions for all future trials can be determined.

Theorem 5.2.1   *If* $\mathbf{p}^{(0)} = (p_1^{(0)}, p_2^{(0)}, \ldots, p_n^{(0)})$ *is the initial probability distribution of a Markov chain with transition matrix P, then the probability distribution of the system after one trial is* $\mathbf{p}^{(1)} = \mathbf{p}^{(0)}P$. *More generally, the mth-step probability distribution is* $\mathbf{p}^{(m)} = \mathbf{p}^{(m-1)}P = \mathbf{p}^{(0)}P^m$.

Proof: By definition, $\mathbf{p}^{(1)} = (p_1^{(1)}, p_2^{(1)}, \ldots, p_n^{(1)})$, where $p_i^{(1)}$ is the probability that the system is in the state E_i after the first trial. Initially, the system is in the states $E_1, E_2, \ldots, E_n$ with probabilities $p_1^{(0)}, p_2^{(0)}, \ldots, p_n^{(0)}$. The probability that the system starts in E_1 and goes to E_i on the first trial is $p_1^{(0)} p_{1i}$. Therefore, the probability of the outcome E_i on the first trial is the sum of n probabilities corresponding to the n possible initial states.

$$p_i^{(1)} = p_1^{(0)} p_{1i} + p_2^{(0)} p_{2i} + \cdots + p_n^{(0)} p_{ni}, \qquad i = 1, 2, \ldots, n. \qquad (5.2)$$

Equivalently, we have proved $\mathbf{p}^{(1)} = \mathbf{p}^{(0)} P$. To complete the proof, we use mathematical induction. Assume that, after $m - 1$ trials, the system has probability distribution $\mathbf{p}^{(m-1)} = \mathbf{p}^{(0)} P^{m-1}$. This distribution $\mathbf{p}^{(m-1)}$ can be considered as an initial probability distribution and, therefore, the probability distribution on the next trial is $\mathbf{p}^{(m)} = \mathbf{p}^{(m-1)} P$. We conclude that $\mathbf{p}^{(m)} = \mathbf{p}^{(0)} P^{m-1} P = \mathbf{p}^{(0)} P^m$, and the induction proof is complete.

Example 5.2.2   Consider the mouse experiment of Example 5.1.3. If the mouse begins in the first compartment, calculate the probability distributions after the first and second trials.

Solution: In this example, the transition matrix is

$$P = \begin{pmatrix} 0 & \frac{1}{2} & \frac{1}{2} \\ \frac{1}{2} & 0 & \frac{1}{2} \\ \frac{1}{2} & \frac{1}{2} & 0 \end{pmatrix}$$

and, therefore,

$$P^2 = \begin{pmatrix} \frac{1}{2} & \frac{1}{4} & \frac{1}{4} \\ \frac{1}{4} & \frac{1}{2} & \frac{1}{4} \\ \frac{1}{4} & \frac{1}{4} & \frac{1}{2} \end{pmatrix}.$$

The initial probability distribution is $\mathbf{p}^{(0)} = (1, 0, 0)$. Therefore, $\mathbf{p}^{(1)} = \mathbf{p}^{(0)} P = (0, \frac{1}{2}, \frac{1}{2})$ and $\mathbf{p}^{(2)} = \mathbf{p}^{(0)} P^2 = (\frac{1}{2}, \frac{1}{4}, \frac{1}{4})$. This implies, for example, that the probability the mouse returns to compartment I on the second trial is $\frac{1}{2}$.

It is reasonable to ask whether, in the long run, the probability that the system is in a particular state will approach a constant independent of the initial probability distribution. In the mouse experiment, for example, we suspect that the long-run probability of each of the three states is $\frac{1}{3}$. This can be proved for the regular Markov chains.

Definition 5.2.3   Regular Probability Matrix   *The probability matrix $P = (p_{ij})$ is regular if some power P^m of P has all components strictly positive.*

Definition 5.2.4   Regular Markov Chain   *A Markov chain is regular if its transition matrix $P = (p_{ij})$ is regular.*

Example 5.2.3   Which of the following probability matrices are regular?

1.
$$P_1 = \begin{pmatrix} \frac{1}{2} & \frac{1}{2} \\ 0 & 1 \end{pmatrix}.$$

2.
$$P_2 = \begin{pmatrix} 0 & \frac{1}{2} & \frac{1}{2} \\ \frac{1}{2} & 0 & \frac{1}{2} \\ \frac{1}{2} & \frac{1}{2} & 0 \end{pmatrix}.$$

3.
$$P_3 = \begin{pmatrix} \frac{3}{5} & \frac{1}{5} & \frac{1}{5} \\ \frac{1}{4} & \frac{1}{2} & \frac{1}{4} \\ \frac{1}{4} & \frac{1}{4} & \frac{1}{2} \end{pmatrix}.$$

Solution:

1.
$$P_1^2 = \begin{pmatrix} \frac{1}{4} & \frac{3}{4} \\ 0 & 1 \end{pmatrix}, \; P_1^3 = \begin{pmatrix} \frac{1}{8} & \frac{7}{8} \\ 0 & 1 \end{pmatrix}, \ldots, P_1^m = \begin{pmatrix} \frac{1}{2^m} & 1 - \frac{1}{2^m} \\ 0 & 1 \end{pmatrix}.$$

Every power of P_1 has a component 0 and, therefore, P_1 is not regular.

2.
$$P_2^2 = \begin{pmatrix} \frac{1}{2} & \frac{1}{4} & \frac{1}{4} \\ \frac{1}{4} & \frac{1}{2} & \frac{1}{4} \\ \frac{1}{4} & \frac{1}{4} & \frac{1}{2} \end{pmatrix} \quad \text{(regular)}.$$

3. P_3 is regular, since $P_3^1 = P_3$ has all components strictly positive.

The following theorem can be proved for regular probability matrices, although the proof will not be given here.

Theorem 5.2.2   *If $P = (p_{ij})$ is a regular probability matrix, then P has a unique fixed probability vector $\mathbf{t}$ whose components are all positive. The limit matrix $T = \lim_{m \to \infty} P^m$ is defined and each of its rows is equal to $\mathbf{t}$, the fixed probability vector. For any probability vector $\mathbf{p}$, $\lim_{m \to \infty} \mathbf{p}P^m = \mathbf{t}$. (See problem 5.1.9 for an explanation of notation.)*

Example 5.2.4   The regular probability matrix

$$P = \begin{pmatrix} \frac{3}{4} & \frac{1}{4} \\ \frac{1}{4} & \frac{3}{4} \end{pmatrix}$$

has fixed probability vector $\mathbf{t} = (\frac{1}{2}, \frac{1}{2})$. To illustrate Theorem 5.2.2, we calculate the following powers of P.

$$P^2 = \begin{pmatrix} \frac{5}{8} & \frac{3}{8} \\ \frac{3}{8} & \frac{5}{8} \end{pmatrix}, \qquad P^4 = P^2 P^2 = \begin{pmatrix} \frac{17}{32} & \frac{15}{32} \\ \frac{15}{32} & \frac{17}{32} \end{pmatrix}, \qquad P^8 = P^4 P^4 = \begin{pmatrix} \frac{257}{512} & \frac{255}{512} \\ \frac{255}{512} & \frac{257}{512} \end{pmatrix}.$$

It is evident that, as m increases, P^m approaches

$$\begin{pmatrix} \frac{1}{2} & \frac{1}{2} \\ \frac{1}{2} & \frac{1}{2} \end{pmatrix}.$$

We are interested in applying Theorem 5.2.2 to the study of Markov chains. If $P = (p_{ij})$ is the transition matrix of a regular Markov chain, then P has a fixed probability vector $\mathbf{t}$ with all components positive. Also, if $\mathbf{p}^{(0)}$ is the initial probability distribution, then $\lim_{m \to \infty} \mathbf{p}^{(0)} P^m = \mathbf{t}$. This means that, after many trials, the probability distribution of the system is approximately $\mathbf{t}$. The fixed probability vector $\mathbf{t}$ of the regular transition matrix P is called the *stationary distribution* of the regular Markov chain. Since $\mathbf{t}$ is defined by $\mathbf{t}P = \mathbf{t}$, it has the property that, if the probability distribution on some trial is given by $\mathbf{t}$, then the probability distribution on all future trials is also given by $\mathbf{t}$.

Example 5.2.5 What are the long-run probabilities of good, indifferent, and bad weather in the model of Example 5.1.4?

Solution: The transition matrix

$$P = \begin{pmatrix} \frac{3}{5} & \frac{1}{5} & \frac{1}{5} \\ \frac{1}{4} & \frac{1}{2} & \frac{1}{4} \\ \frac{1}{4} & \frac{1}{4} & \frac{1}{2} \end{pmatrix}$$

is regular and Theorem 5.2.2 can be applied. The fixed probability vector or stationary distribution is $\mathbf{t} = (\frac{5}{13}, \frac{4}{13}, \frac{4}{13})$ (see Problem 5.1.5). This implies that the long-run probability of good weather is $\frac{5}{13}$. For indifferent or bad weather, the corresponding probabilities are both equal to $\frac{4}{13}$.

Example 5.2.6 Random Walk with Reflecting Barriers Consider a game in which two players each begin with two marbles. On each play of the game, the first player has probability p of winning one marble and probability

$q = 1 - p$ of losing one marble. Suppose that when either player has lost all his marbles, the other player gives up a marble on the next play in order that the game can continue. Describe this game as a Markov chain.

Solution: The five states E_0, E_1, E_2, E_3, and E_4 in this game correspond to the first player having 0, 1, 2, 3, and 4 marbles, respectively. According to the rules of the game, the transition matrix is

$$P = \begin{pmatrix} p_{00} & p_{01} & p_{02} & p_{03} & p_{04} \\ p_{10} & p_{11} & p_{12} & p_{13} & p_{14} \\ p_{20} & p_{21} & p_{22} & p_{23} & p_{24} \\ p_{30} & p_{31} & p_{32} & p_{33} & p_{34} \\ p_{40} & p_{41} & p_{42} & p_{43} & p_{44} \end{pmatrix} = \begin{pmatrix} 0 & 1 & 0 & 0 & 0 \\ q & 0 & p & 0 & 0 \\ 0 & q & 0 & p & 0 \\ 0 & 0 & q & 0 & p \\ 0 & 0 & 0 & 1 & 0 \end{pmatrix}.$$

The initial probability distribution is $\mathbf{p}^{(0)} = (0, 0, 1, 0, 0)$, corresponding to the system beginning in state E_2. After one play, the probability distribution is $\mathbf{p}^{(1)} = \mathbf{p}^{(0)}P = (0, q, 0, p, 0)$. After two plays, the probability distribution is $\mathbf{p}^{(2)} = \mathbf{p}^{(1)}P = (q^2, 0, 2pq, 0, p^2)$. Similarly, $\mathbf{p}^{(3)} = \mathbf{p}^{(2)}P = (0, q^2 + 2pq^2, 0, 2p^2q + p^2, 0)$. This is an example of a *random walk process*. The states E_0 and E_4 are said to be *reflecting barriers*, since, if the system arrives in either state, it moves in the "opposite direction" or "reflects" on the next move.

Problems 5.2

1. For the regular transition matrix

$$P = \begin{pmatrix} \frac{3}{4} & \frac{1}{4} \\ \frac{1}{4} & \frac{3}{4} \end{pmatrix},$$

prove by induction that

$$P^n = \begin{pmatrix} \dfrac{1}{2} + \dfrac{1}{2^{n+1}} & \dfrac{1}{2} - \dfrac{1}{2^{n+1}} \\ \dfrac{1}{2} - \dfrac{1}{2^{n+1}} & \dfrac{1}{2} + \dfrac{1}{2^{n+1}} \end{pmatrix}$$

and deduce that

$$\lim_{n \to \infty} P^n = \begin{pmatrix} \frac{1}{2} & \frac{1}{2} \\ \frac{1}{2} & \frac{1}{2} \end{pmatrix}.$$

2. An examination consists of 100 true-or-false questions. For an average student, the examination is such that, if a question is answered correctly, the probability that the next question will be answered correctly is $\frac{3}{4}$. Similarly, if a question is answered incorrectly, the probability that the next question will be answered correctly is $\frac{1}{4}$. Estimate the average score on this examination.

3. A laboratory animal has a choice of three foods available in standard units. After lengthy observation, it is found that, if the animal chooses one food on one trial, it will choose the same food on the next trial with probability 50 per cent and it will choose the other foods on the next trial with equal probabilities 25 per cent. Describe this process as a Markov chain and determine the transition matrix. Prove that, in the long run, equal quantities of the three foods are consumed.

4. For the regular transition matrix

$$P = \begin{pmatrix} \frac{3}{5} & \frac{2}{5} \\ \frac{2}{5} & \frac{3}{5} \end{pmatrix},$$

prove by mathematical induction that

$$P^n = \begin{pmatrix} \frac{1}{2} + \frac{1}{2\cdot 5^n} & \frac{1}{2} - \frac{1}{2\cdot 5^n} \\ \frac{1}{2} - \frac{1}{2\cdot 5^n} & \frac{1}{2} + \frac{1}{2\cdot 5^n} \end{pmatrix}.$$

What is $\lim_{n\to\infty} P^n$?

5. For the regular transition matrix

$$P = \begin{pmatrix} \frac{4}{7} & \frac{3}{7} \\ \frac{3}{7} & \frac{4}{7} \end{pmatrix},$$

determine an exact formula for P^n. Use this formula to evaluate $\lim_{n\to\infty} P^n$.

6. Consider the transition matrix

$$P = \begin{pmatrix} 0 & 0 & 1 \\ 0 & 1 & 0 \\ 1 & 0 & 0 \end{pmatrix}.$$

(a) Describe the Markov chain with this transition matrix. Is P regular?

(b) Calculate a fixed probability vector for P.

(c) Determine the m-step transition matrix for this Markov chain.

7. Consider the transition matrix

$$P = \begin{pmatrix} 0 & \frac{1}{2} & \frac{1}{2} \\ \frac{1}{2} & 0 & \frac{1}{2} \\ \frac{1}{2} & \frac{1}{2} & 0 \end{pmatrix}.$$

(a) Describe the Markov chain with this transition matrix. Is P regular?

(b) Calculate a fixed probability vector for P.

(c) Determine the three-step transition matrix for this Markov chain.

8. Calculate the two-step probability distributions for the following transition matrices and initial probability distributions.

(a)

$$\begin{pmatrix} \frac{1}{2} & \frac{1}{2} & 0 \\ 0 & \frac{1}{2} & \frac{1}{2} \\ \frac{1}{2} & \frac{1}{2} & 0 \end{pmatrix}, \quad \mathbf{p}^{(0)} = (\tfrac{1}{2}, 0, \tfrac{1}{2}).$$

(b)

$$\begin{pmatrix} \frac{1}{3} & \frac{1}{3} & \frac{1}{3} \\ 0 & 1 & 0 \\ \frac{1}{3} & \frac{1}{3} & \frac{1}{3} \end{pmatrix}, \quad \mathbf{p}^{(0)} = (1, 0, 0).$$

(c)

$$\begin{pmatrix} \frac{1}{2} & \frac{1}{4} & \frac{1}{4} & 0 \\ 0 & \frac{1}{2} & \frac{1}{4} & \frac{1}{4} \\ 0 & 0 & 1 & 0 \\ \frac{1}{4} & \frac{1}{4} & \frac{1}{4} & \frac{1}{4} \end{pmatrix}, \quad \mathbf{p}^{(0)} = (\tfrac{1}{4}, \tfrac{1}{4}, \tfrac{1}{4}, \tfrac{1}{4}).$$

9. There are five tests in a certain course. The possible grades on each test are A, B, C, D, and E. It is estimated that the probability is 60 per cent that a student will obtain the same grade as on the previous test and 10 per cent for each of the alternative possibilities. Describe this process as a Markov chain. What is the transition matrix?

10. (a) If a student receives an A on the first test of Problem 9, what is the probability that he will receive a grade of C on the third test?

(b) If a student receives a B on the second test, what is the probability that all his grades on the remaining three tests are B?

11. Two subspecies of birds are competing for territory. Initially, the two subspecies each occupy 10 territorial units. Competition proceeds by a series of encounters. On each encounter, a subspecies either gains or

loses a unit of territory with equal probabilities. Describe this competition as a Markov chain with 21 states. What is the transition matrix? What are the probability distributions of the territorial units after one encounter and after two encounters? Assume that the competition is terminated if one of the subspecies loses all of its territory.

12. Consider the following mating experiment. An individual of unknown genotype AA, Aa, or aa (see Section 9.3 for an explanation of terminology) is mated to a heterozygous individual Aa. One offspring chosen at random is again mated to a heterozygous individual. After repeating this procedure for many generations, what is the probability that an offspring chosen at random is heterozygous?

5.3 Absorbing Markov Chains

In the mouse experiment of Section 5.1 (Figure 5.1), we may introduce the modification that the mouse is removed from the box when it reaches compartment III. This new experimental procedure can be described as a Markov chain if we agree to say that the system remains in the third state on all trials after it reaches the third state. States with this property are very common in applications of Markov chains. For this reason, we introduce the following definitions.

Definition 5.3.1   Absorbing State   *A state E_i of a Markov chain is absorbing if, once the system reaches the state E_i on some trial, the system remains in the state E_i on all future trials.*

Definition 5.3.2   Absorbing Markov Chain   *A Markov chain is absorbing if it has one or more absorbing states and if it is possible to reach an absorbing state from every nonabsorbing state.*

If the state E_i is absorbing, the probability of transition from E_i to E_i is 1. In other words, the state E_i is absorbing if and only if $p_{ii} = 1$. The number of absorbing states of an absorbing Markov chain is equal to the number of ones on the diagonal of its transition matrix. The nonabsorbing states of an absorbing Markov chain are called the *transient states*. The probability that the system is in a transient state decreases as the number of trials increases.

Example 5.3.1   If the mouse (Example 5.1.3) remains in the third compartment once it reaches there, the transition matrix is

$$P = \begin{pmatrix} 0 & \frac{1}{2} & \frac{1}{2} \\ \frac{1}{2} & 0 & \frac{1}{2} \\ 0 & 0 & 1 \end{pmatrix}.$$

E_1 and E_2 are transient states and E_3 is an absorbing state. We note that it is possible to reach E_3 from either E_1 or E_2, and therefore the Markov chain is absorbing.

Example 5.3.2   The matrix

$$P = \begin{pmatrix} \frac{1}{2} & 0 & \frac{1}{2} \\ 0 & 1 & 0 \\ \frac{1}{3} & \frac{1}{3} & \frac{1}{3} \end{pmatrix}$$

is the transition matrix of an absorbing Markov chain with three states. The second state is absorbing. It can be reached from E_3 on one trial and from E_1 after two trials (first go from E_1 to E_3 with probability $\frac{1}{2}$ and then from E_3 to E_2 with probability $\frac{1}{3}$).

Example 5.3.3   Random Walk with Absorbing Barriers   As in Example 5.2.6, consider a game in which two players each begin with two marbles. On each play of the game, the first player has probability p of winning one marble and probability $q = 1 - p$ of losing one marble. The game ends when either player has lost all his marbles. Describe this game as a Markov chain.

Solution: This game has five states, E_0, E_1, E_2, E_3, and E_4, corresponding to the first player having $0, 1, 2, 3, 4$ marbles. The states E_0 and E_4 are absorbing and the states E_1, E_2, and E_3 are nonabsorbing or transient. The transition matrix is

$$P = \begin{pmatrix} p_{00} & p_{01} & p_{02} & p_{03} & p_{04} \\ p_{10} & p_{11} & p_{12} & p_{13} & p_{14} \\ p_{20} & p_{21} & p_{22} & p_{23} & p_{24} \\ p_{30} & p_{31} & p_{32} & p_{33} & p_{34} \\ p_{40} & p_{41} & p_{42} & p_{43} & p_{44} \end{pmatrix} = \begin{pmatrix} 1 & 0 & 0 & 0 & 0 \\ q & 0 & p & 0 & 0 \\ 0 & q & 0 & p & 0 \\ 0 & 0 & q & 0 & p \\ 0 & 0 & 0 & 0 & 1 \end{pmatrix}.$$

The system begins in state E_2; that is, the initial probability distribution is $\mathbf{p}^{(0)} = (0, 0, 1, 0, 0)$. On succeeding plays, the successive probability distributions are $\mathbf{p}^{(1)} = \mathbf{p}^{(0)}P = (0, q, 0, p, 0)$, $\mathbf{p}^{(2)} = (q^2, 0, 2pq, 0, p^2)$, $\mathbf{p}^{(3)} = (q^2, 2pq^2, 0, 2p^2q, p^2)$, $\mathbf{p}^{(4)} = (q^2 + 2pq^3, 0, 4p^2q^2, 0, p^2 + 2p^3q), \ldots$. After many trials, the probability that the system is in one of the transient states becomes very small. This means that, after a large number of plays, it is almost certain that one of the players has lost all his marbles. A more complicated random walk process with absorbing barriers will be studied in Chapter 9 as a model of the survival and extinction of species.

Consider a general absorbing Markov chain with k transient states $E_1, E_2, \ldots, E_k$ and $n - k$ absorbing states $E_{k+1}, E_{k+2}, \ldots, E_n$. The transition matrix for this Markov chain has the general form

$$P = \begin{pmatrix} p_{11} & p_{12} & \cdots & p_{1k} & p_{1k+1} & p_{1k+2} & \cdots & p_{1n} \\ p_{21} & p_{22} & \cdots & p_{2k} & p_{2k+1} & p_{2k+2} & \cdots & p_{2n} \\ \vdots & \vdots & & \vdots & \vdots & \vdots & & \vdots \\ p_{k1} & p_{k2} & \cdots & p_{kk} & p_{kk+1} & p_{kk+2} & \cdots & p_{kn} \\ 0 & 0 & \cdots & 0 & 1 & 0 & \cdots & 0 \\ 0 & 0 & \cdots & 0 & 0 & 1 & \cdots & 0 \\ \vdots & \vdots & & \vdots & \vdots & \vdots & & \vdots \\ 0 & 0 & \cdots & 0 & 0 & 0 & \cdots & 1 \end{pmatrix}.$$

Suppose that initially the system is in a nonabsorbing state E_i. Then we may be interested in how long the system "survives," that is, on average, how many trials are required before the system reaches an absorbing state. For the initial state E_i, let q_{ij} be the expected number of times the system will be in a nonabsorbing state E_j before it reaches an absorbing state. If $i \neq j$, the system can reach the state E_j on the first trial with probability p_{ij}. It may reach the state E_j on the second trial by passing through any intermediate transient state E_l on the first trial. The probability of reaching E_j on the second trial is therefore $\sum_{l=1}^{k} p_{il}p_{lj}$. Similarly, the probability of reaching E_j on the third trial is $\sum_{l_1=1}^{k} \sum_{l_2=1}^{k} p_{il_1}p_{l_1l_2}p_{l_2j}$, where E_{l_1} and E_{l_2} are the intermediate states on the first and second trials. We conclude that, if $i \neq j$,

$$q_{ij} = p_{ij} + \sum_{l_1=1}^{k} p_{il_1}p_{l_1j} + \sum_{l_1=1}^{k} \sum_{l_2=1}^{k} p_{il_1}p_{l_1l_2}p_{l_2j} + \cdots. \tag{5.3}$$

If $i = j$, the system begins in state E_i and, therefore,

$$q_{ii} = 1 + p_{ii} + \sum_{l_1=1}^{k} p_{il_1} p_{l_1 i} + \sum_{l_1=1}^{k} \sum_{l_2=1}^{k} p_{il_1} p_{l_1 l_2} p_{l_2 i} + \cdots. \qquad (5.4)$$

Define the two $k \times k$ matrices

$$Q = \begin{pmatrix} q_{11} & q_{12} & \cdots & q_{1k} \\ q_{21} & q_{22} & \cdots & q_{2k} \\ \vdots & \vdots & & \vdots \\ q_{k1} & q_{k2} & \cdots & q_{kk} \end{pmatrix}, \qquad R = \begin{pmatrix} p_{11} & p_{12} & \cdots & p_{1k} \\ p_{21} & p_{22} & \cdots & p_{2k} \\ \vdots & \vdots & & \vdots \\ p_{k1} & p_{k2} & \cdots & p_{kk} \end{pmatrix}.$$

Then equations (5.3) and (5.4) can be summarized in the matrix equation

$$Q = I + R + R^2 + R^3 + \cdots, \qquad (5.5)$$

where I is the $k \times k$ identity matrix. This expression for Q as an infinite sum of powers of R can be studied in the same way as the infinite sums of Section 1.6. The series in (5.5) can be summed to yield $Q = (I - R)^{-1}$. To give a rough justification, note that

$$RQ = R + R^2 + R^3 + R^4 + \cdots = Q - I$$

or

$$(I - R)Q = I \qquad \text{and} \qquad Q = (I - R)^{-1}.$$

Our original problem was to determine the average number of trials required before the system reaches an absorbing state. Define m_i to be the expected number of trials before the system reaches an absorbing state when it begins in the transient state E_i. Clearly, m_i is the sum of the expected number of trials the system will be in each transient state.

$$m_i = q_{i1} + q_{i2} + \cdots + q_{ik}. \qquad (5.6)$$

In other words, m_i is the sum of the components in the ith row of the matrix $Q = (I - R)^{-1}$.

Example 5.3.1 (continued) In this example, E_1 and E_2 are transient states and E_3 is absorbing.

$$P = \begin{pmatrix} 0 & \frac{1}{2} & \frac{1}{2} \\ \frac{1}{2} & 0 & \frac{1}{2} \\ 0 & 0 & 1 \end{pmatrix} \quad \text{and} \quad R = \begin{pmatrix} 0 & \frac{1}{2} \\ \frac{1}{2} & 0 \end{pmatrix};$$

$$R^2 = \begin{pmatrix} \frac{1}{4} & 0 \\ 0 & \frac{1}{4} \end{pmatrix}, \quad R^3 = \begin{pmatrix} 0 & \frac{1}{8} \\ \frac{1}{8} & 0 \end{pmatrix}, \cdots,$$

$$R^{2n} = \begin{pmatrix} \frac{1}{2^{2n}} & 0 \\ 0 & \frac{1}{2^{2n}} \end{pmatrix}, \quad R^{2n+1} = \begin{pmatrix} 0 & \frac{1}{2^{2n+1}} \\ \frac{1}{2^{2n+1}} & 0 \end{pmatrix}.$$

Therefore,

$$Q = I + R + R^2 + R^3 + \cdots$$

$$= \begin{pmatrix} 1 & 0 \\ 0 & 1 \end{pmatrix} + \begin{pmatrix} 0 & \frac{1}{2} \\ \frac{1}{2} & 0 \end{pmatrix} + \begin{pmatrix} \frac{1}{4} & 0 \\ 0 & \frac{1}{4} \end{pmatrix} + \begin{pmatrix} 0 & \frac{1}{8} \\ \frac{1}{8} & 0 \end{pmatrix} + \begin{pmatrix} \frac{1}{16} & 0 \\ 0 & \frac{1}{16} \end{pmatrix} + \cdots$$

$$= \begin{pmatrix} 1 + \frac{1}{4} + \frac{1}{16} + \cdots & \frac{1}{2} + \frac{1}{8} + \frac{1}{32} + \cdots \\ \frac{1}{2} + \frac{1}{8} + \frac{1}{32} + \cdots & 1 + \frac{1}{4} + \frac{1}{16} + \cdots \end{pmatrix}$$

$$= \begin{pmatrix} \frac{4}{3} & \frac{2}{3} \\ \frac{2}{3} & \frac{4}{3} \end{pmatrix}.$$

Note that

$$1 + \tfrac{1}{4} + \tfrac{1}{16} + \cdots = 1 + \tfrac{1}{4} + (\tfrac{1}{4})^2 + (\tfrac{1}{4})^3 + \cdots = \frac{1}{1 - \frac{1}{4}} = \frac{1}{\frac{3}{4}} = \frac{4}{3}.$$

(See Example 1.6.8). Also,

$$\tfrac{1}{2} + \tfrac{1}{8} + \tfrac{1}{32} + \cdots = \tfrac{1}{2}(1 + \tfrac{1}{4} + \tfrac{1}{16} + \cdots) = \tfrac{1}{2} \cdot \tfrac{4}{3} = \tfrac{2}{3}.$$

The meaning of $q_{12} = \frac{2}{3}$, for example, is that if the system begins in state E_1, the expected number of times the system will be in state E_2 is $\frac{2}{3}$. Similarly, $m_1 = q_{11} + q_{12} = \frac{4}{3} + \frac{2}{3} = 2$. In other words, if the mouse is in compartment I on the first move, the expected number of moves (including the first move) before reaching compartment III is 2.

Example 5.3.3 (continued)   In this example, the transient states are E_1, E_2 and E_3 and

$$R = \begin{pmatrix} 0 & p & 0 \\ q & 0 & p \\ 0 & q & 0 \end{pmatrix}.$$

For simplicity, consider the case $p = q = \frac{1}{2}$. Then, from Problem 3.4.2,

$$I - R = \begin{pmatrix} 1 & -\frac{1}{2} & 0 \\ -\frac{1}{2} & 1 & -\frac{1}{2} \\ 0 & -\frac{1}{2} & 1 \end{pmatrix} \quad \text{and} \quad (I - R)^{-1} = \begin{pmatrix} \frac{3}{2} & 1 & \frac{1}{2} \\ 1 & 2 & 1 \\ \frac{1}{2} & 1 & \frac{3}{2} \end{pmatrix} = Q.$$

This tells us that, for example, if the first player begins with two marbles, then he can expect the game to last $q_{21} + q_{22} + q_{23} = 1 + 2 + 1 = 4$ moves before he either wins everything or loses everything.

Example 5.3.3 is a game whose outcome is not predictable. The result of each play or move of the game is determined by chance. In the remaining sections of this chapter, we study other types of games in which the skill of the player can affect the outcome.

Problems 5.3

1. Prove that a regular Markov chain is not an absorbing Markov chain. Give an example of a Markov chain that is not regular and not absorbing.

2. Which of the following matrices are the transition matrices of absorbing Markov chains? Determine the number of absorbing states in each example.

(a) $\begin{pmatrix} 1 & 0 & 0 \\ 0 & 0 & 1 \\ 0 & 1 & 0 \end{pmatrix}.$
    (b) $\begin{pmatrix} 0 & 1 & 0 \\ 1 & 0 & 0 \\ 0 & 0 & 1 \end{pmatrix}.$
    (c) $\begin{pmatrix} \frac{1}{3} & \frac{1}{3} & \frac{1}{3} \\ 0 & 0 & 1 \\ \frac{1}{3} & \frac{1}{3} & \frac{1}{3} \end{pmatrix}.$

(d) $\begin{pmatrix} \frac{1}{2} & \frac{1}{2} & 0 & 0 \\ 1 & 0 & 0 & 0 \\ 0 & 1 & 0 & 0 \\ 0 & 0 & \frac{1}{2} & \frac{1}{2} \end{pmatrix}.$
    (e) $\begin{pmatrix} \frac{1}{2} & \frac{1}{2} & 0 & 0 \\ \frac{1}{2} & \frac{1}{2} & 0 & 0 \\ 0 & 0 & 1 & 0 \\ 0 & 0 & 0 & 1 \end{pmatrix}.$

3. An experiment must be repeated until exactly two successful experiments have been completed. Suppose that the probability of a successful experiment on one trial is $\frac{1}{3}$. Describe this process as an absorbing Markov chain with three states. What is the transition matrix? If the first experiment is a failure, what is the probability that the first success occurs on the fourth trial?

4. What is the expected number of times that the experiment of the previous problem will be repeated?

5. A laboratory animal must complete a certain task to receive a unit of food. The probability of successful completion of the task on any trial is $\frac{4}{5}$. Suppose that the animal repeats the task until he receives a total of four units of food. Describe this process as an absorbing Markov chain with five states. What is the transition matrix?

6. What is the expected number of times that the task of Problem 5 will be repeated until four successful completions of the task?

7. Two gamblers G_1 and G_2 are playing a game; the probability that G_1 wins on each move is $\frac{3}{7}$. Suppose that G_1 begins with $7 and G_2 begins with $1. On each play, $1 is bet by both players and the game continues until one player has lost all his money. Describe this game as an absorbing Markov chain with nine states. What are the probability distributions of the states after one play and after two plays. (For a biological interpretation of this game, see Section 9.2.)

8. In the game of Problem 7, suppose that G_2 bets all his money on every play of the game. Describe this game as a Markov chain with five states. What is the expected number of plays of this game?

9. A laboratory animal must choose one of four panels to receive food. Panels I and II, if chosen, result in a very small amount of food. Panels III and IV both yield much larger amounts of food. If either panel III or panel IV is chosen on a trial, it is observed that the same panel is chosen on all future trials. If either panel I or panel II is chosen on a trial, then any of the four panels may be chosen on the next trial with equal probability. Describe this process as an absorbing Markov chain with four states. If panel I is chosen on the first trial, what is the expected number of trials before either panel III or panel IV is chosen?

10. (a) Consider a game played between team A and team B with a total of n players on the two teams. On each play of the game, one team gains a player from the other team and the game continues until a team has lost all its members. If there are k players on team A and $n - k$ players on team B, the probability that team A gains one player on

the next play is $(k/n)^2$. Describe this game as an absorbing Markov chain. Are the rules "fair"? (This is an elementary model of species competition.)

(b) For the above game, suppose that $k = 3$ and $n = 6$. What is the probability distribution of the teams after one play? After two plays?

11. Referring to Problem 5.2.12, suppose that the offspring chosen at random are mated to recessive individuals (aa) in each generation. After many generations, what is the probability that an offspring chosen at random is recessive?

5.4 The Theory of Games

The modern theory of games was developed in the 1940s to provide a general mathematical framework for economics. The principal ideas of this theory were abstracted from ordinary games such as chess, bridge, solitaire, dominoes, and checkers. The general theory was developed without direct reference to any particular game. The theory of games can be applied to the analysis of any competitive behavior including ordinary games, economics, warfare, and biological competition. In the study of biological competition, the theory of games provides a useful conceptual framework for understanding behavior.[2,3]

Many very familiar games have opponents or competitors who make a sequence of moves according to the rules of the game. In some games, the successive moves are made with complete information about the opponent's opportunities (chess). In other games, moves are made with incomplete information (bridge). A player may decide his moves purely by chance (by tossing a coin, for example) or he may choose his moves deliberately from all possible moves. The game may terminate after a finite number of moves with a winner and a loser. There is generally a return to the winner of the game, which may be a cash payment or merely the satisfaction of winning. (The reward to a species playing the game of ecology is to be allowed to continue to play.)

A game is characterized by its rules. In some cases, the game may be so complicated that discovering its rules may be a considerable achievement. Consider the problem of determining the rules of chess by watching it being played. After four or five games, the principal rules would be apparent,

[2] R. C. Lewontin, "Evolution and the Theory of Games," *Journal of Theoretical Biology*, 1:382–403 (1961).

[3] L. B. Slobodkin, "The Strategy of Evolution," *American Scientist*, 52:342–357 (1964).

but it would be necessary to observe many more games to determine all the rules. Analogously, the complex interactions of a human social system or of an ecosystem can be thought of as the unfolding of a game with a very large number of players, whose rules are not completely understood.

Once the rules of the game are known, the problem is to determine how the players should choose their moves and what the consequences of these moves are. In other words, the players must determine their strategies by analyzing the rules of the game. The final outcome of a game will usually be critically dependent on the choices of moves of all players. For complex games, it may be impossible to analyze all possibilities, and, in this case, the players must rely on experience, intuition, or simple trial and error to determine their moves.

In this chapter, we study a simple two-person game in considerable detail. The concepts introduced and the results proved for this game form a model for the analysis of more general games.

Definition 5.4.1   Matrix Game   *Let $A = (a_{ij})$ be an $m \times n$ matrix. Consider a game determined by A played between two competitors R and C (rows and columns) according to the following rules.*

1. *On each move of the game, R chooses one of the m rows of A and C chooses one of the n columns of A. These choices are made simultaneously, and neither competitor knows in advance the choice (or move) of the other competitor.*
2. *If R chooses the ith row and C chooses the jth column, then C pays R an amount a_{ij}. If a_{ij} is negative, this is interpreted to mean that C receives an amount $-a_{ij}$ from R.*

This game is the $m \times n$ matrix game determined by the $m \times n$ matrix $A = (a_{ij})$.

The matrix game may terminate after one move or it may continue for any number of moves. The matrix $A = (a_{ij})$ of the game is called the *game matrix* or the *payoff matrix*.

The following examples illustrate how matrix games are analyzed.

Example 5.4.1   Describe the matrix games whose payoff matrices are

1.
$$A = \begin{pmatrix} 1 & 2 \\ -2 & 3 \end{pmatrix}.$$

2.
$$B = \begin{pmatrix} 1 & 0 & 1 & 0 \\ 0 & -1 & 2 & 0 \\ -1 & 0 & 3 & 1 \end{pmatrix}.$$

Solution:

1. In this 2×2 matrix game, R and C each have two choices. If R chooses the first row, he gains one unit if C chooses the first column and two units if C chooses the second column. If R chooses the second row, he loses two units if C chooses the first column and he gains three units if C chooses the second column. If C is playing rationally, C will choose the first column. In this case, R should choose the first row. With these choices, R guarantees that he will win at least one unit and C guarantees that he will lose no more than one unit.

2. In this 3×4 matrix game, R has three choices and C has four choices. By analyzing all possible choices, it is clear that C will do best by choosing the second column. By this choice, C guarantees that no loss will be suffered; R will do best by choosing the first row. If these choices are made, there is no payment between the players.

The general $m \times n$ matrix game is an example of a *two-person, zero-sum game*, since there are two competitors and the sum of their winnings is zero. The gains of one competitor are the losses of the other. These matrix games are completely analyzed in the remaining sections of this chapter. The game of Example 5.3.3 is a two-person, zero-sum game that is not a matrix game.

Example 5.4.2 In experiments, ravens and parakeets have learned to recognize numbers up to seven. The following experiment is proposed. The diet of a raven R and a parakeet C is to be determined by a matrix game. Each bird is shown three cards, labeled with 2, 4, and 7 dots. If each bird chooses the same card, then R receives from the diet of C a number of worms equal to twice the number of dots on the card. If the cards chosen are different, then C receives from the diet of R a number of worms equal to the difference of the number of dots on the cards. Assuming that the moves are made independently (perhaps by having two sets of cards), describe this experiment as a matrix game.

Solution: This is a 3×3 matrix game with game matrix

$$A = \begin{array}{c} \\ 2 \\ 4 \\ 7 \end{array} \begin{array}{ccc} \overset{2}{} & \overset{4}{} & \overset{7}{} \\ \left(\begin{array}{ccc} 4 & -2 & -5 \\ -2 & 8 & -3 \\ -5 & -3 & 14 \end{array} \right). \end{array}$$

The numbers on the top and left side indicate the possible choices of R and C. The significance of this matrix should be clear. For example, if R (the raven) chooses the card with four dots and C (the parakeet) chooses the card with two dots, then the diet of C is supplemented by $4 - 2 = 2$ worms from the diet of R. If the raven always chooses four, its maximum loss is three worms. If the parakeet always chooses two, its maximum loss is four worms. If both birds play this way, then the raven will lose two worms on each play of the game. Clearly, the raven should occasionally take the risk of losing more than four worms to have a chance at winning. Knowing this, the parakeet should also vary its choice of cards. The theory of matrix games will tell how the raven and parakeet should vary their choices of cards. This is the subject of the final sections of this chapter. Somewhat surprisingly, we will find that the rules of this matrix game are slightly favorable to the raven.

Problems 5.4

1. Describe the matrix games with the following payoff matrices.

(a)
$$\begin{pmatrix} 1 & 0 \\ 0 & 1 \end{pmatrix}.$$

(b)
$$\begin{pmatrix} -1 & 2 \\ -3 & 0 \\ -2 & 3 \end{pmatrix}.$$

(c)
$$\begin{pmatrix} 0 & 0 & 0 \\ 1 & 1 & 1 \\ 2 & 2 & 2 \end{pmatrix}.$$

2. Consider a matrix game with payoff matrix

$$\begin{pmatrix} 1 & -1 & 2 \\ -1 & 0 & -2 \end{pmatrix}.$$

Suppose that R moves first by choosing a row and then C chooses a column, knowing the choice that R has made. What effect will this change in the rules have on the strategies of the competitors?

3. As we have defined a matrix game, the two competitors R and C both have full knowledge of the possible payoffs before the game begins. They move at the same time and do not know what move the competitor has chosen. Are these features of matrix games likely to be true in biological competition? What effects would uncertain knowledge of the outcome have on the choice of strategies?

4. In a two-person, zero-sum game, there are two competitors; the sum of their winnings is zero. The winnings of one competitor are the losses of the other competitor. Some types of biological competition can be

described as non-zero-sum games. In these games, both competitors may win if they cooperate in appropriate ways. Discuss non-zero-sum games in a general way and suggest rules for a game of this type.

5. In a gambling game, the player must choose one of three alternatives to receive a prize. Two of the alternatives give prizes worth 5 cents; the third alternative gives a prize worth 50 cents. Before each play of the game the three prizes are distributed at random among the three alternatives. If the player pays 25 cents to play this game, what is the expected amount that he will win? Is this a two-person, zero-sum game?

5.5 Strategies for Matrix Games

The two players of the $m \times n$ matrix game $A = (a_{ij})$ must analyze the alternative moves and decide which rows or columns to play on successive moves. A *pure strategy* for R (or C) is the decision to play the same row (or column) on every move of the game. Player R (or C) is said to be using a *mixed strategy* if more than one row (or column) is chosen on different moves of the game. If both players employ pure strategies, the outcome of each move is exactly the same and the game is completely predictable. For example, if R always chooses the ith row and C always chooses the jth column, then on every play of the game R receives from C a_{ij} units. When mixed strategies are used by one or both players, the game is more complicated. For example, if R decides to play a mixed strategy, he will randomize his choice of rows in a way that will increase his return. These ideas lead to the following definition.

Definition 5.5.1   Strategy   *A strategy for R in the $m \times n$ matrix game $A = (a_{ij})$ is an m-component probability vector $\mathbf{p} = (p_1, p_2, \ldots, p_m)$, where p_i is the probability that R plays the ith row for $i = 1, 2, \ldots, m$. A strategy for C is an n-component probability vector $\mathbf{q} = (q_1, q_2, \ldots, q_n)$, where q_j is the probability that C plays the jth column for $j = 1, 2, \ldots, n$.*

The players R and C must choose their strategies $\mathbf{p}$ and $\mathbf{q}$. In other words, they must determine the probabilities p_i and q_j that will determine how often they play the various rows and columns. For example, if R and C play the first row and first column of A on every move, they are playing the pure strategies $\mathbf{p} = (1, 0, \ldots, 0)$ and $\mathbf{q} = (1, 0, \ldots, 0)$. If R and C play all rows and columns with equal probabilities, they are playing the mixed strategies

$\mathbf{p} = \left(\dfrac{1}{m}, \dfrac{1}{m}, \ldots, \dfrac{1}{m}\right)$ and $\mathbf{q} = \left(\dfrac{1}{n}, \dfrac{1}{n}, \ldots, \dfrac{1}{n}\right)$. Every m-component probability vector is a possible strategy for R, and every n-component probability vector is a possible strategy for C.

If R and C choose the strategies $\mathbf{p} = (p_1, p_2, \ldots, p_m)$ and $\mathbf{q} = (q_1, q_2, \ldots, q_n)$, then $p_i q_j$ is the probability that, on a given move, R plays the ith row and C plays the jth column. On such a move, the return to R from C is a_{ij} units. The *expected return* $E(\mathbf{p}, \mathbf{q})$ to R from C is defined by

$$E(\mathbf{p}, \mathbf{q}) = \sum_{i=1}^{m} \sum_{j=1}^{n} a_{ij} p_i q_j.$$

This is the expected value of the return to R and C on each move of the matrix game if R plays the strategy $\mathbf{p}$ and C plays the strategy $\mathbf{q}$. If we use vector and matrix notation, the expected return can be written

$$E(\mathbf{p}, \mathbf{q}) = (p_1, p_2, \ldots, p_m) \begin{pmatrix} a_{11} & a_{12} & \cdots & a_{1n} \\ a_{21} & a_{22} & \cdots & a_{2n} \\ \vdots & \vdots & & \vdots \\ a_{m1} & a_{m2} & \cdots & a_{mn} \end{pmatrix} \begin{pmatrix} q_1 \\ q_2 \\ \vdots \\ q_n \end{pmatrix}.$$

In the following, $\mathbf{p}$ will be an m-component row probability vector and $\mathbf{q}$ will be an n-component column probability vector. The expected return to R from C on each move of the matrix game can then be written $E(\mathbf{p}, \mathbf{q}) = \mathbf{p}A\mathbf{q}$.

Example 5.5.1   What are the expected returns to R and C in the 2×2 matrix game

$$A = \begin{pmatrix} -1 & 2 \\ 1 & 0 \end{pmatrix}$$

for the following pairs of strategies?

1. $\mathbf{p} = (\tfrac{1}{2}, \tfrac{1}{2})$, $\mathbf{q} = (1, 0)$.
2. $\mathbf{p} = (0, 1)$, $\mathbf{q} = (\tfrac{3}{4}, \tfrac{1}{4})$.

Solution: For a general 2×2 matrix game,

$$E(\mathbf{p}, \mathbf{q}) = (p_1, p_2)\begin{pmatrix} a_{11} & a_{12} \\ a_{21} & a_{22} \end{pmatrix}\begin{pmatrix} q_1 \\ q_2 \end{pmatrix} = a_{11}p_1q_1 + a_{12}p_1q_2 + a_{21}p_2q_1 + a_{22}p_2q_2.$$

In this example, $E(\mathbf{p}, \mathbf{q}) = -p_1q_1 + 2p_1q_2 + p_2q_1$. Therefore, in case (1), we have $E(\mathbf{p}, \mathbf{q}) = -\frac{1}{2}(1) + 2(\frac{1}{2})(0) + \frac{1}{2}(1) = 0$. This indicates that if C always plays the first column and R plays the two rows with equal probabilities, then the expected return to R is zero. In case (2), we have $E(\mathbf{p}, \mathbf{q}) = -(0)(\frac{3}{4}) + 2(0)(\frac{1}{4}) + 1(\frac{1}{4}) = \frac{1}{4}$. With these strategies, the average return to R is $\frac{1}{4}$ on each play of the game.

Example 5.5.2   In Example 5.4.2, what are the expected returns to R for the following pairs of strategies?

   1. $\mathbf{p} = (\frac{1}{3}, \frac{1}{3}, \frac{1}{3})$, $\mathbf{q} = (1, 0, 0)$.   2. $\mathbf{p} = (\frac{1}{3}, \frac{1}{3}, \frac{1}{3})$, $\mathbf{q} = (\frac{1}{2}, \frac{1}{4}, \frac{1}{4})$.

Solution:

1. With these strategies, the raven R chooses a card at random and the parakeet C always chooses the 2. Then

$$E(\mathbf{p}, \mathbf{q}) = (\tfrac{1}{3}, \tfrac{1}{3}, \tfrac{1}{3})\begin{pmatrix} 4 & -2 & -5 \\ -2 & 8 & -3 \\ -5 & -3 & 14 \end{pmatrix}\begin{pmatrix} 1 \\ 0 \\ 0 \end{pmatrix} = (\tfrac{1}{3}, \tfrac{1}{3}, \tfrac{1}{3})\begin{pmatrix} 4 \\ -2 \\ -5 \end{pmatrix} = -1.$$

    Thus the raven will lose an average of one worm per move to the parakeet.

2. In this case, R chooses a card at random and C chooses the 2 with twice the probability of either 4 or 7. Then

$$E(\mathbf{p}, \mathbf{q}) = (\tfrac{1}{3}, \tfrac{1}{3}, \tfrac{1}{3})\begin{pmatrix} 4 & -2 & -5 \\ -2 & 8 & -3 \\ -5 & -3 & 14 \end{pmatrix}\begin{pmatrix} \tfrac{1}{2} \\ \tfrac{1}{4} \\ \tfrac{1}{4} \end{pmatrix} = (\tfrac{1}{3}, \tfrac{1}{3}, \tfrac{1}{3})\begin{pmatrix} \tfrac{1}{4} \\ \tfrac{1}{4} \\ \tfrac{1}{4} \end{pmatrix} = \tfrac{1}{4}.$$

With these strategies, the raven gains an average of $\frac{1}{4}$ worm per move from the parakeet. Of course, the players are free to change their strategies. For example, in case (1), if the raven observes that the parakeet always chooses the 2, the raven should alter its strategy to take advantage of this fact. If the raven does this, then the parakeet should alter its strategy.

In the $m \times n$ matrix game $A = (a_{ij})$, R should adopt a strategy $\mathbf{p}$ that makes $E(\mathbf{p}, \mathbf{q})$ as large as possible. Of course, R cannot control the strategy $\mathbf{q}$ of C and, therefore, R should choose the strategy $\mathbf{p}_0$, which assures him of the largest possible gain independent of the strategy $\mathbf{q}$ of C. This means that R should choose the strategy $\mathbf{p}_0$, which maximizes the minimum value of $E(\mathbf{p}_0, \mathbf{q})$ over all possible choices of $\mathbf{q}$. In other words, R should choose the strategy $\mathbf{p}_0$, which ensures that $E(\mathbf{p}_0, \mathbf{q})$ is at least u for any strategy $\mathbf{q}$ of C where $\mathbf{p}_0$ is the strategy for which u is as large as possible. The strategy $\mathbf{p}_0$ satisfies $E(\mathbf{p}_0, \mathbf{q}) \geq u$ for every strategy $\mathbf{q}$. Acting in this way, R has an expected return of at least u independent of the strategy of C.

The competitor C should adopt the strategy $\mathbf{q}_0$ which minimizes the maximum loss. This strategy $\mathbf{q}_0$ is chosen to ensure that $E(\mathbf{p}, \mathbf{q}_0)$ is at most w for all $\mathbf{p}$, where w is as small as possible. This means that $E(\mathbf{p}, \mathbf{q}_0) \leq w$ for every strategy $\mathbf{p}$.

Theorem 5.5.1   *The expected return $E(\mathbf{p}_0, \mathbf{q})$ is greater than or equal to u for every strategy $\mathbf{q}$ of C if and only if the components of the n-component row vector $\mathbf{p}_0 A$ are all greater than or equal to u. Similarly, the expected return $E(\mathbf{p}, \mathbf{q}_0)$ is less than or equal to w for every strategy $\mathbf{p}$ of R if and only if the components of the m-component column vector $A\mathbf{q}_c$ are all less than or equal to w.*

Proof: If $\mathbf{p}_0 A$ has all components greater than or equal to u, then $E(\mathbf{p}_0, \mathbf{q}) = \mathbf{p}_0 A \mathbf{q} \geq u$, since the components of $\mathbf{q}$ add to 1. Conversely, if $E(\mathbf{p}_0, \mathbf{q}) \geq u$ for every strategy $\mathbf{q}$ of C, then this is true for the pure strategy $\mathbf{q} = (1, 0, \ldots, 0)$. But, in this case, $E(\mathbf{p}_0, \mathbf{q})$ is equal to the first component of $\mathbf{p}_0 A$. Using all possible pure strategies for C, we conclude that all components of $\mathbf{p}_0 A$ are at least u. The proof of the second part of the theorem is the same and is left as an exercise for the reader.

Theorem 5.5.1 gives a method for deciding the best strategies for R and C. It leads to the following fundamental result of the theory of matrix games due to von Neumann.

Theorem 5.5.2   *For any $m \times n$ matrix game $A = (a_{ij})$, there exist strategies $\mathbf{p}_0$ and $\mathbf{q}_0$ and numbers u and w such that $E(\mathbf{p}_0, \mathbf{q}) \geq u$ for all $\mathbf{q}$ and $E(\mathbf{p}, \mathbf{q}_0) \leq w$ for all $\mathbf{p}$, where u and w are, respectively, the largest and smallest such numbers. Furthermore, the numbers u and w are equal.*

The proof of this theorem is quite difficult and lengthy. Instead of giving a proof, we will develop in the next section a method for explicitly calculating $\mathbf{p}_o$ and $\mathbf{q}_o$. The strategies $\mathbf{p}_o$ and $\mathbf{q}_o$ are called the *optimal strategies* for R and C. The number v, which equals u and w, is called the *value of the game*. It is the expected return to R from C on each move of the game when R and C play the optimal strategies $\mathbf{p}_o$ and $\mathbf{q}_o$.

$$v = E(\mathbf{p}_o, \mathbf{q}_o).$$

If $v = 0$, the game is said to be *fair*. If the matrix game is fair, the average payment is zero when optimal strategies are played.

Example 5.5.3   Verify that the pure strategies $\mathbf{p}_o = (0, 1, 0)$ and $\mathbf{q}_o = (0, 0, 1)$ are optimal for the 3×3 matrix game

$$A = \begin{pmatrix} -2 & -3 & 2 \\ 5 & 4 & 3 \\ 0 & 6 & -1 \end{pmatrix}.$$

Solution:

$$\mathbf{p}_o A = (0, 1, 0) \begin{pmatrix} -2 & -3 & 2 \\ 5 & 4 & 3 \\ 0 & 6 & -1 \end{pmatrix} = (5, 4, 3) \geq (3, 3, 3),$$

$$A\mathbf{q}_o = \begin{pmatrix} -2 & -3 & 2 \\ 5 & 4 & 3 \\ 0 & 6 & -1 \end{pmatrix} \begin{pmatrix} 0 \\ 0 \\ 1 \end{pmatrix} = \begin{pmatrix} 2 \\ 3 \\ -1 \end{pmatrix} \leq \begin{pmatrix} 3 \\ 3 \\ 3 \end{pmatrix}.$$

In this example, $u = w = 3$ or the value v of the game is 3.

If the optimal strategies for R and C are pure strategies, then the matrix game is said to be *strictly determined*. The matrix game in Example 5.5.3 is strictly determined. In this example, the component $a_{23} = 3$ of the game matrix A is the value of the game. Note that this component is a minimum in its row and a maximum in its column. By playing the second row, R guarantees a gain of at least three units and, by playing the third column, C guarantees a loss of no more than three units. In general, a component of the

$m \times n$ game matrix $A = (a_{ij})$ is said to be a *saddle point* if it is a minimum in its row and a maximum in its column.

Theorem 5.5.3   *If $A = (a_{ij})$ is an $m \times n$ game matrix with saddle point a_{kl}, then the optimal strategies for R and C are the pure strategies of playing the kth row and the lth column. The value of the game is $v = a_{kl}$.*

Proof: We must verify that $\mathbf{p}_o = (0, \ldots, 0, 1_k, 0, \ldots, 0)$ and $\mathbf{q}_o = (0, \ldots, 0, 1_l, 0, \ldots, 0)$ are optimal strategies. But $\mathbf{p}_o A = (a_{k1}, a_{k2}, \ldots, a_{kl}, \ldots, a_{kn})$ and, since a_{kl} is the minimum in its row, each component of $\mathbf{p}_o A$ is greater than or equal to a_{kl}. Similarly, each component of $A\mathbf{q}_o$ is less than or equal to a_{kl}. Therefore, $\mathbf{p}_o$ and $\mathbf{q}_o$ are optimal strategies and $v = a_{kl}$ is the value of the game.

The following theorem states that, if the game matrix has more than one saddle point, then the saddle points are equal.

Theorem 5.5.4   *If a_{ij} and a_{kl} are two saddle points for the $m \times n$ game matrix $A = (a_{ij})$, then $a_{ij} = a_{kl}$.*

Proof: Since a_{ij} and a_{kl} are saddle points, $a_{ij} \leq a_{il} \leq a_{kl}$. This follows because a_{ij} and a_{il} are in the same row (a_{ij} is a minimum in its row) and a_{il} and a_{kl} are in the same column (a_{kl} is a maximum in its column). We therefore have $a_{ij} \leq a_{kl}$. By similar reasoning, $a_{kl} \leq a_{kj} \leq a_{ij}$ and $a_{kl} \leq a_{ij}$. For the inequalities $a_{ij} \leq a_{kl}$ and $a_{kl} \leq a_{ij}$ to be valid, we must have $a_{ij} = a_{kl}$, and the proof is complete.

Example 5.5.4   Determine optimal strategies and the values of the following matrix games.

1.
$$A = \begin{pmatrix} 1 & 2 & 3 \\ 0 & 3 & 5 \\ -1 & 2 & -2 \end{pmatrix}.$$

2.
$$B = \begin{pmatrix} 1 & 0 & 5 \\ 1 & -5 & 0 \end{pmatrix}.$$

Solution:

1. The component $a_{11} = 1$ is a saddle point. Optimal strategies are $\mathbf{p}_o = (1, 0, 0)$ and $\mathbf{q}_o = (1, 0, 0)$. The value of the game is $v = a_{11} = 1$.

2. The component $b_{12} = 0$ is a saddle point (a minimum in its row and a maximum in its column). Optimal strategies are $\mathbf{p}_o = (1, 0)$ and $\mathbf{q}_o = (0, 1, 0)$. The value of the game is $v = b_{12} = 0$. This is a fair game.

In Example 5.5.4(2), every component in the first row of B is larger than or equal to the corresponding component in the second row. Clearly, R should never play the second row. For the general $m \times n$ game matrix $A = (a_{ij})$, a *row* is *recessive* if every component is less than or equal to the corresponding component in another row. A *column* is *recessive* if every component is greater than or equal to the corresponding component in another column. The recessive rows and columns can be eliminated from a matrix game since R and C can do as well or better by choosing other rows and columns.

Example 5.5.5   Determine the recessive rows and columns of the following matrix games.

1.
$$A = \begin{pmatrix} 1 & 2 & 3 \\ 4 & 2 & 5 \\ -1 & 2 & 3 \end{pmatrix}.$$

2.
$$B = \begin{pmatrix} -1 & 0 & 1 & 0 \\ 0 & 1 & 2 & 1 \\ 0 & 2 & 3 & 2 \end{pmatrix}.$$

Solution:

1. The first and third rows are recessive and the third column is recessive. The game matrix reduces to $A_1 = (4 \quad 2)$. In this 1×2 matrix, the first column is recessive and the game matrix reduces to $A_2 = (2)$. Note that $a_{22} = 2$ is a saddle point and the optimal strategies are $\mathbf{p}_o = (0, 1, 0)$ and $\mathbf{q}_o = (0, 1, 0)$.
2. The first and second rows and the second, third, and fourth columns are recessive. The game matrix reduces to $B_1 = (0)$, and the optimal strategies are $\mathbf{p}_o = (0, 0, 1)$ and $\mathbf{q}_o = (1, 0, 0, 0)$.

We now consider 2×2 matrix games in general. If

$$A = \begin{pmatrix} a_{11} & a_{12} \\ a_{21} & a_{22} \end{pmatrix}$$

is the game matrix, then we have the following theorems.

Theorem 5.5.5   *The above 2×2 matrix game is not strictly determined if and only if either* (1) a_{11} *and* a_{22} *are greater than both* a_{12} *and* a_{21} *or* (2) a_{11} *and* a_{22} *are less than both* a_{12} *and* a_{21}.

Proof: To prove that the game is not strictly determined, we must show that there is no saddle point, that is, no component of A which is a minimum in its row and a maximum in its column. Consider a_{11}. In case (1), it is a maximum in its column but not a minimum in its row. In case (2), it is a minimum in its row but not a maximum in its column. Similarly, a_{12}, a_{21}, and a_{22} are not saddle points in either case (1) or case (2).

Theorem 5.5.6 *If the above 2 × 2 matrix game is not strictly determined, then the optimal strategies for R and C are*

$$\mathbf{p}_o = \left(\frac{a_{22} - a_{21}}{a_{11} + a_{22} - a_{12} - a_{21}}, \frac{a_{11} - a_{12}}{a_{11} + a_{22} - a_{12} - a_{21}} \right)$$

and

$$\mathbf{q}_o = \left(\frac{a_{22} - a_{12}}{a_{11} + a_{22} - a_{12} - a_{21}}, \frac{a_{11} - a_{21}}{a_{11} + a_{22} - a_{12} - a_{21}} \right).$$

The value of the game is

$$v = \frac{a_{11}a_{22} - a_{12}a_{21}}{a_{11} + a_{22} - a_{12} - a_{21}}.$$

Proof: Applying Theorem 5.5.2, we calculate $\mathbf{p}_o A$ and $A\mathbf{q}_o$.

$$\mathbf{p}_o A = \left(\frac{a_{22} - a_{21}}{a_{11} + a_{22} - a_{12} - a_{21}}, \frac{a_{11} - a_{12}}{a_{11} + a_{22} - a_{12} - a_{21}} \right) \begin{pmatrix} a_{11} & a_{12} \\ a_{21} & a_{22} \end{pmatrix}$$

$$= \left(\frac{a_{11}a_{22} - a_{12}a_{21}}{a_{11} + a_{22} - a_{12} - a_{21}}, \frac{a_{11}a_{22} - a_{12}a_{21}}{a_{11} + a_{22} - a_{12} - a_{21}} \right)$$

$$A\mathbf{q}_o = \begin{pmatrix} a_{11} & a_{12} \\ a_{21} & a_{22} \end{pmatrix} \begin{pmatrix} \dfrac{a_{22} - a_{12}}{a_{11} + a_{22} - a_{12} - a_{21}} \\[2ex] \dfrac{a_{11} - a_{21}}{a_{11} + a_{22} - a_{12} - a_{21}} \end{pmatrix} = \begin{pmatrix} \dfrac{a_{11}a_{22} - a_{12}a_{21}}{a_{11} + a_{22} - a_{12} - a_{21}} \\[2ex] \dfrac{a_{11}a_{22} - a_{12}a_{21}}{a_{11} + a_{22} - a_{12} - a_{21}} \end{pmatrix}.$$

We conclude that each component of $\mathbf{p}_o A$ is equal to v and each component of $A\mathbf{q}_o$ is equal to v where

$$v = \frac{a_{11}a_{22} - a_{12}a_{21}}{a_{11} + a_{22} - a_{12} - a_{21}}.$$

This verifies that $\mathbf{p}_o$ and $\mathbf{q}_o$ are optimal strategies and v is the value of the game.

Example 5.5.6 Determine optimal strategies and values of the following 2×2 matrix games.

1. $A = \begin{pmatrix} 2 & -1 \\ -1 & 2 \end{pmatrix}.$

2. $B = \begin{pmatrix} 1 & 3 \\ 2 & -1 \end{pmatrix}.$

Solution:

1. $\mathbf{p}_o = (\frac{1}{2}, \frac{1}{2})$, $\mathbf{q}_o = (\frac{1}{2}, \frac{1}{2})$, $v = \frac{1}{2}$.
2. $\mathbf{p}_o = (\frac{4}{5}, \frac{1}{5})$, $\mathbf{q}_o = (\frac{3}{5}, \frac{2}{5})$, $v = \frac{7}{5}$.

The problem of calculating the optimal strategies of R and C for a general $m \times n$ matrix will be considered in the next section. We conclude this section by considering the following examples of a learning experiment and a decision problem that can be interpreted as matrix games.

Example 5.5.7   A Learning Experiment[4] Experiments with monkeys, rats, pigeons, turtles, and fish were designed to test qualitative differences in intelligence. In these experiments, the animal must choose between two panels on which various colors or patterns are projected. A "correct" choice is rewarded with a unit of food. If the alternative choices are I and II, and if the choice I is always rewarded, then the animals learn after a characteristic average number of errors to always choose I. This can be described as a simple 2×2 matrix game with payoff matrix

$$A = \begin{pmatrix} 1 & 0 \\ 0 & 1 \end{pmatrix}.$$

The animal chooses the first row (panel I) and is rewarded when the experimenter chooses the first column (panel I). If choice I is always rewarded,

---

[4] M. E. Bitterman, "The Evolution of Intelligence," *Scientific American*, 212:92–100 (Jan. 1965).

the experimenter is playing the strategy $\mathbf{q} = (1, 0)$. In this case, the best strategy for the animal is $\mathbf{p} = (1, 0)$.

Other types of learning are involved when the experimenter begins to reward the two choices randomly. Suppose that I is rewarded on 70 per cent of the trials chosen randomly and II is rewarded on the remaining trials. This is the strategy $\mathbf{q} = (.7, .3)$. The best strategy for the animal is again $\mathbf{p} = (1, 0)$. If the strategy $\mathbf{p} = (p_1, 1 - p_1)$ is used, the expected return is

$$E(\mathbf{p}, \mathbf{q}) = (p_1, 1 - p_1)\begin{pmatrix} 1 & 0 \\ 0 & 1 \end{pmatrix}\begin{pmatrix} .7 \\ .3 \end{pmatrix} = (p_1, 1 - p_1)\begin{pmatrix} .7 \\ .3 \end{pmatrix} = .7p_1 + .3 - .3p_1$$

$$= .3 + .4p_1.$$

This is as large as possible when $p_1 = 1$. Some of the experimental animals did adopt this strategy, but others consistently chose the alternative that had been rewarded on the previous trial. This procedure results in a strategy $\mathbf{p} = (.7, .3)$ with expected return $E(\mathbf{p}, \mathbf{q}) = (.7)(.7) + (.3)(.3) = .58$.

If both alternatives are rewarded on 50 per cent of the trials chosen randomly, then $\mathbf{q} = (.5, .5)$ and every strategy $\mathbf{p} = (p_1, 1 - p_1)$ produces the same expected return: $E(\mathbf{p}, \mathbf{q}) = (.5)p_1 + (.5)(1 - p_1) = .5$. The experiments reported by Bitterman indicated interesting qualitative differences in the responses of the different species.

Example 5.5.8   Assessing Risk in Medical Procedures   As is well known, many medical procedures involve substantial risk to the patient and should only be undertaken when the patient is exposed to greater risk if no treatment is given. How do we decide in a given situation which risk is greater? This problem is complicated further when it is not completely certain that the patient does have the disease suspected. For example, surgery is often undertaken to remove tumors even when there is only a relatively small probability that the tumor will be found to be malignant. How large must this probability be before surgery can be recommended?

To analyze this question, suppose that the probability that a patient has a particular disease is q_1. (This probability has been arrived at by performing various tests.) The treatment for this disease is a serious operation. If the patient has the disease but does not have the operation, he can expect to live 5 years, but if he has the operation he can expect to live 20 years. If the patient does not have the disease, he can expect to live 25 years if he has the operation and 30 years if he does not have the operation. The decision to have the operation or not clearly depends on q_1, the probability that the patient has the disease. If $q_1 = 0$, the patient does not have the disease and should not have the operation. If $q_1 = 1$, the patient has the disease and should have the operation. What is the smallest value of q_1 for which the operation is advisable?

This problem can be analyzed as a matrix game. Define

$$A = \begin{pmatrix} 20 & 25 \\ 5 & 30 \end{pmatrix}$$

to be the game matrix. The patient "plays" the rows; row I corresponding to having the operation and row II corresponding to not having the operation. His opponent, nature, plays the columns; column I corresponding to the disease and column II to no disease. The strategy of nature is $\mathbf{q} = (q_1, q_2)$, where q_1 is the probability that the patient has the disease. The patient must play a pure strategy, but for the moment let us suppose that his strategy is $\mathbf{p} = (p_1, p_2)$. The expected return (in years of life) is

$$\begin{aligned} E(\mathbf{p}, \mathbf{q}) &= 20p_1q_1 + 25p_1q_2 + 5p_2q_1 + 30p_2q_2 \\ &= 20p_1q_1 + 25p_1(1 - q_1) + 5(1 - p_1)q_1 + 30(1 - p_1)(1 - q_1) \\ &= 20p_1q_1 - 5p_1 - 25q_1 + 30. \end{aligned}$$

If the patient has the operation, then $\mathbf{p} = (1, 0)$ and $E(\mathbf{p}, \mathbf{q}) = 25 - 5q_1$. If he does not have the operation, then $\mathbf{p} = (0, 1)$ and $E(\mathbf{p}, \mathbf{q}) = 30 - 25q_1$. The patient should have the operation if $25 - 5q_1 > 30 - 25q_1$, that is, if $20q_1 > 5$ or $q_1 > .25$. This means that the patient should have the operation if the probability that he has the disease is greater than 25 per cent. If the information available indicates that the probability of the disease is, say, 15 per cent, the operation should not be performed. More information must be obtained before the operation can be recommended.

Problems 5.5

1. If all components of a game matrix are positive, prove that the value of the game is positive.

2. Given two $m \times n$ matrix games $A = (a_{ij})$ and $B = (b_{ij})$ such that $b_{ij} = a_{ij} + k$ for all i and j, prove that the value of the game B is equal to the value of the game A plus the constant k. Prove that the optimal strategies for R and C are the same for game B as for game A. (The games A and B are said to be *equivalent matrix games*.)

3. What are the optimal strategies and values of the equivalent matrix games

$$A = \begin{pmatrix} 3 & 2 \\ 2 & -3 \end{pmatrix} \quad \text{and} \quad B = \begin{pmatrix} 5 & 4 \\ 4 & -1 \end{pmatrix}?$$

4. What are the optimal strategies and values of the equivalent matrix games

$$A = \begin{pmatrix} 1 & 1 & 1 \\ 2 & 2 & 3 \\ -2 & 4 & 5 \end{pmatrix} \quad \text{and} \quad B = \begin{pmatrix} 0 & 0 & 0 \\ 1 & 1 & 2 \\ -3 & 3 & 4 \end{pmatrix}?$$

5. Determine the saddle points and optimal strategies of the following matrix games.

(a) $\begin{pmatrix} 1 & 0 & 0 \\ 0 & -1 & 0 \end{pmatrix}$. (b) $\begin{pmatrix} 1 & 1 & 1 \\ 0 & 2 & 2 \end{pmatrix}$. (c) $\begin{pmatrix} 1 & -1 & 4 \\ 0 & -1 & 2 \\ 3 & -2 & 0 \end{pmatrix}$.

6. Determine optimal strategies for the following 2 × 2 matrix games.

(a) $\begin{pmatrix} 1 & 0 \\ -2 & 2 \end{pmatrix}$. (b) $\begin{pmatrix} \frac{1}{2} & -\frac{1}{2} \\ -\frac{1}{2} & \frac{1}{2} \end{pmatrix}$. (c) $\begin{pmatrix} -1 & -1 \\ 0 & -2 \end{pmatrix}$. (d) $\begin{pmatrix} \frac{1}{2} & \frac{3}{2} \\ \frac{1}{2} & 1 \end{pmatrix}$.

7. Prove that the 2 × 2 matrix game

$$\begin{pmatrix} \sin t & \cos t \\ -\cos t & \sin t \end{pmatrix}$$

is strictly determined for $0 \le t \le \pi/4$ but not strictly determined for $\pi/4 < t \le \pi/2$. Determine the value of the game as a function of t for $0 \le t \le \pi/2$.

8. As the variable t increases from 0 to 1, how do the optimal strategies change in the following games?

(a) $\begin{pmatrix} t & 0 \\ 0 & 1-t \end{pmatrix}$. (b) $\begin{pmatrix} 1 & 2t \\ 0 & t \end{pmatrix}$. (c) $\begin{pmatrix} t & t^2 \\ \frac{1}{2} & \frac{1}{4} \end{pmatrix}$.

9. Determine the values of the 2 × 2 matrix games of Problem 8 as functions of t for $0 \le t \le 1$. For which values of t are these games fair?

10. Prove that the 2×2 matrix game

$$\begin{pmatrix} t & 1 - t \\ 1 - t & t \end{pmatrix}$$

is not strictly determined for any value of t. Prove that the value of this game is a constant independent of t.

11. Which of the following 2×2 matrix games are fair?

(a) $\begin{pmatrix} 0 & 1 \\ -1 & 0 \end{pmatrix}$.   (b) $\begin{pmatrix} 1 & -1 \\ -1 & 1 \end{pmatrix}$.   (c) $\begin{pmatrix} 5 & -3 \\ -5 & 3 \end{pmatrix}$.

12. In an experiment, a monkey must choose one of three panels to receive a banana. On each trial of the experiment, the experimenter places either two bananas behind panel I or one banana behind both panel II and panel III. (These are the two "moves" of the experimenter.) Describe this experiment as a 3×2 matrix game. What are the optimal strategies of the monkey and the experimenter?

13. In a certain farming region, the average weather during the growing season is either cool or hot. Two crops are to be planted on a farm of 1500 acres. If the growing season is cool, the expected profits are $20 per acre from crop I and $10 per acre from crop II. If the growing season is hot, the expected profits are $10 per acre from crop I and $30 per acre from crop II. Describe the competition between the farmer and the weather as a matrix game. If no information is available concerning the probabilities of hot or cool weather, what is the optimal strategy of the farmer?

14. Suppose that the weather in Problem 13 is equally likely to be hot or cool. How many acres of each crop should the farmer plant?

15. Two possible strategies of foraging for food are (a) to minimize the time spent foraging and (b) to maximize the net energy return from foraging. What are some possible situations where one strategy would obviously be better than the other? For example, which strategy would be better for an animal exposed to predators during foraging?

16. Certain species of butterflies are distasteful to predators and are therefore avoided as food sources. By natural selection, other butterfly species have evolved to appear very similar to distasteful species. This mimicry of a "model" species can be thought of as a "strategy" for survival. When is this strategy a good one? What can be expected to happen when the mimic species becomes much more numerous than the model species?

17. Suppose that the game matrix in Example 5.5.8 is

$$A = \begin{pmatrix} 10 & 39 \\ 5 & 40 \end{pmatrix}.$$

What is the minimum value of the probability that the patient has the disease for which the operation can be recommended?

18. The model of medical decision making in Example 5.5.8 can be studied more generally. Define the game matrix

$$A = \begin{pmatrix} a_{11} & a_{12} \\ a_{21} & a_{22} \end{pmatrix},$$

where a_{11}, a_{12}, a_{21}, and a_{22} are the expected lengths of life if the patient has the operation and the disease, the operation and no disease, the disease and no operation, or no disease and no operation. Assuming that the patient wishes to maximize his expected length of life, prove that the operation can be recommended if the probability that the patient has the disease is greater than

$$\frac{a_{22} - a_{12}}{a_{11} + a_{22} - a_{12} - a_{21}}.$$

(*Hint:* Compare the patient's expected life when he has the operation to his expected length of life when he does not have the operation.)

5.6 Matrix Games and Linear Programming

The problem of this section is to determine the optimal strategies of the two competitors, R and C, in a general $m \times n$ matrix game with payoff matrix $A = (a_{ij})$. We may assume that the game matrix A has no recessive rows or columns, since these will never be chosen when optimal strategies are used. To avoid a small problem, we will also assume that all components of A are positive. If this is not the case, define a game matrix $B = (b_{ij}) = (a_{ij} + k)$ by adding a positive constant k to all components of A. The constant k is chosen so that the components of B are positive. The value of the $m \times n$ matrix game determined by B is positive (Problems 5.5.1 and 5.5.2).

Theorems 5.5.1 and 5.5.2 imply that the optimal strategy $\mathbf{q}_0$ of C satisfies the condition that all components of $A\mathbf{q}_0$ are less than or equal to v, the value of the game. Equivalently, all components of $(1/v)A\mathbf{q}_0$ are less than or equal to 1. For $i = 1, 2, \ldots, n$, define x_i to be the ith component of $(1/v)\mathbf{q}_0$. Note that $x_i \geq 0$, since $v > 0$ and $\mathbf{q}_0$ is a probability vector. Define the objective function $f = x_1 + x_2 + \cdots + x_n$. Then $f = 1/v$, since $\mathbf{q}_0$ is a probability vector. The competitor C attempts to minimize v, the expected return to R from C. This is done by maximizing $f = 1/v$ subject to the constraints. Therefore, the problem that C must solve is the following. Maximize

$$f = x_1 + x_2 + \cdots + x_n$$

subject to

$$a_{11}x_1 + a_{12}x_2 + \cdots + a_{1n}x_n \leq 1$$
$$a_{21}x_1 + a_{22}x_2 + \cdots + a_{2n}x_n \leq 1$$
$$\vdots \qquad \vdots \qquad\qquad \vdots \qquad \vdots$$
$$a_{m1}x_1 + a_{m2}x_2 + \cdots + a_{mn}x_n \leq 1$$
$$x_1 \geq 0, \quad x_2 \geq 0, \ldots, x_n \geq 0.$$

We recognize this problem as a maximum problem in linear programming. By a similar argument, the reader may verify that R must solve the dual minimum problem to determine his optimal strategy. The initial simplex tableau for these problems is

x_1	x_2		x_n	s_1	s_2		s_m		
a_{11}	a_{12}	$\cdots$	a_{1n}	1	0	$\cdots$	0	1	s_1
a_{21}	a_{22}	$\cdots$	a_{2n}	0	1	$\cdots$	0	1	s_2
$\vdots$	$\vdots$		$\vdots$	$\vdots$	$\vdots$		$\vdots$	$\vdots$	
a_{m1}	a_{m2}	$\cdots$	a_{mn}	0	0	$\cdots$	1	1	s_m
1	1	$\cdots$	1	0	0	$\cdots$	0	f	

The positive indicators in the last row are eliminated successively by the simplex method. By this method, the initial simplex tableau is reduced to a terminal tableau. The value v of the game is the reciprocal of the maximum value of f. The solutions $\mathbf{x}^* = (x_1, x_2, \ldots, x_n)$ and $\mathbf{y}^* = (y_1, y_2, \ldots, y_m)$ of the maximum problem and dual minimum problem can be read off the last column and last row of the terminal tableau. The optimal strategies for C and R are then $\mathbf{q}_0 = v\mathbf{x}^* = (vx_1, vx_2, \ldots, vx_n)$ and $\mathbf{p}_0 = v\mathbf{y}^* = (vy_1, vy_2, \ldots, vy_m)$.

Example 5.6.1 Determine the optimal strategies of the 2×2 matrix game with payoff matrix

$$A = \begin{pmatrix} 1 & 2 \\ 2 & 3 \end{pmatrix}.$$

Solution: The components of A are all positive, and the methods of this section can be applied. The associated maximum problem is to maximize $f = x_1 + x_2$ subject to $x_1 + 2x_2 \le 1$, $2x_1 + 3x_2 \le 1$, $x_1 \ge 0$, and $x_2 \ge 0$. The initial simplex tableau of this problem is

x_1	x_2	s_1	s_2		
1	2	1	0	1	s_1
②	3	0	1	1	s_2
1	1	0	0	f	

Pivoting on the circled component, we obtain the following tableau:

x_1	x_2	s_1	s_2		
0	$\frac{1}{2}$	1	$-\frac{1}{2}$	$\frac{1}{2}$	s_1
1	$\frac{3}{2}$	0	$\frac{1}{2}$	$\frac{1}{2}$	x_1
0	$-\frac{1}{2}$	0	$-\frac{1}{2}$	$f - \frac{1}{2}$	

This is a terminal tableau, since there are no positive indicators. We conclude that the maximum f subject to the constraints is $\frac{1}{2}$ and, therefore, the value of the game is $1/\frac{1}{2} = 2$. The maximum of f occurs at $\mathbf{x}^* = (\frac{1}{2}, 0)$ and the minimum of the dual problem occurs at $\mathbf{y}^* = (0, \frac{1}{2})$. The optimal strategies for R and C are $\mathbf{p}_0 = v\mathbf{y}^* = (0, 1)$ and $\mathbf{q}_0 = v\mathbf{x}^* = 2(\frac{1}{2}, 0) = (1, 0)$.

Example 5.6.1 could be solved by the methods of the previous section for general 2×2 matrix games or by observing that the component in the first column and second row is a saddle point. The methods of this section can be applied to all $m \times n$ matrix games. To end this section, we consider two examples of game matrices with negative components.

Example 5.6.2 Determine the optimal strategies of R and C in the 2×2 matrix game with payoff matrix

$$A = \begin{pmatrix} 2 & 1 \\ -2 & 2 \end{pmatrix}.$$

Solution: By adding 3 to each component of A, define a new game matrix

$$B = \begin{pmatrix} 5 & 4 \\ 1 & 5 \end{pmatrix}$$

with all components positive. The optimal strategies in the game B are the same as the optimal strategies in the game A. To determine these optimal strategies, we write the initial simplex tableau.

x_1	x_2	s_1	s_2		
⑤	4	1	0	1	s_1
1	5	0	1	1	s_2
1	1	0	0	f	

Pivoting on the circled components leads to the following tableaux:

x_1	x_2	s_1	s_2		
1	$\frac{4}{5}$	$\frac{1}{5}$	0	$\frac{1}{5}$	x_1
0	$\left(\frac{21}{5}\right)$	$-\frac{1}{5}$	1	$\frac{4}{5}$	s_2
0	$\frac{1}{5}$	$-\frac{1}{5}$	0	$f - \frac{1}{5}$	

x_1	x_2	s_1	s_2		
1	0	$\frac{5}{21}$	$-\frac{4}{21}$	$\frac{1}{21}$	x_1
0	1	$-\frac{1}{21}$	$\frac{5}{21}$	$\frac{4}{21}$	x_2
0	0	$-\frac{4}{21}$	$-\frac{1}{21}$	$f - \frac{5}{21}$	

The final tableau is terminal, since all indicators are nonpositive. The maximum of f is $\frac{5}{21}$, which occurs at $\mathbf{x}^* = (\frac{1}{21}, \frac{4}{21})$. The value of the game B is, therefore, $\frac{21}{5}$. The value of the original game A is $\frac{21}{5} - 3 = \frac{6}{5}$. The optimal strategies for R and C are

$$\mathbf{p}_o = \tfrac{21}{5}(\tfrac{4}{21}, \tfrac{1}{21}) = (\tfrac{4}{5}, \tfrac{1}{5}) \quad \text{and} \quad \mathbf{q}_o = \tfrac{21}{5}(\tfrac{1}{21}, \tfrac{4}{21}) = (\tfrac{1}{5}, \tfrac{4}{5}).$$

Example 5.6.3 Determine the optimal strategies of R and C in the 3×3 matrix game with payoff matrix

$$A = \begin{pmatrix} 4 & -2 & -5 \\ -2 & 8 & -3 \\ -5 & -3 & 14 \end{pmatrix}.$$

Solution: This is the matrix game of Example 5.4.2. Since it is not obvious that the value of this game is positive, we define an equivalent game matrix

$$B = \begin{pmatrix} 10 & 4 & 1 \\ 4 & 14 & 3 \\ 1 & 3 & 20 \end{pmatrix}$$

with all components positive by adding 6 to each component of A. The optimal strategies for R and C are the same in the equivalent games A and B. The initial simplex tableau associated with the matrix game B is

I

x_1	x_2	x_3	s_1	s_2	s_3		
⑩	4	1	1	0	0	1	s_1
4	14	3	0	1	0	1	s_2
1	3	20	0	0	1	1	s_3
1	1	1	0	0	0	f	

Pivoting on the circled components in succession leads to the following tableaux:

II

x_1	x_2	x_3	s_1	s_2	s_3		
1	$\frac{4}{10}$	$\frac{1}{10}$	$\frac{1}{10}$	0	0	$\frac{1}{10}$	x_1
0	$\left(\frac{124}{10}\right)$	$\frac{26}{10}$	$-\frac{4}{10}$	1	0	$\frac{6}{10}$	s_2
0	$\frac{26}{10}$	$\frac{199}{10}$	$-\frac{1}{10}$	0	1	$\frac{9}{10}$	s_3
0	$\frac{6}{10}$	$\frac{9}{10}$	$-\frac{1}{10}$	0	0	$f - \frac{1}{10}$	

III

x_1	x_2	x_3	S_1	S_2	S_1		
1	0	$\frac{1}{62}$	$\frac{7}{62}$	$\frac{13}{155}$	0	$\frac{5}{62}$	x_1
0	1	$\frac{13}{62}$	$-\frac{1}{31}$	$\frac{5}{62}$	0	$\frac{3}{62}$	x_2
0	0	$\left(\frac{600}{31}\right)$	$-\frac{1}{62}$	$-\frac{13}{62}$	1	$\frac{24}{31}$	S_3
0	0	$\frac{24}{31}$	$-\frac{5}{62}$	$-\frac{3}{62}$	0	$f-\frac{4}{31}$	

IV

x_1	x_2	x_3	S_1	S_2	S_3		
1	0	0	*	*	*	$\frac{2}{25}$	x_1
0	1	0	*	*	*	$\frac{1}{25}$	x_2
0	0	1	$-\frac{1}{1200}$	$-\frac{13}{1200}$	$\frac{31}{600}$	$\frac{1}{25}$	x_3
0	0	0	$-\frac{2}{25}$	$-\frac{1}{25}$	$-\frac{1}{25}$	$f-\frac{4}{25}$	

The reader is invited to complete the terminal tableau (IV). These numbers are not needed in the final step. The maximum of the objective function f is $\frac{4}{25}$ and, therefore, the value of the game B is $\frac{25}{4}$. The value of the original game A is $\frac{25}{4} - 6 = \frac{1}{4}$. The optimal strategy for C is

$$\mathbf{q}_0 = v\mathbf{x}^* = \tfrac{25}{4}\left(\tfrac{2}{25}, \tfrac{1}{25}, \tfrac{1}{25}\right) = \left(\tfrac{1}{2}, \tfrac{1}{4}, \tfrac{1}{4}\right).$$

The optimal strategy for R is

$$\mathbf{p}_0 = v\mathbf{y}^* = \tfrac{25}{4}\left(\tfrac{2}{25}, \tfrac{1}{25}, \tfrac{1}{25}\right) = \left(\tfrac{1}{2}, \tfrac{1}{4}, \tfrac{1}{4}\right).$$

To interpret this result in terms of the experiment of Example 5.4.2, the raven can ensure an average return of at least one-fourth worm per move if it plays the strategy $(\tfrac{1}{2}, \tfrac{1}{4}, \tfrac{1}{4})$; that is, choose rows at random but choose the first row twice as frequently as either of the other two rows. The parakeet by playing the columns with the strategy $(\tfrac{1}{2}, \tfrac{1}{4}, \tfrac{1}{4})$ can ensure an average loss of at most one-fourth worm per move.

In Example 5.5.2(2), we observed that, if the parakeet plays the optimal strategy $\mathbf{q}_0 = (\tfrac{1}{2}, \tfrac{1}{4}, \tfrac{1}{4})$, the raven can do just as well by choosing a card at random, that is, by using the strategy $\mathbf{p} = (\tfrac{1}{3}, \tfrac{1}{3}, \tfrac{1}{3})$. The expected return from these strategies is $E(\mathbf{p}, \mathbf{q}_0) = \tfrac{1}{4}$. However, the strategy $\mathbf{p} = (\tfrac{1}{3}, \tfrac{1}{3}, \tfrac{1}{3})$ is not optimal for the raven, since the parakeet can improve its position by playing a different (nonoptimal) strategy. For example, if $q = (1, 0, 0)$, then

$$E(\mathbf{p}, \mathbf{q}) = (\tfrac{1}{3}, \tfrac{1}{3}, \tfrac{1}{3}) \begin{pmatrix} 4 & -2 & -5 \\ -2 & 8 & -3 \\ -5 & -3 & 14 \end{pmatrix} \begin{pmatrix} 1 \\ 0 \\ 0 \end{pmatrix} = -1.$$

Similarly, if the parakeet adopts the strategy $q = (1, 0, 0)$, the raven can take advantage of this by adopting the strategy $\mathbf{p} = (0, 0, 1)$. This illustrates that, in the long run, the two competitors will do best by playing their optimal strategies.

Problems 5.6

1. By the methods of linear programming, determine the optimal strategies and values of the following 2×2 matrix games.

(a) $\begin{pmatrix} 2 & 4 \\ 3 & -2 \end{pmatrix}$.   (b) $\begin{pmatrix} 1 & 0 \\ -1 & 1 \end{pmatrix}$.   (c) $\begin{pmatrix} -2 & 1 \\ 1 & -2 \end{pmatrix}$.

2. By the methods of linear programming, determine the optimal strategies and values of the following 3×3 matrix games.

(a) $\begin{pmatrix} 1 & 0 & 1 \\ -1 & 2 & 0 \\ 2 & -1 & -1 \end{pmatrix}$.   (b) $\begin{pmatrix} 2 & 3 & 2 \\ 1 & -2 & -1 \\ 4 & -1 & 0 \end{pmatrix}$.

3. The linear programming method of this section determines the optimal strategies when the value of the matrix game is positive. Why does the method not work when the value of the game is zero or negative?

4. The values of the matrix games of Examples 5.6.2 and 5.6.3 were found to be $\tfrac{6}{5}$ and $\tfrac{1}{4}$. Since the values are positive, it was not necessary to define equivalent matrix games with components all positive. Apply the linear programming method to the original game matrices to determine the optimal strategies for both examples.

5. Determine the optimal strategies and value of the matrix game

$$A = \begin{pmatrix} 2 & 4 & 0 & 0 \\ -1 & 7 & 4 & -1 \\ 2 & -1 & 17 & -1 \\ 5 & 2 & -1 & 0 \end{pmatrix}.$$

6. Determine the optimal strategies and value of the matrix game

$$A = \begin{pmatrix} 1 & -1 & 1 & -1 & 1 \\ 0 & 1 & -2 & 2 & -2 \\ 0 & 0 & 1 & -3 & 3 \\ 0 & 0 & 0 & 1 & -4 \\ 0 & 0 & 0 & 0 & 1 \end{pmatrix}.$$

7. The strategies adopted by animals in their struggle for survival are extremely complex. They include migration, choice of foraging periods, camouflage, hibernation, mimicry, and other physiological and behavioral responses to the conditions of competition. What features of matrix games are relevant to the study of general competition processes? How would concepts such as strategy, fairness, payoff, outcome, and value be defined in the general case?

8. Develop a game theory model of a learning experiment in which an animal must choose between four alternatives to receive a reward. Describe some of the types of nonoptimal behavior that may be observed. In what ways may behavior change as the differences in the rewards for the various alternatives become (a) smaller and (b) larger?

9. Determine the optimal strategies of the matrix game

$$\begin{pmatrix} 10 & 9 \\ 0 & 100 \end{pmatrix}.$$

The optimal strategies guarantee a minimum return to R over many moves of the game. If the game is to be played only once, how does R determine his move? Under what conditions will R choose the first row to guarantee a return of at least nine units? When will R gamble by choosing the second row?

Difference Equations 6

6.1 Introduction

In this and the following chapter we develop some of the mathematical techniques capable of describing the dynamics of biological systems. How does one biological variable change when another variable changes? We may be interested in studying the changes over time of the population of a particular species in a certain environment. The biological variable in which we are interested may be a function of variables such as temperature, humidity, or food abundance. The mathematical methods introduced in this chapter can be applied to these problems. However, for definiteness, the biological variable in most cases will be the population of a species in a given environment as a function of time.

We can construct both discrete and continuous models of time-dependent processes. In a discrete model, time is recorded as a discrete variable and observations are made only at certain fixed intervals of time. For example, a census may be taken hourly, annually, or every 10 years. In a continuous model, time is a continuous variable and the population is thought of as changing continuously with time. In this chapter, we develop the mathematics appropriate for discrete models. Continuous models are discussed in Chapter 7.

In the discrete models of population growth of this chapter, x_n will represent the population at the end of the nth time period. The initial or starting population which is usually assumed to be known, is denoted x_0. After one time period, the population is x_1. The population after two time periods is x_2, and so on. The development over time of the population is described by the sequence of numbers $x_0, x_1, x_2, x_3, \ldots, x_n, x_{n+1}, \ldots$.

Example 6.1.1   Suppose that a population grows according to the formula $x_n = 1000 + 500(1 - 2^{-n})$. When $n = 0$, the initial population is $x_0 = 1000 + 500(1 - 2^{-0}) = 1000$. When $n = 1$, the population is $x_1 = 1000 + 500 \times (1 - 2^{-1}) = 1000 + 250 = 1250$. After two time intervals, the population is $x_2 = 1000 + 500(1 - 2^{-2}) = 1375$. Since 2^{-n} becomes small as n gets large, the term $1 - 2^{-n}$ approaches 1 as n increases. This implies that, as n becomes large, x_n approaches a limiting or equilibrium population of 1500. The population growth in the nth time period is $x_n - x_{n-1} = [1000 + 500(1 - 2^{-n})] - [1000 + 500(1 - 2^{-n+1})] = 500(2)^{-n}$. This approaches zero as n increases. This growth process is illustrated in Figure 6.1.

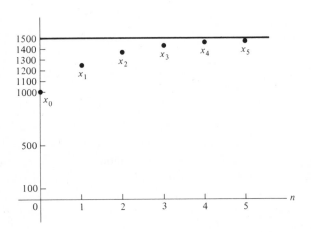

Figure 6.1. $x_n = 1000 + 500\,(1 - 2^{-n})$

In Example 6.1.1, we are given the formula for x_n and, therefore, we know the population at the end of every time period. More often, we know only the initial population and we may have some information about the rates of growth of the population in the various time periods. The problem is then to use this information to determine an explicit formula for x_n. For example, we may have an estimate for the growth $x_n - x_{n-1}$ of the population in the nth period. Can we use this information to determine x_n? These ideas lead to the following definition.

Definition 6.1.1 Difference Equation *A difference equation is an equation that relates the values of x_n for different values of n. If N_1 and N_2 are the largest and smallest values of n that occur in the equation, then the order of the difference equation is $N_1 - N_2$.*

Example 6.1.2 The following are examples of difference equations.

1. $x_n - x_{n-1} = 2^{-n}$. Here $N_1 = n$ and $N_2 = n - 1$. The order is $N_1 - N_2 = n - (n - 1) = 1$. This is a first order difference equation.
2. $x_{n+1} = 2^{-n}x_n + (x_{n-1})^2$. This is a second order equation, since $N_1 = n + 1$, $N_2 = n - 1$, and $N_1 - N_2 = 2$.
3. $2x_{n+2} + 3x_{n+1} = \sin(x_{n+1})$ first order.
4. $2x_{n+2} + 3x_{n+1} + x_n = 0$ second order.
5. $(x_{n+3})^2 + x_n = 5$ third order.

Example 6.1.3 An insect population grows in such a way that the growth in the nth time period is twice the growth in the preceding time period. Describe this growth process by means of a difference equation. What is the order of this equation?

Solution: Define x_n to be the population after n time periods. The growth in the nth time period is $x_n - x_{n-1}$ and the growth in the $(n - 1)$st time period is $x_{n-1} - x_{n-2}$. We are given that $x_n - x_{n-1} = 2(x_{n-1} - x_{n-2})$. This is a second order difference equation, which can also be written $x_n - 3x_{n-1} + 2x_{n-2} = 0$.

Example 6.1.4 Cattle are fed in order to maximize their weight at the time of slaughter. Under certain conditions, the weight of an average cow increases by 5 per cent each week. Describe this weight increase by means of a difference equation. What is the order of this equation?

Solution: Define w_n to be the weight of an average cow after n weeks. After $(n + 1)$ weeks, the weight w_n has increased by 5 per cent. Therefore, we have the equation $w_{n+1} = (1.05)w_n$. This is a first order difference equation.

In the remaining sections of this chapter, we develop methods for solving several important types of difference equations. Given a difference equation, can we determine an explicit formula for x_n? If we can determine such a formula, it will be called the *solution* of the difference equation.

Problems 6.1

1. By drawing graphs, illustrate the discrete growth processes from the initial time $n = 0$, given by the following formulas. Does x_n approach a limiting or equilibrium size as n becomes large?

   (a) $x_n = 100 + 100(2)^{-n}$.

   (b) $x_n = 100 + 100\left(\dfrac{1}{1 + n^2}\right)$.

   (c) $x_n = 100 + 5n^2$.

   (d) $x_n = \sqrt{9 + n}$.

2. Determine the order of each of the following difference equations.

   (a) $x_n = x_{n+1} + x_{n+3}$.

   (b) $x_{n+1} = (x_n)^2 + (x_{n-1})^3$.

   (c) $x_n + nx_{n-1} = n^2$.

   (d) $x_{n+2} + 4x_{n+1} + 4x_n = 2^n$.

   (e) $x_{n+4} = x_n + 1$.

   (f) $x_n = x_{n-2} + x_{n-3}$.

3. Verify that $x_n = n^2 + n$ is a solution of the difference equation $x_{n+1} = x_n + 2n + 2$. Verify that $x_n = n^2 + n + k$ is also a solution for any value of the constant k.

4. Verify that $x_n = ca^n$ is a solution of the difference equation $x_{n+1} = ax_n$ for every value of the constant c. If it is known that $x_2 = 3$ and $x_3 = 5$, determine the constants a and c.

5. Verify that $x_n = 2^{n(n+1)/2}$ is a solution of the difference equation $x_n = 2^n x_{n-1}$. Verify that $x_n = k2^{n(n+1)/2}$ is also a solution for any value of the constant k.

6. The growth of a bacteria culture in a nutrient medium is observed every 2 hours. At each measurement, it is found that the bacteria population is 25 per cent larger than at the previous measurement.

   (a) Describe this growth process by means of a difference equation for x_n, the population after n hours.

   (b) What is the order of this difference equation?

   (c) If $x_0 = 1600$, determine x_2 and x_4.

7. The per capita production of garbage in the United States is estimated to be approximately 5 pounds per day and to be increasing at the rate of 4 per cent per year. Define x_n to be the average daily production of garbage per capita in n year's time from now.

   (a) Describe this growth process by means of a difference equation.

(b) What is the order of this difference equation?

(c) Determine x_1, x_2, x_3, and x_4.

6.2 First Order Linear Difference Equations

The first difference equation we will consider can be thought of as a simple model of population growth. Consider a population which is growing in such a way that, as the population increases, the rate of growth of the population also increases. More explicitly, we assume that the rate of growth of the population in any time period is proportional to the size of the population at the beginning of the time period.

To express this assumption in mathematical form, let x_n represent the population at the end of the nth time period. Then, $x_{n+1} - x_n$ is the population growth in the next time period; that is, it is the rate of growth or the growth per unit time in the $(n + 1)$st time period. This is proportional to x_n. If a is the constant of proportionality, then $x_{n+1} - x_n = ax_n$. Rearranging terms, we have the first order difference equation

$$x_{n+1} = (1 + a)x_n. \tag{6.1}$$

In order to solve this equation, we must know the initial population x_0. Then, using Equation (6.1), we can determine successively x_1, x_2, x_3, and so on. We have

$$x_1 = (1 + a)x_0,$$

$$x_2 = (1 + a)x_1 = (1 + a)(1 + a)x_0 = (1 + a)^2 x_0,$$

$$x_3 = (1 + a)x_2 = (1 + a)(1 + a)^2 x_0 = (1 + a)^3 x_0.$$

The pattern should be clear. In going from one time to the next, the population is multiplied by the factor $(1 + a)$. Therefore, the general solution or the general formula for x_n is $x_n = (1 + a)^n x_0$. If x_0 is known, this formula determines x_n.

If the constant of proportionality a is positive $(a > 0)$, then $1 + a > 1$ and $(1 + a)^n$ increases without bound as n increases. If $a = 0$, the population remains at the constant level x_0. This is the case of no growth. If a is negative (but larger than -1), then $0 < 1 + a < 1$ and x_n approaches zero as n increases. In this case, the population eventually becomes extinct. Note that, if $a = -1$, the population becomes extinct after the first time period. In this model, we are not interested in values of a less than -1, since they would lead to negative populations.

Example 6.2.1 Solve the first order difference equation $x_{n+1} - x_n = 2x_n$. Assuming that $x_0 = 10$, what is x_3?

Solution: The equation can be written $x_{n+1} = 3x_n$. This is an example of the above type with $a = 2$ and $1 + a = 3$. The general solution is $x_n = (1 + a)^n x_0 = 3^n x_0$. Since $x_0 = 10$, we have $x_n = 3^n(10)$ and $x_3 = 3^3(10) = 270$.

Example 6.2.2 A bacteria population is initially 1000 and is growing steadily with a 50 per cent increase every hour. What is the population after 10 hours?

Solution: Let x_n represent the population after n hours. We are given that $x_{n+1} = 1.5x_n$ and $x_0 = 1000$. Therefore, $x_1 = (1.5)(1000)$, $x_2 = (1.5)^2(1000)$, and so on. The general solution is $x_n = (1.5)^n x_0$. After 10 hours, the population is $x_{10} = (1.5)^{10}(1000) \approx 57,700$.

The equation $x_{n+1} = (1 + a)x_n$ is an example of a *linear* first order difference equation. The terms involving x_n and x_{n+1} contain only terms of the form $a(n)x_n$ and $b(n)x_{n+1}$ where $a(n)$ and $b(n)$ depend only on n. Terms such as $x_n^2, x_{n+1}^3, 2^{x_n}, x_n x_{n+1}, 1/x_n$, and so on, do not appear. When such terms appear, the difference equation is said to be *nonlinear*.

Example 6.2.3 The following difference equations are linear.

1. $x_{n+1} = 5x_n - 4x_{n-1}$.
2. $x_{n+1} = n^2 x_n$.
3. $x_{n+2} - x_n = 0$.
4. $x_{n+1} - nx_n = n^2$.

Example 6.2.4 The following difference equations are nonlinear.

1. $x_{n+1} = x_n^2$.
2. $x_{n+1} x_n = x_{n-1}$.
3. $x_{n+2} = x_{n+1}(1 + x_n)$.
4. $x_{n+1} = \sqrt{x_n}$.

The most general first order linear difference equation is

$$x_{n+1} = f(n)x_n + g(n), \tag{6.2}$$

where $f(n)$ and $g(n)$ are known functions of n. If x_n is known, we can determine

x_{n+1} from the equation. Equation (6.2) is said to be *homogeneous* if $g(n) = 0$. Otherwise the equation is *inhomogeneous*.

Consider first the homogeneous equation $x_{n+1} = f(n)x_n$. To solve, we set $n = 0$ to obtain $x_1 = f(0)x_0$. For $n = 1$, we find $x_2 = f(1)x_1 = f(1)f(0)x_0$. Similarly, for $n = 2$, $x_3 = f(2)x_2 = f(2)f(1)f(0)x_0$. This suggests the general solution $x_n = f(n - 1)f(n - 2)\ldots f(1)f(0)x_0$. This solution can be verified by substitution in the original equation $x_{n+1} = f(n)x_n$. From the formula for the general solution we have

$$x_{n+1} = f(n)f(n - 1)f(n - 2)\cdots f(2)f(1)f(0)x_0 = f(n)x_n,$$

which proves that we do have a solution.

Example 6.2.5 Consider a bacteria population that grows from an initial size of 1000 in such a way that the population after $n + 1$ hours is $(n + 3)/(n + 2)$ times the population after n hours. What is the population after 10 hours?

Solution: As before, let x_n represent the population size after n hours. We know that $x_0 = 1000$ and $x_{n+1} = [(n + 3)/(n + 2)]x_n$. This is a homogeneous equation with $f(n) = (n + 3)/(n + 2)$. The general solution is

$$x_n = f(n - 1)f(n - 2)\cdots f(2)f(1)f(0)x_0 = \frac{n + 2}{n + 1}\cdot\frac{n + 1}{n}\cdots\frac{5}{4}\cdot\frac{4}{3}\cdot\frac{3}{2}(1000).$$

Canceling, we obtain $x_n = [(n + 2)/2](1000) = 500(n + 2)$. After 10 hours, the population is $x_{10} = 500(10 + 2) = 6000$. In this growth process, the population increases by 500 every hour. Ultimately, this growth model will be unrealistic, since resources necessary for growth are always limited. However, it may be a good description of some types of growth over a limited number of hours.

Let us now return to the general first order linear difference equation $x_{n+1} = f(n)x_n + g(n)$. The method of solution is the same as for the special cases considered above, although the form of the solution is more complicated. Setting $n = 0$, we have $x_1 = f(0)x_0 + g(0)$. With $n = 1$, the equation implies $x_2 = f(1)x_1 + g(1) = f(1)f(0)x_0 + f(1)g(0) + g(1)$. Similarly, $x_3 = f(2)x_2 + g(2) = f(2)f(1)f(0)x_0 + f(2)f(1)g(0) + f(2)g(1) + g(2)$. This suggests the general solution

$$x_n = f(n-1)f(n-2)\cdots f(1)f(0)x_0 + f(n-1)f(n-2)\cdots f(1)g(0)$$
$$+ f(n-1)f(n-2)\cdots f(2)g(1) + \cdots + f(n-1)g(n-2) + g(n-1).$$
$$(6.3)$$

The reader should verify by substitution in Equation (6.2) that this is the solution. The formula (6.3) should not be memorized, since it is probably much easier to derive it from the above argument.

Example 6.2.6   A bacteria population grows from an initial size of 1000 in such a way that the population increase between n hours and $(n+1)$ hours is $500(2)^{-n}$. What is the population after 10 hours?

Solution: We are given that $x_{n+1} = x_n + 500(2)^{-n}$ and $x_0 = 1000$, where x_n is the population after n hours. In this example, $f(n) = 1$ and $g(n) = 500(2)^{-n}$. After 1 hour, $x_1 = x_0 + 500(2)^{-0} = 1000 + 500 = 1500$. After 2 hours, $x_2 = x_1 + 500(2)^{-1} = 1500 + \frac{500}{2} = 1750$. After 3 hours, $x_3 = x_2 + 500(2)^{-2} = 1750 + \frac{500}{4} = 1875$. The general solution is

$$x_n = 1000 + 500 + \frac{500}{2} + \frac{500}{2^2} + \cdots + \frac{500}{2^{n-1}} = 1000 + 500\frac{1 - (\frac{1}{2})^n}{1 - \frac{1}{2}}.$$

Therefore, $x_n = 1000 + 1000[1 - (\frac{1}{2})^n] = 2000 - 1000(2)^{-n}$. We have used the sum of a finite geometric series in this calculation (see Problem 1.6.14). After 10 hours, the population is $x_{10} = 2000 - 1000(2)^{-n} \approx 1999$. As time increases, the bacteria population grows to a limiting or equilibrium size of 2000. In terms of the growth model, this would be interpreted as the size of population that can be supported by the available resources.

Problems 6.2

1. Determine the general solutions of the following first order difference equations.

   (a) $x_{n+1} - x_n = 2^{-n}$.   (b) $x_{n+1} - x_n = 2x_n$.

   (c) $x_{n+1} = \frac{n+5}{n+3}x_n$.   (d) $x_{n+1} = nx_n$.

   (e) $x_{n+1} - 3x_n = 3x_{n+1} - x_n$.   (f) $x_{n+1} = nx_n + 1$.

2. Determine the particular solutions of the following first order difference equations, which satisfy the initial condition $x_0 = 1$.

(a) $2x_{n+1} = x_n$.

(b) $x_{n+1} = x_n + e^{-n}$.

(c) $(n + 1)x_{n+1} = (n + 2)x_n$.

(d) $x_{n+1} - x_n = 2x_n + 2$.

The reader is referred to Appendix D for a discussion of the number e.

3. A study of the cost of repeating an experiment n times found that the total cost x_n satisfied the equation $x_n = x_{n-1} + cn^{-1/2}$, where c is a constant.

(a) What is the cost of the nth experiment in terms of c?

(b) Assuming that $x_0 = 0$, determine the formula for x_n.

4. A fair coin is marked 1 on one side and 2 on the other side. The coin is tossed repeatedly and a cumulative score of the outcomes is recorded. Define p_n to be the probability that the cumulative score takes on the value n. Prove that $p_n = 1 - \frac{1}{2}p_{n-1}$. Assuming that $p_0 = 1$, derive the formula for p_n.

5. In a study of fasting, the weight of a volunteer decreased from 140 pounds to 110 pounds in 30 days. It was observed that the weight loss per day was proportional to the weight of the volunteer. Define w_n to be the weight after n days of fasting. What is the difference equation satisfied by w_n? Determine the weight of the volunteer after 15 days.

6. The equation $x_{n+1} = f(n)x_n + g(n)$ can be interpreted as describing the growth of a population from one generation to the next. The term $f(n)x_n$ may represent the growth of the population present in the nth generation. The term $g(n)$ may represent the increase due to immigration (or the decrease due to emigration). Consider a population whose initial size is $x_0 = 10$. Suppose that the population would double in each generation if no outside influences were present. Suppose further that nine individuals are removed from the population in each generation. Determine the population x_n in the nth generation.

7. In Problem 6, prove that if $x_n = g(n)/[1 - f(n)]$ for some n, then $x_{n+1} = x_n$. Deduce that, if $g(n)/[1 - f(n)]$ is a constant independent of n and if $x_0 = g(0)/[1 - f(0)]$, then the population remains at a constant size in every generation.

8. Radium transmutes at the rate of 1 per cent every 25 years. Consider a sample of r_0 grams of radium. Define r_n to be the amount of radium remaining in the sample after $25n$ years. Determine a difference equation for r_n and determine its solution. How much is left after 100 years?

9. A zoologist stocks a lake with 100 fish of a certain species every spring. This program is continued for several years until the stock of fish increases to 2000. Define x_n to be the fish population after the nth stocking. If it is found that $x_{n+1} = 1.5x_n + 100$ and if $x_0 = 0$, how many years will the stocking program continue?

10. In constructing a mathematical model of a population, it is assumed that the probability p_n that a couple produces exactly n offspring satisfies the equation $p_n = .7p_{n-1}$. Determine p_n in terms of p_0. Since $p_0 + p_1 + p_2 + \cdots = 1$, determine p_0 and p_n.

11. An alternative model in Problem 10 is given by $p_n = (1/n)p_{n-1}$. In this model, determine the formula for p_n in terms of p_0 and prove that $p_0 = 1/e$. (*Hint:* Use the fact, discussed in Appendix D, that $e = 1 + \dfrac{1}{1!} + \dfrac{1}{2!} + \dfrac{1}{3!} + \dfrac{1}{4!} + \cdots + \dfrac{1}{n!} + \cdots$)

12. Which of the two models of Problems 10 and 11 is a more reasonable description of human populations? Why? How would you test both models? How would you modify either model to make it more realistic?

13. The reaction to a drug n hours after injection is r_n measured in appropriate units. Suppose that the reaction satisfies the difference equation $r_{n+1} = (.8)r_n + (.4)^n$. If $r_0 = 1$, determine r_n. Draw a graph of r_n.

14. In Problem 13, after how many hours does the maximum reaction to the drug occur? After how many hours is the reaction less than 50 per cent of its initial value?

15. Define p_n to be the probability that a critically ill patient will live at least n more days. If $p_0 = 1$ and $p_n = \frac{2}{3}p_{n-1}$, determine the probability that this patient will live at least 5 days. What is the expected number of days that this patient will live?

6.3 Second Order Linear Difference Equations

The general second order linear difference equation is

$$a(n)x_{n+2} + b(n)x_{n+1} + c(n)x_n = d(n), \tag{6.4}$$

where $a(n)$, $b(n)$, $c(n)$, and $d(n)$ are given functions. If $d(n) = 0$, the equation is said to be *homogeneous*. If $a(n)$, $b(n)$, and $c(n)$ are constants, the equation is said to have *constant coefficients*. Equation (6.4) can be solved by methods very similar to those for first order equations. Setting $n = 0$, we can determine

x_2 in terms of x_0 and x_1. Then, setting $n = 1$, x_3 is determined in terms of x_2 and x_1 and, therefore, in terms of x_0 and x_1. Theoretically, we can determine x_n for any n in terms of x_0 and x_1, but the calculations can be extremely tedious and it is usually very difficult to establish a general formula for x_n. The constant coefficients case can, however, be solved by general methods that involve relatively little calculation.

In this section, we study the second order homogeneous linear difference equation with constant coefficients

$$ax_{n+2} + bx_{n+1} + cx_n = 0. \qquad (6.5)$$

We are assuming that a, b, and c are constants and $a \neq 0$. In the next section, we study the corresponding inhomogeneous equation. Equation (6.5) arises in models of population growth and competition as well as in many other biological contexts. In the final chapter, a model of the survival and extinction of species leads to a difference equation of this type (Section 9.2.)

Equation (6.5) can be rewritten $x_{n+2} = -(b/a)x_{n+1} - (c/a)x_n$ and then solved by successively setting $n = 0, 1, 2, 3, \ldots$. For $n = 0$, we have $x_2 = -(b/a)x_1 - (c/a)x_0$. Similarly,

$$x_3 = -\frac{b}{a}x_2 - \frac{c}{a}x_1 = \frac{b^2 - ac}{a^2}x_1 + \frac{bc}{a^2}x_0.$$

This leads to an extremely complex formula for x_n in terms of x_0 and x_1. A better method is suggested by the form of the solution of the first order equation $x_{n+1} = (1 + a)x_n$. In the previous section, we found solutions of the form $x_n = \lambda^n$, where $\lambda = 1 + a$. By analogy, we look for solutions of Equation (6.5) of the form $x_n = \lambda^n$ for certain values of λ.

If $x_n = \lambda^n$ is to satisfy Equation (6.5), we must have

$$a\lambda^{n+2} + b\lambda^{n+1} + c\lambda^n = 0$$

for $n = 0, 1, 2, 3, \ldots$. In particular, setting $n = 0$, we obtain

$$a\lambda^2 + b\lambda + c = 0. \qquad (6.6)$$

This equation is called the *auxiliary equation* of the difference equation. It is a quadratic equation whose roots are

$$\lambda_1 = \frac{-b + \sqrt{b^2 - 4ac}}{2a} \quad \text{and} \quad \lambda_2 = \frac{-b - \sqrt{b^2 - 4ac}}{2a}. \qquad (6.7)$$

There are three possibilities that arise in solving the auxiliary equation. The two roots may be real and distinct ($b^2 - 4ac > 0$); they may be complex numbers ($b^2 - 4ac < 0$); or they may be real and equal ($b^2 - 4ac = 0$).

If $b^2 - 4ac > 0$, the above method produces two solutions of Equation (6.5), $x_n = \lambda_1^n$ and $x_n = \lambda_2^n$. The general solution is

$$x_n = k_1\lambda_1^n + k_2\lambda_2^n, \tag{6.8}$$

where k_1 and k_2 are arbitrary constants. Every solution is of this form for some values of the constants k_1 and k_2. To verify that we have a solution, we substitute it in Equation (6.5).

$$ax_{n+2} + bx_{n+1} + cx_n = a(k_1\lambda_1^{n+2} + k_2\lambda_2^{n+2}) + b(k_1\lambda_1^{n+1} + k_2\lambda_2^{n+1})$$
$$+ c(k_1\lambda_1^n + k_2\lambda_2^n)$$
$$= k_1\lambda_1^n(a\lambda_1^2 + b\lambda_1 + c) + k_2\lambda_2^n(a\lambda_2^2 + b\lambda_2 + c).$$

Since λ_1 and λ_2 are roots of the quadratic equation $a\lambda^2 + b\lambda + c = 0$, the above expression is equal to zero and $x_n = k_1\lambda_1^n + k_2\lambda_2^n$ is a solution of Equation (6.5).

The constants k_1 and k_2 can be determined in terms of the values of x_n at $n = 0$ and $n = 1$. Setting $n = 0$ and $n = 1$ in the general solution (6.8), we have $x_0 = k_1 + k_2$ and $x_1 = k_1\lambda_1 + k_2\lambda_2$. This is a simple system of linear equations for the constants k_1 and k_2. By using any of the methods of Chapter 3, we solve and find that

$$k_1 = \frac{x_1 - x_0\lambda_2}{\lambda_1 - \lambda_2} \quad \text{and} \quad k_2 = \frac{\lambda_1 x_0 - x_1}{\lambda_1 - \lambda_2}. \tag{6.9}$$

Therefore, given x_0 and x_1, we have determined a unique solution of Equation (6.5).

Example 6.3.1   Determine the general solution of the second order difference equation $x_{n+2} - 3x_{n+1} + 2x_n = 0$. If $x_0 = 1000$ and $x_1 = 1500$, determine a general formula for x_n. What is x_5?

Solution: In this example, the auxiliary equation is $\lambda^2 - 3\lambda + 2 = 0$. The two roots are

$$\lambda_1 = \frac{3 + \sqrt{9 - 4(2)}}{2} = 2 \quad \text{and} \quad \lambda_2 = \frac{3 - \sqrt{9 - 4(2)}}{2} = 1.$$

According to Equation (6.8), the general solution is

$$x_n = k_1(2^n) + k_2(1^n) = k_1(2^n) + k_2.$$

But $x_0 = k_1 + k_2 = 1000$ and $x_1 = 2k_1 + k_2 = 1500$. Solving for k_1 and k_2, we find $k_1 = 500$ and $k_2 = 500$. The unique solution is

$$x_n = 500(2^n) + 500 = 500(1 + 2^n).$$

When $n = 5$, we find $x_5 = 500(1 + 2^5) = 16,500$.

Let us now examine the case in which $b^2 - 4ac = 0$. In this case, the roots of the auxiliary equation are equal, $\lambda_1 = \lambda_2 = -b/2a$, and our method produces only one solution, $x_n = \lambda_1^n$. We will show that $x_n = n\lambda_1^{n-1}$ is another solution of Equation (6.5) in this case. The general solution can be written

$$x_n = k_1\lambda_1^n + k_2 n\lambda_1^{n-1}, \tag{6.10}$$

where $\lambda_1 = -b/2a$ and k_1 and k_2 are arbitrary constants. To verify that this is a solution, we substitute it in Equation (6.5).

$$
\begin{aligned}
ax_{n+2} + bx_{n+1} + cx_n &= a(k_1\lambda_1^{n+2} + k_2(n + 2)\lambda_1^{n+1}) \\
&\quad + b(k_1\lambda_1^{n+1} + k_2(n + 1)\lambda_1^n) \\
&\quad + c(k_1\lambda_1^n + k_2 n\lambda_1^{n-1}) \\
&= k_1\lambda_1^n(a\lambda_1^2 + b\lambda_1 + c) + k_2 n\lambda_1^{n-1}(a\lambda_1^2 + b\lambda_1 + c) \\
&\quad + k_2\lambda_1^n(2a\lambda_1 + b).
\end{aligned}
$$

The first two terms are zero, since $a\lambda_1^2 + b\lambda_1 + c = 0$. The third term is zero, since $\lambda_1 = -b/2a$ and $2a\lambda_1 + b = 2a(-b/2a) + b = 0$. We conclude that Equation (6.10) is a solution of Equation (6.5). The constants k_1 and k_2 can be determined in terms of x_0 and x_1. Setting $n = 0$ and $n = 1$ in the general solution (6.10), we have $x_0 = k_1$ and $x_1 = k_1\lambda_1 + k_2$, and therefore $k_1 = x_0$ and $k_2 = x_1 - k_1\lambda_1$. This case is illustrated in the following example.

Example 6.3.2   Determine the general solution of the second order difference equation $x_{n+2} - 4x_{n+1} + 4x_n = 0$. If $x_0 = 500$ and $x_1 = 1000$, determine x_n. What is x_5?

Solution: The auxiliary equation is $\lambda^2 - 4\lambda + 4 = 0$. In this example, $b^2 - 4ac = 0$ and the unique root is $\lambda_1 = -b/2a = \frac{4}{2} = 2$. The general solution is therefore $x_n = k_1 2^n + k_2 n 2^{n-1}$. Setting $n = 0$ and $n = 1$, we have $x_0 = 500 = k_1$ and $x_1 = 1000 = 2k_1 + k_2$. We conclude that $k_1 = 500$ and $k_2 = 0$. Therefore, $x_n = (500)2^n$. In particular, $x_5 = (500)2^5 = 16,000$.

The third and final case to examine occurs when $b^2 - 4ac < 0$. In this case, the roots λ_1 and λ_2 are conjugate complex numbers.

$$\lambda_1 = -\frac{b}{2a} + i\frac{\sqrt{4ac - b^2}}{2a} \qquad \text{and} \qquad \lambda_2 = -\frac{b}{2a} - i\frac{\sqrt{4ac - b^2}}{2a}.$$

(For a brief review of complex numbers and their properties, see Appendix E.) These numbers can be written in polar form, $\lambda_1 = re^{i\theta}$ and $\lambda_2 = re^{-i\theta}$, where $r = \sqrt{c/a}$ and $\tan \theta = -\sqrt{4ac - b^2}/b$. Therefore, $\lambda_1^n = r^n e^{in\theta} = r^n(\cos n\theta + i \sin n\theta) = (c/a)^{n/2}(\cos n\theta + i \sin n\theta)$. Similarly, $\lambda_2^n = (c/a)^{n/2} \times (\cos n\theta - i \sin n\theta)$.

As in the two previous cases, any linear combinations of the basic solutions λ_1^n and λ_2^n are also solutions. In particular, $\frac{1}{2}(\lambda_1^n + \lambda_2^n) = (c/a)^{n/2} \cos n\theta$ and $(1/2i)(\lambda_1^n - \lambda_2^n) = (c/a)^{n/2} \sin n\theta$ are solutions. Using these two solutions, we can write the general solution of the difference equation (6.5) as

$$x_n = k_1\left(\frac{c}{a}\right)^{n/2} \cos n\theta + k_2\left(\frac{c}{a}\right)^{n/2} \sin n\theta. \tag{6.11}$$

Again, the constants k_1 and k_2 can be determined in terms of x_0 and x_1.

Solution (6.11) of Equation (6.5) in the case of complex roots has some very interesting properties. Since $\cos n\theta$ and $\sin n\theta$ oscillate between the values $+1$ and -1 as n increases, the solution x_n oscillates in a complicated way. In applications, populations may oscillate in size from one year to the next because of changes in food supply or for other reasons. The properties of the solution are best illustrated by a number of examples.

Example 6.3.3 Determine the general solution of the second order difference equation $x_{n+2} + x_n = 0$. If $x_0 = 0$ and $x_1 = 1000$, determine x_n.

Solution: In this example, the auxiliary equation is $\lambda^2 + 1 = 0$ with complex roots $\lambda_1 = +i$ and $\lambda_2 = -i$. In polar form, $r = (c/a)^{1/2} = 1$ and $\theta = \pi/2$. The general solution is therefore $x_n = k_1 \cos (n\pi/2) + k_2 \sin (n\pi/2)$. Setting $n = 0$ and $n = 1$, we have $x_0 = 0 = k_1$ and $x_1 = 1000 = k_1 \cos (\pi/2) + k_2 \sin (\pi/2) = k_2$. The required solution is $x_n = 1000 \sin (n\pi/2)$. For example,

$x_2 = 0$, $x_3 = -1000$, $x_4 = 0$, $x_5 = 1000$, and so forth. This could represent the seasonal variations of a population about its average value.

Example 6.3.4 **A Model of the Influence of Preceding Generations on Population Growth** Consider a population which is growing from generation to generation. Let x_n represent the population in the nth generation. Clearly, the population in the nth generation depends on the population in the previous generation, and it may also depend on the populations of other preceding generations. For example, preceding generations may have exhausted most of the available resources making reproduction difficult or impossible. The equation $x_{n+2} = rx_{n+1} + sx_n$ can be thought of as a model of population growth from generation to generation. The population in the $(n + 2)$nd generation is made up of two contributions, one from the $(n + 1)$st generation and one from the nth generation. The constants r and s measure the relative importance of these two terms.

The auxiliary equation is $\lambda^2 - r\lambda - s = 0$ with roots $\lambda_1 = (r + \sqrt{r^2 + 4s})/2$ and $\lambda_2 = (r - \sqrt{r^2 + 4s})/2$. If $r^2 + 4s > 0$, these roots are real and distinct and the general solution is $x_n = k_1\lambda_1^n + k_2\lambda_2^n$. If $r^2 + 4s < 0$ (this occurs if s is a large negative number), the general solution has the very different form $x_n = (-s)^{n/2}(k_1 \cos n\theta + k_2 \sin n\theta)$, where $\tan \theta = -\sqrt{-(r^2 + 4s)}/r$. In this very simple model, we have the interesting result that, if the mortality in one generation is sufficiently affected by the individuals of two generations before, the population will oscillate from generation to generation.

Example 6.3.5 Determine the general solution of the second order difference equation $x_{n+2} + x_{n+1} + x_n = 0$. If $x_0 = 100$ and $x_1 = 0$, determine x_n.

Solution: The auxiliary equation is $\lambda^2 + \lambda + 1 = 0$ with complex roots $\lambda_1 = (-1 + i\sqrt{3})/2$ and $\lambda_2 = (-1 - i\sqrt{3})/2$. In polar form, $r = (c/a)^{1/2} = 1$ and $\theta = 2\pi/3$ (since $\tan \theta = -\sqrt{3}$). The general solution is $x_n = k_1 \cos (2n\pi/3) + k_2 \sin (2n\pi/3)$. Setting $n = 0$ and $n = 1$, we have $x_0 = 100 = k_1$ and $x_1 = 0 = -\frac{1}{2}k_1 + k_2(\sqrt{3}/2)$. Therefore, $k_1 = 100$ and $k_2 = 100/\sqrt{3}$. The required solution is $x_n = 100 \cos (2n\pi/3) + (100/\sqrt{3}) \sin (2n\pi/3)$.

Problems 6.3

1. Determine the general solutions of the following second order difference equations.

(a) $x_{n+2} = x_n + x_{n-1}$.

(b) $x_{n+2} + 2x_{n+1} - 3x_n = 0$.

(c) $4x_n + 4x_{n+1} + x_{n+2} = 0$.

(d) $3x_{n+2} - 2x_{n+1} - x_n = 0$.

2. Determine the solutions of the following difference equations, which
   satisfy the initial conditions $x_0 = 1, x_1 = 2$.
   (a) $x_{n+2} + x_{n+1} = 6x_n$. (b) $x_{n+4} + 6x_{n+3} + 9x_{n+2} = 0$.
   (c) $x_{n+2} + 2x_{n+1} + 2x_n = 0$. (d) $x_n = x_{n+2} - x_{n+1}$.

3. Determine the solutions of the following difference equations, which
   satisfy the initial conditions $x_0 = 0, x_1 = 1$.
   (a) $x_{n+2} + x_n = 0$. (b) $6x_{n+2} + 5x_{n+1} + x_n = 0$.
   (c) $x_{n+2} + 4x_{n+1} + 6x_n = 0$. (d) $x_{n+2} + x_n = 2x_{n+1}$.

4. Determine the general solution of the second order equation $x_{n+2} - 4x_{n+1} + 4x_n = 0$. Determine the particular solution which satisfies $x_0 = 160$ and $x_1 = 300$. If this difference equation represents the development of a population recorded at yearly intervals from an initial population of 160, when does the population become extinct?

5. Determine a second order difference equation of the form $ax_{n+2} + bx_{n+1} + cx_n = 0$ with particular solutions $x_n = 2^n$ and $x_n = 4^{2n}$.

6. Determine a second order difference equation of the form $ax_{n+2} + bx_{n+1} + cx_n = 0$ with particular solutions $x_n = (1.5)^n$ and $x_n = n(1.5)^n$.

7. Determine a second order difference equation of the form $ax_{n+2} + bx_{n+1} + cx_n = 0$ with particular solutions $x_n = 4^n \cos(n\pi/2)$ and $x_n = 4^n \sin(n\pi/2)$.

8. (a) In a simple model of the growth of a population, it is assumed that the population x_n in the nth generation is the sum of the populations in the two preceding generations. If $x_0 = 0$ and $x_1 = 1$, determine x_n for all generations n. (The sequence x_n is known as the Fibonacci sequence.)
   (b) For this growth model, evaluate the limit of the ratio x_{n+1}/x_n as n tends to infinity. (This model is a special case of Example 6.3.4.)

9. Prove that all solutions of $4x_{n+2} + 4x_{n+1} + x_n = 0$ approach zero as n tends to infinity.

10. Under what conditions on the coefficients a, b, and c do all solutions of $ax_{n+2} + bx_{n+1} + cx_n = 0$ approach zero as n tends to infinity?

11. Determine solutions of the following equations satisfying the given initial conditions.
   (a) $x_{n+2} - x_n = 0, x_0 = 1, x_1 = 3$.
   (b) $x_{n+2} - 3x_{n+1} + 2x_n = 0, x_0 = 2, x_1 = 1$.
   (c) $x_{n+2} - 6x_{n+1} + 8x_n = 0, x_0 = 3, x_1 = 6$.

12. Determine solutions of the following equations satisfying the given initial conditions.

(a) $x_{n+2} + 6x_{n+1} + 9x_n = 0$, $x_0 = 1$, $x_1 = 2$.
(b) $x_{n+2} + 2x_{n+1} + x_n = 0$, $x_0 = -2$, $x_1 = -4$.
(c) $4x_{n+2} + 20x_{n+1} + 25x_n = 0$, $x_0 = 0$, $x_1 = 1$.

13. Determine solutions of the following equations satisfying the given initial conditions.

(a) $x_{n+2} + x_{n+1} + x_n = 0$, $x_0 = 1$, $x_1 = 0$.
(b) $3x_{n+2} + 3x_{n+1} + x_n = 0$, $x_0 = 0$, $x_1 = 2$.
(c) $x_{n+2} + \frac{1}{2}x_{n+1} + \frac{1}{4}x_n = 0$, $x_0 = 1$, $x_1 = 1$.

14. In a study of fasting, the weights of two volunteers decreased from 140 and 170 pounds to 110 and 125 pounds, respectively, in 30 days. It was observed that the daily weight loss of each volunteer was proportional to his weight. Define x_n to be the total weight of the two volunteers after n days of fasting. Determine a second order difference equation satisfied by x_n. What is the total weight after 15 days of fasting?

15. Two competing species of drosophila are growing under favorable conditions. In each generation, species I increases its population by 60 per cent and species II increases by 40 per cent. If initially there are 1000 flies of each species, what is the total population after n generations? Determine a second order difference equation for the total population.

16. In Problem 15, suppose that 500 flies of each species are removed in each generation. In how many generations does the second species become extinct. What is the total population after n generations?

17. Discuss the properties of the solutions of (a) $x_{n+2} = x_{n+1} + \frac{1}{2}x_n$ and (b) $x_{n+2} = x_{n+1} - \frac{1}{2}x_n$ as special cases of the model of population growth of Example 6.3.4.

18. If x_n and y_n satisfy the first order difference equations $x_{n+1} = ax_n$ and $y_{n+1} = by_n$, prove that $z_n = x_n + y_n$ satisfies the second order difference equation $z_{n+2} - (a + b)z_{n+1} + abz_n = 0$. (How is this result useful in Problems 14 and 15?)

19. If x_n and y_n satisfy the first order difference equations $x_{n+1} = f(n)x_n$ and $y_{n+1} = h(n)y_n$, determine a second order difference equation satisfied by $z_n = x_n + y_n$.

20. In a study of the spread of infectious diseases, a record is kept of outbreaks of measles in a particular school. It is estimated that the probability p_n of at least one new case in the nth week after an outbreak satisfies the equation $p_n = p_{n-1} - \frac{1}{5}p_{n-2}$. If $p_0 = 0$ and $p_1 = 1$, determine p_n. After how many weeks is the probability of a new case of measles less than 10 per cent?

6.4 The Variation of Constants Method for Second Order Difference Equations

In this section, we derive a method for solving the equation

$$ax_{n+2} + bx_{n+1} + cx_n = f(n) \tag{6.12}$$

with constant coefficients a, b, and c (with $a \neq 0$). This is a second order linear inhomogeneous difference equation with constant coefficients. Of course, if $f(n) = 0$ for all n, the equation is homogeneous. As in the previous section, there are three cases to consider corresponding to the roots of the auxiliary equation being real and distinct, real and equal, or complex. The methods of this section are applicable to all three cases but, for simplicity, we consider only the case of two distinct real roots, λ_1 and λ_2. Then $x_n = k\lambda_1^n + l\lambda_2^n$ is the general solution of the homogeneous equation (6.5), where k and l are arbitrary constants. The idea of the variation of constants method is to allow the constants k and l to vary with n in such a way that we produce a solution of the inhomogeneous equation (6.12). Accordingly, let us look for solutions of the form

$$x_n = k_n\lambda_1^n + l_n\lambda_2^n, \tag{6.13}$$

where k_n and l_n can vary with n. We now substitute equation (6.13) into (6.12) and attempt to determine functions k_n and l_n that yield a solution.

Using Equation (6.13), we can calculate x_{n+1} and x_{n+2}. We have $x_{n+1} = k_{n+1}\lambda_1^{n+1} + l_{n+1}\lambda_2^{n+1}$. To produce a more useful expression, we add and subtract $k_n\lambda_1^{n+1} + l_n\lambda_2^{n+1}$ to produce the formula

$$x_{n+1} = k_n\lambda_1^{n+1} + l_n\lambda_2^{n+1} + (k_{n+1} - k_n)\lambda_1^{n+1} + (l_{n+1} - l_n)\lambda_2^{n+1}.$$

In order to simplify this expression, we will assume that

$$(k_{n+1} - k_n)\lambda_1^{n+1} + (l_{n+1} - l_n)\lambda_2^{n+1} = 0 \tag{6.14}$$

for every n. With this assumption, we have (for every n)

$$x_{n+1} = k_n\lambda_1^{n+1} + l_n\lambda_2^{n+1}. \tag{6.15}$$

Similarly, from Equation (6.15), we have $x_{n+2} = k_{n+1}\lambda_1^{n+2} + l_{n+1}\lambda_2^{n+2}$ or

$$x_{n+2} = k_n\lambda_1^{n+2} + l_n\lambda_2^{n+2} + (k_{n+1} - k_n)\lambda_1^{n+2} + (l_{n+1} - l_n)\lambda_2^{n+2}. \tag{6.16}$$

We now substitute Equations (6.13), (6.15), and (6.16) into Equation (6.12) to obtain

$$ax_{n+2} + bx_{n+1} + cx_n = a[k_n\lambda_1^{n+2} + l_n\lambda_2^{n+2} + (k_{n+1} - k_n)\lambda_1^{n+2}$$
$$+ (l_{n+1} - l_n)\lambda_2^{n+2}] + b[k_n\lambda_1^{n+1} + l_n\lambda_2^{n+1}]$$
$$+ c[k_n\lambda_1^n + l_n\lambda_2^n].$$

Rearranging terms and equating to $f(n)$, we obtain

$$f(n) = k_n[a\lambda_1^{n+2} + b\lambda_1^{n+1} + c\lambda_1^n] + l_n[a\lambda_2^{n+2} + b\lambda_2^{n+1} + c\lambda_2^n]$$
$$+ a[(k_{n+1} - k_n)\lambda_1^{n+2} + (l_{n+1} - l_n)\lambda_2^{n+2}]. \qquad (6.17)$$

Since λ_1 and λ_2 are roots of the auxiliary equation (6.6), the first two terms of (6.17) are equal to zero. We conclude that

$$ax_{n+2} + bx_{n+1} + cx_n = a[(k_{n+1} - k_n)\lambda_1^{n+2} + (l_{n+1} - l_n)\lambda_2^{n+2}] = f(n).$$

This gives us a second equation for k_n and l_n, the first equation being (6.14). Writing these two equations together, we have

$$(k_{n+1} - k_n)\lambda_1^{n+1} + (l_{n+1} - l_n)\lambda_2^{n+1} = 0$$
$$(k_{n+1} - k_n)\lambda_1^{n+2} + (l_{n+1} - l_n)\lambda_2^{n+2} = \frac{f(n)}{a}. \qquad (6.18)$$

This gives a system of two equations for the two unknowns $k_{n+1} - k_n$ and $l_{n+1} - l_n$. To solve, multiply the first equation by λ_2 and subtract from the second equation. This determines $k_{n+1} - k_n$; then $l_{n+1} - l_n$ can be determined from the first equation. The solutions are

$$k_{n+1} - k_n = \frac{f(n)}{a\lambda_1^{n+1}(\lambda_1 - \lambda_2)}, \qquad (6.19)$$

$$l_{n+1} - l_n = \frac{f(n)}{a\lambda_2^{n+1}(\lambda_2 - \lambda_1)}. \qquad (6.20)$$

We recognize Equations (6.19) and (6.20) as first order linear difference equations that can be solved by the methods of Section 6.2. To solve (6.19), we have

$$k_1 = k_0 + \frac{f(0)}{a\lambda_1(\lambda_1 - \lambda_2)},$$

$$k_2 = k_1 + \frac{f(1)}{a\lambda_1^2(\lambda_1 - \lambda_2)} = k_0 + \frac{f(0)}{a\lambda_1(\lambda_1 - \lambda_2)} + \frac{f(1)}{a\lambda_1^2(\lambda_1 - \lambda_2)},$$

$$k_3 = k_2 + \frac{f(2)}{a\lambda_1^3(\lambda_1 - \lambda_2)} = k_0 + \frac{f(0)}{a\lambda_1(\lambda_1 - \lambda_2)} + \frac{f(1)}{a\lambda_1^2(\lambda_1 - \lambda_2)}$$
$$+ \frac{f(2)}{a\lambda_1^3(\lambda_1 - \lambda_2)}.$$

The general solution of (6.19) is easily seen to be

$$k_n = k_0 + \frac{1}{a\lambda_1(\lambda_1 - \lambda_2)}\left[f(0) + \frac{f(1)}{\lambda_1} + \frac{f(2)}{\lambda_1^2} + \cdots + \frac{f(n-1)}{\lambda_1^{n-1}}\right]. \quad (6.21)$$

Similarly, for (6.20), the general solution is

$$l_n = l_0 + \frac{1}{a\lambda_2(\lambda_2 - \lambda_1)}\left[f(0) + \frac{f(1)}{\lambda_2} + \frac{f(2)}{\lambda_2^2} + \cdots + \frac{f(n-1)}{\lambda_2^{n-1}}\right]. \quad (6.22)$$

Finally, the general solution of Equation (6.12) is

$$x_n = k_n\lambda_1^n + l_n\lambda_2^n, \quad (6.23)$$

where k_n and l_n are given by Equations (6.21) and (6.22). The constants k_0 and l_0 can be determined if x_0 and x_1 are known. Setting $n = 0$ and $n = 1$ in (6.23), we have $x_0 = k_0 + l_0$ and $x_1 = k_1\lambda_1 + l_1\lambda_2$.

This very lengthy calculation has yielded a formula for the general solution that is not difficult to apply. This can be best illustrated by a number of examples.

Example 6.4.1   Determine the general solution of $x_{n+2} - x_n = 1$. Determine the formula for x_n if $x_0 = 50$ and $x_1 = 100$.

Solution: In this example, $f(n) = 1$. The auxiliary equation $\lambda^2 - 1 = 0$ has roots $\lambda_1 = 1$ and $\lambda_2 = -1$. The general solution is $x_n = k_n(1)^n + l_n(-1)^n = k_n + (-1)^n l_n$, where k_n and l_n are given by Equations (6.21) and (6.22).

$$k_n = k_0 + \tfrac{1}{2}(1 + 1 + \cdots + 1) = k_0 + \frac{n}{2},$$

$$l_n = l_0 + \tfrac{1}{2}[1 - 1 + 1 - \cdots + (-1)^{n-1}] = l_0 + \tfrac{1}{4}[1 - (-1)^n].$$

The general solution can therefore be written

$$x_n = k_0 + \frac{n}{2} + (-1)^n[l_0 + \tfrac{1}{4}(1 - (-1)^n)].$$

If $x_0 = 50$ and $x_1 = 100$, we have

$$x_0 = 50 = k_0 + l_0 \qquad \text{and} \qquad x_1 = 100 = k_0 + \tfrac{1}{2} - (l_0 + \tfrac{1}{2}) = k_0 - l_0.$$

Therefore, $2k_0 = 150$ or $k_0 = 75$ and $l_0 = -25$. This yields the particular solution

$$x_n = 75 + \frac{n}{2} + (-1)^n[-25 + \tfrac{1}{4}(1 - (-1)^n)].$$

Example 6.4.2   If left to grow undisturbed, the increase of a fish population in the $(n + 1)$st year would be double the increase in the nth year. However, for research purposes, 100 fish are added to the population each year. If $x_0 = 1000$ and $x_1 = 1200$, determine x_n, the fish population in the nth year.

Solution: The fish population x_n satisfies the second order difference equation

$$x_{n+2} - x_{n+1} = 2(x_{n+1} - x_n) + 100 \qquad \text{or} \qquad x_{n+2} - 3x_{n+1} + 2x_n = 100.$$

In this example, $f(n) = 100$. The auxiliary equation is $\lambda^2 - 3\lambda + 2 = 0$ with roots $\lambda_1 = 2$ and $\lambda_2 = 1$. The general solution is $x_n = k_n(2)^n + l_n(1)^n$. From Equations (6.21) and (6.22),

$$k_n = k_0 + \tfrac{1}{2}\left(100 + \frac{100}{2} + \frac{100}{2^2} + \cdots + \frac{100}{2^{n-1}}\right)$$

$$= k_0 + \frac{100}{2}\frac{1 - (\tfrac{1}{2})^n}{1 - \tfrac{1}{2}} = k_0 + 100\left(1 - \frac{1}{2^n}\right),$$

(See Problem 1.6.14.)

$$l_n = l_0 - (100 + 100 + \cdots + 100) = l_0 - (100)n.$$

Therefore, $x_n = k_0 2^n + 100(2^n - 1) + l_0 - (100)n$. The constants k_0 and l_0 can be determined in terms of x_0 and x_1.

$$x_0 = 1000 = k_0 + l_0 \quad \text{and} \quad x_1 = 1200 = 2k_0 + 100 + l_0 - 100.$$

Therefore, $k_0 = 200$ and $l_0 = 800$ and the required solution is

$$x_n = (200)2^n + (100)2^n - 100 + 800 - (100)n = (300)2^n + 700 - (100)n.$$

For example, $x_2 = 1700$, $x_3 = 2800$, and $x_4 = 5100$. Clearly, the fish population continues to grow very rapidly.

Problems 6.4

1. Determine the general solutions of the following second order difference equations.
   (a) $x_{n+2} = x_n + 2$.
   (b) $x_{n+2} + 5x_{n+1} - 6x_n = n$.
   (c) $x_{n+2} - 4x_n = (1.5)^n$.
   (d) $x_{n+2} - 3x_{n+1} + 2x_n = 3^n$.

2. Determine the solutions of the following equations, which satisfy the initial conditions $x_0 = 0$ and $x_1 = 1$.
   (a) $x_{n+2} - 9x_n = 1$.
   (b) $x_{n+2} - 4x_n + 3x_n = -1$.
   (c) $4x_{n+2} + 5x_{n+1} + x_n = 2^{2n}$.
   (d) $x_{n+2} + 2x_{n+1} - 3x_n = (-2)^n$.

3. Consider the second order equation $ax_{n+2} + bx_{n+1} + cx_n = r^n$, where a, b, c, and r are constants. Under what conditions on the coefficients a, b, and c is there a solution of the form $x_n = kr^n$ for some constant k?

4. Use the result of Problem 3 to determine solutions of the four second order difference equations of Problem 2.

5. Verify that $x_n = (n/2)2^n$ is a particular solution of the difference equation $x_{n+2} - 3x_{n+1} + 2x_n = 2^n$.

6. Verify that $x_n = (n^2/8)2^n$ is a particular solution of the difference equation $x_{n+2} - 4x_{n+1} + 4x_n = 2^n$.

7. (a) If y_n and z_n satisfy the difference equations $ay_{n+2} + by_{n+1} + cy_n = f(n)$ and $az_{n+2} + bz_{n+1} + cz_n = g(n)$, verify that $x_n = y_n + z_n$ satisfies the difference equation $ax_{n+2} + bx_{n+1} + cx_n = f(n) + g(n)$.
   (b) What equation does $k_1 y_n + k_2 z_n$ satisfy when k_1 and k_2 are arbitrary constants?

8. Use the results of Problems 3 and 7 to determine particular solutions of the following second order equations.
   (a) $x_{n+2} + 2x_{n+1} - 3x_n = 2^n + 1$. (b) $x_{n+2} + 3x_{n+1} - 4x_n = 2^n + 3^n$.
   (c) $x_{n+2} + 3x_{n+1} - 4x_n = 3^n - 2^n$. (d) $x_{n+2} - x_n = 2^{2n} + 3^{2n}$.

9. Suppose that w_n and y_n are any solutions of the second order equation $ax_{n+2} + bx_{n+1} + cx_n = f(n)$. Prove that $z_n = w_n - y_n$ is a solution of the homogeneous equation $ax_{n+2} + bx_{n+1} + cx_n = 0$.

10. If left to grow undisturbed, a bird population would grow in such a way that its size in each generation would be the sum of its sizes in the preceding two generations. However, because there is a constant population of predators, 1000 birds are lost in each generation. The population x_n in the nth generation satisfies $x_n = x_{n-1} + x_{n-2} - 1000$. If $x_0 = 2000$ and $x_1 = 2500$, determine the population in the nth generation.

11. Suppose that, in Problem 10, 20 per cent of the birds are harvested in each generation. The population x_n in the nth generation satisfies $x_n = .8(x_{n-1} + x_{n-2} - 1000)$. If $x_0 = 2000$ and $x_1 = 2500$, determine the population in the nth generation.

12. If x_n and y_n satisfy the first order difference equations $x_{n+1} = f(n)x_n + g(n)$ and $y_{n+1} = h(n)x_n + k(n)$, determine a second order difference equation satisfied by $z_n = x_n + y_n$.

6.5 Systems of First Order Difference Equations

In the preceding sections, we have considered difference equations which could have arisen in describing the population growth of a species in a given environment. The presence of competing species and other aspects of the environment were not taken into account explicitly in these equations. To have a more detailed description and a better understanding of population growth, we must find ways to include in the growth equations factors corresponding to the important features of the environment.

As an example, we may be interested in studying the interaction of two or more species. If two species coexist in an environment, their populations in the $(n + 1)$st time period (year, generation, etc.) will depend on their populations in the nth time period. If the two species are predator and prey, a large prey population in the nth period will lead to an increased predator population in the $(n + 1)$st time period. In turn, this may lead to a decreased prey population in the next time period. On the other hand, two species in a symbiotic relationship will tend to increase and decrease together. Of course, a real ecosystem is a complicated web of competition and cooperation among many different species. We will restrict ourselves here to the two-species case, although, in principle, the methods can be generalized to any number of species.

In this section, we will consider the system of first order difference equations

$$x_{n+1} = a_{11}x_n + a_{12}y_n + f(n)$$

$$y_{n+1} = a_{21}x_n + a_{22}y_n + g(n),$$

(6.24)

where a_{11}, a_{12}, a_{21}, and a_{22} are constants and $f(n)$ and $g(n)$ are known functions. The problem is to determine functions x_n and y_n that satisfy these equations. The system is said to be *homogeneous* if $f(n) = g(n) = 0$ for all n. Otherwise, the system is *inhomogeneous*.

System (6.24) can be thought of as a model or description of the interaction of two species in an environment if x_n and y_n are the populations of species I and species II at the end of the nth time period. If both species compete for the same resources, this can be modeled by the coefficients a_{12} and a_{21} being negative. If a_{12} is negative, for example, then the population of species I will decrease if the population of species II is large. This can be best illustrated by an example.

Example 6.5.1 A Model of Species Competition The homogeneous linear system

$$x_{n+1} = 2x_n - y_n$$

$$y_{n+1} = -x_n + 2y_n$$

(6.25)

describes the influence of two competing species on their populations. Suppose that the initial populations are $x_0 = 100$ and $y_0 = 150$. Determine the populations of both species at all future times.

Solution: We will solve (6.25) by deriving a second order linear difference equation with constant coefficients for x_n. From the first equation of (6.25),

$$x_{n+2} = 2x_{n+1} - y_{n+1} = 2x_{n+1} - (-x_n + 2y_n) = 2x_{n+1} + x_n - 2y_n.$$

But, again from the first equation, $y_n = 2x_n - x_{n+1}$, and we have

$$x_{n+2} = 2x_{n+1} + x_n - 2(2x_n - x_{n+1}) = 4x_{n+1} - 3x_n.$$

Therefore, $x_{n+2} - 4x_{n+1} + 3x_n = 0$. The auxiliary equation is $\lambda^2 - 4\lambda + 3 = 0$ with roots $\lambda_1 = 3$ and $\lambda_2 = 1$. The general solution is $x_n = k_1(3)^n +$

$k_2(1)^n = k_1(3)^n + k_2$. From the first equation of (6.25), $y_n = 2x_n - x_{n+1} = 2k_1(3)^n + 2k_2 - k_1(3)^{n+1} - k_2 = k_2 - k_1(3)^n$. To determine the constants k_1 and k_2, we have $x_0 = 100 = k_1 + k_2$ and $y_0 = 150 = k_2 - k_1$. This implies that $k_1 = -25$ and $k_2 = 125$. The required solution is

$$x_n = 125 - 25(3)^n \qquad \text{and} \qquad y_n = 125 + 25(3)^n.$$

The first species rapidly becomes extinct ($x_0 = 100$, $x_1 = 50$, $x_2 = \cdots = 0$), while the second species increases ($y_0 = 150$, $y_1 = 200$, $y_2 = 350, \dots$).

We now return to the general system (6.24) and determine its general solution by the methods of the above example. From the first equation, we have

$$x_{n+2} = a_{11}x_{n+1} + a_{12}y_{n+1} + f(n+1)$$
$$= a_{11}x_{n+1} + a_{12}(a_{21}x_n + a_{22}y_n + g(n)) + f(n+1)$$
$$= a_{11}x_{n+1} + a_{12}a_{21}x_n + a_{12}a_{22}y_n + a_{12}g(n) + f(n+1).$$

Again, from the first equation, $a_{12}y_n = x_{n+1} - a_{11}x_n - f(n)$. Therefore,

$$x_{n+2} = a_{11}x_{n+1} + a_{12}a_{21}x_n + a_{22}(x_{n+1} - a_{11}x_n - f(n))$$
$$+ a_{12}g(n) + f(n+1)$$
$$= (a_{11} + a_{22})x_{n+1} - (a_{11}a_{22} - a_{12}a_{21})x_n - a_{22}f(n)$$
$$+ a_{12}g(n) + f(n+1).$$

This implies that x_n satisfies the second order difference equation

$$x_{n+2} - (a_{11} + a_{22})x_{n+1} + (a_{11}a_{22} - a_{12}a_{21})x_n = h(n), \qquad (6.26)$$

where $h(n) = -a_{22}f(n) + a_{12}g(n) + f(n+1)$. This equation can be solved by the variation of constants method of Section 6.4. Then y_n can be determined from the first equation of (6.24).

Example 6.5.2   Solve the first order linear system

$$x_{n+1} = 2x_n - y_n - 1$$
$$y_{n+1} = -x_n + 2y_n + 2.$$

Difference Equations [Ch. 6

Solution: In this example, $f(n) = -1$ and $g(n) = 2$. Equation (6.26) becomes $x_{n+2} - 4x_{n+1} + 3x_n = -1$. This equation is Problem 6.4.2(b). The general solution is $x_n = (n/2) + k_1(3)^n + k_2$. From the first equation of the system,

$$y_n = 2x_n - x_{n+1} - 1 = 2\left[\frac{n}{2} + k_1(3)^n + k_2\right]$$
$$-\left[\frac{(n+1)}{2} + k_1(3)^{n+1} + k_2\right] - 1.$$

Therefore, $y_n = [(n-1)/2] - k_1(3)^n + k_2 - 1$. The constants k_1 and k_2 could be determined if the initial populations x_0 and y_0 were known.

6. Birds of a certain species reproduce in such a way that their population would double each year if predators were not present. The presence of predators has the effect of decreasing the bird population in each year by an amount equal to 10 times the number of predators. Assume that the predator population remains constant independent of the size of the bird population. Describe the development from year to year of the bird and predator populations by a system of first order difference equations. If initially there are 1000 birds and 50 predators, determine the solution of the system.

7. In Problem 6, suppose that the predator population doubles each year. Describe the development of the two populations by a system of first-order difference equations. If initially there are 1000 birds and 50 predators, determine the solution of the system. Does the bird population become extinct?

Differential Equations | 7

7.1 Introduction

In this chapter, we continue the development of the mathematical techniques capable of describing the dynamics of biological systems. Again, the biological variables in which we are interested may be functions of variables such as temperature, food abundance, hours of sunlight, and frequency of disease. However, as in Chapter 6, our development will be motivated by considering the population of a species in a given environment as a function of time. Other examples are given in the problems following each section.

In the discrete models of Chapter 6, the biological variable (usually population) was measured at discrete time intervals. In this chapter, we will consider the corresponding continuous models in which the biological variable is a continuous function of time. These two points of view in describing biological systems are both useful. Continuous models are more appropriate for describing the growth of very large populations such as bacteria populations. On the other hand, whooping crane populations are better described by discrete models corresponding to a daily, monthly, or annual census.

In the continuous models, $x(t)$ will represent the population of a certain species in a given environment as a function of time. The initial or starting population is $x(0)$ and the development of the population over time is described by the function $x(t)$.

Example 7.1.1   The function $x(t) = 1000 + 500(1 - 2^{-t})$ represents the continuous growth of a population from the initial size $x(0) = 1000$ to a limiting or equilibrium size $\lim_{t \to \infty} x(t) = 1500$. If this population is observed at $t = 1, t = 2, t = 3, \ldots$, the populations at these times are $x(1) = 1250$, $x(2) = 1375$, $x(3) = 1438, \ldots$ (This is the discrete population growth of Example 6.1.1.) This continuous growth is illustrated in Figure 7.1.

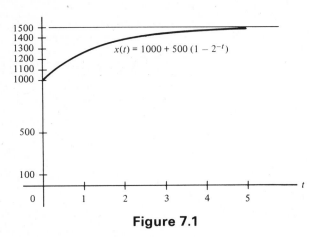

Figure 7.1

In studying the continuous growth of a population $x(t)$, we may have information about the growth rate $dx(t)/dt$. For example, the growth rate may be zero, $dx(t)/dt = 0$. In this case, the population remains at a constant size. This suggests the following definition.

Definition 7.1.1   Differential Equation   *A differential equation for the function $x(t)$ is an equation containing derivatives of $x(t)$ with respect to t. The* order *of the differential equation is defined as the order of the highest derivative occurring in the equation.*

Example 7.1.2   The following are differential equations.

1. $\dfrac{dx}{dt} = 2x$ first order.

2. $\dfrac{dx}{dt} + 2tx = e^{-t^2}$ first order.

3. $\dfrac{d^2x}{dt^2} + 4\left(\dfrac{dx}{dt}\right)^3 + 4x = 0$ second order.

4. $\dfrac{d^3x}{dt^2} + x^2(1 + t^4) = 0$ third order.

Example 7.1.3 A bacteria population grows in such a way that its growth rate at time t (measured in hours) is equal to its population divided by 10. Describe this growth process by a differential equation. What is the order of this equation?

Solution: Define $x(t)$ to be the size of the bacteria population at time t (measured in hours). Then the growth rate $dx(t)/dt$ at time t is equal to $(.1)x(t)$; that is, $dx(t)/dt = (.1)x(t)$. This is a first order differential equation for $x(t)$.

In the rest of this chapter, we will develop methods for solving some important types of differential equations. Given a differential equation for $x(t)$, can we determine all functions $x(t)$ which satisfy the equation? The corresponding biological problem can be phrased: Given information about the initial population and about growth rates, can we determine what the population will be at all future times? Any differentiable function $x(t)$ that satisfies the differential equation is called a *solution* of the equation.

Problems 7.1

1. By drawing graphs, illustrate the continuous growth processes from time zero given by the following functions. For each example, determine the initial populations $x(0)$ and $\lim_{t \to \infty} x(t)$ when this limit is defined.

(a) $x(t) = 100 + 100\dfrac{1}{1 + t^2}$.

(b) $x(t) = 100 + 100e^{-t}$.

(c) $x(t) = 90 + 10t^2$.

(d) $x(t) = 10e^{t/10}$.

2. Determine the order of each of the following differential equations.

(a) $\dfrac{dx}{dt} + x^2 = \sin x$.

(b) $\dfrac{d^2x}{dt^2} + t^2x = e^t$.

(c) $\dfrac{d^2x}{dt^2} + 4\dfrac{dx}{dt} + 4x = 5$.

(d) $\left(\dfrac{d^3x}{dt^3}\right)^2 + \left(\dfrac{dx}{dt}\right)^4 = x + t$.

3. Determine first order differential equations satisfied by the following functions.

   (a) $x(t) = t^2$.   (b) $x(t) = e^t$.   (c) $x(t) = te^t$.

4. Verify that $x(t) = \sqrt{te^t}$ satisfies the differential equation

$$\frac{dx}{dt} = x\left(1 + \frac{1}{2t}\right).$$

5. Verify that $x(t) = \tan t$ satisfies the differential equation

$$\frac{d^2x}{dt^2} = 2x\frac{dx}{dt}.$$

6. (a) Verify that $x(t) = c_1 e^t + c_2 e^{2t}$ is a solution of the differential equation $(d^2x/dt^2) - 3(dx/dt) + 2x = 0$ for any values of the constants c_1 and c_2.

   (b) Determine a solution $x(t)$ of the differential equation in (a) which satisfies $x(0) = 1$ and $dx(0)/dt = 2$.

7. The growth rate of a bacteria population at time t (measured in hours) is observed to be equal to the size $x(t)$ of the population divided by 5. Describe this growth process by a differential equation for $x(t)$. What is the order of the differential equation?

8. Yeast is growing in a sugar solution in such a way that the weight of the yeast is growing at a rate equal to half the weight at time t (when time is measured in hours). Describe the change of the weight of yeast by a differential equation. What is the order of the differential equation?

7.2 First Order Linear Differential Equations

The first differential equation we will consider can be thought of as a simple model of population growth. If the resources available to a population are very abundant, it is reasonable to assume that the rate of growth will be proportional to the size of the population. If $x(t)$ is the population at time t, this assumption is expressed by the differential equation

$$\frac{d}{dt}x(t) = ax(t), \qquad (7.1)$$

where a is a constant. The assumption that leads to this equation is that the growth rate per unit of population (the average growth rate) is constant;

that is,

$$\frac{1}{x(t)} \frac{d}{dt} x(t) = a.$$

In Section 6.2, the difference equation $x_{n+1} = (1 + a)x_n$ can be derived from the assumption that the average increase in population in the $(n + 1)$st time period is a constant; that is,

$$\frac{x_{n+1} - x_n}{x_n} = a.$$

This is the discrete version of the continuous model.

There are two ways to solve Equation (7.1). The first method is to make an educated guess, also known as the method of trial solutions. Do we know any function which has the property that its derivative is equal to the function multiplied by the constant a? You may recall that the exponential function $x(t) = e^{at}$ has this property. More generally, for any value of the constant c, the function $x(t) = ce^{at}$ has the required property.

$$\frac{d}{dt} x(t) = \frac{d}{dt}(ce^{at}) = cae^{at} = ax(t).$$

If the method of trial solutions seems mysterious, there is a more systematic way to determine solutions of Equation (7.1). The equation can be written

$$\frac{1}{x(t)} \frac{d}{dt} x(t) = a. \tag{7.2}$$

But

$$\frac{1}{x(t)} \frac{d}{dt} x(t) = \frac{d}{dt}[\log_e x(t)],$$

where $\log_e x(t)$ is the natural logarithm or the logarithm to the base e. Therefore, Equation (7.2) is equivalent to

$$\frac{d}{dt}[\log_e x(t)] = a. \tag{7.3}$$

Integrating both sides of Equation (7.3) with respect to the variable t, we have

$$\log_e x(t) = at + k, \tag{7.4}$$

where k is a constant of integration. Taking exponentials, we obtain $x(t) = e^{at+k} = e^k e^{at}$. Define the constant $c = e^k$. This gives the general solution of Equation (7.1) in the form

$$x(t) = ce^{at}. \tag{7.5}$$

By general solution we mean that every solution of (7.1) can be written in this form for some value of the constant c. We observe that the general solution derived by this systematic method is exactly the same as that obtained by the method of trial solutions.

If the population $x(t)$ is known for some time $t = t_0$, the constant c in the solution can be uniquely determined. Since $x(t) = ce^{at}$, we have $x(t_0) = ce^{at_0}$ and, therefore, $c = x(t_0)e^{-at_0}$. The solution becomes

$$x(t) = x(t_0)e^{-at_0}e^{at} = x(t_0)e^{a(t-t_0)}. \tag{7.6}$$

When the initial or starting population $x(0)$ is known, we have

$$x(t) = x(0)e^{at}. \tag{7.7}$$

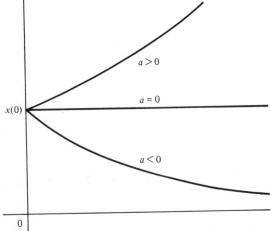

Figure 7.2. Exponential Growth $x(t) = x(0)e^{at}$

The solution at time t is completely determined by knowing the population at $t = 0$ or at $t = t_0$. The solution (7.6) or (7.7) is the equation of *exponential growth*. When $a > 0$, the population increases with time; when $a = 0$, the population remains at the constant level $x(t) = x(0)$; when $a < 0$, the population decreases with time to zero. These three cases are illustrated in Figure 7.2.

Example 7.2.1   Determine the solution of the differential equation $dx/dt = (.1)x$ that satisfies the initial condition $x(0) = 1000$. If $x(t)$ represents the size of a bacteria population after t hours, what is the population after 10 hours?

Solution: In this example $a = .1$. The general solution is $x(t) = x(0)e^{(.1)t}$. The solution satisfying the initial condition is $x(t) = 1000e^{(.1)t}$. After 10 hours, the population is $x(1) = 1000e \approx 2718$.

The equation $dx/dt = ax$ is an example of a linear differential equation. A differential equation is *linear* if terms involving $x(t)$ and its derivatives contain $x(t)$ or a derivative to the first degree only and if there are no product terms such as $x(dx/dt)$. For example, terms such as x^2, $x^3(d^2x/dt^2)$, $\sin x$, and so on, are absent. When such terms occur, the equation is said to be *nonlinear*.

Example 7.2.2   The following differential equations are linear.

1. $\dfrac{d^2x}{dt^2} = 8x.$

2. $\dfrac{dx}{dt} - t^2x = 0.$

3. $\dfrac{dx}{dt} + 2tx = e^{-t^2}.$

4. $\dfrac{d^3x}{dt^3} + 3\dfrac{d^2x}{dt^2} + 3\dfrac{dx}{dt} = 1 + \cos t.$

Example 7.2.3   The following differential equations are nonlinear.

1. $\dfrac{dx}{dt} = x^2.$

2. $x\dfrac{d^2x}{dt^2} = t^3.$

3. $\left(\dfrac{dx}{dt}\right)^2 + xe^x = 1 + t.$

4. $\dfrac{d^2x}{dt^2} + \dfrac{1}{1 + x^2} = x.$

The most general first order linear differential equation is

$$\frac{dx}{dt} + a(t)x = f(t). \tag{7.8}$$

This equation is said to be *homogeneous* if $f(t) = 0$ for all t. Otherwise, it is said to be *inhomogeneous*. If the coefficient $a(t)$ is a constant, the equation is said to have *constant coefficients*. Equation (7.1) is a special case of Equation (7.8) with $f(t) = 0$ and $a(t) = -a$.

Before solving Equation (7.8), the following property of linear homogeneous equations should be understood. Suppose that $x_1(t)$ and $x_2(t)$ are solutions of $(dx/dt) + a(t)x = 0$. Then, $k_1 x_1(t) + k_2 x_2(t)$ is also a solution for any values of the constants k_1 and k_2. To see this, note that

$$\frac{d}{dt}[k_1 x_1(t) + k_2 x_2(t)] + a(t)[k_1 x_1(t) + k_2 x_2(t)] = k_1 \left[\frac{dx_1}{dt} + a(t)x_1 \right]$$

$$+ k_2 \left[\frac{dx_2}{dt} + a(t)x_2 \right]$$

$$= k_1(0) + k_2(0) = 0.$$

This property of linear homogeneous equations (of any order) is often used to define the word "linear."

We now consider the homogeneous equation

$$\frac{dx}{dt} + a(t)x = 0. \tag{7.9}$$

To solve this equation, we write it in the form

$$\frac{1}{x}\frac{dx}{dt} = -a(t) \qquad \text{or} \qquad \frac{d}{dt} \log_e x(t) = -a(t).$$

Integrating on both sides, we conclude that

$$\log_e x(t) = -\int^t a(s)\, ds + k, \tag{7.10}$$

where k is a constant of integration. Taking exponentials, we have

$$x(t) = e^{-\int^t a(s)\, ds + k} = e^k e^{-\int^t a(s)\, ds}.$$

If we define a constant $c = e^k$, the solution of Equation (7.9) can be written

$$x(t) = ce^{-\int^t a(s)\, ds}. \tag{7.11}$$

The constant c can be determined if an initial condition is known. This is illustrated in the following examples.

Example 7.2.4 Determine the general solution of the differential equation $(dx/dt) + x/(1 + t) = 0$. What is the solution that satisfies the initial condition $x(0) = 10$?

Solution: In this example, $a(t) = 1/(1 + t)$ and $\int^t a(s)\, ds = \int^t [1/(1 + s)]\, ds = \log_e (1 + t)$. The general solution (7.11) is $x(t) = ce^{-\log_e(1+t)} = c/(1 + t)$ [since $e^{\log_e f(t)} = f(t)$ for any function $f(t)$]. If $x(0) = 10$, we have $10 = c/(1 + 0)$ and $c = 10$. The solution that satisfies the given initial condition is $x(t) = 10/(1 + t)$. As t increases, $x(t)$ decreases steadily to zero.

Example 7.2.5 A bacteria population is growing in such a way that the average growth rate of the bacteria in the population at time t (measured in hours) is $1/(1 + 2t)$. Suppose that the initial population is $x(0) = 1000$. What is the population after 4 hours? After 12 hours?

Solution: The average growth rate is $(1/x)(dx/dt) = 1(1 + 2t)$. This is a homogeneous first order linear equation with $a(t) = -1/(1 + 2t)$. Integrating the differential equation, we have $\log_e x(t) = \frac{1}{2} \log_e (1 + 2t) + k$, where k is the constant of integration. Taking exponentials, we have $x(t) = e^k e^{(1/2)\log_e (1 + 2t)} = e^k(1 + 2t)^{1/2}$. Defining $c = e^k$, we obtain the solution $x(t) = c(1 + 2t)^{1/2}$. Since $x(0) = 1000$, we have $1000 = c$, and the required solution is $x(t) = 1000(1 + 2t)^{1/2}$. The population after 4 hours is $x(4) = 1000(1 + 8)^{1/2} = 3000$. After 12 hours the population has grown to $x(12) = 1000(25)^{1/2} = 5000$. The general solution $x(t) = c(1 + 2t)^{1/2}$ can be verified by substitution in the original equation. For this solution, we have

$$\frac{1}{x(t)}\frac{dx}{dt} = \frac{1}{c(1 + 2t)^{1/2}}\frac{c}{(1 + 2t)^{1/2}},$$

which is equal to $1/(1 + 2t)$.

Example 7.2.6 A Model of Seasonal Growth The first order differential equation $dx/dt = rx(t) \cos (t)$, where r is a positive constant, can be thought of as a simple model of seasonal growth. As t increases, the growth rate dx/dt of the population $x(t)$ is alternately positive and negative and, therefore, the population alternately increases and decreases. This can be caused by seasonal factors such as the availability of food.

To solve the equation, we note that $a(t) = -r \cos(t)$ in this example and that $\int^t a(s)\, ds = -r \int^t \cos(s)\, ds = -r \sin(t) + k$. The general solution can be written $x(t) = ce^{r \sin(t)}$. Setting $t = 0$, we have $c = x(0)$ and the population at time t is $x(t) = x(0)e^{r \sin(t)}$. The maximum size of the population $e^r x(0)$ occurs at $t = \pi/2, 5\pi/2, 9\pi/2, \ldots$, where $\sin(t) = 1$. The minimum size of the population $e^{-r}x(0)$ occurs at $t = 3\pi/2, 7\pi/2, 11\pi/2, \ldots$, where $\sin(t) = -1$.

In this model, the population oscillates between $e^{-r}x(0)$ and $e^r x(0)$ with period 2π. The times $t = 0, 2\pi, 4\pi, \ldots$ can be thought of as the midpoints of the seasons of greatest food availability (summer) and $t = \pi, 3\pi, 5\pi, \ldots$ corresponds to the midpoints of the seasons of greatest food scarcity (winter). The length of the year is 2π time units. This is illustrated in Figure 7.3.

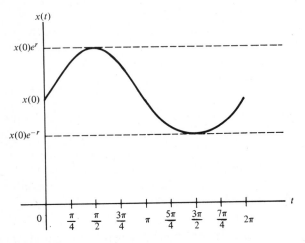

Figure 7.3. Seasonal Growth $x(t) = x(0)e^{r \sin t}$

We now return to the general inhomogeneous equation (7.8). Our method of solution is to find an *integrating factor* for this equation. An integrating factor is a function $g(t)$ that has the property

$$g(t)\frac{dx}{dt} + g(t)a(t)x(t) = \frac{d}{dt}(g(t)x(t)).$$

But, from the product rule of differentiation,

$$\frac{d}{dt}(g(t)x(t)) = g(t)\frac{dx}{dt} + x(t)\frac{dg}{dt}.$$

Therefore, the integrating factor $g(t)$ must be chosen to satisfy the first-order homogeneous equation $dg/dt = a(t)g(t)$. One solution of this equation is $g(t) = e^{\int^t a(s)\,ds}$. Therefore, we have

$$\frac{d}{dt}[e^{\int^t a(s)\,ds}x(t)] = e^{\int^t a(s)\,ds}\,\frac{dx}{dt}(t) + e^{\int^t a(s)\,ds}a(t)x(t).$$

We conclude that Equation (7.8) can be written

$$\frac{d}{dt}[e^{\int^t a(s)\,ds}x(t)] = e^{\int^t a(s)\,ds}f(t). \tag{7.12}$$

Integrating both sides of Equation (7.12) with u as the variable of integration, we have

$$e^{\int^t a(s)\,ds}x(t) = \int^t e^{\int^u a(s)\,ds}f(u)\,du + c, \tag{7.13}$$

where c is the constant of integration. Finally, we multiply both sides of Equation (7.13) by $e^{-\int^t a(s)\,ds}$ to obtain

$$x(t) = ce^{-\int^t a(s)\,ds} + e^{-\int^t a(s)\,ds}\int^t e^{\int^u a(s)\,ds}f(u)\,du. \tag{7.14}$$

This is the general solution of Equation (7.8). It is a formidable formula but, with practice, it is not difficult to apply. Since $f(t)$ and $a(t)$ are given functions, we can evaluate the integrals in (7.14) and determine the general solution.

For convenience, we outline the steps that lead to the solution of Equation (7.8).

1. Write the equation in the form (7.8) and identify $a(t)$ and $f(t)$.
2. Calculate $\int^t a(s)\,ds$ and $e^{\int^t a(s)\,ds}$.
3. Multiply the equation by the integrating factor $e^{\int^t a(s)\,ds}$ and write the equation in the form (7.12).
4. Determine $I(t)$, the indefinite integral of the function $e^{\int^t a(s)\,ds}f(t)$.
5. The general solution is then $x(t) = ce^{-\int^t a(s)\,ds} + e^{-\int^t a(s)\,ds}I(t)$.
6. Finally, if $x(t)$ is known for some time t_0, the constant c can be determined in terms of $x(t_0)$.

Example 7.2.7 Determine the general solution of $(dx/dt) + x = e^{-t}$. What is the solution which satisfies $x(0) = 1000$?

Solution: In this example, $a(t) = 1$ and $f(t) = e^{-t}$. Then $\int^t a(s)\, ds = \int^t 1\, ds = t$ and $e^{\int^t a(s)\, ds} = e^t$. Multiplying the differential equation by the integrating factor e^t, we have

$$\frac{d}{dt}(e^t x(t)) = e^t e^{-t} = 1.$$

Integrating, we obtain $x(t)e^t = c + t$ or $x(t) = ce^{-t} + te^{-t}$. This is the general solution, which, of course, can be verified by differentiation. When $t = 0$, $x(0) = c = 1000$. The required solution is $x(t) = (1000 + t)e^{-t}$.

Example 7.2.8   Determine the solution of $(dx/dt) + 2tx = te^{-t^2}$ that satisfies the initial condition $x(0) = 1$.

Solution: In this example, we have $a(t) = 2t$ and $f(t) = te^{-t^2}$. The integrating factor $e^{\int^t a(s)\, ds} = e^{\int^t 2s\, ds} = e^{t^2}$. Multiplying by the integrating factor, the differential equation becomes

$$\frac{d}{dt}(e^{t^2} x(t)) = te^{-t^2} e^{t^2} = t.$$

Integrating, we obtain $e^{t^2} x(t) = c + t^2/2$. The general solution is $x(t) = ce^{-t^2} + (t^2/2)e^{-t^2}$. The initial condition gives $x(0) = c = 1$. Therefore, the required solution is $x(t) = e^{-t^2} + (t^2/2)e^{-t^2}$.

Example 7.2.9   Determine the solution of $(dx/dt) + (1/t)x = t$ that satisfies the condition $x(6) = 20$.

Solution: The integrating factor is $e^{\int^t a(s)\, ds} = e^{\int^t (1/s)\, ds} = e^{\log_e t} = t$. The differential equation becomes $d(tx(t))/dt = t^2$. Integrating, we have $tx(t) = c + t^3/3$, where c is the constant of integration. The general solution is $x(t) = (c/t) + (t^2/3)$. To determine c, we know that $x(6) = 20 = (c/6) + \frac{36}{3} = (c/6) + 12$. Therefore, $c = 48$ and the required solution is $x(t) = (48/t) + (t^2/3)$.

Example 7.2.10   Intravenous Feeding of Glucose   Infusion of glucose into the bloodstream is an important medical technique. To study this process, define $G(t)$ to be the amount of glucose in the bloodstream of a patient at time t. Suppose that glucose is infused into the bloodstream at the constant rate c grams per minute. At the same time, the glucose is converted and re-moved from the bloodstream at a rate proportional to the amount of glucose present. Therefore, the function $G(t)$ satisfies the first order differential equation

$$\frac{dG}{dt} = c - aG,$$

where a is a positive constant. This is a nonhomogeneous first order linear differential equation of the form (7.8) with $a(t) = a$ and $f(t) = c$.

To solve, we write the equation in the form $(dG/dt) + aG = c$ and multiply by the integrating factor e^{at} to obtain $d(e^{at}G(t))/dt = ce^{at}$. Integrating, we have $e^{at}G(t) = k + (c/a)e^{at}$, where k is the constant of integration. Finally, the general solution is $G(t) = ke^{-at} + c/a$. The constant k can be determined in terms of the initial amount $G(0)$ of glucose in the bloodstream. We have $G(0) = k + c/a$. The solution can be written in the form

$$G(t) = \frac{c}{a} + \left[G(0) - \frac{c}{a} \right] e^{-at}.$$

As time increases, $G(t)$ approaches the limit c/a. This will be the equilibrium amount of glucose in the bloodstream.

Problems 7.2

1. Determine the general solutions of the following first order equations.

(a) $\dfrac{dx}{dt} = 5x.$

(b) $\dfrac{dx}{dt} = \cos t.$

(c) $\dfrac{dx}{dt} = 3t^2 + 5.$

(d) $\dfrac{dx}{dt} = x + t.$

2. Determine the solutions of the following first order equations that satisfy the initial condition $x(0) = 1$.

(a) $\dfrac{dx}{dt} - 3x = 0.$

(b) $\dfrac{dx}{dt} = 1 + 2t + t^2.$

(c) $\dfrac{dx}{dt} = e^{-2t}.$

(d) $\dfrac{dx}{dt} = x + 1.$

3. In a study of fasting, the weight of a volunteer decreased from 140 pounds to 110 pounds in 30 days. It was observed that the weight loss per day

was proportional to the weight of the volunteer. What is the differential equation for the weight as a function of time? Determine the weight of the volunteer after 15 days.

4. Under ideal conditions, human lice populations will increase exponentially with constant a equal to 0.10 when time is measured in days. Suppose that initially there are 100 lice growing under ideal conditions. Determine the lice population after (a) 10 days, (b) 20 days, and (c) 50 days.

5. Determine the solutions of the following first order equations that satisfy the initial condition $x(0) = 0$.

(a) $\dfrac{dx}{dt} = \tfrac{1}{2}x.$

(b) $\dfrac{dx}{dt} = t - \dfrac{x}{t+1} + 1.$

(c) $\dfrac{dx}{dt} = (1+t)^3.$

(d) $\dfrac{dx}{dt} = 2 \sin t \cos t + \cos t.$

6. Suppose that $T(t)$ is the temperature difference at time t between an object and its surrounding medium. Newton's law of cooling states that the rate of change of this temperature difference is proportional to the temperature difference. This implies that $T(t)$ satisfies the first order differential equation $dT/dt = -kT$, where k is a positive constant characteristic of the object and the surrounding medium.
   (a) In terms of k, calculate the length of time for the temperature difference to decrease to one half of its original value.
   (b) Calculate the length of time for the temperature difference to decrease to one fourth of its original value.

7. In Problem 6, suppose that $k = 0.05$, when t is measured in hours, and suppose that the initial temperature difference is $T(0) = 100°C$.
   (a) Calculate $T(t)$ for $t = 1, 10, 24,$ and 48 hours.
   (b) Calculate the lengths of time for the temperature difference to decrease to 50°C and to 25°C.

8. Determine the solutions of the differential equation $(dx/dt) + 2tx = t^{2k+1}e^{-t^2}$, where k is any positive number. Determine the particular solution that satisfies the initial condition $x(1) = 1$.

9. Determine the solutions of the following first order equations.

(a) $\dfrac{dx}{dt} + \dfrac{1}{t}x = 6t^2.$

(b) $\dfrac{dx}{dt} + \dfrac{1}{t^2}x = \dfrac{1}{t^2}.$

(c) $\dfrac{dy}{dt} - y \tan t = \sin t.$

(d) $\dfrac{dy}{dx} + x^2 = -xy.$

10. Suppose that $x_1(t)$ and $x_2(t)$ are two solutions of the first order differential equation $(dx/dt) + a(t)x = f(t)$. Define $y(t) = x_1(t) - x_2(t)$. Prove that $y(t) = y(0)e^{-\int_0^t a(s)\,ds}$.

11. The first order differential equation $(dx/dt) + a(t)x = b(t)x^n$ is called the *Bernoulli equation*. If $n \neq 1$, define the change of variable $y = x^{1-n}$.
    (a) Prove that y satisfies a first order linear equation and determine its solution.
    (b) Determine the general solutions for the special cases $n = 0$ and $n = 1$.

12. Use the method of Problem 11 to solve the following equations.

    (a) $\dfrac{dx}{dt} - x = tx^2.$

    (b) $\dfrac{dx}{dt} - \dfrac{2}{t}x = \dfrac{x^2}{t^2}.$

    (c) $\dfrac{dx}{dt} + \dfrac{1}{t}x = x^3.$

    (d) $\dfrac{dx}{dt} = e^t x^5.$

13. Determine the solutions of the equations of Problem 12 that satisfy the initial condition $x(1) = 1$.

14. Consider a gene with two alleles A and a, which occur in the proportions $p(t)$ and $q(t) = 1 - p(t)$ at time t in a population. Suppose that the allele A mutates to the allele a at the rate μ per unit of time. This means that $dp/dt = -\mu p$. The constant μ is called the *mutation rate*.
    (a) If initially $p(0) = q(0) = 0.5$, determine $p(t)$ and $q(t)$ in terms of μ.
    (b) In terms of μ, determine the length of time for $p(t)$ to decrease to 0.3.

15. Mutations may occur in both directions between the alleles A and a with forward mutation rate μ and backward mutation rate v. This implies that $dp/dt = -\mu p + vq = -\mu p + v(1 - p) = v - (\mu + v)p$.
    (a) Determine $p(t)$ and $q(t)$ in terms of $p(0)$, $q(0)$, μ, and v.
    (b) Prove that $\lim_{t \to \infty} p(t) = v/(\mu + v)$ and $\lim_{t \to \infty} q(t) = \mu/(\mu + v)$. (These are the equilibrium gene frequencies.)

16. A chemical substance S is produced at the rate of r moles per minute in a chemical reaction. At the same time, it is consumed at the rate of c moles per minute per mole of S. Define $S(t)$ to be the number of moles of the chemical present at time t.
    (a) Determine the differential equation satisfied by $S(t)$.
    (b) By solving the differential equation, determine $S(t)$ in terms of $S(0)$.
    (c) Prove that $\lim_{t \to \infty} S(t) = r/c$. (This is the equilibrium amount of the chemical.)

17. If the equation $dx/dt = 2x + e^{3t}$ represents the growth rate at time t of a population $x(t)$, give a biological interpretation for each term on the right-hand side. If the initial population is $x(0) = 50$, determine the populations at times $t = 0.1$, $t = 0.2$, and $t = 0.5$.

18. An infectious disease is introduced to a large population. The proportion of people who have been exposed to the disease increases with time. Suppose that $p(t)$ is the proportion of people who have been exposed to the disease within t years of its introduction. If $p'(t) = [1 - p(t)]/3$ and if $p(0) = 0$, determine $p(t)$ for all $t > 0$. After how many years has the proportion increased to 90 per cent?

7.3 First Order Nonlinear Equations: Separation of Variables

We now turn briefly to nonlinear first order differential equations. Consider the very general first order equation

$$\frac{dx}{dt} = g(t, x), \tag{7.15}$$

where $g(t, x)$ is a given function that is usually assumed to be continuous. The biological interpretation of this equation may be that the growth rate of a population is a function of the time and the size of the population. Equation (7.15) is linear if $g(t, x) = -a(t)x + f(t)$ for some functions $a(t)$ and $f(t)$. In general, it is not possible to find an explicit formula for the solution $x(t)$, although there are theorems that guarantee the existence of at least one solution for every initial value $x(t_0)$. It is usually necessary to use numerical methods to determine approximate solutions.

There are a number of special types of first order nonlinear equations that can be solved explicitly. Discussions of the special methods involved in finding solutions can be found in any elementary textbook on differential equations. We will restrict ourselves here to one type of nonlinear equation that can always be solved by elementary methods. Fortunately, it is this type of equation which is most useful in applications.

We say that the variables x and t in Equation (7.15) are *separable* if $g(t, x) = h(x)k(t)$, where $h(x)$ is a function only of x and $k(t)$ is a function only of t. If the variables are separable in Equation (7.15), the equation can be solved by the method of *separation of variables*. In this case, Equation (7.15) can be written

$$\frac{1}{h(x)}\frac{dx}{dt} = k(t) \qquad\qquad (7.16)$$

or, less formally, $[1/h(x)]\,dx = k(t)\,dt$. In this form, the left-hand side can be integrated over the variable x (which, of course, depends on t) and the right-hand side can be integrated over the variable t. Performing these two integrations, we have the general solution

$$\int \frac{dx}{h(x)} = \int k(t)\,dt + c. \qquad\qquad (7.17)$$

If $h(x)$ and $k(t)$ are sufficiently simple, these integrals can be evaluated to determine the solution $x(t)$ explicitly. The method of separation of variables is best illustrated by a number of examples.

Example 7.3.1   Determine the solution of $dx/dt = (1 + x^2)(1 + 2t)$ that satisfies the initial condition $x(0) = 0$.

Solution: This is a first order nonlinear equation with variables separable. We have $h(x) = 1 + x^2$ and $k(t) = 1 + 2t$. After separating the variables, the equation becomes $dx/(1 + x^2) = (1 + 2t)\,dt$. Integrating both sides, we obtain $\tan^{-1} x = t + t^2 + c$, where c is the constant of integration. Equivalently, we can write $x(t) = \tan(c + t + t^2)$. This is the general solution. To determine the solution that satisfies the initial condition, we have $x(0) = 0 = \tan(c)$. Therefore, $c = 0$, and the required solution is $x(t) = \tan(t + t^2)$.

Example 7.3.2   Determine the general solution of $dx/dt = (1 + x)e^{-t}$. What is the solution that satisfies $x(0) = 1$?

Solution: We write the equation in the form $dx/(1 + x) = e^{-t}\,dt$ and integrate both sides. This gives $\log_e(1 + x) = c - e^{-t}$, where c is the constant of integration. This is the general solution. To determine c, we have the condition $x(0) = 1$. This implies that $\log_e 2 = c - 1$ or $c = 1 + \log_e 2$. The solution satisfying $x(0) = 1$ is $\log_e(1 + x) = 1 + \log_e 2 - e^{-t}$.

Example 7.3.3 Logistic Growth   The growth rate per individual in a population is the difference between the average birth rate and the average death rate. We will assume that the average birth rate is a positive constant β, independent of time t and population size $x(t)$. Let us assume that the average death rate is proportional to the size of the population and is, therefore, $\delta x(t)$, where δ is a positive constant. This increasing mortality with increasing population may be due to the effects of crowding or to increased competition for the available food resources. In this model, the equation governing population growth is

$$\frac{1}{x}\frac{dx}{dt} = \beta - \delta x \quad \text{or} \quad \frac{dx}{dt} = x(\beta - \delta x). \tag{7.18}$$

Separating variables, we have $\int [1/x(\beta - \delta x)]\,dx = \int dt$. Noting that we can write

$$\frac{1}{x(\beta - \delta x)} = \frac{1}{\beta x} + \frac{\delta}{\beta(\beta - \delta x)},$$

we can integrate to obtain

$$\frac{1}{\beta}\int \frac{1}{x}\,dx + \frac{\delta}{\beta}\int \frac{1}{\beta - \delta x}\,dx = \int dt,$$

$$\frac{1}{\beta}\log_e x + \frac{\delta}{\beta}\left(-\frac{1}{\delta}\right)\log_e(\beta - \delta x) = t + c.$$

Therefore,

$$\frac{1}{\beta}[\log_e x - \log_e (\beta - \delta x)] = \frac{1}{\beta}\log_e \frac{x}{\beta - \delta x} = t + c.$$

If the initial population is $x(0)$, we have $1/\beta \log_e \{x(0)/[\beta - \delta x(0)]\} = c$. The solution becomes

$$\log_e \frac{x(t)}{\beta - \delta x(t)} - \log_e \frac{x(0)}{\beta - \delta x(0)} = \beta t \quad \text{or} \quad \log_e \frac{x(t)(\beta - \delta x(0))}{x(0)(\beta - \delta x(t))} = \beta t.$$

Taking exponentials, we have $x(t)(\beta - \delta x(0))/x(0)(\beta - \delta x(t)) = e^{\beta t}$. Finally, by solving for $x(t)$. the solution can be written

$$x(t) = \frac{x(0)\beta e^{\beta t}}{\beta - \delta x(0) + \delta x(0)e^{\beta t}}. \tag{7.19}$$

The growth process described by this function is called *logistic growth* and Equation (7.18) is called the *logistic equation.* In logistic growth, as time increases the population approaches a limiting or equilibrium size. The equilibrium population is

$$\lim_{t \to \infty} x(t) = \lim_{t \to \infty} \frac{x(0)\beta e^{\beta t}}{\beta - \delta x(0) + \delta x(0)e^{\beta t}}.$$

Multiplying numerator and denominator of $x(t)$ by $e^{-\beta t}$, we have

$$\lim_{t \to \infty} x(t) = \lim_{t \to \infty} \frac{x(0)\beta}{\beta e^{-\beta t} - \delta x(0)e^{-\beta t} + \delta x(0)} = \frac{x(0)\beta}{x(0)\delta} = \frac{\beta}{\delta},$$

since as $t \to \infty$, $e^{-\beta t} \to 0$. This formula tells us that the equilibrium population is directly proportional to the average birth rate and inversely proportional to the average death rate per unit of population. Logistic growth is illustrated in Figure 7.4.

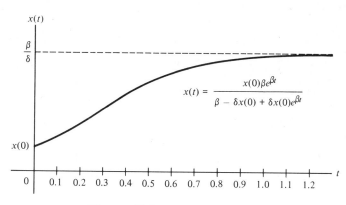

Figure 7.4. Logistic Growth

Example 7.3.4 Determine the solution of $dx/dt = x((1/10) - (x/1000))$ that satisfies the initial condition $x(0) = 10$. What is the equilibrium population $\lim_{t \to \infty} x(t)$?

Solution: This is the logistic equation with $\beta = .1$ and $\delta = .001$. From Equation (7.19), the solution satisfying $x(0) = 10$ is

$$x(t) = \frac{e^{(.1)t}}{(.1) - (.01) + (.01)e^{(.1)t}} = \frac{e^{(.1)t}}{(.09) + (.01)e^{(.1)t}}.$$

As $t \to \infty$, the population $x(t)$ approaches the equilibrium value $\beta/\delta = 100$. This represents the population size for which the birth rate exactly balances the death rate.

Problems 7.3

1. Determine the general solutions of the following first order differential equations.

   (a) $\dfrac{dx}{dt} = x \cos t.$ (b) $\dfrac{dx}{dt} = x^2(1 + t^2).$

   (c) $\dfrac{dx}{dt} = t^2(1 + x^2).$ (d) $\dfrac{dx}{dt} + x = xte^{(t-1)^2}.$

2. Determine the particular solutions of the following equations that satisfy the initial condition $x(0) = 1$.

   (a) $\dfrac{dx}{dt} = x(\cos t + \sin t).$ (b) $\dfrac{dx}{dt} = x^2(1 + t^2).$

   (c) $\dfrac{dx}{dt} = t^2(1 + x^2).$ (d) $\dfrac{dx}{dt} + 2x = xt^2.$

3. Cigarette consumption in the United States increased from 50 per capita in 1900 to 3900 per capita in 1960 (approximate figures). Assume that the growth in cigarette consumption is governed by the logistic equation and assume that the limiting consumption is 4000 cigarettes per capita. Estimate the consumptions in 1910, 1920, 1930, 1940, and 1950.

4. Consider a population $x(t)$ that is growing according to the equation $dx/dt = x(\beta - \delta x)$ of logistic growth. Prove that the growth rate is a maximum when the population is equal to half its equilibrium size. (If the population is to be harvested, it should be maintained at this level to maximize the yield.)

5. A bacteria population grows from an initial size of 100 to a limiting or equilibrium size of 100,000. Suppose that, in the first hour, the population increases to 120. Assuming that the growth is governed by the logistic equation, determine the population as a function of time.

6. In how many hours will the bacteria population of Problem 5 increase to 1000? To 10,000? To 50,000?

7. One possible shortcoming of the logistic model of population growth is that the average growth rate $(1/x)(dx/dt)$ is largest when the population x is small. However, there are species which may become extinct if their

population becomes too small. Suppose that m is the minimum viable population size for such a species. Populations smaller than m will become extinct. Prove that the modification of the logistic equation $dx/dt = x(\beta - \delta x)(1 - m/x)$ has the required property that the population becomes extinct if $x < m$. (The term $1 - m/x$ is a correction term which takes into account a factor ignored in the original model of population growth.)

8. (a) Solve the modified logistic equation $dx/dt = (\beta - \delta x)(x - m)$ by writing

$$\frac{1}{(\beta - \delta x)(x - m)} = \frac{1}{\beta - \delta m}\left(\frac{\delta}{\beta - \delta x} + \frac{1}{x - m}\right)$$

and integrating.

(b) If $\beta = 100$, $\delta = 1$, and $m = 10$, plot the solutions $x(t)$ for $t > 0$ when $x(0) = 20$ and when $x(0) = 5$.

9. Bacteria are supplied as food to a protozoan population at a constant rate w. It is observed that the bacteria are consumed at a rate that is proportional to the square of their numbers. The concentration $c(t)$ of bacteria at time t satisfies the first order equation $dc/dt = w - rc^2$, where r is a positive constant.

(a) Determine the bacteria concentration $c(t)$ in terms of $c(0)$.

(b) What is the equilibrium concentration of bacteria; that is, for what value of c is dc/dt equal to zero?

10. In a chemical reaction, two substances, C_1 and C_2, combine in equal amounts to produce a compound, C_3. Suppose that a and b are the initial concentrations (at $t = 0$) of C_1 and C_2. Define $x(t)$ to be the concentration of C_3 at time t. The rate of increase of the concentration of C_3 is $dx/dt = r(a - x)(b - x)$, where r is a positive constant.

(a) If $x(0) = 0$, determine the concentration of C_3 as a function of time for $t > 0$.

(b) If $a = 10$ and $b = 15$ (in appropriate units), determine the limiting concentration of C_3.

11. In some chemical reactions, certain products may catalyze their own formation. If $x(t)$ is the amount of such a product at time t, a possible model for such a reaction is given by the differential equation $dx/dt = rx(c - x)$, where r and c are positive constraints. In this model, the reaction is completed when $x = c$, presumably because one of the chemicals of the reaction is used up.

(a) Determine the general solution in terms of the constants r, c, and $x(0)$.

(b) For $r = 1$, $c = 100$, and $x(0) = 20$, draw a graph of $x(t)$ for $t > 0$.

12. Determine the solutions of the following differential equations that satisfy the given initial conditions.

(a) $\dfrac{dx}{dt} = ax^5, x(0) = 4.$ (b) $x\dfrac{dy}{dx} = y, y(2) = 6.$

(c) $y\dfrac{dy}{dx} = x, y(1) = 2.$ (d) $\dfrac{dy}{dt} = \dfrac{2y}{t}, y(1) = 1.$

13. Nutrients flow across cell walls to determine the growth, survival, and reproduction of the cells. This suggests that, during the early stages of a cell's growth, the rate of increase of the weight of the cell will be proportional to its surface area. If the shape and density of the cell do not change during growth, the weight $x(t)$ of the cell at time t will be proportional to the cube of a radius, while the surface area is proportional to the square of a radius.
   (a) Verify that $x(t)$ satisfies the first order equation $dx/dt = cx^{2/3}$ during the early stages of growth (c is a positive constant).
   (b) In terms of the constant c and the initial weight $x(0)$, determine the weight $x(t)$ of the cell at time t.
   (c) If $c = 3$ and $x(0) = 1$ (in appropriate units), determine the length of time for the weight of the cell to double.

14. What are the limitations of the model of cell growth of Problem 13? If there is a maximum weight which the cell cannot exceed, develop a differential equation model which takes this into account.

15. In a model of epidemics,[1] a single infected individual is introduced into a community containing n individuals susceptible to the disease. Define $x(t)$ to be the number of uninfected individuals in the population at time t. If we assume that the infection spreads to all those susceptible, then $x(t)$ decreases from its initial value $x(0) = n$ to zero. A possible equation for $x(t)$ is $dx/dt = -rx(n + 1 - x)$, where r is a positive constant which measures the rate of infection. Determine the solution of this first order equation. When is the infection rate a maximum?

7.4 Second Order Linear Differential Equations

In this section, we return to linear differential equations, but now we consider second order equations. Our discussion of these equations will closely parallel the discussion of second order difference equations in Section 6.3.

[1] N. Bailey, "A Simple Stochastic Epidemic," *Biometrika*, 37:193–202 (1950).

The most general second order linear differential equation is

$$a(t)x''(t) + b(t)x'(t) + c(t)x(t) = f(t), \qquad (7.20)$$

where $a(t)$, $b(t)$, $c(t)$, and $f(t)$ are given functions of t and $a(t) \neq 0$ for all t. For convenience, we have used the notation $x''(t)$ and $x'(t)$ to denote d^2x/dt^2 and dx/dt. [Another common notation is $\ddot{x}(t)$ and $\dot{x}(t)$.] If the functions $a(t)$, $b(t)$, and $c(t)$ are constants, Equation (7.20) is said to have *constant coefficients*. If $f(t) = 0$ for all t, the equation is said to be *homogeneous*. In this section, we will study the homogeneous, constant coefficient equation

$$ax''(t) + bx'(t) + cx(t) = 0, \qquad (7.21)$$

where a, b, and c are constants and $a \neq 0$.

The first order equation that most closely resembles (7.21) is the equation $x'(t) + ax(t) = 0$, which has the solution $x(t) = e^{-at}$. Using a method of trial solutions, we will look for solutions of (7.21) of the form $x(t) = e^{\lambda t}$ for some values of λ. Now if $x(t) = e^{\lambda t}$, then $x'(t) = \lambda e^{\lambda t}$ and $x''(t) = \lambda^2 e^{\lambda t}$. Substituting into Equation (7.21), we obtain

$$a\lambda^2 e^{\lambda t} + b\lambda e^{\lambda t} + ce^{\lambda t} = 0. \qquad (7.22)$$

Multiplying by $e^{-\lambda t}$, we conclude that λ must satisfy

$$a\lambda^2 + b\lambda + c = 0. \qquad (7.23)$$

Equation (7.23) is called the *auxiliary equation* of Equation (7.21). We recognize it as the same equation as appeared in Section 6.3. The roots of the auxiliary equation are

$$\lambda_1 = \frac{-b + \sqrt{b^2 - 4ac}}{2a} \quad \text{and} \quad \lambda_2 = \frac{-b - \sqrt{b^2 - 4ac}}{2a}. \qquad (7.24)$$

There are three cases to be considered, according to whether $b^2 - 4ac$ is greater than, equal to, or less than zero.

The simplest case occurs when $b^2 - 4ac > 0$, or when the auxiliary equation has two real, distinct roots, λ_1 and λ_2. Then the differential equation has two distinct solutions, $x_1(t) = e^{\lambda_1 t}$ and $x_2(t) = e^{\lambda_2 t}$. The general solution is

$$x(t) = k_1 e^{\lambda_1 t} + k_2 e^{\lambda_2 t}, \qquad (7.25)$$

where k_1 and k_2 are arbitrary constants. This solution can be verified by substitution into Equation (7.21). If $x(t) = k_1 e^{\lambda_1 t} + k_2 e^{\lambda_2 t}$, then

$$x'(t) = k_1 \lambda_1 e^{\lambda_1 t} + k_2 \lambda_2 e^{\lambda_2 t} \quad \text{and} \quad x''(t) = k_1 \lambda_1^2 e^{\lambda_1 t} + k_2 \lambda_2^2 e^{\lambda_2 t}.$$

Then,

$$ax''(t) + bx'(t) + cx(t) = k_1 e^{\lambda_1 t}(a\lambda_1^2 + b\lambda_1 + c)$$
$$+ k_2 e^{\lambda_2 t}(a\lambda_2^2 + b\lambda_2 + c) = 0.$$

To determine the constants k_1 and k_2, we must know two conditions that the solution $x(t)$ must satisfy. If $x(0)$ and $x'(0)$ are given, then k_1 and k_2 can be determined. We have, from (7.25), $x(0) = k_1 + k_2$ and $x'(0) = \lambda_1 k_1 + \lambda_2 k_2$. These two equations determine k_1 and k_2.

Example 7.4.1   Determine the general solution of $x''(t) - 4x'(t) + 3x(t) = 0$. What is the particular solution that satisfies the initial conditions $x(0) = 1$ and $x'(0) = 0$?

Solution: The auxiliary equation is $\lambda^2 - 4\lambda + 3 = 0$ with roots $\lambda_1 = 3$ and $\lambda_2 = 1$. The general solution is $x(t) = k_1 e^t + k_2 e^{3t}$. The initial conditions imply that $x(0) = 1 = k_1 + k_2$ and $x'(0) = 0 = k_1 + 3k_2$. Solving, we have $k_1 = \frac{3}{2}$ and $k_2 = -\frac{1}{2}$. The required solution is $x(t) = \frac{3}{2}e^t - \frac{1}{2}e^{3t}$.

Example 7.4.2   Determine the general solution of $x''(t) + x'(t) - 6x(t) = 0$. What is the particular solution that satisfies the conditions $x(0) = 0$ and $x(1) = 1$?

Solution: The auxiliary equation is $\lambda^2 + \lambda - 6 = 0$ with roots $\lambda_1 = 2$ and $\lambda_2 = -3$. The general solution is $x(t) = k_1 e^{2t} + k_2 e^{-3t}$. The two conditions $x(0) = 0$ and $x(1) = 1$ imply that $0 = k_1 + k_2$ and $1 = k_1 e^2 + k_2 e^{-3}$. Therefore, $k_2 = -k_1$ and $k_1(e^2 - e^{-3}) = 1$ or

$$k_1 = \frac{1}{e^2 - e^{-3}} \quad \text{and} \quad k_2 = \frac{-1}{e^2 - e^{-3}}.$$

The required solution is

$$x(t) = \frac{e^{2t}}{e^2 - e^{-3}} - \frac{e^{-3t}}{e^2 - e^{-3}} = \frac{e^{2t} - e^{-3t}}{e^2 - e^{-3}}.$$

The second case we will examine occurs when $b^2 - 4ac = 0$. In this case, the roots of the auxiliary equation are equal, $\lambda_1 = \lambda_2 = -b/2a$. As in Chapter 6, one solution is $x_1(t) = e^{\lambda_1 t}$ and a second solution is $x_2(t) = te^{\lambda_1 t}$. To verify the second solution, $x_2'(t) = \lambda_1 te^{\lambda_1 t} + e^{\lambda_1 t}$ and $x_2''(t) = \lambda_1^2 te^{\lambda_1 t} + 2\lambda_1 e^{\lambda t}$. Substituting in Equation (7.21), we have

$$ax_2''(t) + bx_2'(t) + cx_2(t) = te^{\lambda_1 t}(a\lambda_1^2 + b\lambda_1 + c) + e^{\lambda_1 t}(2a\lambda_1 + b) = 0,$$

since $2a\lambda_1 + b = 2a(-b/2a) + b = 0$. The general solution is

$$x(t) = k_1 e^{\lambda_1 t} + k_2 te^{\lambda_2 t}, \tag{7.26}$$

where k_1 and k_2 are arbitrary constants. The constants can be determined if the values of $x(t)$ and $x'(t)$ are known for some t.

Example 7.4.3 Determine the general solution of the equation $x''(t) - 4x'(t) + 4x(t) = 0$. What is the particular solution that satisfies $x(0) = 1$ and $x'(0) = 0$?

Solution: The auxiliary solution is $\lambda^2 - 4\lambda + 4 = 0$ with equal roots $\lambda_1 = \lambda_2 = 2$. The general solution is $x(t) = k_1 e^{2t} + k_2 te^{2t}$. The first derivative of the general solution is $x'(t) = (2k_1 + k_2)e^{2t} + 2k_2 te^{2t}$. The initial conditions imply that $x(0) = 1 = k_1$ and $x'(0) = 0 = 2k_1 + k_2$. Therefore, $k_1 = 1$ and $k_2 = -2$. The required solution is $x(t) = e^{2t} - 2te^{2t}$.

Example 7.4.4 Determine the solution of the equation $x''(t) - 6x'(t) + 9x(t) = 0$ that satisfies the conditions $x(1) = 0$ and $x'(1) = 2$.

Solution: The auxiliary equation is $\lambda^2 - 6\lambda + 9 = 0$ with equal roots $\lambda_1 = \lambda_2 = 3$. The general solution is $x(t) = k_1 e^{3t} + k_2 te^{3t}$. Since $x(1) = k_1 e^3 + k_2 e^3 = 0$, we conclude that $k_2 = -k_1$. Also $x'(t) = (3k_1 + k_2)e^{3t} + 3k_2 te^{3t} = 2k_1 e^{3t} - 3k_1 te^{3t}$. Since $x'(1) = 2 = -k_1 e^3$, we conclude that $k_1 = -2e^{-3}$ and $k_2 = 2e^{-3}$. The required solution is $x(t) = -2e^{3(t-1)} + 2te^{3(t-1)} = 2(t-1)e^{3(t-1)}$.

The last case to examine is the most difficult and, in many ways, the most interesting. When $b^2 - 4ac < 0$, the roots of the auxiliary equation are conjugate complex numbers

$$\lambda_1 = -\frac{b}{2a} + i\frac{\sqrt{4ac - b^2}}{2a} \quad \text{and} \quad \lambda_2 = -\frac{b}{2a} - i\frac{\sqrt{4ac - b^2}}{2a}.$$

To simplify notation, we write $\lambda_1 = \alpha + i\beta$ and $\lambda_2 = \alpha - i\beta$, where $\alpha = -b/2a$ and $\beta = \sqrt{4ac - b^2}/2a$. The two solutions are $x_1(t) = e^{\lambda_1 t}$ and $x_2(t) = e^{\lambda_2 t}$. From Euler's formula, (see Appendix E)

$$x_1(t) = e^{\lambda_1 t} = e^{(\alpha + i\beta)t} = e^{\alpha t}e^{i\beta t} = e^{\alpha t}(\cos \beta t + i \sin \beta t),$$

$$x_2(t) = e^{\lambda_2 t} = e^{(\alpha - i\beta)t} = e^{\alpha t}e^{-i\beta t} = e^{\alpha t}(\cos \beta t - i \sin \beta t).$$

Since any linear combination of two solutions is a solution, the combinations

$$\frac{x_1(t) + x_2(t)}{2} = e^{\alpha t} \cos \beta t$$

and

$$\frac{x_1(t) - x_2(t)}{2i} = e^{\alpha t} \sin \beta t$$

are solutions. The general solution of (7.21) in this case can be written

$$x(t) = k_1 e^{\alpha t} \cos \beta t + k_2 e^{\alpha t} \sin \beta t, \qquad (7.27)$$

where k_1 and k_2 are arbitrary constants. Again, the constants can be determined if the values of $x(t)$ and $x'(t)$ are known for some t. The solution (7.27) represents an oscillatory motion, where the magnitude of the oscillations increases if $\alpha > 0$, remains constant if $\alpha = 0$, and decreases if $\alpha < 0$. This is illustrated by the following examples.

Example 7.4.5 Determine the solution of $\frac{1}{2}x''(t) + 3x'(t) + 17x(t) = 0$ that satisfies the initial conditions $x(0) = 1$ and $x'(0) = 0$.

Solution: The auxiliary equation is $\frac{1}{2}\lambda^2 + 3\lambda + 17 = 0$ with roots $\lambda_1 = -3 + 5i$ and $\lambda_2 = -3 - 5i$. In this example, $\alpha = -3$ and $\beta = 5$. This leads to the general solution $x(t) = k_1 e^{-3t} \cos 5t + k_2 e^{-3t} \sin 5t$. The first derivative is $x'(t) = (5k_2 - 3k_1)e^{-3t} \cos 5t + (-5k_1 - 3k_2)e^{-3t} \sin 5t$. Applying the initial conditions, we obtain $x(0) = 1 = k_1$ and $x'(0) = 0 = 5k_2 - 3k_1$. Therefore, $k_1 = 1$, $k_2 = \frac{3}{5}$ and the required solution is $x(t) = e^{-3t} \cos 5t + \frac{3}{5}e^{-3t} \sin 5t$. This solution oscillates but tends rapidly to zero as t increases. This is illustrated in Figure 7.5.

Example 7.4.6 Determine the solution of $x''(t) - 2x'(t) + 5x(t) = 0$ that satisfies the initial conditions $x(0) = 2$ and $x'(0) = 10$.

Solution: The auxiliary equation is $\lambda^2 - 2\lambda + 5 = 0$ with roots $\lambda_1 = 1 + 2i$ and $\lambda_2 = 1 - 2i$. Since $\alpha = 1$ and $\beta = 2$, the general solution is $x(t) = k_1 e^t \cos 2t + k_2 e^t \sin 2t$. The first derivative is $x'(t) = (k_1 + 2k_2)e^t \cos 2t + (k_2 - 2k_1)e^t \sin 2t$. The initial conditions imply that $x(0) = 2 = k_1$ and $x'(0) = 10 = k_1 + 2k_2$. Therefore, $k_1 = 2$ and $k_2 = 4$, and the required solution is $x(t) = 2e^t \cos 2t + 4e^t \sin 2t$. This solution oscillates as in the previous example, except that now the magnitude of the oscillation increases as t increases.

Oscillatory motion occurs in biological processes in many different ways. The complex daily, monthly, and annual biological rhythms of many plants and animals are examples of natural oscillations. Many phenomena connected with the nervous system and its disorders provide further examples. Oscillatory motions can be classified into three distinct types. In Example 7.4.5, the magnitude or *amplitude* of the oscillations decreases with time. In Example 7.4.6, the amplitude increases with time. Finally, in Example 7.4.7, the amplitude of the oscillations does not change with time.

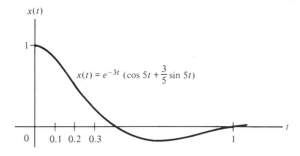

Figure 7.5. $x(t) = e^{-3t}(\cos 5t + \frac{3}{5} \sin 5t)$

Example 7.4.7 The Harmonic Oscillator The equation $x''(t) + \omega^2 x(t) = 0$ is the equation of simple harmonic motion. It occurs frequently in biological problems dealing with periodic or oscillatory phenomena. The auxiliary equation $\lambda^2 + \omega^2 = 0$ has the roots $\lambda_1 = i\omega$ and $\lambda_2 = -i\omega$ (ω is a real positive constant). In this example, $\alpha = 0$ and $\beta = \omega$. The general solution is $x(t) = k_1 \cos \omega t + k_2 \sin \omega t$. If $x(0)$ and $x'(0)$ are known, k_1 and k_2 can be determined. We have $x(0) = k_1$, $x'(0) = k_2 \omega$ and the solution can be written $x(t) = x(0) \cos \omega t + [x'(0)/\omega] \sin \omega t$. The solution represents periodic motion with period $2\pi/\omega$. This is illustrated in Figure 7.6.

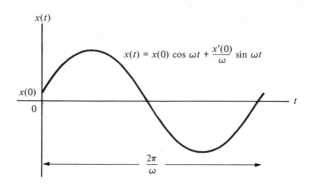

Figure 7.6. Harmonic Oscillator

Problems 7.4

1. Determine the general solutions of the following second order differential equations.
   (a) $x''(t) + 3x'(t) + 2x(t) = 0$.   (b) $x''(t) + 4x'(t) - 5x(t) = 0$.
   (c) $4x''(t) + x(t) = 0$.   (d) $x''(t) + 4x'(t) + 4x(t) = 0$.

2. Determine the solutions of the following differential equations that satisfy the initial conditions $x(0) = 1$, $x'(0) = 1$.
   (a) $x''(t) + x'(t) - 6x(t) = 0$.   (b) $x''(t) + x'(t) + x(t) = 0$.
   (c) $x''(t) + 9x(t) = 0$.   (d) $4x''(t) + 4x'(t) + x(t) = 0$.

3. Determine the solutions of the following differential equations that satisfy the initial conditions $x(0) = 0$, $x'(0) = 1$.
   (a) $x''(t) + x(t) = 0$.   (b) $x''(t) + x'(t) = 0$.
   (c) $x''(t) + 2x'(t) + x(t) = 0$.   (d) $x''(t) + 2x'(t) + 2x(t) = 0$.

4. Determine a second order differential equation of the form $ax''(t) + bx'(t) + cx(t) = 0$ with particular solutions $x(t) = e^{-t}$ and $x(t) = e^{9t}$.

5. Determine a second order differential equation of the form $ax''(t) + bx'(t) + cx(t) = 0$ with particular solutions $x(t) = e^{2t}$ and $x(t) = te^{2t}$.

6. Determine a second order differential equation of the form $ax''(t) + bx'(t) + cx(t) = 0$ with particular solutions $x(t) = e^{-3t} \cos t$ and $x(t) = e^{-3t} \sin t$.

7. Prove that all solutions of $x''(t) + 2x'(t) + 2x(t) = 0$ approach zero as t tends to infinity.

8. Under what conditions on the coefficients a, b, and c do all solutions of $ax''(t) + bx'(t) + cx(t) = 0$ tend to zero as t tends to infinity?

9. Determine solutions of the following equations satisfying the given initial conditions.

   (a) $x''(t) = x(t)$, $x(0) = 1$, $x'(0) = 0$.
   (b) $x''(t) - 6x'(t) + 5x(t) = 0$, $x(0) = 1$, $x'(0) = 1$.
   (c) $x''(t) - 6x'(t) + 8x(t) = 0$, $x(0) = 4$, $x'(0) = 8$.

10. Determine solutions of the following equations satisfying the given initial conditions.

   (a) $x''(t) - 4x'(t) + 4x(t) = 0$, $x(0) = -1$, $x'(0) = 2$.
   (b) $9x''(t) + 6x'(t) + x(t) = 0$, $x(0) = 1$, $x'(0) = -1$.
   (c) $x''(t) = 0$, $x(0) = 5$, $x'(0) = 3$.

11. Determine solutions of the following equations satisfying the given initial conditions.

   (a) $x''(t) + 5x'(t) + 7x(t) = 0$, $x(0) = 2$, $x'(0) = 0$.
   (b) $x''(t) + 3x'(t) + 3x(t) = 0$, $x(0) = -1$, $x'(0) = 1$.
   (c) $8x''(t) + 4x'(t) + x(t) = 0$, $x(0) = 0$, $x'(0) = 4$.

12. In a study of fasting, the weights of two volunteers decreased from 140 and 170 pounds to 110 and 125 pounds, respectively, in 30 days. It was observed that the rate of weight loss for each volunteer was proportional to his weight. Define $x(t)$ to be the total weight of the two volunteers after t days of fasting. Determine a second order differential equation satisfied by $x(t)$. What is the total weight after 15 days of fasting?

13. For a certain bird species, the total time spent foraging varies from a minimum of 2 hours per day in summer to a maximum of 8 hours per day in winter. Assuming that the change in foraging times are described by the equation of the harmonic oscillator, determine the daily foraging period as a function of the time of year.

14. Two human lice populations are growing under favorable conditions. For population I, the growth rate per unit of population is 0.10 when time is measured in days. For population II, the corresponding growth rate is 0.08. Define $x(t)$ to be the total lice population at time t. Determine a second order differential equation satisfied by $x(t)$.

15. If, in Problem 14, both populations are initially equal to 1000, determine the total population after 10 days and after 20 days.

16. If $x(t)$ and $y(t)$ satisfy the first order differential equations $dx/dt = ax$ and $dy/dt = by$, prove that $z(t) = x(t) + y(t)$ satisfies the second order differential equation $d^2z/dt^2 - (a + b)(dz/dt) + abz = 0$. (How is this result useful in Problems 12 and 14?)

17. If $x(t)$ and $y(t)$ satisfy the first order differential equations $dx/dt = f(t)x$ and $dy/dt = g(t)y$, prove that $z(t) = x(t) + y(t)$ satisfies a second order differential equation and determine the equation.

7.5 The Variation of Constants Method for Second Order Differential Equations

In this section, we derive a method for solving the equation

$$ax''(t) + bx'(t) + cx(t) = f(t) \tag{7.28}$$

with constant coefficients a, b, and c (with $a \neq 0$). The method of solution will be analogous to the method of Section 6.4.

We first note that it is only necessary to find one solution of (7.28) in order to obtain the general solution. If $y(t)$ and $z(t)$ are two solutions, then $w(t) = y(t) - z(t)$ is a solution of the homogeneous equation (7.21). To see this, we note that

$$
\begin{aligned}
aw''(t) + bw'(t) + cw(t) &= a(y''(t) - z''(t)) + b(y'(t) - z'(t)) + c(y(t) - z(t)) \\
&= ay''(t) + by'(t) + cy(t) - az''(t) - bz'(t) - cz(t) \\
&= f(t) - f(t) = 0.
\end{aligned}
$$

Therefore, any two solutions of (7.28) differ by a solution of (7.21). If we know one solution of (7.28), the general solution can be written as the sum of the particular solution and of the general solution of (7.21). We now give a method for finding one solution of (7.28).

Suppose that $g_1(t)$ and $g_2(t)$ are independent solutions of the homogeneous equation $ax''(t) + bx'(t) + cx(t) = 0$. By independent, we mean that $g_1(t)$ is not a constant multiple of $g_2(t)$. Then, the general solution is $x(t) = k_1 g_1(t) + k_2 g_2(t)$, where k_1 and k_2 are arbitrary constants. To determine a solution of (7.28), we will allow the constants k_1 and k_2 to vary and we will look for a solution of the form

$$x(t) = k_1(t)g_1(t) + k_2(t)g_2(t). \tag{7.29}$$

The coefficients $k_1(t)$ and $k_2(t)$ are now functions of t to be determined in such a way that Equation (7.29) gives a solution of (7.28). This is the *variation of constants method*.

From Equation (7.29), we calculate $x'(t)$ and $x''(t)$ and substitute in (7.28).

$$x'(t) = k_1(t)g_1'(t) + k_2(t)g_2'(t) + k_1'(t)g_1(t) + k_2'(t)g_2(t).$$

To simplify this expression, we will assume that $k_1'(t)g_1(t) + k_2'(t)g_2(t) = 0$. This gives one equation for the two unknown functions $k_1(t)$ and $k_2(t)$. Differentiating once more, we have

$$x''(t) = k_1(t)g_1''(t) + k_2(t)g_2''(t) + k_1'(t)g_1'(t) + k_2'(t)g_2'(t).$$

Substitution of our expressions for $x(t)$, $x'(t)$, and $x''(t)$ in Equation (7.28) gives

$$ax''(t) + bx'(t) + cx(t) = k_1(t)[ag_1''(t) + bg_1'(t) + cg_1(t)]$$
$$+ k_2(t)[ag_2''(t) + bg_2'(t) + cg_2(t)] + a[k_1'(t)g_1'(t)$$
$$+ k_2'(t)g_2'(t)] = f(t).$$

But $g_1(t)$ and $g_2(t)$ are solutions of the homogeneous equation. Therefore,

$$ax''(t) + bx'(t) + cx(t) = a[k_1'(t)g_1'(t) + k_2'(t)g_2'(t)] = f(t).$$

This gives a second equation relating $k_1(t)$ and $k_2(t)$. These functions must therefore satisfy the system of first order differential equations

$$g_1(t)k_1'(t) + g_2(t)k_2'(t) = 0,$$
$$g_1'(t)k_1'(t) + g_2'(t)k_2'(t) = \frac{f(t)}{a}. \tag{7.30}$$

This system of equations can be solved for $k_1'(t)$ and $k_2'(t)$ by elementary methods. For example, multiply the first equation by $g_2'(t)$, the second by $g_2(t)$, and subtract to obtain $k_1'(t)$. Similarly, $k_2'(t)$ can be determined. The solutions are

$$k_1'(t) = \frac{-f(t)g_2(t)}{a[g_1(t)g_2'(t) - g_2(t)g_1'(t)]}, \tag{7.31}$$

$$k_2'(t) = \frac{f(t)g_1(t)}{a[g_1(t)g_2'(t) - g_2(t)g_1'(t)]}. \tag{7.32}$$

Define $D(t) = g_1(t)g_2'(t) - g_2(t)g_1'(t)$. To justify our solution, we must verify that $D(t)$ is never equal to zero. To see this, we calculate

$$D'(t) = g_1(t)g_2''(t) - g_2(t)g_1''(t)$$

$$= \frac{1}{a}\{g_1(t)[-bg_2'(t) - cg_2(t)] - g_2(t)[-bg_1'(t) - cg_1(t)]\}$$

$$= -\frac{b}{a}[g_1(t)g_2'(t) - g_2(t)g_1'(t)] = -\frac{b}{a}D(t).$$

We observe that $D(t)$ satisfies a simple first order equation whose solution is $D(t) = D(0)e^{-(b/a)t}$. We can see that $D(t)$ is either always zero or never zero, depending on whether $D(0) = 0$ or $D(0) \neq 0$. Therefore, if $g_1(t)$ and $g_2(t)$ are chosen so that $D(0) \neq 0$, the solutions (7.31) and (7.32) are defined for all t.

We recognize (7.31) and (7.32) as simple first order equations for $k_1(t)$ and $k_2(t)$. We can integrate these equations to determine functions $k_1(t)$ and $k_2(t)$, which give a solution $x(t) = k_1(t)g_1(t) + k_2(t)g_2(t)$ of the nonhomogeneous equation (7.28).

Example 7.5.1   Determine the solution of $x''(t) - x(t) = e^t$ that satisfies $x(0) = 1$ and $x'(0) = 0$.

Solution: The corresponding homogeneous equation is $x''(t) - x(t) = 0$ with auxiliary equation $\lambda^2 - 1 = 0$ and roots $\lambda_1 = +1$, $\lambda_2 = -1$. Two independent solutions of the homogeneous equation are $g_1(t) = e^t$ and $g_2(t) = e^{-t}$. For this example, $D(t) = g_1(t)g_2'(t) - g_2(t)g_1'(t) = e^t(-e^{-t}) - e^{-t}(e^t) = -2 \neq 0$. Equations (7.31) and (7.32) become $k_1'(t) = \frac{1}{2}e^te^{-t} = \frac{1}{2}$, and $k_2'(t) = -\frac{1}{2}e^te^t = -e^{2t}/2$. Upon integration, $k_1(t) = (t/2) + c_1$ and $k_2(t) = -(e^{2t}/4) + c_2$, where c_1 and c_2 are constants of integration. The general solution is

$$x(t) = k_1(t)g_1(t) + k_2(t)g_2(t) = \frac{t}{2}e^t - \frac{e^t}{4} + c_1e^t + c_2e^{-t}.$$

Defining $b_1 = c_1 - \frac{1}{4}$ and $b_2 = c_2$, we find that the general solution is $x(t) = (t/2)e^t + b_1e^t + b_2e^{-t}$. The reader should verify that $(t/2)e^t$ is a particular solution of the nonhomogeneous equation. The initial conditions $x(0) = 1$ and $x'(0) = 0$ determine the constants b_1 and b_2. We have $x(0) = 1 = b_1 + b_2$ and $x'(0) = \frac{1}{2} + b_1 - b_2 = 0$. Therefore, $b_1 = \frac{1}{4}$ and $b_2 = \frac{3}{4}$, and the required solution is

$$x(t) = \frac{t}{2}e^t + \frac{1}{4}e^t + \frac{3}{4}e^{-t}.$$

Example 7.5.2 Determine the solution of $x''(t) - 6x'(t) + 9x(t) = e^{3t}$ that satisfies the initial conditions $x(0) = 0$, $x'(0) = 1$.

Solution: The corresponding homogeneous equation $x''(t) - 6x'(t) + 9x(t) = 0$ has two independent solutions $g_1(t) = e^{3t}$ and $g_2(t) = te^{3t}$. Then

$$D(t) = g_1(t)g_2'(t) - g_2(t)g_1'(t) = e^{3t}(3t + 1)e^{3t} - 3e^{3t}te^{3t} = e^{6t} \neq 0.$$

From Equations (7.31) and (7.32),

$$k_1'(t) = -\frac{e^{3t}te^{3t}}{e^{6t}} = -t, \qquad k_2'(t) = \frac{e^{3t}e^{3t}}{e^{6t}} = 1.$$

Integrating, we obtain $k_1(t) = -(t^2/2) + c_1$ and $k_2(t) = t + c_2$, where c_1 and c_2 are constants of integration. The general solution can be written

$$x(t) = -\frac{t^2}{2}e^{3t} + t^2e^{3t} + c_1e^{3t} + c_2te^{3t} = \frac{t^2}{2}e^{3t} + c_1e^{3t} + c_2te^{3t}.$$

The initial conditions imply that $x(0) = 0 = c_1$ and $x'(0) = 3c_1 + c_2 = 1$. Therefore, $c_1 = 0$ and $c_2 = 1$, and the required solution is

$$x(t) = \frac{t^2}{2}e^{3t} + te^{3t}.$$

Example 7.5.3 Determine the solution of $x''(t) + x(t) = 2 \cos(t)$ that satisfies the initial conditions $x(0) = 5$ and $x'(0) = 2$.

Solution: The homogeneous equation is $x''(t) + x(t) = 0$ with two independent solutions $g_1(t) = \cos(t)$ and $g_2(t) = \sin(t)$. Then

$$D(t) = g_1(t)g_2'(t) - g_2(t)g_1'(t) = \cos^2(t) + \sin^2(t) = 1 \neq 0.$$

From Equations (7.31) and (7.32),

$$k_1'(t) = -2 \cos(t) \sin(t) \qquad \text{and} \qquad k_2'(t) = 2 \cos^2(t).$$

Since $d \sin^2(t)/dt = 2 \sin(t) \cos(t)$, we have $k_1(t) = -\sin^2(t) + c_1$. To integrate $2 \cos^2(t)$, we use the trigonometric identity $\cos^2(t) = [1 + \cos(2t)]/2.$
Then

$$k_2(t) = t + \frac{\sin(2t)}{2} + c_2 = t + \sin(t) \cos(t) + c_2,$$

since $\sin(2t) = 2 \sin(t) \cos(t)$. The general solution is

$$x(t) = k_1(t)g_1(t) + k_2(t)g_2(t)$$

$$= -\sin^2(t) \cos(t) + t \sin(t) + \sin^2(t) \cos(t) + c_1 \cos(t) + c_2 \sin(t)$$

$$= t \sin(t) + c_1 \cos(t) + c_2 \sin(t).$$

Since $x(0) = c_1 = 5$ and $x'(0) = c_2 = 2$, the required solution is $x(t) = (t + 2) \sin(t) + 5 \cos(t)$. The reader should verify that this function satisfies the given differential equation and both initial conditions.

The method of variation of constants gives a general method of solving Equation (7.28). We will make use of this method in our study of systems of first order differential equations in the next section.

Problems 7.5

1. Determine the general solutions of the following second order differential equations.
   (a) $x''(t) - x(t) = t.$
   (b) $x''(t) + 4x'(t) + 4x(t) = e^{-2t}.$
   (c) $x''(t) + x(t) = \sin(t).$
   (d) $3x''(t) - 2x'(t) + x(t) = t + 1.$

2. Determine the solutions of the following second order differential equations which satisfy the initial conditions $x(0) = 0$ and $x'(0) = 1$.
   (a) $x''(t) + 4x(t) = 3.$
   (b) $x''(t) + x'(t) + x(t) = e^t.$
   (c) $x''(t) - x'(t) + 2x(t) = e^{3t}.$
   (d) $x''(t) - 2x'(t) + x(t) = 5e^{2t}.$

3. Consider the second order equation $ax''(t) + bx'(t) + cx(t) = e^{rt}$, where $a, b, c,$ and r are constants. Under what conditions on the coefficients $a, b,$ and c is there a solution of the form $x(t) = ke^{rt}$ for some constant k?

4. Use the result of Problem 3 to determine solutions of the four second order differential equations of Problem 2.

5. Verify that $x(t) = te^{2t}$ is a particular solution of the differential equation $x''(t) - 3x'(t) + 2x(t) = e^{2t}.$

6. Verify that $x(t) = (t^2/2)e^{2t}$ is a particular solution of the differential equation $x''(t) - 4x'(t) + 4x(t) = e^{2t}.$

7. (a) If $y(t)$ and $z(t)$ satisfy the differential equations $ay''(t) + by'(t) + cy(t) = f(t)$ and $az''(t) + bz'(t) + cz(t) = g(t)$, verify that $x(t) = y(t) + z(t)$ satisfies the differential equation $ax''(t) + bx'(t) + cx(t) = f(t) + g(t)$.

   (b) If k_1 and k_2 are arbitrary constants, what equation does $k_1 y(t) + k_2 z(t)$ satisfy?

8. Use the results of Problem 7 to solve the following second order equations.

   (a) $x''(t) + 6x'(t) + 9x(t) = 1 + e^{2t}$.

   (b) $x''(t) - 3x'(t) - 4x(t) = e^t + e^{2t}$.

   (c) $x''(t) + 4x'(t) + 8x(t) = 4 + e^t$.

   (d) $x''(t) + x(t) = e^{-t} + e^{-2t}$.

9. (a) The height $x(t)$ at time t of an object falling freely, due to gravity, satisfies the equation $x''(t) = -g$, where g is the constant acceleration due to gravity. Determine $x(t)$ in terms of the initial height $x(0)$ and the initial velocity $x'(0)$.

   (b) If an object falls from a height h with initial velocity zero, what is its velocity when it reaches the earth?

10. The equilibrium size of a population of a certain species in an environment is estimated to be 1000 individuals. The population fluctuates about this average value according to the equation $x''(t) = 4\pi^2[1000 - x(t)]$, where $x(t)$ is the population at time t when t is measured in years. If $x(0) = 1500$ and $x'(0) = 0$, determine the population after 6 months, after 12 months, and after 18 months. Draw a graph of the population $x(t)$ as a function of t.

7.6 Systems of First Order Differential Equations

How do we extend the mathematical techniques developed in the previous sections to the description of the interactions of several biological variables? For example, if two species coexist in the same environment and compete for the same resources, we must take into account the effect of the presence of the competing species in describing population growth. It is reasonable to assume that the growth rate of each species will depend on the population of the other species, as well as on its own population. More than one differential equation is required to describe this type of interaction.

The systems of differential equations that we will study are of the form

$$x'(t) = a_{11}x(t) + a_{12}y(t) + f'(t)$$
$$y'(t) = a_{21}x(t) + a_{22}y(t) + g(t),$$

$$(7.33)$$

where a_{11}, a_{12}, a_{21}, and a_{22} are constants and $f(t)$ and $g(t)$ are given functions. To solve the linear system (7.33), we must determine two functions, $x(t)$ and $y(t)$, which satisfy both equations. The system is said to be *linear*, since both equations are linear.

To solve the linear system, we proceed in the following way. Differentiating the first equation gives $x''(t) = a_{11}x'(t) + a_{12}y'(t) + f'(t)$. But, substituting the second equation for $y'(t)$, we have

$$x''(t) = a_{11}x'(t) + a_{12}[a_{21}x(t) + a_{22}y(t) + g(t)] + f'(t)$$
$$= a_{11}x'(t) + a_{12}a_{21}x(t) + a_{12}a_{22}y(t) + a_{12}g(t) + f'(t).$$

From the first equation of (7.33), we have an equation for $a_{12}y(t)$. Then,

$$x''(t) = a_{11}x'(t) + a_{12}a_{21}x(t) + a_{22}[x'(t) - a_{11}x(t) - f(t)] + a_{12}g(t) + f'(t).$$

Finally, we conclude that $x(t)$ satisfies

$$x''(t) - (a_{11} + a_{22})x(t) + (a_{11}a_{22} - a_{12}a_{21})x(t)$$
$$= a_{12}g(t) - a_{22}f(t) + f'(t). \qquad (7.34)$$

But this is a second order linear nonhomogeneous differential equation with constant coefficients that can be solved by the methods of the previous section. Having solved for $x(t)$, then we can determine $y(t)$ from the system (7.33).

The linear system (7.33) can be thought of as a model of species interaction if $x(t)$ and $y(t)$ represent the populations at time t of two species present in a given environment. If both species compete for the same resources, this can be modeled by the coefficients a_{12} and a_{21} being negative. For example, if a_{12} is negative, the growth rate of the first species will decrease as the population of the second species increases. The possible types of interactions can best be illustrated by a number of examples.

Example 7.6.1   A Model of Species Competition   The homogeneous linear system

$$x'(t) = 2x(t) - y(t)$$
$$y'(t) = -x(t) + 2y(t)$$

describes the influence of the populations of two competing species on their growth rates. Suppose that the initial populations are $x(0) = 100$ and $y(0) = 200$. Determine the populations of both species at all future times.

Solution: Differentiating the first equation, we have $x''(t) = 2x'(t) - y'(t)$. But $y'(t) = -x(t) + 2y(t) = -x(t) + 2[2x(t) - x'(t)]$. Therefore, $x(t)$ satisfies the second order equation $x''(t) - 4x'(t) + 3x(t) = 0$. The general solution is $x(t) = k_1 e^{3t} + k_2 e^t$, where k_1 and k_2 are constants. From the first equation $y(t) = 2x(t) - x'(t) = -k_1 e^{3t} + k_2 e^t$. The initial populations are $x(0) = 100$ and $y(0) = 200$. This implies that $k_1 + k_2 = 100$ and $-k_1 + k_2 = 200$ or $k_1 = -50$, $k_2 = 150$. The solution is $x(t) = 150e^t - 50e^{3t}$ and $y(t) = 150e^t + 50e^{3t}$. The first species becomes extinct when $150e^t - 50e^{3t} = 0$ or when $e^{2t} = 3$. This occurs when $t = \frac{1}{2} \log_e 3 \approx .552$ time unit. After .552 time unit, the second species continues to grow according to the equation $y'(t) = 2y(t)$. The general solution is $y(t) = y(t_0)e^{2(t-t_0)}$. With $t_0 = \frac{1}{2} \log_e 3$ and $y(t_0) = 150e^{t_0} + 50e^{3t_0}$, this gives the time development of the second species after the extinction of the first species. This is illustrated in Figure 7.7.

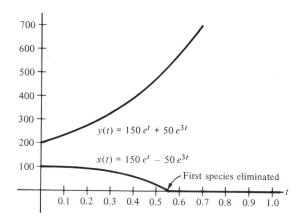

Figure 7.7

Example 7.6.2 Predator–Prey Interaction Suppose that $x(t)$ is the population of a predator species at time t and that $y(t)$ is the population of the prey species. Suppose further that their growth rates are given by the linear system

$$x'(t) = x(t) + y(t)$$
$$y'(t) = -x(t) + y(t).$$

If the initial populations are $x(0) = y(0) = 1000$, determine the populations at all future times. When does the prey species become extinct?

Solution: In this example, the minus sign in the second equation indicates that the growth rate $y'(t)$ of the prey population decreases when the predator population $x(t)$ increases. By the usual method, we find that $x''(t) - 2x'(t) + 2x(t) = 0$. The auxiliary equation is $\lambda^2 - 2\lambda + 2 = 0$ with complex roots $\lambda_1 = 1 + i$ and $\lambda_2 = 1 - i$. The general solution is $x(t) = k_1 e^t \cos t + k_2 e^t \sin t$. From the first equation of the system, $y(t) = x'(t) - x(t) = k_2 e^t \cos t - k_1 e^t \sin t$. The initial conditions are $x(0) = y(0) = 1000$. This implies that $k_1 = 1000$ and $k_2 = 1000$. The required solution of the system is $x(t) = 1000e^t(\cos t + \sin t)$ and $y(t) = 1000e^t(\cos t - \sin t)$. The prey species becomes extinct when $\cos t - \sin t = 0$, that is, after $t = \pi/4$ time units. The predator population may continue to grow on other resources in the environment according to the equation $x'(t) = x(t)$. It usually happens, however, that the assumptions made in deriving a model of species interaction are less accurate when one of the species becomes extinct. The development of the two populations until the extinction of the prey species is illustrated in Figure 7.8.

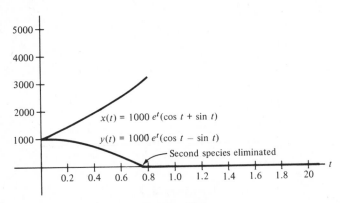

Figure 7.8

Example 7.6.3 A Model of Species Cooperation Suppose that two species coexist in a symbiotic relationship; that is, the population of each increases proportionally to the population of the other and decreases proportionally to its own population. A possible model for this behavior is given by the linear system

$$x'(t) = -2x(t) + 4y(t)$$
$$y'(t) = x(t) - 2y(t).$$

If the initial populations are $x_1(0) = 100$ and $x_2(0) = 300$, determine the populations at all future times.

Solution: By differentiation and elimination, we find that $x''(t) + 4x'(t) = 0$. The auxiliary equation is $\lambda^2 + 4\lambda = 0$ with roots $\lambda_1 = 0$ and $\lambda_2 = -4$. The general solution is $x(t) = k_1 e^{0t} + k_2 e^{-4t} = k_1 + k_2 e^{-4t}$. Also, $y(t) = \frac{1}{4}[x'(t) + 2x(t)] = (k_1/2) - (k_2/2)e^{-4t}$. Applying the initial conditions, $x(0) = k_1 + k_2 = 100$, $y(0) = (k_1/2) - (k_2/2) = 300$, we find $k_1 = 350$ and $k_2 = -250$. The required solution is $x(t) = 350 - 250e^{-4t}$ and $y(t) = 175 + 125e^{-4t}$. The population of the first species increases from an initial size of 100 to a limiting size of 350. The second species decreases from an initial size of 300 to a limiting size of 175. These limiting sizes are the equilibrium populations of the two species that the environment can support. This is illustrated in Figure 7.9.

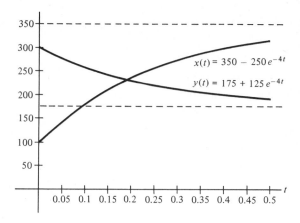

$$x(t) = 350 - 250e^{-4t}$$
$$y(t) = 175 + 125e^{-4t}$$

Figure 7.9

Calculated Values

t	e^{-4t}	$x(t)$	$y(t)$
0	1	100	300
.05	.8187	145	277
.10	.6703	182	259
.15	.5488	213	244
.20	.4493	238	231
.25	.3679	258	221
.30	.3012	275	213
.35	.2466	288	206
.40	.2019	300	200
.45	.1653	309	196
.50	.1353	316	192

Problems 7.6

1. Determine the general solutions of the following systems of first order differential equations.

(a) $x'(t) = x(t) - y(t)$
   $y'(t) = -x(t) + y(t).$

(b) $x' = x - \frac{1}{2}y$
   $y' = -\frac{1}{2}x + y.$

(c) $x' = \quad -y + e^{-t}$
   $y' = -x \quad + e^{-t}.$

(d) $x' = x + y + te^t$
   $y' = \quad y.$

2. Determine the solutions of the systems in Problem 1 that satisfy the initial conditions $x(0) = 100$, $y(0) = 200$.

3. Describe a method of solution of the general system of three first order differential equations with constant coefficients.

$$x'(t) = a_{11}x + a_{12}y + a_{13}z + f(t)$$

$$y'(t) = a_{21}x + a_{22}y + a_{23}z + g(t)$$

$$z'(t) = a_{31}x + a_{32}y + a_{33}z + h(t).$$

4. Give an interpretation of the linear system of Problem 3 as a model of the interaction of three coexisting species.

5. Determine the solutions of the following systems of first order differential equations that satisfy the initial conditions $x(0) = 1$, $y(0) = 0$.

(a) $x'(t) = x + y$
   $y'(t) = x - y.$

(b) $x' = 2x \quad + e^{2t}$
   $y' = \quad 2y + e^{2t}.$

(c) $x' = \quad -y + \sin t$
   $y' = x \quad + \cos t.$

(d) $x' = 2x + \quad y + t$
   $y' = \quad x + 2y + t^2.$

6. A salt solution flows from one container at a rate proportional to its volume into a second container from which it flows at a constant rate. Define $V_1(t)$ and $V_2(t)$ to be the volumes of the salt solution in the first and second containers at time t. Then $V_1(t)$ and $V_2(t)$ satisfy the equations $dV_1/dt = -aV_1$ and $dV_2/dt = aV_1 - b$, where a and b are positive constants.

(a) If $V_1(0) = 1000$ and $V_2(0) = 100$, determine $V_1(t)$ and $V_2(t)$ for $t > 0$.
(b) Prove that, if $b/a > V_1(0)$, the volume of solution in the second container decreases as t increases.
(c) If $b/a < V_1(0)$, what is the maximum accumulation of the solution in the second container and when does it occur?

7. A species population at time t consists of $x(t)$ males and $y(t)$ females. The system proposed as a model of the growth of this population is $x'(t) = -ax + by$ and $y'(t) = cy$, where a, b, and c are positive constants.

(a) Determine the general solution in terms of the initial populations $x(0)$ and $y(0)$ of males and females.

(b) What happens if $x(0) = 0$? If $y(0) = 0$?

(c) In what ways is this model oversimplified? How could it be made more realistic?

8. A community of n individuals is exposed to a rare infectious disease.[2] At time t, the community is divided into $x(t)$ susceptibles, $y(t)$ infectious cases in circulation, and $z(t)$ who are isolated, dead, or immune. Suppose that initially $y(t)$ and $z(t)$ are both small compared to $x(t)$. A model of the spread of this infection is given by

$$\frac{dx}{dt} = -\beta x(0)y, \qquad \frac{dy}{dt} = \beta x(0)y - \gamma y, \qquad \frac{dz}{dt} = \gamma y,$$

where β and γ are positive constants measuring the rates at which susceptibles become infected and infected individuals become isolated, dead, or immune.

(a) Determine the solution in terms of $x(0)$, $y(0)$, and $z(0) = n - x(0) - y(0)$.

(b) Prove that if $\beta x(0) < \gamma$, the disease will not produce an epidemic.

(c) What happens if $\beta x(0) > \gamma$?

---

[2] N. Bailey, "The Total Size of a General Stochastic Epidemic," *Biometrika*, 40:177–185 (1953).

Continuous Probability 8

8.1 Continuous Random Variables

In this chapter, we extend the theory of probability to experiments with a continuous range of possible outcomes. As a simple example, consider a perfect spinner that can point to any number from 0 to 1 (see Figure 8.1).

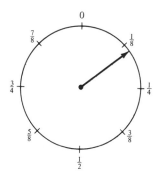

Figure 8.1. Spinner Experiment

If each outcome is equally likely (a reasonable assumption), then the probability of the outcome $\frac{1}{2}$ is, by the laws of equiprobable spaces, equal to

1/(number of possible outcomes) $= 1/\infty = 0$. Since there is nothing exceptional about the point $\frac{1}{2}$, it is clear that the probability of every outcome is zero. The sample space S for this experiment is made up of all numbers between 0 and 1. On a trial of the experiment, some outcome must occur $[P(S) = 1]$, but, for every x from 0 to 1, we have $P(x) = 0$.

Another example of this phenomenon is given by the annual rainfall over a certain region. The amount of rainfall in one year is a random variable that can take any value from a continuous range of values. The probability that there will be exactly 30.2 inches of rain in a given year is clearly extremely small (in fact, equal to zero).

These two examples illustrate the fact that we will not progress very far if we attempt to assign nonzero probabilities to every possible value of a continuous random variable. A more promising method is to study the probabilities that the value of the random variable falls between certain limits. For example, it is easy to see that the probability that the spinner points to numbers between $\frac{1}{3}$ and $\frac{1}{2}$ is $\frac{1}{2} - \frac{1}{3} = \frac{1}{6}$. We could also estimate the probability that the rainfall in a given year will be between 30 and 35 inches. This leads to the following definition.

Definition 8.1.1   Distribution Function of a Continuous Random Variable   *The distribution function $F(x)$ of a continuous random variable X is a continuous function which represents the probability that X takes on values less than or equal to x.*

$$F(x) = P(X \le x).$$

It is important to note that $P(X \le x) = P(X < x)$ for a continuous random variable X, since $P(X = x) = 0$.

In the examples, if X represents the outcome of the spinner experiment, then $P(\frac{1}{3} \le X \le \frac{1}{2}) = F(\frac{1}{2}) - F(\frac{1}{3}) = \frac{1}{2} - \frac{1}{3} = \frac{1}{6}$. The distribution function in this case is $F(x) = x$ for $0 \le x \le 1$. If X represents the annual rainfall, then the probability above is $F(35) - F(30)$, where F is the distribution function of X. In general, the probability that X takes on values between a and b is $P(X \le b) - P(X \le a)$; that is,

$$P(a \le X \le b) = F(b) - F(a).$$

The distribution function $F(x)$ of a continuous random variable X increases from 0 to 1 continuously as x increases from $-\infty$ to $+\infty$. Figure 8.2 shows the graph of a typical distribution function.

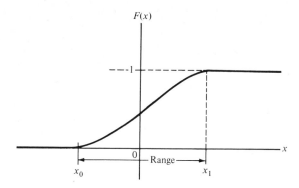

Figure 8.2. Typical Distribution Function (as x increases from x_0 to x_1, $F(x)$ increases from 0 to 1)

To specify $F(x)$, we need to know only the interval of the x axis in which $F(x)$ lies between 0 and 1. This interval is called the *range* of the continuous random variable X.

Example 8.1.1 $F(x) = x/R$ on the interval $0 \le x \le R$. This defines the distribution function of a continuous random variable X with range $0 \le x \le R$ (Figure 8.3). In this case, X is said to be *uniformly distributed* on its range.

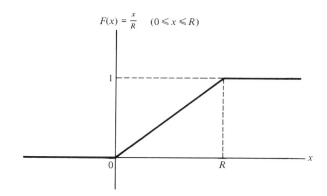

Figure 8.3. Uniform Distribution on $0 \le x \le R$

In this example, the probability that X takes a value between $\frac{1}{3}R$ and $\frac{2}{3}R$ is

$$P\left(\frac{R}{3} \le X \le \frac{2}{3}R\right) = F\left(\frac{2}{3}R\right) - F\left(\frac{R}{3}\right) = \frac{2}{3} - \frac{1}{3} = \frac{1}{3}.$$

The probability that X takes a value greater than $\frac{3}{5}R$ is

$$P(\tfrac{3}{5}R \le X \le R) = F(R) - F(\tfrac{3}{5}R) = 1 - \tfrac{3}{5} = \tfrac{2}{5}.$$

Example 8.1.2   A laboratory rat requires at least 2 minutes to complete a certain task but never requires more than 10 minutes. Define T to be the time required to complete the task. If all times between 2 and 10 minutes are equally likely, then T is a continuous random variable with distribution function $F(t) = P(T \le t) = (t - 2)/8$ for $2 \le t \le 10$. The probability that the task is completed in less than 5 minutes is $F(5) = (5 - 2)/8 = \tfrac{3}{8}$. The probability that at least 3 minutes are required is $1 - F(3) = 1 - \tfrac{1}{8} = \tfrac{7}{8}$, since $F(3) = \tfrac{1}{8}$ is the probability that no more than 3 minutes will be required.

Example 8.1.3   For any positive constant a, the function $F(x) = 1 - e^{-ax^2}$ is the distribution function of a continuous random variable X with range $0 \le x < \infty$. As x increases from 0 to ∞, $F(x)$ increases from 0 to 1. The probability that the random variable X takes on a value between 1 and 2, for example, is $P(1 \le X \le 2) = F(2) - F(1) = (1 - e^{-4a}) - (1 - e^{-a}) = e^{-a} - e^{-4a}$.

Example 8.1.4   A mathematical model of the distribution of bird nests in a certain environment is to be developed. The probability that a circle of radius r (centered on any point chosen at random) contains at least one nest is estimated to be $1 - e^{-5r^2}$ when r is measured in kilometers. In this model, what is the probability that there is at least one nest within 100 meters of an arbitrarily chosen point?

Solution: If we define R to be the distance from the point to the nearest nest, then R is a continuous random variable with probability distribution function $F(r) = P(R \le r) = 1 - e^{-5r^2}$. The probability that there is a nest within 100 meters of the arbitrarily chosen point is $F(.1) = P(R \le .1) = 1 - e^{-.05} \approx .049$ or slightly less than 5 per cent.

Problems 8.1

1. Which of the following functions represent the distribution function of a continuous random variable? Draw a graph of each of the distribution functions and determine the range of the random variable.
   (a) $F(x) = e^x$ for $x \le 0$
   $\quad\;\; = 1$ for $x > 0$.
   (b) $F(x) = \tfrac{1}{2}e^x$ for $x \le 0$
   $\quad\;\; = 1 - \tfrac{1}{2}e^{-x}$ for $x \ge 0$.

(c) $F(x) = 0$ for $x \leq 0$
   $= \sin x$ for $0 < x \leq \pi/2$
   $= 1$ for $x > \pi/2$.

(d) $F(x) = 0$ for $x \leq -\frac{1}{2}$
   $= x + \frac{1}{2}$ for $-\frac{1}{2} < x < \frac{1}{2}$
   $= 1$ for $x \geq \frac{1}{2}$.

2. For each of the distribution functions in Problem 1, determine the probability that the corresponding random variable takes on values between 0 and 2.

3. Define a function $F(x)$ to be zero for x negative and to be equal to $1 - ke^{-x/2}$ for x nonnegative.
   (a) For what value of the constant k is this the distribution function of a continuous random variable? Define X to be this random variable.
   (b) Determine $P(X \leq 2)$, $P(X \geq 4)$, and $P(2 \leq X \leq 4)$.

4. Define a function $F(x) = Ae^x$ for $x < 0$ and $F(x) = 1 - e^{-2x}/2$ for $x \geq 0$.
   (a) For what value of the constant A is this the distribution function of a continuous random variable? Define X to be this random variable.
   (b) Determine $P(X \geq -1)$, $P(X \leq 2)$, and $P(-2 \leq X \leq 2)$.

5. The probability that a certain task will be learned in x minutes is estimated to be $1 - 1/(1 + x)$ for an average person. What is the probability that the task will be learned in fewer than 10 minutes? In less than 1 hour?

6. The probability that a bird will find a suitable nesting site in x days of searching is estimated to be $1 - [1/(1 + x)^2]$. If the nesting site must be found within a 1-week period, what proportion of the birds do not find suitable nesting sites?

7. A student is allowed exactly 30 minutes to answer each of two questions on a 1-hour exam. For each question, the probability that he will be able to solve the equation in x minutes is given by $1 - 10/(10 + x)$. Assuming that the questions are unrelated, what is the probability that both questions are answered correctly? That only one is answered correctly? That none are answered correctly?

8. In 250 grams of a glucose solution of unknown concentration, define the random variable X to be the weight of glucose in the solution.
   (a) What is the range of the random variable X?
   (b) Define the events $\{50 \leq X \leq 100\}$ and $\{X \geq 75\}$.
   (c) Suppose that, on its range, the distribution function of X is $F(x) = x(500 - x)/(250)^2$. Draw a graph of $F(x)$. Determine the probabilities $P(50 \leq X \leq 100)$ and $P(X \geq 75)$.

9. Suppose that the probability that there is a bird nest within r kilometers of a given point is $1 - e^{-8r^2}$. If it is found that there is no nest within 100

meters of the given point, what is the probability that there is at least one nest within 200 meters of the point?

10. An experimental drug has the effect of increasing the pulse rate. The drug is tested on a large number of volunteers by injecting each volunteer with a standard amount of the drug and observing the change in pulse rate 5 minutes later. Suppose that the proportion of volunteers whose pulse is increased by more than x beats per minute is found to be $(1 + e^{x/10})^{-1}$. What proportion of the volunteers have their pulse rate increased by between 5 and 10 beats per minute?

11. In Problem 10, if it is known that the pulse rate of a volunteer has increased by more than 5 beats per minute, what is the probability that it has increased by more than 10 beats per minute?

8.2 Density Functions

In the previous section, we have seen that, in general, it is not possible to assign nonzero probabilities to particular values of a continuous random variable X. However, we would like to be able to estimate the probability that X takes on values "close" to a given value x. The probability that the random variable X takes on a value between x and $x + \Delta x$ is

$$P(x \leq X \leq x + \Delta x) = F(x + \Delta x) - F(x) \equiv \Delta F(x).$$

This is the probability associated with an interval of length Δx. If we divide this probability by Δx, we obtain the probability per unit length (or the probability density) in the interval. If we now let Δx tend to zero, we are led to the following definition, which is central to the study of continuous random variables.

Definition 8.2.1   Probability Density Function   *Consider a continuous random variable X that has the differentiable probability distribution function $F(x)$ on $-\infty < x < \infty$. The probability density function of X at x is the function $f(x)$ defined by*

$$f(x) = \lim_{\Delta x \to 0} \frac{F(x + \Delta x) - F(x)}{\Delta x} = \lim_{\Delta x \to 0} \frac{\Delta F(x)}{\Delta x} = \frac{d}{dx}F(x) \qquad (-\infty < x < \infty).$$

The probability density function represents the probability per unit distance along the x axis. The continuous random variable X can be specified by its density function $f(x)$, since, if $f(x)$ is known, we can determine the distribution function $F(x)$ by the formula

$$F(x) = \int_{-\infty}^{x} f(t)\, dt.$$

A function $f(x)$ is a probability density function if it satisfies the properties

1. $f(x) \geq 0 \qquad (-\infty < x < \infty)$.

2. $\displaystyle\int_{-\infty}^{\infty} f(x)\, dx = 1$.

The second property is the requirement that the total probability along the x axis is 1. The first property must be satisfied because $f(x)$ is the derivative of a nondecreasing function.

   If X is a continuous random variable with probability density function $f(x)$, then the probability that X takes on values between a and b is

$$P(a \leq X \leq b) = F(b) - F(a) = \int_{a}^{b} f(t)\, dt.$$

In the spinner example given earlier, the outcome of the experiment was a continuous random variable X with range $0 \leq x \leq 1$. The distribution function was $F(x) = x$ and, therefore, the density function is $f(x) = 1$ for $0 \leq x \leq 1$. The probability that X takes on values between $\frac{1}{3}$ and $\frac{1}{2}$, for example, is

$$P(\tfrac{1}{3} \leq X \leq \tfrac{1}{2}) = F(\tfrac{1}{2}) - F(\tfrac{1}{3}) = \int_{1/3}^{1/2} 1\, dt = \tfrac{1}{2} - \tfrac{1}{3} = \tfrac{1}{6}.$$

The following examples are typical probability density functions.

Example 8.2.1   Uniform Distribution $f(x) = 1/R$ on $0 \leq x \leq R$ and $f(x) = 0$ elsewhere. This function clearly satisfies the two properties that define a probability density function. It is the density function corresponding to the random variable X that has uniform distribution on $0 \leq x \leq R$ (see Figure 8.4).

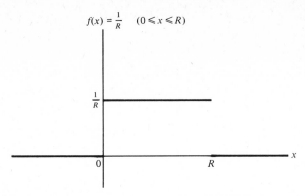

Figure 8.4. Density Function of Uniform Distribution on $0 \leqslant x \leqslant R$

Example 8.2.2 $f(x) = \frac{3}{4}x(2 - x)$ on $0 \leq x \leq 2$ and $f(x) = 0$ elsewhere (Figure 8.5). It is easy to see that $f(x) \geq 0$ and $\int_{-\infty}^{\infty} f(x)\,dx = \int_{0}^{2} f(x)\,dx = 1$. Therefore, $f(x)$ is a probability density function for a continuous random variable X with range $0 \leq x \leq 2$.

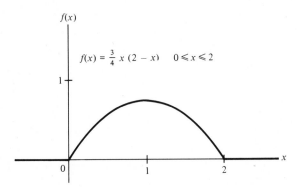

Figure 8.5. Density Function $f(x) = \frac{3}{4}x(2 - x)$ $(0 \leqslant x \leqslant 2)$

Example 8.2.3 Exponential Distribution   Here, $f(x) = (1/a)e^{-x/a}$ for $x \geq 0$ and $f(x) = 0$ for $x < 0$ (Figure 8.6), where a is any positive number. We must verify the second property defining a probability density function.

$$\int_{-\infty}^{\infty} f(x)\,dx = \int_{0}^{\infty} \frac{1}{a}e^{-x/a}\,dx = -e^{-x/a}\Big|_{0}^{\infty} = 1.$$

Therefore, $f(x)$ defines a continuous random variable X with range $0 \leq x < \infty$. The random variable X is said to have the exponential distribution with parameter a.

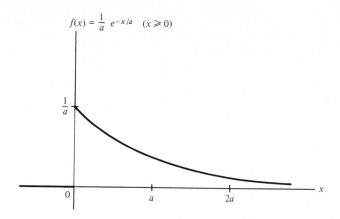

$$f(x) = \frac{1}{a} e^{-x/a} \quad (x \geqslant 0)$$

Figure 8.6. Density Function of Exponential Distribution

Example 8.2.4 Unit Normal Distribution In this example,

$$f(x) = (1/\sqrt{2\pi})e^{-x^2/2} \quad (-\infty < x < \infty)$$

(Figure 8.7). It is obvious that $f(x) \geq 0$, and it can be shown that $\int_{-\infty}^{\infty} f(x)\, dx = 1$. The range of the corresponding continuous random variable X is the entire real line. We shall study this example in the next section of this chapter.

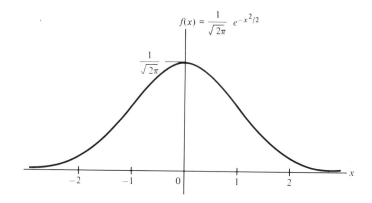

$$f(x) = \frac{1}{\sqrt{2\pi}} e^{-x^2/2}$$

Figure 8.7. Unit Normal Distribution

For a continuous random variable, we have the following definition of the mean and variance in terms of the probability density function.

Definition 8.2.2   Expected Value and Variance   *Let X be a continuous random variable with density function f(x). Then, the expected value μ and the variance σ² of X are defined by*

$$\mu = E(X) = \int_{-\infty}^{\infty} xf(x)\, dx,$$

$$\sigma^2 = E(X - \mu)^2 = \int_{-\infty}^{\infty} (x - \mu)^2 f(x)\, dx.$$

The square root σ of the variance is called the *standard deviation* of X. As before (Theorem 2.8.1), we have $E(X - \mu)^2 = E(X^2) - \mu^2 = E(X^2) - (E(X))^2$.

For the example of the spinner, the density function is $f(x) = 1$ on $0 \le x \le 1$. Therefore, $E(X) = \int_0^1 x \cdot 1\, dx = x^2/2 \big|_0^1 = \frac{1}{2}$. This can be thought of as the average outcome of the experiment. If the experiment is repeated many times, the average of the outcomes is expected to be close to $\frac{1}{2}$. To calculate the variance, $E(X^2) = \int_0^1 x^2 \cdot 1\, dx = \frac{1}{3}$. Therefore, $\sigma^2 = \frac{1}{3} - (\frac{1}{2})^2 = \frac{1}{12}$.

Example 8.2.2 (continued)   For this example,

$$E(X) = \int_0^2 x \frac{3}{4} x(2 - x)\, dx = \frac{3}{4} \int_0^2 x^2(2 - x)\, dx = \frac{3}{4} \left[\frac{2x^3}{3} - \frac{x^4}{4} \right]_0^2 = 1.$$

This is what one would guess by looking at Figure 8.5. $E(X^2) = \frac{3}{4} \int_0^2 (2x^3 - x^4)\, dx = \frac{6}{5}$. Therefore, $\sigma^2 = \frac{6}{5} - 1^2 = \frac{1}{5}$.

Example 8.2.3 (continued)   Here,

$$E(X) = \frac{1}{a} \int_0^{\infty} xe^{-x/a}\, dx.$$

This integral can be evaluated by integrating by parts. Define $u = x/a$ and $dv = e^{-x/a}\, dx$. Then,

$$E(X) = \int_0^{\infty} u\, dv = uv \Big|_0^{\infty} - \int_0^{\infty} v\, du = -xe^{-x/a} \Big|_0^{\infty} + \int_0^{\infty} e^{-x/a}\, dx$$

$$= -ae^{-x/a} \Big|_0^{\infty} = a.$$

Similarly, after integrating by parts twice, we obtain $E(X^2) = 2a^2$ and so

$\sigma^2 = 2a^2 - a^2 = a^2$. This gives the interesting property of the exponential distribution with parameter a that $\mu = \sigma = a$.

The following examples illustrate how density functions can be applied in practical problems. Other applications are given in the problems at the end of this section.

Example 8.2.5   Distribution of Plant Lifetimes   The length of life of plants of a given species in a certain environment is a continuous random variable X. Suppose that the probability density function of X is $f(x) = \frac{1}{120}e^{-x/120}$.

1. What is the distribution function of X?
2. What proportion of plants of this species die within 100 days?
3. If an individual plant lives for 100 days, what is the probability that it will live another 100 days?

Solution: We recognize $f(x)$ as the probability density function of the exponential distribution with parameter $a = 120$ (Example 8.2.3). The average length of life for plants of this species is $E(X) = 120$ days.

1. The distribution function of X is

$$F(x) = \int_0^x \frac{1}{120}e^{-t/120}\, dt = 1 - e^{-x/120}.$$

2. The proportion of plants that die within 100 days is $P(0 \le X \le 100) = F(100) = 1 - e^{-100/120} \approx .70$.
3. The required probability is

$$P(X \ge 200 | X \ge 100) = \frac{P(X \ge 200)}{P(X \ge 100)} = \frac{1 - F(200)}{1 - F(100)} = \frac{e^{-200/120}}{e^{-100/120}}$$

$$= e^{-100/120} \approx .30.$$

In other words, about 30 per cent of the plants that live 100 days will live at least another 100 days.

Example 8.2.6   In a study of the diffusion of insect populations, a large number of ants are released at a given point. After 1 minute, it is observed that the proportion of ants at a distance at least r meters from the point of release is approximately e^{-2r}.

1. What proportion of the ants travel more than 1 meter from the point of release?
2. What is the average distance traveled by the ants from the point of release?

Solution: Define R to be the distance traveled by an ant from the point of release in 1 minute. Then, R is a continuous random variable with distribution function $F(r) = P(R \le r) = 1 - e^{-2r}$.

1. The proportion of ants that travel more than 1 meter is $P(R \ge 1) = 1 - F(1) = e^{-2} \approx .135$.
2. The probability density function of the random variable R is $f(r) = dF(r)/dr = 2e^{-2r}$. Therefore, the average distance is

$$E(R) = \int_0^\infty rf(r)\,dr = \int_0^\infty 2re^{-2r}\,dr = \tfrac{1}{2}.$$

We conclude that the average distance traveled in 1 minute from the point of release is .5 meter.

Problems 8.2

1. Which of the following functions represent the probability density function of a continuous random variable? Draw a graph of each of the probability density functions and determine the range of the random variable.

(a) $f(x) = e^x, \quad -\infty < x \le 0,$
$\quad\;\; = 0, \quad x > 0.$

(b) $f(x) = 1, \quad -\tfrac{1}{2} \le x \le \tfrac{1}{2},$
$\quad\;\; = 0, \quad$ elsewhere.

(c) $f(x) = \tfrac{1}{2}\sin x, \quad 0 \le x \le \pi,$
$\quad\;\; = 0, \quad$ elsewhere.

(d) $f(x) = \dfrac{1}{\pi}\dfrac{1}{1 + x^2},$
$\quad\; -\infty < x < \infty.$

(e) $f(x) = \tfrac{1}{2}e^{-|x|}, \quad -\infty < x < \infty.$

(f) $f(x) = \tfrac{3}{4}(1 - x^2), \quad -1 \le x \le 1,$
$\quad\;\; = 0, \quad$ elsewhere.

2. For each of the probability density functions in Problem 1,
(a) Determine the corresponding distribution functions.

   (b) Determine the probabilities that the random variables take on values between 0 and 1.
   (c) Determine the expected values and variances of the random variables.

3. Define a function $f(x)$ to be zero for x negative and to be equal to ke^{-3x} for x nonnegative.
   (a) For what value of the constant k is this function the probability density function of a continuous random variable? Define X to be this random variable.
   (b) Determine $P(X \leq 1)$, $P(X \geq 2)$, and $P(1 \leq X \leq 2)$.

4. The life span (measured in days) of a certain kind of bacteria is a continuous random variable whose probability distribution is approximately an exponential distribution. If the average life span is 12 hours, calculate
   (a) The probability that a particular bacterium will die within 12 hours.
   (b) The probability that a particular bacterium, which has lived for one day, will die within the next day.

5. The distribution function of the exponential distribution is $F(x) = 1 - e^{-x/a}$ for $x \geq 0$. The corresponding random variable X has mean a and standard deviation a. If b and h are any positive numbers, prove that

$$P(0 \leq X \leq h) = \frac{P(b \leq X \leq b + h)}{P(X \geq b)}.$$

   How is this result useful in Problem 4(b)?

6. Define the function $f(x) = 0$ for $x < 0$ and $f(x) = 4xe^{-2x}$ for $x \geq 0$.
   (a) Verify that this is the probability density function of a continuous random variable. Define X to be this random variable.
   (b) Determine $P(X \leq 1)$, $P(X \geq 2)$, and $P(1 \leq X \leq 2)$.
   (c) Determine the expected value and standard deviation of X.

7. The digestion time (measured in hours) of a unit of food is a random variable whose probability density function is $f(x) = 4xe^{-2x}$ in a mathematical model. What is the probability that a unit of food is not completely digested after 1 hour? What is the average time required for the digestion of a unit of food?

8. The probability that an insect born at time $t = 0$ will be eaten before time t (measured in days) is $F(t) = 1 - e^{-\gamma t}$ for a constant γ.
   (a) In terms of the constant γ, what is the expected length of time before an insect is eaten?
   (b) Assume that $\gamma = .5$. What is the probability that the insect will be eaten before it is 2 days old?

(c) Suppose that the insect must live 10 days before it is able to reproduce. When $\gamma = .5$, what is the probability that it will be eaten before it will be able to reproduce?

9. The life span (in days) of a certain plant is a random variable X with probability density function $f(x) = x/20,000$ for $0 \le x \le 200$ and $f(x) = 0$ elsewhere.
   (a) What is the expected length of life of this plant?
   (b) What is the probability that it will die during the first 30 days?
   (c) If the plant lives for 100 days, what is the probability that it will die during the next 30 days?

10. An experimental operation requires at least 4 minutes to complete but never requires more than 10 minutes. Define a random variable T to be the length of time required to complete the operation and assume that the probability density function of T is of the form $f(t) = k(t - 4)(10 - t)$ on $4 \le t \le 10$.
   (a) For what value of the constant k is this function a probability density function?
   (b) Determine the expected value and standard deviation of T.

11. For a certain instrument, the time (measured in years) until the first repairs are required is a random variable X whose probability density function is $f(x) = 2xe^{-x^2}$.
   (a) What is the average time until the first repairs?
   (b) What is the probability that no repairs will be required in the first 2 years?

12. (a) Verify that $f(x) = 6x/(1 + x)^4$ for x nonnegative is the probability density function of a continuous random variable X with range the positive x axis.
   (b) Draw a graph of $f(x)$ and determine its maximum.
   (c) Calculate $P(X \le \frac{1}{3})$, $P(X \le 1)$, and $P(\frac{1}{3} \le X \le 1)$.

13. The probability that a cat will catch a mouse increases with the time t of pursuit and is given by $F(t) = 1 - e^{-t/10}$, where t is measured in minutes. If the cat always becomes bored after 15 minutes and abandons the pursuit, what percentage of mice escape? What percentages of mice are captured within 5 minutes, within 10 minutes, and within 15 minutes?

14. An ecologist is interested in studying the dispersal of seeds from a certain plant. Suppose that the seeds travel an average distance of 1 meter from the plant and that the probability distribution of this distance is exponential. What proportion of the seeds are dispersed more than 2 meters from the plant?

8.3 The Normal Distribution

We now introduce a continuous random variable X with range the real line whose probability distribution comes up in every area of probability theory. It can be considered to be an approximation to many other probability distributions, and it is an important distribution in its own right.

Definition 8.3.1   The Normal Distribution   *The continuous random variable X has the normal distribution with mean μ and variance σ^2 if it has the probability density function*

$$f(x) = \frac{1}{\sigma\sqrt{2\pi}} e^{-(x-\mu)^2/2\sigma^2}$$

on the range $-\infty < x < \infty$ (Figure 8.8).

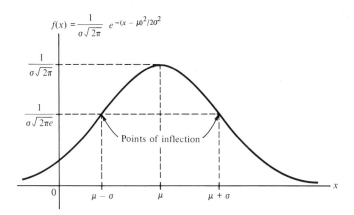

Figure 8.8. The Normal Distribution

It can be verified that this function $f(x)$ satisfies (1) $\int_{-\infty}^{\infty} f(x)\,dx = 1$, (2) $\int_{-\infty}^{\infty} xf(x)\,dx = \mu$, and (3) $\int_{-\infty}^{\infty} (x-\mu)^2 f(x)\,dx = \sigma^2$. The special case with $\mu = 0$ and $\sigma = 1$, introduced in Example 8.2.4, is particularly useful and is called the *unit normal distribution*.

The graph of $f(x)$ is shown in Figure 8.8. This is the famous "bell-shaped" curve. To study the shape of this curve, note that $f(\mu + y) = f(\mu - y)$ and, therefore, the curve is symmetric about the line $x = \mu$. The slope of the curve is

$$f'(x) = -\frac{(x-\mu)}{\sigma^3\sqrt{2\pi}} e^{-(x-\mu)^2/2\sigma^2}.$$

This is positive for $x < \mu$ and negative for $x > \mu$. Therefore, $f(x)$ increases to its maximum as x increases to μ and then decreases as x increases. The second derivative is

$$f''(x) = \frac{e^{-(x-\mu)^2/2\sigma^2}}{\sigma^3\sqrt{2\pi}}\left[\frac{(x-\mu)^2}{\sigma^2} - 1\right].$$

The points of inflection occur when $f''(x) = 0$ at $x = \mu + \sigma$ and $x = \mu - \sigma$. It is clear that $f(x) > 0$ for all values of x and that $f(x)$ tends to 0 as x tends to infinity.

In order to use the normal distribution, it is necessary to calculate many of its values. This would be very difficult to do for all possible values of μ and σ, but, because of the following theorem, we need only to calculate values of the unit normal distribution.

Theorem 8.3.1   *If X is a normal random variable with mean μ and variance σ^2, then $Y = (X - \mu)/\sigma$ is a random variable with unit normal distribution.*

Proof: X has distribution function $F(x) = P(X \le x) = (1/\sigma\sqrt{2}) \int_{-\infty}^{x} \times e^{-(t-\mu)^2/2\sigma^2}\, dt$.   If $y = (X - \mu)/\sigma$,   then $P(Y \le y) = P(X \le \mu + \sigma y) = F(\mu + \sigma y)$. Therefore, $P(Y \le y) = (1/\sigma\sqrt{2\pi}) \int_{-\infty}^{\mu+\sigma y} e^{-(t-\mu)^2/2\sigma^2}\, dt$. Introducing the change of variables $s = (t - \mu)/\sigma$ in the integral, we have $P(Y \le y) = (1/\sqrt{2\pi}) \int_{-\infty}^{y} e^{-s^2/2}\, dt$. This proves that Y has the unit normal distribution.

The converse to Theorem 8.3.1 is that, if Y has the unit normal distribution, then $X = \sigma Y + \mu$ is normally distributed with mean μ and variance σ^2. If we have a table of values of the density function and the distribution function for the unit normal distribution, by the above theorem, this table can be used to calculate probabilities for normal random variables with arbitrary mean and variance. If X and Y are defined as before, then

$$P(a \le X \le b) = P\left(\frac{a-\mu}{\sigma} \le Y \le \frac{b-\mu}{\sigma}\right) = F\left(\frac{b-\mu}{\sigma}\right) - F\left(\frac{a-\mu}{\sigma}\right),$$

where F is the distribution function of the unit normal distribution. In applying the tables, we use the result that $F(x) + F(-x) = 1$. This follows from the symmetry of the density function $f(x)$ about $x = 0$. If X is a unit normal random variable with distribution function $F(x)$, then

$$F(-x) = P(X \le -x) = P(X \ge x) = 1 - P(X \le x) = 1 - F(x).$$

Because of this result, the values of the distribution function are tabulated for positive values of the argument only.

Example 8.3.1 Suppose that the random variable X is normally distributed with mean 5 and standard deviation 2. What is the probability that X takes on values between 4 and 7? What is the probability that it takes a value greater than 10?

Solution: Define the unit normal random variable $Y = (X - 5)/2$. Then the first probability is $P(4 \leq X \leq 7) = P(-\frac{1}{2} \leq Y \leq 1) = F(1) - F(-\frac{1}{2})$. But $F(-\frac{1}{2}) = 1 - F(\frac{1}{2})$. Therefore, $P(4 \leq X \leq 7) = F(1) - 1 + F(\frac{1}{2}) = .8413 - 1 + .6915 = .5328$. Also, $P(X \geq 10) = P(Y \geq \frac{5}{2}) = 1 - P(Y \leq \frac{5}{2}) = 1 - F(\frac{5}{2}) = 1 - .9938 = .0062$.

Example 8.3.2 The time T required for digestion of a unit of food by a certain protozoan is a normal random variable with mean 31 minutes and standard deviation 5 minutes.

1. What is the probability that a unit of food will be digested in less than 35 minutes?
2. If a particular unit of food is observed to be incompletely digested after 30 minutes, what is the probability that it will be digested in less than 35 minutes?

Solution: Define the unit normal random variable $X = (T - 31)/5$.

1. $P(T \leq 35) = P(X \leq .8) = F(.8) \approx .79$.
2. To calculate the second probability, we must use conditional probability. The required probability is

$$P(T \leq 35 | T \geq 30) = \frac{P(30 \leq T \leq 35)}{P(T \geq 30)}.$$

But $P(30 \leq T \leq 35) = P(-.2 \leq X \leq .8) = F(.8) - F(-.2) = F(.8) - 1 + F(.2) = .3674$ and $P(T \geq 30) = P(X \geq -.2) = F(-.2) = 1 - F(.2) = .5793$. Therefore,

$$P(T \leq 35 | T \geq 30) = \frac{.3674}{.5793} \approx .63.$$

This means that approximately 63 per cent of the units of food not completely digested after 30 minutes will be digested after 35 minutes.

The normal distribution can be thought of as an approximation to many other distributions. For example, the binomial probability

$$f(n, k, p) = \binom{n}{k} p^k q^{n-k}$$

tends, as n increases, to

$$f(k) = \frac{1}{\sqrt{npq}} \frac{1}{\sqrt{2\pi}} e^{-(k-np)^2/2npq}.$$

We recognize this as the normal distribution with mean $\mu = np$ and variance $\sigma^2 = npq$. This approximation is sufficiently accurate for many purposes when $np \geq 10$. This is illustrated for $n = 20$, $p = .5$ in Figure 8.9.

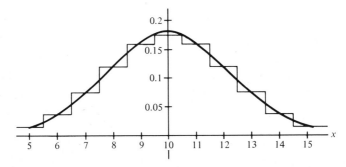

Figure 8.9. Comparison of Binomial Distribution with $n = 20$, $p = 0.5$, and Normal Distribution with $\mu = 10$, $\sigma^2 = 5$

Calculated Values

x	$\binom{20}{x} \dfrac{1}{2^{20}}$	$\dfrac{1}{\sqrt{10\pi}} e^{-(x-10)^2/10}$
10	.174	.179
9	.160	.161
8	.120	.119
7	.074	.073
6	.037	.036
5	.015	.015

In Section 2.9, we proved that the Poisson distribution with $\mu = np$ approximates the binomial distribution for n large and p small. Therefore, the normal distribution can be used as an approximation to the Poisson distribution when μ is sufficiently large. The normal approximation to $P(k) = (\mu^k/k!)e^{-\mu}$ is $f(k) = (1/\sqrt{2\pi\mu})e^{-(k-\mu)^2/2\mu}$.

Example 8.3.3 A binomial experiment with $p = 0.02$ is repeated 2500 times. The expected number of successes is $\mu = np = 50$. The probability of exactly k successes is

$$\binom{2500}{k}(.02)^k(.98)^{2500-k}.$$

The Poisson approximation is $[(50)^k/k!]e^{-50}$ and the normal approximation is $[1/\sqrt{(49)(2\pi)}]e^{-(k-50)^2/98}$. For example, when $k = 40$, these probabilities are .0212, .0215, and .0205, respectively. (To obtain the last probability, we have evaluated $f(x)$, the normal density function, at $x = 40$.)

Example 8.3.4 It is known that 20 per cent of the people in a large population are left-handed. Suppose that 10,000 people are chosen at random from this population and the number of left-handed people in this sample is determined. By using a normal approximation, estimate the probability that at least 1900 people in the sample are left-handed. What is the probability that at least 1960, but no more than 2040, people in the sample are left-handed?

Solution: Define X to be the number of people in a sample of 10,000 who are left-handed. Then X is a binomial random variable with $n = 10,000$ and $p = .2$. The expected number of left-handed people in the sample is $\mu = E(X) = np = 2000$. The standard deviation of X is $\sigma = \sqrt{npq} = 40$. If we define $Y = (X - 2000)/40$, then Y is approximately a unit normal random variable. Therefore, the required probabilities are $P(X \geq 1900) = P(Y \geq -2.5) \approx .994$ and $P(1960 \leq X \leq 2040) = P(-1 \leq Y \leq 1) \approx .68$.

Example 8.3.5 On an average day, there are 100 emergency cases admitted to a large hospital.

1. Estimate the probability that at least 90 emergency cases will be admitted to the hospital on a given day.

2. If the hospital provides facilities adequate for the number of emergency cases on 95 per cent of all days, what number of emergency patients is the hospital prepared to handle each day?

Solution: Define X to be the number of emergency cases admitted to the hospital on a given day. Then X is a random variable with a Poisson distribution (why?). The mean and variance are both equal to $\mu = E(X) = 100$. Therefore, the standard deviation is $\sigma = \sqrt{\mu} = 10$. If we define $Y = (X - 100)/10$, then Y is approximately a unit normal random variable.

1. The first probability is $P(X \geq 90) = P(Y \geq -1) \approx .84$.
2. Suppose that k is the number of emergency patients that the hospital is prepared to handle each day. Then k is determined by

$$P(X \leq k) = P\left(Y \leq \frac{k - 100}{10}\right) = .95.$$

From the tables, we conclude that $(k - 100)/10 \approx 1.65$ or $k \approx 116.5$. The hospital should have facilities for 117 emergency patients each day to meet the demand on at least 95 per cent of all days.

Problems 8.3

1. Suppose that X is a normal random variable with mean 5 and variance 4.
   (a) Define the corresponding unit normal random variable.
   (b) Determine $P(3 \leq X)$, $P(X \leq 6)$, and $P(2 \leq X \leq 8)$.
   (c) Determine numbers a and b such that $P(X \leq a) = .95$ and $P(X \leq b) = .99$.

2. (a) If X is a normal random variable with mean μ and variance σ^2, the 95 and 99 per cent *probability intervals* for X are the ranges $\mu - a \leq X \leq \mu + a$ and $\mu - b \leq X \leq \mu + b$, where a and b are chosen so that $P(\mu - a \leq X \leq \mu + a) = .95$ and $P(\mu - b \leq X \leq \mu + b) = .99$. Prove that $a \approx 1.96\sigma$ and $b \approx 2.58\sigma$.
   (b) Suppose that X is a normal random variable with mean 25 and variance 4. What are the 95 and 99 per cent probability intervals for X?

3. The acidity of human blood measured on the pH scale is a normal random variable with mean 7.4 and standard deviation 0.2. What is the probability that the pH level is greater than 7.43? Between 7.35 and 7.45?

4. The annual rainfall in a certain region is a normally distributed random variable with mean 30 inches and standard deviation 2 inches. What is

the probability of more than 31 inches of rain in a given year? What are the 95 and 99 per cent probability intervals for the annual rainfall?

5. The adult weight of a certain mammal is a normally distributed random variable with mean 100 pounds and standard deviation 8 pounds. What are the probabilities that this animal weighs (a) less than 90 pounds, (b) between 95 and 105 pounds, and (c) more than 110 pounds?

6. The diastolic blood pressure in hypertensive women has been estimated to have an average value of 98 millimeters with a standard deviation of 15 millimeters. Assuming that the diastolic blood pressure is a normal random variable, estimate the probabilities that it is (a) less than 89 millimeters, (b) more than 104 millimeters, and (c) between 86 and 100 millimeters. If it is known that the diastolic blood pressure is greater than 104 millimeters, what is the probability that it is more than 110 millimeters?

7. The probability of success on a single trial of a binomial experiment is $p = .3$. By using a normal approximation, estimate the probability of obtaining more than 40 successes in 100 trials of the experiment.

8. A rare disease affects 0.1 per cent of a large population. Define X to be the number of people who have the disease in a group of 100,000 people chosen at random from the population. By using a normal approximation, estimate the probability that at least 80 people have the disease and the probability that no more than 130 people have the disease.

9. In a large population of drosophila, 25 per cent have a wing mutation. A sample of 300 flies is chosen at random. By using a normal approximation, estimate the probability that at least 60 and no more than 90 flies in the sample have the wing mutation.

10. The frequency of tuberculosis in a large population is estimated to be 0.04 per cent. What is the probability that, among 1,000,000 people chosen at random from the population, at least 360 and no more than 440 have tuberculosis?

11. The average length of adult fish of a certain species has been estimated to have an average value of 65 centimeters with a standard deviation of 5 centimeters. Assuming a normal distribution, what is the probability that a given fish is longer than 70 centimeters? Shorter than 55 centimeters?

12. (a) An observation is made of the normal random variable X with mean μ and variance σ^2. If it is known that the observed value is less than μ, what is the probability that it is less than $\mu - \sigma$?

(b) If a fish of Problem 11 is known to be shorter than the average, what is the probability that it is shorter than 60 centimeters?

13. In two large school systems in neighboring cities, the intelligence quotients of equal numbers of third-grade schoolchildren were measured by standard tests. In system I, the average was 100 with standard deviation 10; in system II, the average was 105 with standard deviation 12. A child chosen at random was found to have an intelligence quotient over 120. Assuming normal distributions, what is the probability that this child came from the second school system? (*Hint:* Use Bayes' theorem.)

14. About 1 child in 700 is born with Down's syndrome. In one state, there are 34,300 births recorded in 1 year. Estimate the probability that at least 56 cases of Down's syndrome occur among these recorded births.

8.4 Chebyshev's Inequality and Confidence Intervals

We have studied probability distributions by calculating their means and variances. In this section, we prove a useful inequality that emphasizes the importance of these two numbers associated with probability distributions.

Suppose that X is a continuous random variable with mean μ and variance σ^2. We may want to estimate the probability that X takes on values which differ considerably from the mean. The following theorem allows us to do this.

Theorem 8.4.1   Chebyshev's Inequality   *Let X be a continuous random variable with mean μ and standard deviation σ. Then, for any $t > 0$, the probability that X takes on values which differ from μ by $t\sigma$ or more is less than $1/t^2$; that is, $P(|X - \mu| \geq t\sigma) \leq 1/t^2$.*

Proof: Divide the range of X into two sets, $S_1 = \{x : (x - \mu)^2 \geq t^2\sigma^2\}$ and $S_2 = \{x : (x - \mu)^2 < t^2\sigma^2\}$. By definition,

$$\sigma^2 = \int_{-\infty}^{\infty} (x - \mu)^2 f(x)\, dx = \int_{S_1} (x - \mu)^2 f(x)\, dx + \int_{S_2} (x - \mu)^2 f(x)\, dx.$$

Therefore, $\sigma^2 \geq \int_{S_1} (x - \mu)^2 f(x)\, dx$, since the second integral is nonnegative.

But, on S_1, we have $(x - \mu)^2 \geq t^2\sigma^2$. Therefore,

$$\sigma^2 \geq \int_{S_1} t^2\sigma^2 f(x) \, dx = t^2\sigma^2 \int_{S_1} f(x) \, dx.$$

This integral is exactly the probability we are interested in. This implies $\sigma^2 \geq t^2\sigma^2 P(|X - \mu| \geq t\sigma)$. Rearranging, we have

$$P(|X - \mu| \geq t\sigma) \leq \frac{1}{t^2}.$$

This is the result.

The theorem is only useful for $t > 1$. It says, for example, that the probability that X differs from its mean by more than 5 standard deviations is less than $1/5^2 = .04$. This is true for every probability distribution. The reader should try to construct a proof following the same argument for a discrete random variable. For particular distributions such as the normal distribution, this probability will be much less than 4 per cent. This is illustrated in the following example.

Example 8.4.1   In a large population of fruit flies, it is known that 40 per cent have a particular mutation. How large should a sample of the population be in order to be 95 per cent certain that between 38 and 42 per cent of the flies in the sample have the mutation?

Solution: If n is the required sample size, we have n repeated trials of a binomial experiment with $p = .4$. Therefore, $\mu = np = (.4)n$ and $\sigma^2 = npq = (.24)n$. Let X be the random variable equal to the number of flies in the sample which have the mutation. Then, X/n is the proportion of flies with the mutation and we are seeking the probability

$$P\left(\left|\frac{X}{n} - \frac{\mu}{n}\right| \geq .02\right) = P(|X - \mu| \geq (.02)n).$$

We require n to be chosen sufficiently large so that

$$P(|X - \mu| \geq (.02)n) \leq .05.$$

By Chebyshev's inequality, we know $P(|X - \mu| \geq t\sigma) \leq 1/t^2$. Therefore,

$t\sigma = (.02)n$ or $t^2 = n^2/\sigma^2(50)^2$. But $\sigma^2 = (.24)n$. Therefore, $t^2 = n/(.24)(50)^2$. Also $1/t^2 = .05 = \frac{1}{20}$. Therefore, $n = 20(.24)(50)^2 = 12{,}000$. A sample size of 12,000 flies is large enough to ensure that between 38 and 42 per cent of the flies in the sample have the mutation in 95 per cent of such samples.

Another method of estimating the sample size required would be to use the normal approximation to the binomial distribution. Assume, then, that X is normally distributed with mean $(.4)n$ and variance $(.24)n$. We must determine n such that

$$P(|X - (.4)n| \geq (.02)n) \leq .05.$$

As before, $(.02)n = t\sigma = t\sqrt{(.24)n}$ and $t^2 = n/600$. From the tables, we determine x such that $1 - F(x) = .025$. (The probability .05 is made up of two equal contributions from the opposite tails of the unit normal distribution.) This occurs for $x \approx 2.0$. Therefore, $t = 2.0$ and $n \approx 600(2.0)^2 = 2400$. This example illustrates that using Chebyshev's inequality gives a very conservative estimate of the sample size required. When possible, a more useful estimate can be obtained by using the normal approximation.

In Example 8.4.1, we have assumed that we know the proportion of the population which has a particular characteristic. In many problems, this proportion is not known; we attempt to estimate it by observing the proportion with the characteristic in a sample drawn from the population. Suppose that, in n trials of a binomial experiment, k successes occur. Then, the proportion k/n should be approximately equal to p, the probability of success on one trial, if n is large. If X is the random variable equal to the number of successes on n trials, then $(X - np)/\sqrt{npq}$ can be approximated by the unit normal random variable. Therefore,

$$P\left(\left|\frac{X - np}{\sqrt{npq}}\right| \leq 2\right) \approx .95.$$

This implies that $P(|(X/n) - p| \leq 2\sqrt{pq/n}) \approx .95$. Since $pq = p(1 - p) \leq \frac{1}{4}$ (why?) and setting $X = k$, we have

$$P\left(\left|\frac{k}{n} - p\right| \leq \frac{1}{\sqrt{n}}\right) > .95.$$

We conclude that

$$\frac{k}{n} - \frac{1}{\sqrt{n}} \le p \le \frac{k}{n} + \frac{1}{\sqrt{n}}$$

in at least 95 per cent of all cases. This range of values for the true proportion p is referred to as the *95 per cent confidence interval*. The *99 per cent confidence interval* is

$$\frac{k}{n} - \frac{1.3}{\sqrt{n}} \le p \le \frac{k}{n} + \frac{1.3}{\sqrt{n}}.$$

This can be derived from the relation

$$P\left(\left|\frac{X - np}{\sqrt{npq}}\right| \le 2.6\right) > .99$$

by the same argument as above.

Example 8.4.2   In a large population of fruit flies, an unknown proportion p have a certain mutation. It is found that, in a sample of 2500 flies, 975 have the mutation. What are the 95 and 99 per cent confidence intervals for p?

Solution: In this example, $n = 2500$, $k = 975$, $k/n = .39$, and $\sqrt{n} = 50$. The 95 per cent confidence interval for p is $.37 \le p \le .41$. The 99 per cent confidence interval is $.36 \le p \le .42$. This means that the exact proportion almost certainly lies between 36 and 42 per cent.

Example 8.4.3   In a study of the relative effects of two weight-reducing diets, two groups of 400 volunteers followed the two diets. After 1 month, 280 of the volunteers on the first diet reported weight losses of at least 10 pounds. On the second diet, 300 of the volunteers lost at least 10 pounds during the month. Comment on the significance of this data.

Solution: The data obtained from these two samples allow us to estimate the true proportions of people who would lose at least 10 pounds in 1 month on these diets. Define p_1 and p_2 to be these true proportions for the first and second diets, respectively. The observed proportions in the two samples of 400 people are $\frac{280}{400} = .7$ and $\frac{300}{400} = .75$. The 95 per cent confidence interval for p_1 is $.7 - \frac{1}{20} \le p_1 \le .7 + \frac{1}{20}$ or $.65 \le p_1 \le .75$. The 99 per cent confidence interval for p_1 is $.7 - (1.3/20) \le p_1 \le .7 + (1.3/20)$ or $.635 \le p_1 \le .765$. Similarly, the 95 per cent confidence interval for p_2 is $.7 \le p_2 \le .8$ and the 99 per cent confidence interval is $.685 \le p_1 \le .815$. We observe that these confidence intervals for p_1 and p_2 overlap considerably. Therefore, on the basis of the given data, we cannot conclude that one diet is more likely than the other to produce a weight loss of more than 10 pounds in 1 month.

Problems 8.4

1. (a) Suppose that X is a continuous random variable with mean 100 and standard deviation 10. Use Chebyshev's inequality to prove that the probabilities that $80 < X < 120$ and $70 < X < 130$ are at least $\frac{3}{4}$ and $\frac{8}{9}$, respectively.

   (b) If X is a normal random variable with mean 100 and standard deviation 10, calculate $P(80 < X < 120)$ and $P(70 < X < 130)$.

2. In a large population of drosophila, 25 per cent have a particular wing mutation. A sample of 300 flies is chosen at random from this population.

   (a) Use Chebyshev's inequality to estimate a lower bound to the probability that between 45 and 105 flies in the sample have the wing mutation.

   (b) What size sample is required to be 95 per cent certain that between 20 and 30 per cent of the flies in the sample have the wing mutation?

3. Let X be any random variable (discrete or continuous) with mean μ and variance σ^2. The 95 and 99 per cent *probability intervals* for X are the ranges of values for the random variable centered about the mean, which have the property that X lies in the ranges with probabilities 95 and 99 per cent, respectively. Using Chebyshev's inequality, prove that the 95 per cent probability interval for X is contained in the interval $\mu - 2\sqrt{5}\sigma \le X \le \mu + 2\sqrt{5}\sigma$ and that the 99 per cent probability interval for X is contained in the interval $\mu - 10\sigma \le X \le \mu + 10\sigma$.

4. An electronic device is claimed to last for at least 1000 hours under normal use. If the average length of life of the device is estimated by the manufacturer to be 1050 hours, what is the maximum value of the standard deviation in order that this claim be correct in at least 99 per cent of all cases? Solve this problem by using (a) Chebyshev's inequality and (b) the normal distribution.

5. It is known that 10 per cent of the people in a large population are left-handed. Samples are to be chosen at random from this population. How large should the samples be so that, in 95 per cent or more of the samples,

the proportion of left-handed people is between 9 and 11 per cent? Estimate the required sample size by (a) Chebyshev's inequality and (b) the normal distribution. Why are these estimates so different?

6. A standard treatment for a disease is known to be completely effective in 20 per cent of the cases where it is used. A modification of this treatment is introduced and, in the next 10,000 cases, the modified treatment produced a cure in 2500 cases. What can be said about the effectiveness of the modification?

7. In a study of the incidence of cardiovascular disease, two groups of 400 veterans were observed over a period of 8 years. One group had a diet high in unsaturated fats and the incidence of disease was 31 per cent. The second group had a normal diet high in saturated fats and the cardiovascular disease rate was 48 per cent. What are the 95 and 99 per cent confidence intervals for the incidence of disease for the two groups?

8. Two groups of 1000 men between the ages of 60 and 70 were studied for evidence of a direct association between cigarette smoking and coronary atherosclerosis (fatty deposits in the arteries of the heart). Group I consisted of individuals who had never smoked regularly and 210 had advanced atherosclerosis. Individuals of group II had smoked between 20 and 40 cigarettes daily and 440 of this group had advanced atherosclerosis. Comment on the significance of these data by comparing confidence intervals.

9. (a) A particular environment can support an average of 1000 animals of a certain species. Because of fluctuations in the availability of food, in the presence of predators, and in other factors, the population of the species fluctuates about this average value. Assuming that the probability distribution of the population is normal with standard deviation $\sigma = 50$, what is the probability that the population is greater than 900? Less than 1100?

(b) It is proposed to harvest 400 animals from the above population. However, it is believed that, if the population falls below 500, it will not be able to restore itself to the average size within a reasonable time. What is the probability that the population will fall below 500?

10. There is evidence to suggest that the death rate from sudden coronary attacks is affected by the hardness of the local water supply. In a study, the annual death rates per 1,000,000 people from sudden coronary attacks were 2000 where the water supply was soft and 1200 where the water supply was hard. Comment on the significance of these results.

11. In a large hospital, 20 per cent of the patients have cancer. Suppose that two samples of 400 patients are chosen at random from the list of patients with one sample consisting of smokers and the second sample consisting of nonsmokers.

   (a) Assuming that smoking has no effect on the incidence of cancer, what are the expected numbers in each sample that have cancer?

   (b) Suppose that 120 of the smokers and 50 of the nonsmokers in the two samples do have cancer. On this evidence, is the hypothesis that smoking is linked with cancer supported?

12. The frequencies of stillbirths have been reported as 36 per 1000 births for children of parents not known to be related and 62 per 1000 births for children of related parents. Assuming that the sample size is 10,000 for both groups, is the hypothesis that stillbirths are more probable when parents are related supported by this evidence?

13. Opinion polls may give incorrect results when a small sample is taken. Suppose that 60 per cent of a group of 1,000,000 people are opposed to a certain action and 40 per cent are in favor. How large should a sample from this population be made in order to be 95 per cent certain that the sample indicates between 55 and 65 per cent of the group are opposed to the action? Estimate the required sample size by using (a) Chebyshev's inequality and (b) the normal approximation. Why are these two estimates so different?

14. In one year, among 12,100 reported deaths from leukemia, 2200 were children under the age of 15. Estimate the proportion of deaths due to leukemia of children under 15. What are the 95 and 99 per cent confidence intervals for this estimate based on these data?

15. Psychokinesis is the theory that the motion of objects can be influenced by conscious thought. Design an experiment to test whether you are psychokinetic. (Use tossing of coins or drawing of cards from well-shuffled decks.) What type of evidence would you require before concluding that you are psychokinetic?

16. Prove Chebyshev's inequality for a discrete random variable X by dividing the range space of X into two sets and following the steps of the proof for a continuous random variable.

Mathematical Models in Biology | 9

9.1 Construction of Models

In the previous chapters, a large number of simple models of biological processes were introduced. The growth models of Chapter 7 are typical example of such models. By making assumptions about the growth rates of the populations being studied, we were able to determine the populations at all future times. A model of a biological process is simply a set of assumptions about that process, together with some analysis of the logical consequences of the assumptions. We have a mathematical model when the assumptions can be expressed in mathematical form. Of course, models are meant to describe real processes, and any predictions based on the analysis of a model should be compared with experimental observations to determine the validity of the assumptions.

The first problem in constructing a model of a biological process is to select the key variables in terms of which the essential properties of the process can be understood. Depending on the particular problem, the key variables may be energy consumption, respiration rate, space available, frequency of disease, and so forth. Other variables, such as mutation rate, humidity, and time of year, for example, may be ignored in the construction of a model if it is believed that they are relatively unimportant for the description of a particular biological process.

Once the key variables have been chosen, it is then necessary to construct a framework of concepts and language capable of describing the significant features of the process. This framework is established to explain experimental observations and it is always subject to modification to take into account further experiments or new interpretations of previous experiments.

If the assumptions of a model can be expressed in mathematical form, then the whole apparatus of mathematics becomes available to deduce the logical consequences of the assumptions. In many cases, these mathematical consequences lead to predictions concerning the biological process that were not suspected or were poorly understood on the basis of experimental studies alone. The model of counting cells under a microscope (Example 2.9.3) leads to the surprising prediction that the total number of cells can be determined simply by counting the number of squares of the grid that contain no cells. The logistic population growth model (Example 7.3.3) generates the prediction that the population approaches a limiting or equilibrium size.

There are some disadvantages in expressing the assumptions of a biological model in mathematical form. It may be necessary to simplify the problem substantially. If this simplification is carried too far, the model will not mirror some important aspects of the reality we are trying to understand. In constructing a mathematical model of an ecosystem, for example, it may be necessary to neglect some potentially important interactions in order to produce a model which is sufficiently simple to study. We should realize that, even when simple models cannot be realistic, the attempt to construct them may lead to a better understanding of key variables and relationships. It may be possible, once the consequences of a simple model have been analyzed, to make the model more realistic by altering the assumptions. It should be pointed out that some types of complexity are no longer a major problem now that high-speed computers are widely available.

In this chapter, the building of models of biological processes is illustrated by means of a number of examples. The models have been chosen to relate to similar types of problems—the origin, survival, and evolution of individuals, populations, and species. The reader should note how the models focus on different questions and how they are related. The assumptions that are made should be analyzed very critically; the reader should decide whether the assumptions are a reasonable approximation to reality. To improve our description of reality, we would continue by altering the assumptions to build a succession of models that provide an increasingly accurate description of the real biological system. In many cases, it is possible to define correction terms that take into account factors which have originally been ignored.

The following quotation is from a survey of the current status (in 1970) of the biological sciences:

"Biology has become a mature science as it has become precise and quantifiable. The biologist is no less dependent upon his apparatus than the physicist. Yet the biologist does not use distinctively biological tools. He is an opportunist who employs a nuclear magnetic resonance spectrometer, a telemetry assembly or an airplane equipped for infrared photography depending on the biological problem he is attacking. In any case, he is always grateful to the physicists, chemists, and engineers who have provided the tools he has adapted to his trade."[1]

We might add that mathematical models have become an indispensable tool in almost every area of biological research.

Problems 9.1

1. Why are models of biological processes necessary? As examples, consider the problems of (a) pest control and (b) the inheritance of characteristics. What limitations would purely verbal or descriptive models of these processes have?

2. What are some of the key variables of the following biological processes?
   (a) The coexistence of species in an environment.
   (b) The productivity of farmland.
   (c) The spread of an infectious disease.

3. Comment on the following quotation. "Models are not intended to be exact copies of the real world but simplifications that reveal the key processes necessary for prediction."[2]

9.2. The Survival and Extinction of Species

If populations of two species occur together in the same geographical region and have the same ecological requirements for food, space, and other resources, the theory of natural selection leads us to expect that the better adapted species will completely displace the other one. This is the idea of the

[1] P. Handler (ed.), *Biology and the Future of Man*, Oxford University Press, New York, 1970, p. 6.

[2] E. P. Odum, *Fundamentals of Ecology*, 3rd ed., W. B. Saunders Company, Philadelphia, 1971.

principle of competitive exclusion. Stated briefly, the principle asserts that complete competitors cannot coexist or that no two species can occupy the same ecological niche.[3,4] The principle has been the subject of some controversy, but it has stimulated some important ecological research.

As a test of the principle of competitive exclusion, MacArthur[5] has studied the coexistence of five species of warblers in a Connecticut forest. These warbler species are found together in the breeding season and are of the same genus; the warblers of the different species have similar sizes and shapes. Their diets are similar, consisting mainly of insects. The coexistence of these warbler species is an apparent contradiction of the principle of competitive exclusion. However, by very detailed observations, MacArthur found that the species behave in ways which expose them to different conditions. Among the important differences are: they feed in different positions on the trees; they have different hawking and hovering strategies in their search for food; and they have the greatest need for food at different times corresponding to different nesting dates. These differences tend to decrease the direct competition between the species and to permit their coexistence.

If, for example, we could distinguish three feeding positions, three hawking strategies, and three nesting dates, then the principle of competitive exclusion would allow the coexistence of as many as $3 \times 3 \times 3 = 27$ warbler species. Clearly, because of the wide range of behavior of species, the observed abundance of coexisting species in nature does not contradict this principle.

The reader may have noticed the use of the word "strategy" in the phrase hawking and hovering strategies. If two competing species both search for insects on the same parts of the trees, then the return to each will be less. On the other hand, they can both expect greater returns if they specialize on different parts of the tree. This can be modeled by a matrix game played between each species and the insects. The game matrix may be the 2×2 matrix

$$A = \begin{pmatrix} \frac{1}{2} & 1 \\ 1 & \frac{1}{2} \end{pmatrix}.$$

---

[3] G. Gause and A. Witt. "Behavior of Mixed Populations and the Problem of Natural Selection," *American Naturalist*, 69:596–609 (1935).

[4] G. Hardin, "The Competitive Exclusion Principle," *Science*, 131:1291–1297 (1960).

[5] R. H. MacArthur, "Population Ecology of Some Warblers of Northeastern Coniferous Forests," *Ecology* **39**:599–619 (1958).

If both species feed in the same position, they each receive $\frac{1}{2}$ insect unit. If both species feed in different positions, they each receive 1 insect unit. By introducing this and more complicated games, we attempt to understand the behavior of the different species in terms of strategies.

The competitive exclusion principle states that, if two species have the same ecological requirements, then the better adapted species will displace the other one. To model this process of extinction, let us consider the case of two competing species with the same ecological requirements, coexisting at some time in an environment that can support exactly N individuals of both species. Suppose that, initially, there are k individuals of species I and $N - k$ individuals of species II in the environment. Assume that the two species compete by means of a succession of encounters which are such that, on each encounter, the probability that species I increases by one individual is p and the probability that species II increases by one individual is $q = 1 - p$. For example, the two species are equally well adapted if $p = q = \frac{1}{2}$ and the first species has a selective advantage if $p > q$. We will assume that p is independent of the populations k, $N - k$ of the two species.

The competition continues according to these rules, until one of the species completely displaces the other. This is the process that we would like to understand. To do this, define p_k to be the probability that the first species will displace the second species if its initial population is k. If the initial population is 0, then $p_0 = 0$, since the first species is already extinct. If the initial population is N, then $p_N = 1$, since the first species has already displaced the second one. After the first encounter, the population of species I is either $k + 1$ or $k - 1$ with probabilities p and q, respectively. The probability p_k is, therefore, the sum of two terms

$$p_k = p p_{k+1} + q p_{k-1}. \tag{9.1}$$

Here, $p p_{k+1}$ is the probability that the first species has population $k + 1$ after one move and then displaces the second species. Similarly, $q p_{k-1}$ is the probability that the first species has population $k - 1$ after one move and then displaces the second species.

We recognize Equation (9.1) as a second-order linear difference equation with constant coefficients. Let us look for solutions of the form $p_k = \lambda^k$. The auxiliary equation is

$$\lambda^k = p\lambda^{k+1} + q\lambda^{k-1} \qquad \text{or} \qquad p\lambda^2 - \lambda + q = 0. \tag{9.2}$$

The roots of this quadratic equation are

$$\lambda = \frac{1 \pm \sqrt{1 - 4pq}}{2p} = \frac{1 \pm \sqrt{1 - 4p(1 - p)}}{2p} = \frac{1 \pm (1 - 2p)}{2p}$$

or

$$\lambda_1 = 1 \quad \text{and} \quad \lambda_2 = \frac{1 - p}{p} = \frac{q}{p}.$$

These roots are equal if $p = q = \frac{1}{2}$. The general solution of Equation (9.1) is, therefore,

$$p_k = c_1 + c_2 \left(\frac{q}{p}\right)^k \quad \text{if} \quad p \neq \tfrac{1}{2},$$

or

$$p_k = c_1 + c_2 k \quad \text{if} \quad p = \tfrac{1}{2},$$

where c_1 and c_2 are constants to be determined. We have seen, from the definition of p_k, that $p_0 = 0$ and $p_N = 1$. Therefore, if $p \neq \frac{1}{2}$,

$$c_1 + c_2 = 0 \quad \text{and} \quad c_1 + c_2 \left(\frac{q}{p}\right)^N = 1.$$

This implies that

$$c_1 = \frac{1}{1 - (q/p)^N} \quad \text{and} \quad c_2 = \frac{-1}{1 - (q/p)^N}.$$

If $p = \frac{1}{2}$, then $c_1 = 0$ and $c_2 N = 1$ or $c_2 = 1/N$. Finally, we conclude that the solution of Equation (9.1) is

$$p_k = \frac{1 - (q/p)^k}{1 - (q/p)^N} \quad \text{if} \quad p \neq \tfrac{1}{2},$$

$$p_k = \frac{k}{N} \quad \text{if} \quad p = \tfrac{1}{2}.$$

$$(9.3)$$

In this simple model, we have been able to calculate explicitly the probability that one species will displace the other. Let us consider the various possibilities for a total population $N = 1000$. If the initial populations of the two species are both 500 and if $p = q = \frac{1}{2}$, then the probability that the first species will displace the second species is $p_{500} = \frac{500}{1000} = \frac{1}{2}$. This indicates that neither species has a competitive advantage in this case. However, if $p = \frac{2}{3}$ and $q = \frac{1}{3}$, then

$$p_{500} = \frac{1 - (\frac{1}{2})^{500}}{1 - (\frac{1}{2})^{1000}} = \frac{1}{1 + (\frac{1}{2})^{500}}.$$

But $(\frac{1}{2})^{500}$ is an extremely small number (more than 100 zeros after the decimal place). Therefore, in this case, it is overwhelmingly probable that species II will become extinct. Even small competitive advantages (p slightly greater than $\frac{1}{2}$) with these initial populations lead to the same result. Note that, if $p = \frac{2}{3}$, $q = \frac{1}{3}$, and the initial populations of species I and II are 1 and 999, respectively, then the probability that the first species displaces the second species is

$$p_1 = \frac{1 - (\frac{1}{2})^1}{1 - (\frac{1}{2})^{1000}} \approx \frac{1}{2}.$$

This implies, for example, that, if one individual of the better adapted species invades the territory of the second species, the probability is $\frac{1}{2}$ that the invader will completely displace the second species. If $p = q = \frac{1}{2}$, then $p_1 = \frac{1}{1000}$. In other words, if the two species are equally well adapted, the probability is very small that the introduction of one individual of one species will lead to the extinction of the other species.

This simple model of the survival and extinction of species could have been developed in the language of Markov chains. The experimental system has $N + 1$ states corresponding to the $N + 1$ possible populations $0, 1, 2, \ldots, N$ of the first species. The transition matrix is an $(N + 1) \times (N + 1)$ matrix with most components equal to zero. If $N = 1000$, this is a 1001×1001 matrix, which indicates that this may not be the most practical method of studying this model. However, if $N = 5$, the transition matrix is a 6×6 probability matrix. In this case, the model may be conveniently studied by the methods of Chapter 5. In fact, the game described in Example 5.3.3 is equivalent to the model of survival and extinction of species developed in this section.

Problems 9.2

1. In the model of survival and extinction of species, assume that the first species has a selective advantage $(p > q)$. Prove that the probability that the first species becomes extinct is

$$1 - p_k = \frac{(p/q)^{N-k} - 1}{(p/q)^N - 1},$$

where k is the initial population of the first species and N is the total population. Deduce that, if N is small, the probability of extinction of the better adapted species is not negligible. Is this a reasonable feature of this model? (This process, by which the better adapted species may become extinct, has been called *sampling error* or *random fixation*.)

2. One of the assumptions of the model is that the probability p of success on each encounter is independent of the population k of the first species. Is this a reasonable assumption for some types of competition? How would you modify this assumption?

3. Because of its simplicity, the model of this section generates exact predictions about the outcome of competition. What difficulties can be anticipated in attempting to compare these predictions with either laboratory or field competition between species?

4. Suppose that, on a small island, there is room for 1000 individuals of a certain species. One year, a favorable mutant appears in this population. In each subsequent generation, the mutant population either increases by one or decreases by one with probabilities .7 and .3, respectively. For example, the probability that the mutant disappears in the first generation is .3. What is the probability that the mutant population displaces the original population?

5. Consider two islands, one large and one small, both at the same distance from a much larger land mass. Suppose that both islands have roughly similar geography and that both have been colonized by species from the mainland. In view of the ideas of this section, why is it reasonable to expect that the large island will support more species than the small island?[6]

6. Referring to the ideas and the model of this section, comment on the following quotation. "In the creative analysis of systems, it becomes

---

[6] For a well-developed theory of this problem, see R. H. MacArthur and E. O. Wilson, *The Theory of Island Biogeography*, Princeton University Press, Princeton, N.J., 1967.

crucial to decide what things about the system are worth knowing. There exists a prejudice . . . that in order to know something we must define it precisely and measure it precisely. In fact in the development of population biology this has not been the case. The theory of the niche begins as a vague heuristic concept, referring to the fact that organisms are coupled into certain kinds of environments. The subsequent development of the concept involves making some things more precise, changing the definitions, applying it to certain test situations."[7]

7. The opening of a canal between two previously separated bodies of water generally leads to a transfer of species between the two bodies of water. Discuss the general features of this process, taking into account the possible existence of unoccupied niches.

9.3 Genetics and the Hardy–Weinberg Law

The modern theory of the inheritance of characteristics has its origins in the experiments of Gregor Mendel on garden peas published in 1865. Mendel's interpretation of his experiments led him to suggest general laws governing the transmission of characteristics from parents to offspring. In particular, heredity was assumed by Mendel to be a result of the transmission of particles (now called *genes*) from parents to children. The exact nature of the genes and the mechanisms by which they determine characteristics have been major problems of biological research to the present day.

A particular gene may occur in several forms or *alleles*. For simplicity, we will consider a gene with two alleles, *A* and *a*. The genes occur in every cell of an organism grouped together on the chromosomes. Except in the reproductive cells, the genes occur in pairs and appear on paired chromosomes. The three possible pairs of this gene, *AA*, *Aa*, and *aa*, determine the three possible *genotypes* of the organism relative to this gene. The genotypes *AA* and *aa* are called *homozygous* or *pure*, and the genotype *Aa* is called *heterozygous* or *hybrid*.

The reproductive cells (the sperm and the egg) have unpaired chromosomes and therefore only one copy of each gene. The genes of an offspring result from the pairing of the genes of two reproductive cells, one from each parent. If both parents are homozygous, the genotype of the offspring is determined. For example, if one parent is genotype *AA* and the other is genotype *aa*,

---

[7] R. Levins, "Complex Systems," in C. H. Waddington, ed., *Towards a Theoretical Biology*, Vol. 3, Aldine-Atherton, Inc., Chicago, 1970.

then the offspring must be genotype Aa. On the other hand, if one or both are heterozygous, the genotype of the offspring is not determined. For example, if both parents are heterozygous, the offspring may be AA, Aa, or aa with probabilities $\frac{1}{4}$, $\frac{1}{2}$, and $\frac{1}{4}$, respectively, assuming equal viability of the offspring.

Many characteristics, such as albinism in human beings, are controlled by a single gene. Other characteristics, such as height or intelligence, are controlled by the cooperative effects of a very large number of genes and are often strongly influenced by environmental factors. One of the two alleles A of a particular gene is said to be *dominant* if the genotypes AA and Aa are indistinguishable from each other. In this case, the allele a is then said to be *recessive* if the genotype aa is observably different from the genotypes AA and Aa. The gene controlling sickle cell anemia in humans is an example of a gene with a dominant and a recessive allele. An aa individual invariably suffers from severe anemia, resulting in premature death.

If it were possible to classify the individuals of a population of a given species according to the genotypes AA, Aa, and aa, then we could determine the proportions of the two alleles in the population. This would not be possible if, for example, AA and Aa were indistinguishable. Let us use the letters u, v, and w to denote the proportions of the three genotypes in the population and assume that these proportions can be determined. Then the proportions p and q of the two alleles A and a in the population satisfy the equations

$$p = u + \tfrac{1}{2}v,$$
$$q = \tfrac{1}{2}v + w. \qquad (9.4)$$

Here we have made use of the fact that the A allele comprises 100 per cent of the AA genotype (with proportion u) and 50 per cent of the Aa genotype and, similarly, for the a allele. We note that the second equation of (9.4) follows from the first equation, since $p + q = 1$ and $u + v + w = 1$. If we assume that the genotypes occur in the same proportions among males and females, then p and q represent (in a large population) the probabilities that the gene is A or a, respectively. A discussion of the case of sex-linked genes is presented later in this section.

Example 9.3.1   In a population, the distribution of genotypes is 50 per cent AA, 30 per cent Aa, and 20 per cent aa. What proportions of the genes in this population are A and a?

Solution: In this example, $u = .50$, $v = .30$, and $w = .20$. Therefore, $p = .50 + \frac{1}{2}(.30) = .65$ and $q = .15 + .20 = .35$. This means that the gene "population" is 65 per cent A and 35 per cent a.

We are often interested in the inverse problem of determining the proportions of the genotypes when the proportions of the alleles are known. This problem does not, in general, have a unique solution. The system of equations (9.4) reduces to one equation in two unknowns, $p = u + \frac{1}{2}v$. To provide a second independent equation, we will make the assumption of *random mating*. This means that the probability that a given individual will become the mate of another individual is not dependent on the genotype of the other individual. In many cases, this is a good approximation. Sometimes it is not. For example, it is known that tall people tend to marry other tall people, and therefore the characteristic of human height cannot be analyzed in this way. On the other hand, it has been shown that the assumption of random mating is applicable to the characteristic of human blood types. Most individuals will choose their spouse without worrying about blood type.

As before, suppose that p and q are the proportions of the alleles A and a among both males and females. Then, if we assume that the population is large, the probability that an offspring will receive allele A from both parents is p^2. Similarly, the probabilities of the genotypes Aa and aa are $2pq$ and q^2, respectively. The $2pq$ term comes from the fact that Aa and aA individuals are identical genotypes. Since the probability of genotype Aa is pq and genotype aA is qp, the probability of Aa is $2pq$. This result leads to the following theorem, discovered independently by Hardy and Weinberg in 1908.

Theorem 9.3.1   (Hardy–Weinberg Law)   *Suppose that, in a large parent population, the alleles A and a of a particular gene are present in the proportions p and $q = 1 - p$. Assuming that these proportions are the same for males and females and assuming that mating is random, the first and all succeeding generations will be composed of the three genotypes, AA, Aa, and aa, in the proportions p^2, $2pq$, and q^2.*

Proof: As we have seen, an individual in the first generation is of genotype AA if both parents contribute the allele A. Since the probability of an allele A coming from either parent is p, the probability of genotype AA in an offspring is p^2. Similarly, the probabilities of the genotypes Aa and aa are $2pq$ and q^2. This implies that the proportions p_1 and q_1 of the alleles A and a

in the first generation are given by

$$p_1 = p^2 + \tfrac{1}{2}(2pq) = p(p + q) = p,$$
$$q_1 = \tfrac{1}{2}(2pq) + q^2 = q(p + q) = q.$$

Therefore, the proportions of the two alleles are unchanged from the initial generation. This continues from one generation to the next. We conclude that, after the initial generation, the proportions of the three genotypes AA, Aa, and aa remain constant at $p^2, 2pq$, and q^2.

Example 9.3.2   The color of sweet pea blossoms is controlled by one pair of genes. The three genotypes AA, Aa, and aa are characterized by red, pink, and white blossoms, respectively. If a field is planted randomly with 60 per cent red flowers and 40 per cent white flowers, what proportions of the three genotypes will be present in the fourth generation?

Solution: In this example, $p = .6$ and $q = .4$. From the Hardy–Weinberg law, the proportions of red, pink, and white flowers in the first and all subsequent generations are $p^2, 2pq$, and q^2 or .36, .48, and .16. We note that the assumption of random planting is equivalent to an assumption of random pollination.

We emphasize that the Hardy–Weinberg law is valid only when mating is random and when the three genotypes are equally viable. In the next section, we consider models in which one allele confers a selective advantage. In some cases, it is quite difficult to verify that mating is random. However, if the genotype proportions remain constant through several generations, and if they satisfy the Hardy–Weinberg law, this can be taken as strong evidence that mating is indeed random. In fact, knowledge that mating is random for human blood types (as well as for many other plant and animal characteristics) has been derived from observations of genotype proportions conforming to this law. We should note, however, that discrepancies in measured genotype proportions do not necessarily mean that mating is not random. These may be attributed to other factors, such as differential mortality.

The reader should attempt to generalize the theorem to the case of a gene with n alleles $A_1, A_2, \ldots, A_n$. If $p_1, p_2, \ldots, p_n$ are the corresponding proportions of the various alleles present in a parent population, then the proportions of the

$$\binom{n + 1}{2}$$

genotypes (see Problem 1.5.13) in the first and all subsequent generations are given by the terms in the multinomial expression $(p_1 + p_2 + \cdots + p_n)^2$. (See Problems 9.3.7 and 9.3.8.)

Sex-linked characteristics are an important group of characteristics which do not follow the Hardy–Weinberg law. As we have observed, genes occur in pairs grouped on paired chromosomes. Sex is determined by one pair of chromosomes: XX for females and XY for males. If a particular gene occurs only on the X chromosome (such as the genes controlling hemophilia or color blindness in humans), the genotypes relative to that gene are different for males and females. A female can still be classified as AA, Aa, or aa, since she will have two X chromosomes. On the other hand, a male has only one X chromosome and therefore belongs to one of two genotypes, A or a. This explains why hemophilia, for example, cannot be passed from father to son, since the son will always inherit the Y chromosome from his father. Of course, the disease can be transmitted from father to daughter to grandson.

To analyze genotype proportions in sex-linked characteristics, we define u, v, and w to be the initial genotype (AA, Aa, aa) proportions in the female population. Then the initial gene proportions in the female population are $p^f = u + \frac{1}{2}v$ and $q^f = \frac{1}{2}v + w$. Similarly, we define p^m and q^m to be the initial gene and genotype proportions of A and a in the male population. Note that $u + v + w = p^f + q^f = p^m + q^m = 1$.

In the first generation, the corresponding proportions are defined to be $u_1, v_1, w_1, p_1^f, q_1^f, p_1^m$, and q_1^m. Our problem is to determine these proportions in the first generation in terms of the proportions in the parent generation. The genotype of a male offspring is determined by the genotype of his mother. This implies that

$$p_1^m = p^f \qquad \text{and} \qquad q_1^m = q^f. \tag{9.5}$$

A female is of genotype AA if she receives the allele A from both mother and father; similarly, for the female genotypes Aa and aa. Therefore, we have

$$u_1 = p^f p^m, \qquad v_1 = p^f q^m + q^f p^m, \qquad w_1 = q^f q^m. \tag{9.6}$$

These equations yield the following expressions for p_1^f and q_1^f:

$$p_1^f = u_1 + \tfrac{1}{2}v_1 = p^f p^m + \tfrac{1}{2}(p^f q^m + q^f p^m),$$
$$q_1^f = \tfrac{1}{2}v_1 + w_1 = \tfrac{1}{2}(p^f q^m + q^f p^m) + q^f q^m.$$

These expressions can be simplified by noting that $p^f + q^f = p^m + q^m = 1$.

$$p_1^f = \tfrac{1}{2}p^f p^m + \tfrac{1}{2}p^f q^m + \tfrac{1}{2}p^f p^m + \tfrac{1}{2}q^f p^m = \tfrac{1}{2}p^f(p^m + q^m) + \tfrac{1}{2}p^m(p^f + q^f).$$

We conclude that

$$p_1^f = \tfrac{1}{2}(p^m + p^f) \quad \text{and} \quad q_1^f = \tfrac{1}{2}(q^m + q^f). \tag{9.7}$$

This implies that the gene proportions in the females of the first generation are the averages of the gene proportions in the female and male parent populations. Combining (9.5) and (9.7), we have

$$p_1^m - p_1^f = \tfrac{1}{2}(p^f - p^m) \quad \text{and} \quad q_1^m - q_1^f = \tfrac{1}{2}(q^f - q^m). \tag{9.8}$$

Therefore, in one generation, the differences in the proportions of the two alleles in the male and female populations have been reduced by one half. In the next generation, these differences will again be reduced by one half. We conclude that, after a few generations, the gene proportions approach equilibrium values with equal proportions in the male and female populations of either allele. The approach to equilibrium is described in greater detail in the following problems.

Problems 9.3

In Problems 1–4, consider a sex-linked characteristic governed by a gene with two alleles A and a. Let p_n^f and p_n^m denote the proportions of allele A in the female and male populations in the nth generation. The corresponding proportions of allele a are denoted by q_n^f and q_n^m.

1. Prove that $p_2^f = \tfrac{3}{4}p^f + \tfrac{1}{4}p^m$ and $q_2^f = \tfrac{3}{4}q^f + \tfrac{1}{4}q^m$.
2. Show, by induction, that

$$p_n^f = \frac{p_{n-1}^f + p_{n-1}^m}{2} = \frac{2}{3}p^f + \frac{1}{3}p^m + (-1)^n \frac{p^f - p^m}{3(2)^n},$$

$$q_n^f = \frac{q_{n-1}^f + q_{n-1}^m}{2} = \frac{2}{3}q^f + \frac{1}{3}q^m + (-1)^n \frac{q^f - q^m}{3(2)^n}.$$

3. From Problem 2, deduce that $\lim_{n \to \infty} p_n^f = \frac{2}{3} p^f + \frac{1}{3} p^m$ and $\lim_{n \to \infty} q_n^f = \frac{2}{3} q^f + \frac{1}{3} q^m$. These are the equilibrium gene proportions for the female population. What are the corresponding equilibrium gene proportions for the male population? (See Problem 5.1.9 or Appendix B for an explanation of this terminology.)

4. From Problems 2 and 3, we conclude that the proportions p_n^f and q_n^f are approximately equal to $\frac{2}{3} p^f + \frac{1}{3} p^m$ and $\frac{2}{3} q^f + \frac{1}{3} q^m$. The terms

$$(-1)^n \frac{p^f - p^m}{3(2)^n} \qquad \text{and} \qquad (-1)^n \frac{q^f - q^m}{3(2)^n}$$

are called *correction terms* for these approximations. Verify that for $n > 9$ the correction terms are both less than .001. Deduce that after 10 generations the fluctuations in the proportions of the two alleles from one generation to the next are less than .001.

5. The ability to taste certain substances is controlled genetically. Phenylthiocarbamide (PTC) has a bitter taste to about 70 per cent of the people in a large population and is tasteless to the remaining 30 per cent. Assuming that the ability or inability to taste PTC is caused by a single gene, estimate the proportions of the dominant taster gene and the recessive nontaster gene in this population. What proportion of individuals in this population are heterozygous for this gene?

6. Many genes affect the ability of the organism to reproduce. The Hardy–Weinberg law may still be used, in some cases, to predict the genotype proportions in the next generation in the stages before selection has occurred. For example, the condition known as vestigial wings, in fruit flies, is controlled by a recessive gene. It is characterized by shortened wings and affects the ability of the recessive genotype to survive and reproduce. Suppose that 1 adult in 6400 in a large population has vestigial wings. Estimate the genotype proportions in the next generation before selection occurs.

7. Consider a gene with four alleles A_1, A_2, A_3, and A_4. Write out the 10 genotypes corresponding to this gene. Suppose that, initially, these four alleles are present in equal proportions in a large population. Assuming that the proportions are the same among both males and females and assuming that mating is random, determine the proportions of the various genotypes in the first and all subsequent generations.

8. Consider a gene with n alleles A_1, A_2, ..., A_n occurring in the proportions $p_1, p_2, \ldots, p_n$. Calculate the ratio of homozygotes to heterozygotes in

this population. If the proportions p_i are all equal, prove that this ratio is equal to $1/(n-1)$.

9.4 Models of Selection and Fitness

The model of Section 9.2 describing the survival and extinction of two competing species does not take into account in a satisfactory way the development of the competing populations in successive generations. Another shortcoming of the model is a failure to relate the probabilities p and q of success and failure on each encounter to genetic or environmental variables. In some circumstances, it is reasonable to assume that p depends mainly on environmental conditions. For example, temperature and humidity are key factors in determining the outcome of competition between two species of beetle.[8] In other circumstances, the outcome of competition between two populations is determined mainly by genetic factors. In Section 9.3, we studied models in which the proportions of the various genotypes in a population approach constants. In this section, we study models of a single population in which one allele of a particular gene may displace another allele.[9]

Consider an environment that supports exactly n reproducing individuals in each successive generation. A particular gene with two alleles, A and a, has $2n$ representatives in each generation. Suppose that, in the mth generation, the allele A occurs i times and the allele a occurs $2n - i$ times. We would like to develop a model to determine the probability that, in the next generation, the allele A occurs j times and the allele a occurs $2n - j$ times for $j = 0, 1, 2, \ldots, 2n$. In the language of Markov chains, the system has $2n + 1$ states $E_0, E_1, \ldots, E_i, \ldots, E_{2n}$ corresponding to the population having $0, 1, 2, \ldots, 2n$ copies of the allele A. The successive trials of the Markov chain correspond to the successive generations of the population. To describe this Markov chain, we must determine the transition probabilities p_{ij} from the ith state to the jth state. Note that the states E_0 and E_{2n} are absorbing states corresponding to populations of all aa and AA individuals, respectively. It is important to note that we are explicitly taking into account the size of the population. In the derivation of the Hardy–Weinberg law in Section 9.3, we merely assumed that the population is large.

---

[8] T. Park, "Beetles, Competition and Populations," *Science*, 138:1369–1375 (1962).

[9] W. Feller, *An Introduction to Probability Theory and Its Applications*, Vol. I, 3rd ed., John Wiley & Sons. Inc.. New York. 1968.

In the absence of other information, we again assume that mating is random and that none of the three genotypes AA, Aa, and aa has a selective advantage which gives an increased probability of successful reproduction. With these assumptions, the population in the $(m + 1)$st generation is determined by $2n$ repeated trials of a binomial experiment with probability $1 - i/2n$ of allele a occurring on a trial. The probability of j "successes" (of j copies of allele A in the next generation) is given by the binomial probability

$$p_{ij} = \binom{2n}{j}\left(\frac{i}{2n}\right)^{j}\left(1 - \frac{i}{2n}\right)^{2n-j}. \tag{9.9}$$

These transition probabilities define a Markov chain. The expected number of copies of allele A in the $(m + 1)$st generation is $\sum_{j=0}^{2n} jp_{ij} = i$. (Why?) The probabilities that allele A or allele a will disappear in the next generation are

$$p_{i0} = \left(1 - \frac{i}{2n}\right)^{2n} \quad \text{and} \quad p_{i,2n} = \left(\frac{i}{2n}\right)^{2n}. \tag{9.10}$$

If i is small relative to $2n$ (that is, if allele A is rare compared to allele a), then $p_{i,2n}$ is extremely small and p_{i0} is approximately equal to e^{-i} (see Appendix D). For example, if there is one copy of allele A in the population in the mth generation, the probability that it will have disappeared in the next generation is approximately equal to $e^{-1} \approx .37$.

In this model, we have assumed that neither allele confers a selective advantage which increases the probability that it will be represented in the next generation. In spite of this assumption, we have observed that there is a finite probability that the gene proportions will change from one generation to the next. This effect is known as *sampling error* or *random drift*. In small populations particularly, this random fluctuation of the gene proportions can lead to the disappearance of one of the alleles from the gene pool.

To model a selective advantage for one of the alleles, we must modify the formula for the binomial probability $i/2n$ that allele A is represented on one of the $2n$ trials. Suppose again that the population in the $(m + 1)$st generation is determined by $2n$ repeated trials of a binomial experiment. Suppose further that the probabilities of alleles A and a occurring on a particular trial are $(i/2n)^{\alpha}$ and $1 - (i/2n)^{\alpha}$, where α is a positive constant. Then the probability that there are exactly j copies of allele A in the next generation is given by

$$p_{ij} = \binom{2n}{j}\left(\frac{i}{2n}\right)^{\alpha j}\left[1 - \left(\frac{i}{2n}\right)^{\alpha}\right]^{2n-j} \tag{9.11}$$

when there were i copies of A in the previous generation. If $\alpha = 1$, this is the same model as before. In this more general model, the probabilities that allele A or allele a will disappear in the next generation are

$$p_{i0} = \left[1 - \left(\frac{i}{2n}\right)^{\alpha}\right]^{2n} \quad \text{and} \quad p_{i,2n} = \left(\frac{i}{2n}\right)^{2n\alpha}. \tag{9.12}$$

If $\alpha > 1$, the allele a has a selective advantage and, if $\alpha < 1$, the allele A has a selective advantage. If $\alpha = 1$, neither allele has a selective advantage. Note that if $\alpha = 0$, for example, then $p_{i,2n} = 1$ and $p_{i0} = 0$. This is the extreme case of complete selective advantage for allele A. In one generation, the allele a has become extinct.

The parameter α could, in principle, be measured by performing the experiment many times and observing the transition probabilities from one generation to the next. The parameter α is one example of a *fitness index*. The value of α is a measure of the fitness of the allele A relative to the fitness of the allele a. In some cases, α will not be constant but will depend on environmental factors such as temperature. The reader may be interested in attempting to include such factors in this model.

We can define a second fitness index, which measures the relative probabilities that the various genotypes will survive to adulthood. We recall that a basic assumption of the Hardy–Weinberg law is that all three genotypes are equally viable. However, it is often the case that, for example, recessive offspring (genotype aa) have a smaller probability of survival than those individuals with the dominant gene. Sickle cell anemia and albinism in humans are examples of such conditions. This smaller probability of survival will, of course, be reflected in the genotype proportions in the next generation. We expect, therefore, that the percentage of adults with sickle cell anemia will be considerably less than that predicted by the Hardy–Weinberg law.

To analyze this situation, we again assume that A is dominant and a is recessive and that mating is random. However, we now assume that the probabilities of survival from fertilization to adulthood are not equal for the three genotypes. Define P to be the probability of survival of an individual of genotype AA or Aa and define $P\beta$ to be the probability that an individual of genotype aa survives to adulthood. The parameter β is a *fitness index*, which measures the fitness of the genotype *aa*. If $\beta = 0$, no recessive individual survives to adulthood. If $\beta = 1$, all three genotypes are equally viable. If $\beta > 1$, the recessive genotype has a selective advantage. We will assume that $0 \leq \beta < 1$, which means that the recessive genotype is less likely to survive.

Define p_n and q_n to be the proportions of alleles A and a in the adults of the nth generation. According to the Hardy–Weinberg law, the genotype proportions in the next generation will be p_n^2, $2p_nq_n$, and q_n^2 for the genotypes AA, Aa, and aa, respectively. However, when AA and Aa have a selective advantage, the proportions of adults in the population will not be given by the Hardy–Weinberg law. To calculate the new proportions, we multiply the Hardy–Weinberg proportions by the probabilities of survival to adulthood. The new proportions are then Pp_n^2, $2Pp_nq_n$, and $P\beta q_n^2$. Since the proportions should add to 1, we divide each one by their sum. The proportions among the adult population in the $(n + 1)$st generation are, therefore,

$$\frac{p_n^2}{p_n^2 + 2p_nq_n + \beta q_n^2}, \qquad \frac{2p_nq_n}{p_n^2 + 2p_nq_n + \beta q_n^2}, \qquad \frac{\beta q_n^2}{p_n^2 + 2p_nq_n + \beta q_n^2} \qquad (9.13)$$

for the three genotypes AA, Aa, and aa. Using Equation (9.4), we can calculate the gene proportions in the $(n + 1)$st generation.

$$p_{n+1} = \frac{p_n^2}{p_n^2 + 2p_nq_n + \beta q_n^2} + \frac{1}{2}\frac{2p_nq_n}{p_n^2 + 2p_nq_n + \beta q_n^2} = \frac{p_n}{p_n^2 + 2p_nq_n + \beta q_n^2},$$

$$q_{n+1} = \frac{1}{2}\frac{2p_nq_n}{p_n^2 + 2p_nq_n + \beta q_n^2} + \frac{\beta q_n^2}{p_n^2 + 2p_nq_n + \beta q_n^2} = \frac{(p_n + \beta q_n)q_n}{p_n^2 + 2p_nq_n + \beta q_n^2}. \qquad (9.14)$$

As a check on these calculations, note that, if $\beta = 1$, we reproduce the results of the Hardy–Weinberg law. The ratio of allele A to allele a in the $(n + 1)$st generation is given by

$$\frac{p_{n+1}}{q_{n+1}} = \frac{1}{p_n + \beta q_n}\frac{p_n}{q_n}. \qquad (9.15)$$

Since $0 \leq \beta < 1$ and $p_n + q_n = 1$, we have that $p_n + \beta q_n < 1$ (unless $p_n = 1$ and $q_n = 0$). Therefore, $p_{n+1}/q_{n+1} > p_n/q_n$. This means that the proportion of allele A increases from one generation to the next. This is reasonable, since we have assumed that the recessive individuals are less likely to survive.

To analyze these ratios in successive generations in greater detail, define $r_n = p_n/q_n$. Then, from Equation (9.15),

$$r_{n+1} = \frac{r_n}{p_n + \beta q_n} = r_n \frac{1/q_n}{(p_n/q_n) + \beta} = r_n \frac{(p_n + q_n)/q_n}{(p_n/q_n) + \beta}.$$

We conclude that

$$r_{n+1} = r_n \frac{r_n + 1}{r_n + \beta}. \tag{9.16}$$

This is a first order, nonlinear difference equation that could be solved by methods similar to those of Chapter 6. Since $0 \le \beta < 1$, $r_{n+1} > r_n$ and the proportion of the dominant allele increases in each generation. It can be proved that $\lim_{n \to \infty} r_n = \infty$. This means that the recessive allele will eventually be eliminated. This is illustrated in the following example.

Example 9.4.1 Suppose that the fitness index of the recessive genotype is $\beta = .5$ in the above model. If the initial proportions of the alleles A and a are $p = .6$ and $q = .4$, determine the ratios r_0, r_1, r_2, and r_3 of the alleles from the initial population to the third generation.

Solution: We are given $r_0 = .6/.4 = 1.5$. From Equation (9.16),

$$r_{n+1} = r_n \frac{r_n + 1}{r_n + \beta}$$

and $\beta = .5$, we have

$$r_1 = r_0 \frac{r_0 + 1}{r_0 + 0.5} = (1.5)\frac{2.5}{2} = 1.875,$$

$$r_2 = (1.875)\frac{2.875}{2.375} = 2.270,$$

$$r_3 = (2.270)\frac{3.270}{2.770} = 2.752.$$

In the fourth generation, allele A will be more than three times as numerous as allele a.

Finally, we may rewrite Equation (9.16) in the form

$$r_{n+1} = r_n \frac{r_n + \beta + 1 - \beta}{r_n + \beta} = r_n + r_n \frac{1 - \beta}{r_n + \beta}. \tag{9.17}$$

If $\beta = 1$, then we again observe that $r_{n+1} = r_n$. The term $r_n[(1 - \beta)/(r_n + \beta)]$ is called a *correction term*. This term gives the correction to the Hardy–Weinberg result when we do not assume that all three genotypes are equally viable. The importance of this correction term is illustrated in the following examples.

Example 9.4.2   Suppose that the fitness index of the recessive genotype is $\beta = .99$ in the above model. Assume that the alleles A and a are initially present in equal numbers. Determine the ratios r_0, r_1, r_2, r_3, and r_4 of allele A to allele a from the initial population to the fourth generation.

Solution: Since $\beta = .99$, the recessive genotype is only slightly less viable than the other two genotypes. Therefore, we expect the correction terms to the Hardy–Weinberg result to be small. From Equation (9.17),

$$r_{n+1} = r_n + r_n \left(\frac{1 - \beta}{r_n + \beta} \right).$$

We are given that $r_0 = 1$ and $\beta = .99$. Therefore,

$$r_1 = r_0 + r_0 \frac{1 - \beta}{r_0 + \beta} = 1 + \frac{.01}{1.99} \approx 1.005,$$

$$r_2 = r_1 + r_1 \frac{1 - \beta}{r_1 + \beta} \approx 1.005 + (1.005)\frac{.01}{1.995} \approx 1.010,$$

$$r_3 = r_2 + r_2 \frac{1 - \beta}{r_2 + \beta} \approx 1.010 + (1.010)\frac{.01}{2.00} \approx 1.015,$$

$$r_4 = r_3 + r_3 \frac{1 - \beta}{r_3 + \beta} \approx 1.015 + (1.015)\frac{.01}{2.005} \approx 1.020.$$

After four generations, the ratio of allele A to allele a has increased by only 2 per cent. Clearly, the correction term in each generation is quite small. Only after many generations will the ratio of allele A to allele a be significantly greater than 1. We may be interested in studying the development of the population over only a few generations. In this case, the error made in assuming that all three genotypes are equally viable is not more than a few per cent. (Would this be true if the fitness index of the recessive genotype was $\beta = .90$?)

Example 9.4.3   Suppose that no individuals with the recessive genotype survive to reproduce. Determine r_n, the ratio of the dominant allele to the recessive allele in the nth generation, in terms of r_0.

Solution: The fitness index of the recessive genotype is $\beta = 0$. In this case, $r_{n+1} = r_n + 1$ from Equation (9.16) or (9.17). This first order difference equation has the general solution $r_n = r_0 + n$. After n generations, the ratio of the dominant allele to the recessive allele has increased by n.

Problems 9.4

1. (a) Consider a recessive genotype with a given fitness index β $(0 \le \beta < 1)$. Prove that

$$r_n\frac{1 - \beta}{r_n + \beta} \le 1 - \beta.$$

Deduce that, if $\beta = .99$, the correction term is never more than 0.01 in any generation.
   (b) Since $\lim_{n \to \infty} r_n = \infty$, we deduce that

$$\lim_{n \to \infty} r_n\frac{1 - \beta}{r_n + \beta} = 1 - \beta.$$

What does this say about the increase in r_n in one generation when n is large, that is, after many generations?
2. Cystic fibrosis is believed to be due to a recessive gene. The incidence of cystic fibrosis is estimated to be 1 in 2500 and, therefore, the frequency of the recessive allele is 1 in 50. Assuming that cystic fibrosis victims do not survive to reproduce and neglecting possible mutations of the dominant gene to the recessive gene, in how many generations will the incidence of cystic fibrosis be 1 in 10,000?
3. In the binomial probability model of selection with $\alpha = 1$, the gene proportions may vary from one generation to the next. This is known as *random drift* of the gene proportions. Is this result of the binomial model

more reasonable than the prediction of the Hardy–Weinberg model that gene proportions remain constant from one generation to the next?

4. In the binomial model of gene selection (with α not necessarily equal to 1), prove that the expected number of copies of allele A in the first generation is $i^{\alpha}(2n)^{1-\alpha}$ if there are i copies of allele A in the parent generation. Deduce that, if $0 \le \alpha < 1$, the expected number is greater than i; if $\alpha = 1$, the expected number is equal to i; and, if $\alpha > 1$, the expected number is less than i.

5. Which of the two fitness models of this section do you expect to be more realistic? (Does selection act on the alleles or on the genotypes?) How would you design experiments to test both of these models?

6. (a) If the fitness index $\beta = 1$, the three genotypes AA, Aa, and aa are equally viable. From the Hardy–Weinberg law, calculate the ratio of heterozygotes to homozygotes. Prove that this ratio is between 0 and 1.
(b) If $0 \le \beta < 1$, calculate the ratio of heterozygotes to homozygotes in the nth generation. Prove that this ratio is equal to $2r_{n-1}/(r_{n-1}^2 + \beta)$.

7. Consider again the binomial probability model of selection with $\alpha = 1$. Suppose that, in a population composed of 200 individuals (100 males and 100 females), the frequency of one allele changed from 40 per cent to 49 per cent in one generation. Can this fluctuation reasonably be considered to be due to random drift of the gene proportions? (Use confidence intervals.)

8. For some genes, the heterozygote has a higher fitness than either homozygote. As an example, the gene for sickle cell anemia confers some resistance to malaria in the heterozygote. Develop a model of selection that describes this possibility.

9. The fitness index β of the recessive genotype may not be a constant in every environment. For example, fruit flies with vestigial wings have a selective advantage in windy environments but not in calm environments. Develop a more general model of gene selection which takes environmental variability into account.

10. (a) We have proved that, if no recessive genotypes survive to reproduce, then $r_n = r_0 + n$, where r_n is the ratio of the dominant allele to the recessive allele in the nth generation. Prove that $q_n = q_0/(1 + nq_0)$, where q_n is the proportion of the recessive allele in the nth generation.
(b) By solving for $n = 1/q_n - 1/q_0$, determine the number of generations required to reduce the proportion of the recessive allele from $q_0 = .01$ to $q_n = .001$.

9.5 The Lotka-Volterra Equations

The logistic equation discussed in Section 7.3,

$$\frac{dx}{dt} = x(\beta - \delta x), \qquad (9.18)$$

is a first order, nonlinear differential equation describing the growth of a population from an initial size $x(0)$ at time $t = 0$ to a size $x(t)$ at time t. The solution of the logistic equation

$$x(t) = \frac{x(0)\beta e^{\beta t}}{\beta - \delta x(0) + \delta x(0) e^{\beta t}} \qquad (9.19)$$

determines the population $x(t)$ at every future time t. The growth curve of the population for some typical values of $x(0)$, β, and δ is drawn in Figure 7.4.

The logistic equation describes the growth of a population in a resource-limited environment. Because one or more essential resources are in limited supply, the population cannot grow indefinitely but instead approaches an equilibrium value $\lim_{t \to \infty} x(t) = \beta/\delta$. This is the stable population size that can be supported by the available resources. If the population is equal to β/δ, then the growth rate of the population is zero and the population remains unchanged at this level. (This population size β/δ corresponds to the population size n considered in the first model of Section 9.4. In that model, we considered an environment which could support exactly n reproducing individuals in each generation.)

If two or more species are competing for the available resources, we expect that their growth can be modeled by a suitable generalization of the logistic equation. Consider the case of two competing species, denoted I and II, whose populations at time t are $x_1(t)$ and $x_2(t)$. Generalizing the ideas that led to the equation of logistic growth, we assume that the growth rates of I and II will be limited by their population sizes. These ideas lead to the following system of equations for the growth rates.

$$\frac{dx_1}{dt} = x_1(b_1 - a_{11}x_1 - a_{12}x_2),$$

$$\frac{dx_2}{dt} = x_2(b_2 - a_{21}x_1 - a_{22}x_2). \qquad (9.20)$$

These equations (known as the *Lotka–Volterra equations*) have been studied extensively in the literature of mathematical ecology.[10] We assume, for the moment, that the constants $a_{11}, a_{12}, a_{21}, a_{22}, b_1$, and b_2 are all positive. (Later in this section, we will consider other possibilities.) The growth rate of the first population is positive if $a_{11}x_1 + a_{12}x_2 < b_1$; zero if $a_{11}x_1 + a_{12}x_2 = b_1$; and negative if $a_{11}x_1 + a_{12}x_2 > b_2$. The growth rate is similar for the second population.

Any populations x_1 and x_2 for which dx_1/dt and dx_2/dt are both zero represent an equilibrium pair of populations in this model. The equations to determine these equilibrium populations are

$$x_1(b_1 - a_{11}x_1 - a_{12}x_2) = 0,$$
$$x_2(b_2 - a_{21}x_1 - a_{22}x_2) = 0. \tag{9.21}$$

Two obvious solutions are $x_1 = 0$, $x_2 = b_2/a_{22}$ and $x_1 = b_1/a_{11}$, $x_2 = 0$. These solutions correspond to one of the species becoming extinct and the other species then reaching its equilibrium population when it is alone in the environment. The third equilibrium point is more interesting, since it may correspond to a stable coexistence of the two species. It is obtained by solving the pair of simultaneous equations

$$a_{11}x_1 + a_{12}x_2 = b_1,$$
$$a_{21}x_1 + a_{22}x_2 = b_2. \tag{9.22}$$

This system of linear equations should be familiar from Chapter 3. Assuming that $a_{11}a_{22} - a_{21}a_{12} \neq 0$ (det $A \neq 0$), we have the unique solution

$$x_1 = \frac{a_{22}b_1 - a_{12}b_2}{a_{11}a_{22} - a_{21}a_{12}}, \qquad x_2 = \frac{a_{11}b_2 - a_{21}b_1}{a_{11}a_{22} - a_{21}a_{12}}. \tag{9.23}$$

If these numbers are positive, they represent populations of the two competing species that can coexist and remain constant.

In the x_1x_2 plane, the equations $a_{11}x_1 + a_{12}x_2 = b_1$ and $a_{21}x_1 + a_{22}x_2 = b_2$ represent straight lines that divide the plane into four distinct regions. This is illustrated in Figure 9.1.

There is another way of looking at the Lotka–Volterra equations. By the chain rule of calculus,

$$\frac{dx_2}{dx_1} = \frac{dx_2}{dt}\frac{dt}{dx_1} = \frac{x_2(b_2 - a_{21}x_1 - a_{22}x_2)}{x_1(b_1 - a_{11}x_1 - a_{12}x_2)}. \tag{9.24}$$

[10] E. C. Pielou, *An Introduction to Mathematical Ecology*, Interscience Publishers, a division of John Wiley & Sons, Inc., New York, 1969.

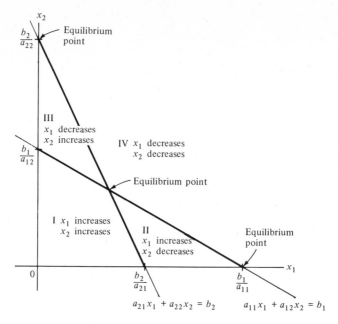

Figure 9.1

We have replaced the two equations of (9.20) by a single first order, nonlinear differential equation. In this equation, we consider x_2 to be a function of x_1. This means that the size of the second population is thought of as depending on the size of the first population. If we could solve for x_2 as a function of x_1, this would tell us exactly how the two species interact. For example, we would know immediately the change in x_2 produced by a change in x_1. Unfortunately, the variables in Equation (9.24) are not separable and the equation cannot be solved, in general, by elementary means. However, there are numerical methods that will give the solution to any desired degree of accuracy.

We should make clear that the Lotka–Volterra equations are indeed a generalization of the equation for logistic growth. For example, if the second population x_2 has no influence on the growth rate of the first population x_1, then $a_{12} = 0$, and the equation for x_1 as a function of t reduces to $dx_1/dt = x_1(b_1 - a_{11}x_1)$. This is exactly the equation of logistic growth for the first population as discussed in Chapter 7. Similarly, if the first population has no influence on the growth rate of the second population, then $dx_2/dt = x_2(b_2 - a_{22}x_2)$. We consider that, if $a_{12} = a_{21} = 0$, then the two populations do not interfere with each other's growth. In this case, they both grow to their equilibrium populations $x_1 = b_1/a_{11}$ and $x_2 = b_2/a_{22}$.

When suitably interpreted, the Lotka–Volterra equations also provide a model for predator–prey interactions. If the first species is a predator of the second species, then the growth rate of the first population can be expected to increase as the second population increases in size. This can be modeled within the framework of the Lotka–Volterra equations if we assume that a_{12} is negative (making $-a_{12}$ positive). It may also be reasonable to assume that the mortality in the second population due to predation is larger than the mortality due to self-crowding when the populations of the two species are approximately equal. This can be modeled by assuming that a_{21} is relatively large compared to a_{22}. Assuming as before that a_{11}, a_{21}, and a_{22} are positive, we conclude that $a_{11}a_{22} - a_{21}a_{12}$ is positive.

From Equations (9.23), we see that there will be an equilibrium point if $a_{11}b_2 - a_{21}b_1 > 0$. (Note that $a_{22}b_1 - a_{12}b_2$ is positive.) But since a_{21} is relatively large compared to other coefficients, it may happen that $a_{11}b_2 - a_{21}b_1$ is negative. This means that there will be no equilibrium populations for which both x_1 and x_2 are positive. In other words, the predator will completely consume the prey. The various possibilities in this case are illustrated in Figure 9.2.

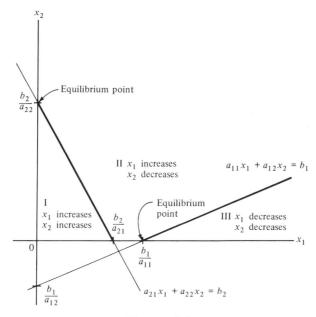

Figure 9.2

The reader may be interested in attempting to model other types of competition by means of the Lotka–Volterra equations. For example, if both species are predators of each other, can this be modeled by assuming that a_{12} and a_{21} are negative? It is also instructive to attempt to construct generalizations of the Lotka–Volterra equations to take into account other interaction terms. The various models that can be constructed in this way give us yet another approach to the description of the survival, reproduction, and extinction of populations.

Problems 9.5

1. What assumptions would lead to the following system of differential equations describing the growth of two competing species?

$$\frac{dx_1}{dt} = x_1^2(k_1 - 2x_1 - x_2),$$

$$\frac{dx_2}{dt} = x_2^2(k_2 - x_1 - 2x_2).$$

Is this system a reasonable model for some types of species competition? What are the equilibrium populations?

2. Describe the four regions of Figure 9.1 when $0 < b_1/a_{11} < b_2/a_{21}$ and $0 < b_2/a_{22} < b_1/a_{12}$, according to whether x_1 and x_2 increase or decrease in each region.

3. Generalize the Lotka–Volterra equations to the case of n competing species. Determine all equilibrium populations.

4. Construct a model of the growth and reproduction of a single population based on the Lotka–Volterra equations which takes into account the age structure of the population. (Divide the population into two or more age groups and make assumptions about the types of competition and cooperation among the age groups.)

5. How would you estimate the constants that appear in the Lotka–Volterra equations? If the populations of the competing species are known at some time, the Lotka–Volterra model determines the populations at all future times. Is this a reasonable approximation to reality? How would you include the effects of seasonal factors? Can you think of other factors not included in the model that may have a significant effect on the outcome of competition?

6. The total population of the two competing species is $x_1 + x_2$; its growth rate is $d(x_1 + x_2)/dt$. In the Lotka–Volterra model, determine the populations x_1 and x_2 for which the growth rate of the total population is a maximum.

7. Suppose that two species are competing for resources as in the Lotka–Volterra model [Equations (9.20)]. Suppose further that average individuals of the first and second species weigh 1 pound and 2 pounds, respectively. Determine the populations x_1 and x_2 for which the rate of increase of weight of the total population is a maximum. (If the two species are to be harvested, they should be maintained at these levels in order to produce the maximum yield.)

8. Competing species may become extinct if their populations become too small. Suppose that species I and species II will become extinct in a certain environment if their populations are less than m_1 and m_2, respectively. How would you extend the Lotka–Volterra model to include this possibility? (See Problem 7.3.7.)

9. Suppose that $x_1(t)$ and $x_2(t)$ are two populations competing according to the Lotka–Volterra system. If individuals of both populations are harvested at rates proportional to their populations, this is described by

$$\frac{dx_1}{dt} = x_1(b_1 - a_{11}x_1 - a_{12}x_2) - k_1x_1,$$

$$\frac{dx_2}{dt} = x_2(b_2 - a_{21}x_1 - a_{22}x_2) - k_2x_2.$$

What are the equilibrium populations in this model?

9.6 The Game of Life

A basic problem of biology is that of understanding the origin of living organisms from nonliving matter. The two fundamental characteristics of living organisms are self-reproduction and evolution. Both of these aspects of life are extremely complex, and it is correspondingly difficult to understand the processes by which nonliving matter acquired these characteristics. This problem has been studied by experiments "contrived to simulate the conditions supposed to have existed on the primitive earth. Several amino acids and even peptides and nucleotides have been thus obtained. Primordial oceans may, then, have contained a very dilute 'broth' of organic compounds. But even if such a 'broth' existed, the origin in it of the first self-reproducing, and hence living, systems remains an unsolved problem."[11]

[11] T. Dobzhansky, *Genetics of the Evolutionary Process*, Columbia University Press, New York, 1970, p. 8.

There have been a number of candidates proposed as possible original self-duplicating forms of matter. We will not examine these possibilities here but will instead study the problem of self-duplication and mutation in a more general way. To do this, we introduce in this section the game of "Life" invented by the mathematician John Conway.[12] This game illustrates in a simple and amusing way many features of processes of self-reproduction and evolution.

The rules of life are surprisingly uncomplicated. The game is played on a plane divided into squares. (Think of an infinite checkerboard.) Each square or cell of the board has eight neighboring cells. These are shown in Figure 9.3. Each square is either occupied by a counter or it is empty. The game begins with a configuration of counters placed on the board with at most one counter to each cell.

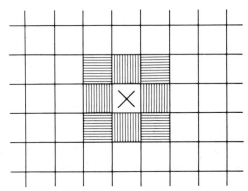

Figure 9.3. Single Counter with Eight Neighboring Squares

The initial configuration or "population" of counters then changes according to the rules of the game. The counters may "die" either from overcrowding or from isolation in subsequent "generations." In other circumstances, the counter population may increase by reproduction. There are two rules governing survival and reproduction from one generation to the next.

1. A counter survives to the next move (or to the next generation) if it has either two or three neighboring counters. Otherwise, it dies and is removed from the board.
2. An empty cell with exactly three neighboring counters becomes alive (and is occupied by a counter) in the next generation.

[12] M. Gardner, "Mathematical Games," *Scientific American*, 223:120–123 (Oct. 1970); 224:112–117 (Feb. 1971).

With these rules, the initial configuration determines the configurations on all subsequent moves. The examples in the figure on the endpapers show the evolution of some simple initial configurations through several generations.

Configurations with only one or two counters die immediately [examples (i) and (ii)]. Example (iii) oscillates from one generation to the next. Since this object appears very frequently in the evolution of other configurations, it is given a name—"blinker." The row of four counters [example (iv)] evolves into a stable configuration called a "beehive" in the third generation. The row of five counters evolves into a stable configuration composed of four blinkers in the sixth generation. The reader may verify that a row of six counters disappears completely. It seems to be impossible to guess the final pattern from the initial configuration. Despite the simplicity of the laws of evolution and despite the fact that the initial configuration determines all future configurations, the final outcome can be very surprising.

The initial configuration of example (vii) with five counters is called a "glider," because it has the remarkable property that it moves across the plane. After four moves, the original shape of the glider has been restored, but it has shifted one square. The existence of the glider makes possible a type of communication between configurations in different parts of the plane.

The game of life becomes really interesting when we start with more complicated initial configurations or when two or more configurations collide. In some cases, a configuration will have the ability to repair itself after a collision with other configurations. In other cases, the collision will give rise to new configurations, which then continue to evolve. Configurations are known which periodically give birth to gliders or to other configurations. Self-reproducing configurations can be constructed. Very generally, it can be said that a complex initial configuration with many thousands or millions of counters can be expected to evolve patterns of increasingly greater stability if the configuration does continue to evolve. In this evolution process, unstable patterns will be eliminated by collision with more stable ones.

The analogies between the processes of growth and reproduction in this game and in real life are extremely suggestive. The primordial broth referred to by Dobzhansky is conceptually very similar to the initial configuration of Conway's game. Of course, the rules by which the primordial broth has evolved are the much more complicated rules of chemistry. However, in principle, the two types of evolution are very similar.

The game of life is an example of a cellular automaton game. The study of these games may provide insights helpful in the understanding of specific biological problems such as cell growth and differentiation, the replication of DNA molecules, the synthesis of molecules in cells, and the origin of life.

Certainly, by studying life and other similar games, we are led to ask some very basic questions concerning the transition from an initial chaos of atoms and molecules to the highly evolved forms of self-reproducing matter that are studied in (and study) the biological sciences.

Trigonometry

One of the oldest branches of mathematics is trigonometry, defined as the study of angles and triangles. We will define trigonometry to be the study of the *trigonometric functions*. The applications of these functions extend far beyond their original applications in surveying and astronomy. To define the trigonometric functions, consider the *unit circle* in the xy plane (Figure A1). This is the set of points (x, y) in the plane whose coordinates satisfy $x^2 + y^2 = 1$.

If 0 is the origin of the xy plane and P is any point (x, y) on the unit circle, define θ to be the angle between the x axis and the line segment $0P$. The angle θ uniquely determines the corresponding point P on the unit circle. By definition, the x and y components of P are written

$$x = \text{cosine } \theta \quad \text{and} \quad y = \text{sine } \theta. \tag{A1}$$

These are the two basic trigonometric functions, usually written $\cos \theta$ and $\sin \theta$. Since $x^2 + y^2 = 1$, $\cos \theta$ and $\sin \theta$ satisfy the equation $\cos^2 \theta + \sin^2 \theta = 1$. As angle θ varies, $\cos \theta$ and $\sin \theta$ oscillate between $+1$ and -1.

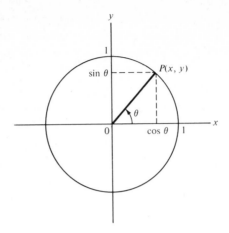

Figure A1. Unit Circle $x^2 + y^2 = 1$

If θ is measured in degrees, then $\theta = 90°$ corresponds to a right angle and $\theta = 360°$ corresponds to a full rotation. It is often more convenient to measure angles in *radians*, defined by the length of the arc on the unit circle from the point $(0, 1)$ to the point $(\cos \theta, \sin \theta)$. Since the circumference of the unit circle is 2π, we define 2π radians to be equal to 360 degrees. Therefore, 1 radian is equal to $360/2\pi = 57.3 \cdots$ degrees. Similarly, $30° = \pi/6$ radians, $45° = \pi/4$ radians, $270° = 3\pi/2$ radians, and so on.

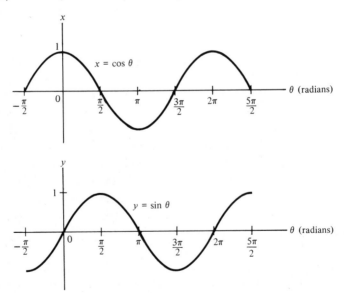

Figure A2. Trigonometric Functions

In measuring angles, we have adopted the convention that positive angles are measured in a counterclockwise direction about the origin. Negative angles are measured in a clockwise direction. One complete rotation corresponds to an angle of 2π radians. Since the point P on the unit circle corresponding to the angle θ also corresponds to the angle $\theta + 2\pi$, we conclude that

$$\cos \theta = \cos (\theta + 2\pi), \qquad \sin \theta = \sin (\theta + 2\pi). \qquad \text{(A2)}$$

We say that these functions are *periodic* with period 2π. Their graphs are drawn in Figure A2.

We emphasize that the trigonometric functions $\sin x$ and $\cos x$ are defined for all real numbers (positive or negative). If x is negative, then it is evident from the definition of these functions that $\sin (-x) = -\sin x$ and $\cos (-x) = \cos x$.

The four other trigonometric functions are the tangent, cotangent, secant, and cosecant. They can be defined most simply in terms of the sine and cosine.

$$\tan \theta = \frac{\sin \theta}{\cos \theta}, \qquad \cot \theta = \frac{\cos \theta}{\sin \theta},$$

$$\sec \theta = \frac{1}{\cos \theta}, \qquad \csc \theta = \frac{1}{\sin \theta}. \qquad \text{(A3)}$$

Table 5 gives values of the three trigonometric functions (sin, cos, tan) for θ ranging from 0 to 90 degrees (0 to $\pi/2$ radians). Other values can be calculated from these. However, there are several values of θ for which the calculation of the trigonometric functions is quite easy. From Figure A1, it is obvious that $\sin 0 = \sin \pi = \cos \pi/2 = 0$, $\cos 0 = \sin \pi/2 = 1$, and $\cos \pi = \sin 3\pi/2 = -1$. Similarly, by the Pythagorean theorem it is clear that $\sin 45° = \sin \pi/4 = \cos \pi/4 = \sqrt{2}/2$ and $\tan \pi/4 = 1$. By the use of elementary geometry, it can be shown that $\sin 30° = \sin \pi/6 = \cos \pi/3 = \frac{1}{2}$ and $\sin \pi/3 = \cos \pi/6 = \sqrt{3}/2$. The reader should know the values of $\sin \theta$ and $\cos \theta$ for $\theta = 0, \pi/6, \pi/4$, and $\pi/2$. Other values can be calculated using the results of Problem 11 and the fact that $\sin (-\theta) = -\sin \theta$ and $\cos (-\theta) = \cos \theta$. (See Problem 12.)

The trigonometric functions satisfy a large number of identities that can be very useful in solving problems. We know that $\sin \theta$ and $\cos \theta$ satisfy the identity $\cos^2 \theta + \sin^2 \theta = 1$ for every angle θ. For all angles θ and ϕ, the following *addition formulas* can be proved.

$$\sin(\theta + \phi) = \sin\theta\cos\phi + \cos\theta\sin\phi,$$

$$\cos(\theta + \phi) = \cos\theta\cos\phi - \sin\theta\sin\phi, \qquad \text{(A4)}$$

$$\tan(\theta + \phi) = \frac{\tan\theta + \tan\phi}{1 - \tan\theta\tan\phi}.$$

To illustrate the applications of the trigonometric functions to the study of triangles, consider the triangle ABC with sides of length a, b, and c (Figure A3). Suppose that θ is the angle between the sides AB and AC and that the point P is the foot of the perpendicular from C to the side AB.

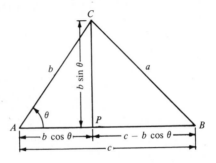

Figure A3

From the definition of the trigonometric functions, if (x, y) is a point on the unit circle, then $x = \cos\theta$ and $y = \sin\theta$. If the circle has radius r instead of radius 1, and (x_1, y_1) is the point on this circle corresponding to the angle θ, then, by similar triangles, $x/1 = x_1/r = \cos\theta$, and $y/1 = y_1/r = \sin\theta$. In general, for any right triangle, let θ be one of the angles (not the right angle). We then could define

$$\cos\theta = \frac{\text{adjacent side}}{\text{hypotenuse}} \qquad \text{and} \qquad \sin\theta = \frac{\text{opposite side}}{\text{hypotenuse}}.$$

This definition is equivalent to the one already given. Now, referring back to Figure A3, the length of PC is $b\sin\theta$ and the length of AP is $b\cos\theta$. Therefore, the length of PB is $c - b\cos\theta$. The triangle CPB is a right triangle and the Pythagorean theorem tells us that

$$(CB)^2 = (CP)^2 + (PB)^2 \qquad \text{or} \qquad a^2 = (b\sin\theta)^2 + (c - b\cos\theta)^2.$$

Therefore,

$$a^2 = b^2(\sin^2 \theta + \cos^2 \theta) + c^2 - 2bc \cos \theta = b^2 + c^2 - 2bc \cos \theta.$$

This result is known as the *law of cosines*.

Another simple geometric argument gives the result that $\sin \theta < \theta < \sin \theta/\cos \theta$ when θ is measured in radians and $0 < \theta < \pi/2$.

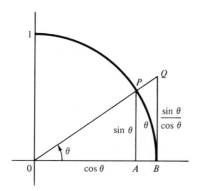

Figure A4

In Figure A4, θ is the length of the arc BP on the unit circle if θ is measured in radians. The length of AP is $\sin \theta$ and (by similar triangles) the length of BQ is $\sin \theta/\cos \theta$. Comparing these three lengths, we have

$$\sin \theta < \theta < \frac{\sin \theta}{\cos \theta} \quad \text{or} \quad 1 < \frac{\theta}{\sin \theta} < \frac{1}{\cos \theta} \quad \text{if} \ \ 0 < \theta < \frac{\pi}{2}. \quad \text{(A5)}$$

This fact will be extremely useful in Appendix B. Since $\cos 0 = 1$, we can conclude that $\theta/\sin \theta$ approaches 1 as θ approaches 0.

Finally, we define the *inverse trigonometric functions*. If $x = \cos \theta$, then θ is the angle whose cosine is x. This is written $\theta = \cos^{-1} x$. For example, if $x = 1/\sqrt{2}$, then $\theta = \cos^{-1} x = \cos^{-1} (1/\sqrt{2}) = \pi/4$ (if θ is measured in radians). Unfortunately, θ is not uniquely determined by x since $\cos \theta = \cos (\theta + 2\pi) = \cos (\theta + 4\pi) = \cdots$. However, if we restrict θ to values between 0 and 2π, the inverse cosine function is better defined. Even in this case, $\cos (\pi/4) = \cos (7\pi/4) = 1/\sqrt{2}$. To make the inverse cosine "function" into a function, we must have a uniquely determined value of $\cos^{-1} x$ for every x. To do this, we simply require that $\theta = \cos^{-1} x$ takes values between 0 and π. In this way, $\cos^{-1} x$ will really be a function. The other inverse trigonometric functions are defined in a similar way.

Problems A

1. Prove the following identities relating the trigonometric functions.
   (a) $1 + \tan^2 \theta = \sec^2 \theta$.   (b) $1 + \cot^2 \theta = \csc^2 \theta$.
   (c) $\sin(-\theta) = -\sin \theta$.   (d) $\cos(-\theta) = \cos \theta$.

2. From the addition formulas, prove the following identities.
   (a) $\sin(\theta - \phi) = \sin \theta \cos \phi - \cos \theta \sin \phi$.
   (b) $\cos(\theta - \phi) = \cos \theta \cos \phi + \sin \theta \sin \phi$.
   (c) $\sin 2\theta = 2 \sin \theta \cos \theta$.
   (d) $\cos 2\theta = \cos^2 \theta - \sin^2 \theta = 2 \cos^2 \theta - 1 = 1 - 2 \sin^2 \theta$.

3. Prove the half-angle formulas.

   (a) $\sin \dfrac{\theta}{2} = \sqrt{\dfrac{1 - \cos \theta}{2}}$.

   (b) $\cos \dfrac{\theta}{2} = \sqrt{\dfrac{1 + \cos \theta}{2}}$.

   (c) $\tan \dfrac{\theta}{2} = \dfrac{\sin \theta}{1 + \cos \theta} = \dfrac{1 - \cos \theta}{\sin \theta}$.

4. By writing $\theta = \left(\dfrac{\theta + \phi}{2}\right) + \left(\dfrac{\theta - \phi}{2}\right)$ and $\phi = \left(\dfrac{\theta + \phi}{2}\right) - \left(\dfrac{\theta - \phi}{2}\right)$, prove

   (a) $\sin \theta - \sin \phi = 2 \cos \left(\dfrac{\theta + \phi}{2}\right) \sin \left(\dfrac{\theta - \phi}{2}\right)$.

   (b) $\cos \theta - \cos \phi = -2 \sin \left(\dfrac{\theta + \phi}{2}\right) \sin \left(\dfrac{\theta - \phi}{2}\right)$.

5. Express in radians:
   (a) $15°$.   (b) $72°$.   (c) $-90°$.   (d) $720°$.
   (e) $-135°$.   (f) $2°$.   (g) $210°$.   (h) $315°$.

6. Show that $\sin \theta < \theta$ for $0 < \theta < \dfrac{\pi}{2}$. (*Hint:* Use Figure A1.)

7. Compute the length of the chord OP in Figure A1 in terms of $\sin \theta$ and $\cos \theta$. Use this result to prove that $1 - \theta^2 < \cos \theta$ for $0 < \theta < \pi/2$. (*Hint:* Use the result of Problem 6.)

8. How far does the tip of the second hand of a clock move in 25 seconds if it is 6 inches long?

9. A wheel of radius 10 inches is rotating at the rate of 90 revolutions per minute. Find the speed of a point on the wheel in feet per second.

10. The three sides of a triangle have lengths 5, 7, and 8. Use the law of cosines to find the three angles.

11. Using Figure A1 or the addition formulas, prove that

    (a) $\sin\left(\dfrac{\pi}{2} - \theta\right) = \cos\theta = \sin\left(\dfrac{\pi}{2} + \theta\right)$.

    (b) $\cos\left(\dfrac{\pi}{2} - \theta\right) = \sin\theta$.

    (c) $\sin(\theta + \pi) = -\sin\theta$.

    (d) $\cos(\theta + \pi) = -\cos\theta$.

    (e) $\cos\left(\theta + \dfrac{\pi}{2}\right) = -\sin\theta$.

12. Compute the values of $\sin\theta$, $\cos\theta$, and $\tan\theta$ for the following angles.

    (a) $\theta = \dfrac{5\pi}{4}$.   (b) $\theta = -\dfrac{2\pi}{3}$.   (c) $\theta = -\dfrac{\pi}{6}$.   (d) $\theta = 6\pi$.

    (e) $\theta = 9\pi$.   (f) $\theta = \dfrac{5\pi}{3}$.   (g) $\theta = \dfrac{3\pi}{2}$.   (h) $\theta = -\dfrac{5\pi}{6}$.

13. Calculate $\sin^{-1}x$ and $\cos^{-1}x$.

    (a) $x = 0$.   (b) $x = 1$.   (c) $x = \tfrac{1}{2}$.   (d) $x = \dfrac{\sqrt{3}}{2}$.

    (e) $x = -1$.   (f) $x = -\dfrac{\sqrt{3}}{2}$.

14. (a) Express $\csc\theta$ in terms of $\cos\theta$.
    (b) Express $\tan\theta$ in terms of $\sin\theta$.
    (c) Express $\cos\theta$ in terms of $\cot\theta$.

15. If $\tan\theta = \tfrac{4}{3}$, find possible values for $\sin\theta$, $\cos\theta$, $\cot\theta$, $\sec\theta$, and $\csc\theta$.

16. Use the half-angle formulas (Problem 3) to find
    (a) $\sin 15°$.   (b) $\cos 22\tfrac{1}{2}°$.   (c) $\tan 67.5°$.

17. Use the addition formulas to find
    (a) $\sin 75°$.   (b) $\cos 15°$.   (c) $\cos 105°$.

Differential Calculus

B

Appendix

B1 Rates of Change and Derivatives

The basic problem of differential calculus is to describe how changes in one variable produce changes in other variables. As an example, consider the equation $y = \frac{5}{9}(x - 32)$ relating the temperature y in Celsius degrees to the temperature x in Fahrenheit degrees. If x increases by 9 degrees, y increases by 5 degrees. If x increases by 1 degree, y increases by $\frac{5}{9}$ degree. We say that the rate of change of y with respect to x is $\frac{5}{9}$. This means that the change in y is $\frac{5}{9}$ the change in x.

More generally, the equation

$$y = mx + b \qquad \text{(B1)}$$

represents a straight line in the xy plane (Figure B1). If (x_1, y_1) and (x_2, y_2) are any two points on this line, then $y_1 = mx_1 + b$ and $y_2 = mx_2 + b$. As x changes from x_1 to x_2, y changes from y_1 to y_2. The *rate of change* of y with respect to x (as x changes from x_1 to x_2) is

$$\frac{y_2 - y_1}{x_2 - x_1} = \frac{(mx_2 + b) - (mx_1 + b)}{x_2 - x_1} = m\frac{x_2 - x_1}{x_2 - x_1} = m.$$

This constant rate of change m is also called the *slope* of the straight line.

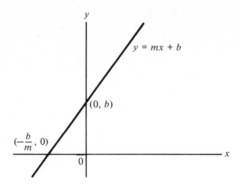

Figure B1

Example B1.1   An iron bar expands in length as it is heated. A bar of length 50 centimeters at temperature 0°C has length $50 + 0.0006T$ centimeters at temperature T°C. What is the rate of change of the length of the iron bar with respect to temperature?

Solution: The length L at temperature T is $L = 50 + 0.0006T$. If the temperature changes from T_1 to T_2, the length changes from L_1 to L_2 and the rate of change is

$$\frac{L_2 - L_1}{T_2 - T_1} = \frac{(50 + 0.0006T_2) - (50 + 0.0006T_1)}{T_2 - T_1} = 0.0006.$$

The length of the iron bar increases by 0.0006 centimeter for each increase of 1°C in the temperature.

We are interested in defining rates of change when y is a more complicated function of x, say $y = f(x)$. In this case, if x changes from x_1 to x_2, then y changes from $f(x_1)$ to $f(x_2)$. The average rate of change of y with respect to x (as x changes from x_1 to x_2) is

$$\frac{y_2 - y_1}{x_2 - x_1} = \frac{f(x_2) - f(x_1)}{x_2 - x_1}. \tag{B2}$$

Unless $f(x)$ is a linear function of x [as in Equation (B1)], this rate of change is not a constant. We are most interested in studying the change in $f(x)$ when x changes by a small amount. Let Δx represent a small change in x (from x to $x + \Delta x$). In Figure B2, the two points $(x, f(x))$ and $(x + \Delta x, f(x + \Delta x))$ on

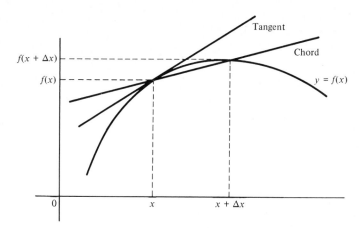

Figure B2

the curve $y = f(x)$ are drawn. The chord or straight line joining these points is also drawn. The average rate of change of y with respect to x between these points is

$$\frac{f(x + \Delta x) - f(x)}{(x + \Delta x) - x} = \frac{f(x + \Delta x) - f(x)}{\Delta x}. \tag{B3}$$

This ratio is the slope of the chord joining the two points. If Δx is made very small, the point $(x + \Delta x, f(x + \Delta x))$ approaches the point $(x, f(x))$ and the chord approaches the tangent line to the curve at $(x, f(x))$. This suggests the definition of the rate of change of y with respect to x at the point $(x, f(x))$ as the slope of the tangent line. This is the limit as Δx tends to zero of the ratio $[f(x + \Delta x) - f(x)]/\Delta x$. This is the idea that leads to the definition of the derivative of $f(x)$. Before defining the derivative, let us illustrate by a number of examples what is meant by $\lim_{\Delta x \to 0}$, the limit as Δx tends to zero.

Example B1.2   Calculate the following limits.

1. $\lim_{t \to 1}(1 + t^2)$.   2. $\lim_{\Delta x \to 0} 2^{\Delta x}$.   3. $\lim_{\Delta x \to 0} \dfrac{3}{3 + \Delta x}$.   4. $\lim_{\Delta x \to 0} \dfrac{5 + 2\Delta x}{10 + \Delta x}$.

Solution:

1. The limit as t approaches 1 of $1 + t^2$ is $1 + 1 = 2$.

2. $\lim_{\Delta x \to 0} 2^{\Delta x} = 2^0 = 1$.

3. $\lim\limits_{\Delta x \to 0} \dfrac{3}{3 + \Delta x} = \dfrac{3}{3} = 1.$

4. $\lim\limits_{\Delta x \to 0} \dfrac{5 + 2\Delta x}{10 + \Delta x} = \dfrac{5}{10} = \dfrac{1}{2}.$

Example B1.3 Calculate $\lim_{\Delta x \to 0} [f(x + \Delta x) - f(x)]/\Delta x$ for the following functions.

1. $f(x) = 3x.$ 2. $f(x) = x^2.$ 3. $f(x) = \dfrac{1}{x}$ $(x \neq 0).$

Solution: If we set $\Delta x = 0$ in the ratio $[f(x + \Delta x) - f(x)]/\Delta x$, we have $0/0$, which is not determined. However, the limit as Δx tends to zero may be defined.

1. $\dfrac{f(x + \Delta x) - f(x)}{\Delta x} = \dfrac{3(x + \Delta x) - 3x}{\Delta x} = 3$

for all Δx not equal to zero. Therefore, the required limit is 3.

2. $\lim\limits_{\Delta x \to 0} \dfrac{f(x + \Delta x) - f(x)}{\Delta x} = \lim\limits_{\Delta x \to 0} \dfrac{(x + \Delta x)^2 - x^2}{\Delta x}$

$= \lim\limits_{\Delta x \to 0} \dfrac{x^2 + 2x\,\Delta x + (\Delta x)^2 - x^2}{\Delta x} = \lim\limits_{\Delta x \to 0} (2x + \Delta x) = 2x.$

3. $\lim\limits_{\Delta x \to 0} \dfrac{f(x + \Delta x) - f(x)}{\Delta x} = \lim\limits_{\Delta x \to 0} \dfrac{1}{\Delta x}\left(\dfrac{1}{x + \Delta x} - \dfrac{1}{x}\right)$

$= \lim\limits_{\Delta x \to 0} \dfrac{1}{\Delta x}\left(\dfrac{x - (x + \Delta x)}{x(x + \Delta x)}\right) = \lim\limits_{\Delta x \to 0} \dfrac{-1}{x(x + \Delta x)} = \dfrac{-1}{x^2}$ $(x \neq 0).$

Having seen that these limits can be calculated for some simple functions, we are now ready to define the concept of a derivative.

Definition B1.1 **Derivative** *The derivative of a function $f(x)$ with respect to the variable x is a function $f'(x) = df(x)/dx$ defined by*

$$f'(x) = \frac{df(x)}{dx} = \lim\limits_{\Delta x \to 0} \frac{f(x + \Delta x) - f(x)}{\Delta x}. \tag{B4}$$

If this limit is defined for $x = x_0$, the function $f(x)$ is said to be differentiable *at $x = x_0$.*

In Example B1.3, we have found that the functions $3x$, x^2, and $1/x$ have the derivatives 3, $2x$, and $-1/x^2$. To gain some experience in calculating derivatives, consider the following example.

Example B1.4   Calculate the derivatives of the following functions.

1. $f(x) = 7(1 + x)$.   ⠀⠀⠀⠀⠀⠀⠀⠀⠀⠀⠀   2. $f(x) = x^2 + x$.

3. $f(x) = \dfrac{1}{x + 1}$ $(x \neq 1)$.   ⠀⠀⠀   4. $f(x) = \dfrac{1}{x^2}$ $(x \neq 0)$.

Solution:

1. $f'(x) = \lim\limits_{\Delta x \to 0} \dfrac{f(x + \Delta x) - f(x)}{\Delta x}$

   $= \lim\limits_{\Delta x \to 0} \dfrac{7(1 + x + \Delta x) - 7(1 + x)}{\Delta x} = 7.$

2. $f'(x) = \lim\limits_{\Delta x \to 0} \dfrac{(x + \Delta x)^2 + (x + \Delta x) - x^2 - x}{\Delta x}$

   $= \lim\limits_{\Delta x \to 0} (2x + 1 + \Delta x) = 2x + 1.$

3. $f'(x) = \lim\limits_{\Delta x \to 0} \dfrac{1}{\Delta x}\left(\dfrac{1}{x + \Delta x + 1} - \dfrac{1}{x + 1}\right)$

   $= \lim\limits_{\Delta x \to 0} \dfrac{-1}{(x + 1)(x + \Delta x + 1)} = \dfrac{-1}{(x + 1)^2}$ $(x \neq -1).$

4. $f'(x) = \lim\limits_{\Delta x \to 0} \dfrac{1}{\Delta x}\left[\dfrac{1}{(x + \Delta x)^2} - \dfrac{1}{x^2}\right]$

   $= \lim\limits_{\Delta x \to 0} \dfrac{-2x - \Delta x}{x^2(x + \Delta x)^2} = \dfrac{-2}{x^3}$ $(x \neq 0).$

In these last two examples, $f(x) = 1/x^2$ is not differentiable at $x = 0$, and $f(x) = 1/(x + 1)$ is not differentiable at $x = -1$.

The derivative $dy/dx = f'(x)$ of a function $y = f(x)$ represents the "instantaneous" rate of change of the function with respect to the variable x. The notation dy/dx indicates that the derivative is the limit of the ratio of a small change in y to a small change in x.

If $p(t)$ represents the size of a bacteria population at time t, then $dp/dt = p'(t)$ represents the growth rate of the population at time t. For example, if the population at time t is $p(t) = 3000 + 100t^2$, where t is measured in hours, the growth rate is

$$\frac{dp}{dt} = p'(t) = \lim_{\Delta t \to 0} \frac{p(t + \Delta t) - p(t)}{\Delta t}$$

$$= \lim_{\Delta t \to 0} \frac{3000 + 100(t + \Delta t)^2 - 3000 - 100t}{\Delta t}$$

$$= \lim_{\Delta t \to 0} (200t + 100\Delta t) = 200t.$$

The growth rate of this population increases with time. When $t = 5$ hours, the growth rate is 1000 per hour. When $t = 10$ hours, the growth rate is 2000 per hour.

To calculate the derivative of every function directly from the definition can be very tedious. We now give a series of results that greatly simplify the calculation of many derivatives.

   I. The derivative of the constant function $f(x) = c$ is $f'(x) = 0$.
  II. If $f(x)$ is a differentiable function and if a is any constant, define the function $g(x) = af(x)$. Then the derivative of $g(x)$ is $g'(x) = af'(x)$. This can be written

$$\frac{d}{dx}(af(x)) = a\frac{df(x)}{dx}.$$

 III. If $f(x)$ and $g(x)$ are differentiable functions, then

$$\frac{d}{dx}(f(x) + g(x)) = \frac{d}{dx}f(x) + \frac{d}{dx}g(x).$$

 IV. The function $f(x) = x^n$, where n is any real number, has the derivative $f'(x) = nx^{n-1}$.

These results can be proved using the definition of the derivative. For example, if $f(x) = c$, then

$$\lim_{\Delta x \to 0} \frac{f(x + \Delta x) - f(x)}{\Delta x} = \lim_{\Delta x \to 0} \frac{c - c}{\Delta x} = \lim_{\Delta x \to 0} 0 = 0.$$

The many applications of these results are illustrated by the following example and by the problems at the end of this section.

Example B1.5   Determine the derivatives of the following functions.

1. $f(x) = x^5$.   2. $x^5 + x^4 + 1$.   3. $x^{4/3} + x^{1/3}$.   4. $4 + \dfrac{4}{x}$.

Solution:

1. From IV with $n = 5$, $f'(x) = 5x^4$.

2. $\dfrac{d}{dx}(x^5 + x^4 + 1) = 5x^4 + 4x^3$, using I, III, and IV.

3. $\dfrac{d}{dx}(x^{4/3} + x^{1/3}) = \frac{4}{3}x^{1/3} + \frac{1}{3}x^{-2/3}$.

4. $\dfrac{d}{dx}\left(4 + \dfrac{4}{x}\right) = \dfrac{d}{dx}\left(\dfrac{4}{x}\right) = \dfrac{d}{dx}(4x^{-1}) = \dfrac{-4}{x^2}$, since $\dfrac{d}{dx}(4) = 0$.

In calculating the derivative, we study the limit of the ratio $[f(x + \Delta x) - f(x)]/\Delta x$. If Δx is small, this ratio must be approximately equal to $f'(x)$. This is written

$$\frac{f(x + \Delta x) - f(x)}{\Delta x} \approx f'(x) \qquad \text{or} \qquad f(x + \Delta x) - f(x) \approx f'(x)\,\Delta x. \quad \text{(B5)}$$

This approximation can be extremely useful. It describes the change in $f(x)$ when x changes by a small amount. For example, the volume of a sphere of radius r is $V(r) = \frac{4}{3}\pi r^3$. If the radius increases by a small amount from r to $r + \Delta r$, the volume increases by the amount $V(r + \Delta r) - V(r) \approx V'(r)\,\Delta r = 4\pi r^2\,\Delta r$. If the radius is $r = 1$ meter and $\Delta r = 0.01$ meter, the volume increases by $\Delta V = V(r + \Delta r) - V(r) \approx 4\pi(1)^2(0.01) \approx .126$ (meter)^3.

Example B1.6   Calculate an approximate value for $(3.01)^2$.

Solution: By multiplying, we find that $(3.01)^2 = 9.0601$. An approximate answer can be found by defining $f(x) = x^2$. Then $f(3) = 9$ and, if we set $\Delta x = .01$, $f(3.01) \approx f(3) + f'(3)(.01)$. But $f'(x) = 2x$. Therefore, $f(3.01) \approx 9 + 6(.01) = 9.06$. This is a very good approximation to the exact value.

Example B1.7   Calculate an approximate value for $(26.5)^{1/3}$.

Solution: Define $f(x) = x^{1/3}$. Since 26.5 is close to 27 and $(27)^{1/3} = 3$, we choose $x = 27$ and $\Delta x = -.5$. Then $f(26.5) \approx f(27) + f'(27)(-.5)$.

But $f'(x) = \frac{1}{3}x^{-2/3}$ and $f'(27) = \frac{1}{3}(27)^{-2/3} = \frac{1}{27}$. Therefore, $(26.5)^{1/3} \approx 3 - .5/27 \approx 2.9815$. The correct value is 2.98137 and our approximation is again very good.

As the last topic in this section, let us calculate the derivative of the trigonometric function $y = \sin x$, where x is measured in radians. In Appendix A, we proved that $1 < \theta/\sin \theta < 1/\cos \theta$ for $0 < \theta < \pi/2$. We conclude that $\lim_{\theta \to 0} \theta/\sin \theta = 1$, since $\lim_{\theta \to 0} 1/\cos \theta = 1/1 = 1$. This is the result we need to evaluate the derivative of $y = \sin x$.

We must evaluate the limit

$$\frac{dy}{dx} = \lim_{\Delta x \to 0} \frac{\sin (x + \Delta x) - \sin x}{\Delta x}.$$

From Problem A.4(a), we have the identity

$$\sin (x + \Delta x) - \sin x = 2 \cos \left(x + \frac{\Delta x}{2} \right) \sin \left(\frac{\Delta x}{2} \right).$$

Using this result, we obtain the limit

$$\frac{dy}{dx} = \lim_{\Delta x \to 0} \frac{2}{\Delta x} \cos \left(x + \frac{\Delta x}{2} \right) \sin \left(\frac{\Delta x}{2} \right) = \cos x \lim_{\Delta x \to 0} \frac{\sin (\Delta x/2)}{\Delta x/2} = \cos x.$$

This proves the result that $d (\sin x)/dx = \cos x$; the derivative of $\sin x$ is $\cos x$. The reader should attempt to prove that $d (\cos x)/dx = -\sin x$. [*Hint:* Use problem A.4(b)].

Problems B1

1. Determine the derivatives of the following functions.

   (a) $f(x) = 3x^2$.

   (b) $f(x) = 1 + x + x^2$.

   (c) $f(x) = \sqrt{x} + \dfrac{1}{\sqrt{x}}$.

   (d) $f(x) = (1 + x)^2$.

   (e) $f(x) = 5x + 5x^3$.

   (f) $f(x) = 10 + 5x^2 - \dfrac{1}{x}$.

2. From the definition of the derivative, determine the derivatives of the following functions.

   (a) $f(x) = \dfrac{1}{1 - x}$.

   (b) $f(x) = \dfrac{1}{(1 - x)^2}$.

   (c) $f(x) = \dfrac{5}{1 + x^2}$.

3. A freely falling object travels a distance $s(t) = ut + \frac{1}{2}at^2$ in time t, where u and a are constants. Calculate the velocity $ds(t)/dt$ at time t.

4. The size of an insect population at time t (measured in days) is given by $p(t) = 10{,}000 - 9000/(1 + t)$. Determine the initial population $p(0)$ and the growth rate $p'(t)$ at time t.

5. Use derivatives to determine approximate values for the following numbers.

   (a) $(5.03)^3$.   (b) $(15.8)^{1/2}$.   (c) $(1.99)^{-2}$.   (d) $(63)^{1/6}$.

6. From the definition of the derivative, prove that (a) the derivative of $af(x)$ is $af'(x)$ when a is a constant, and (b) the derivative of the sum of two functions is the sum of the derivatives.

7. Use the general binomial theorem (Section 1.6) to expand $(x + \Delta x)^n = x^n(1 + \Delta x/x)^n$. Prove that the derivative of $f(x) = x^n$ is $f'(x) = nx^{n-1}$, where n is any real number.

8. Yeast is growing in a sugar solution in such a way that the weight of the yeast increases by 3 per cent every hour. If the initial weight is 1 gram, the weight after t hours is $w(t) = (1.03)^t$. Determine approximate values for the weight after (a) 10 minutes and (b) 20 minutes.

9. Consider a spherical cell of radius r that is continuously growing in volume without changing shape. The volume is $\frac{4}{3}\pi r^3$. Estimate the change in the volume of the cell when the radius of the cell increases from 2.5×10^{-3} to 2.6×10^{-3} centimeters.

10. The size of a bacteria population at time t (measured in hours) is given by $p(t) = 10^6 + 10^4 t - 10^3 t^2$. Determine the growth rate when (a) $t = 1$ hour, (b) $t = 5$ hours, and (c) $t = 10$ hours.

11. On a warm summer evening, the temperature can be estimated by counting the number of chirps a cricket makes in 15 seconds and adding 40. This thermometer is reasonably accurate between 55 and 100°F. Calculate the rate of change of chirps (per 15 seconds) per Fahrenheit degree.

B2 The Rules of Differentiation

We must now develop methods that allow us to calculate efficiently the derivatives of the functions that occur in the many applications of calculus. For example, what is the derivative of the function $y = (x + x^2)(1 + x^3)$? To calculate this derivative, we first multiply the factors $x + x^2$ and $1 + x^3$ to obtain $y = x + x^2 + x^4 + x^5$. Since the derivative of a sum is the sum of the derivatives, the required derivative is $dy/dx = 1 + 2x + 4x^3 + 5x^4$.

Let us consider the general problem of differentiating the product of two differentiable functions. Consider the function $y = f(x) = u(x)v(x)$, where $u(x)$ and $v(x)$ are differentiable functions. From the definition of the derivative, we have

$$\frac{dy}{dx} = \lim_{\Delta x \to 0} \frac{f(x + \Delta x) - f(x)}{\Delta x} = \lim_{\Delta x \to 0} \frac{u(x + \Delta x)v(x + \Delta x) - u(x)v(x)}{\Delta x}.$$

From the identity

$$u(x + \Delta x)v(x + \Delta x) - u(x)v(x) = u(x + \Delta x)[v(x + \Delta x) - v(x)]$$
$$+ v(x)[u(x + \Delta x) - u(x)],$$

we can write the limit as the sum of two terms

$$\frac{dy}{dx} = \lim_{\Delta x \to 0} u(x + \Delta x)\left[\frac{v(x + \Delta x) - v(x)}{\Delta x}\right] + \lim_{\Delta x \to 0} v(x)\left[\frac{u(x + \Delta x) - u(x)}{\Delta x}\right]$$

$$= u(x)\frac{dv}{dx} + v(x)\frac{du}{dx}.$$

This is the *product rule of differentiation.*

$$\frac{d}{dx}(u(x)v(x)) = u(x)\frac{dv}{dx} + v(x)\frac{du}{dx}. \tag{B6}$$

In the example, $y = (x + x^2)(1 + x^3)$ and $u(x) = x + x^2$, $v(x) = 1 + x^3$. Therefore,

$$\frac{dy}{dx} = (x + x^2)\frac{d}{dx}(1 + x^3) + (1 + x^3)\frac{d}{dx}(x + x^2)$$

$$= (x + x^2)(3x^2) + (1 + x^3)(1 + 2x) = 1 + 2x + 4x^3 + 5x^4.$$

This agrees with the result obtained by first multiplying the two factors and then differentiating.

Example B2.1   Calculate the derivatives of the following functions.

1. $y = x(x^2 + 4x)$.   2. $y = (x^3 + x^5)(1 + x^2)$.

3. $y = \dfrac{x^2 + 1}{x}$.

Solution:

1. In this example, $u(x) = x$ and $v(x) = x^2 + 4x$. Therefore, by the product rule of differentiation,

$$\frac{dy}{dx} = x\frac{d}{dx}(x^2 + 4x) + (x^2 + 4x)\frac{dx}{dx} = x(2x + 4) + (x^2 + 4x)$$

$$= 3x^2 + 8x.$$

2. $\dfrac{dy}{dx} = (x^3 + x^5)\dfrac{d}{dx}(1 + x^2) + (1 + x^2)\dfrac{d}{dx}(x^3 + x^5)$

$$= (x^3 + x^5)(2x) + (1 + x^2)(3x^2 + 5x^4)$$

$$= 3x^2 + 10x^4 + 7x^6.$$

3. $\dfrac{dy}{dx} = \dfrac{1}{x}\dfrac{d}{dx}(x^2 + 1) + (x^2 + 1)\dfrac{d}{dx}\left(\dfrac{1}{x}\right) = \dfrac{2x}{x} - \dfrac{x^2 + 1}{x^2}$

$$= 1 - \frac{1}{x^2}.$$

A similar method can be used to calculate the derivative of the ratio of two differentiable functions of x. Suppose that $y = u(x)/v(x)$ is the ratio of the differentiable functions $u(x)$ and $v(x)$ and suppose that $v(x) \neq 0$. The derivative of y is

$$\frac{dy}{dx} = \lim_{\Delta x \to 0} \frac{1}{\Delta x}\left[\frac{u(x + \Delta x)}{v(x + \Delta x)} - \frac{u(x)}{v(x)}\right]$$

$$= \lim_{\Delta x \to 0} \frac{1}{\Delta x} \frac{u(x + \Delta x)v(x) - u(x)v(x + \Delta x)}{v(x + \Delta x)v(x)}$$

$$= \lim_{\Delta x \to 0} \frac{1}{v(x + \Delta x)v(x)}\left[v(x)\frac{u(x + \Delta x) - u(x)}{\Delta x} - u(x)\frac{v(x + \Delta x) - v(x)}{\Delta x}\right]$$

$$= \frac{1}{(v(x))^2}\left[v(x)\frac{du}{dx} - u(x)\frac{dv}{dx}\right].$$

This is the *quotient rule of differentiation*.

$$\frac{d}{dx}\left(\frac{u(x)}{v(x)}\right) = \frac{1}{(v(x))^2}\left[v(x)\frac{du}{dx} - u(x)\frac{dv}{dx}\right]. \tag{B7}$$

Example B2.2   Calculate the derivatives of the following functions.

1. $y = \dfrac{x + 1}{x + 2}$.   2. $y = \dfrac{1}{1 + x^2}$.   3. $y = \dfrac{x + x^3}{1 + x^3}$.

Solution:

1. In this example,

$$\frac{dy}{dx} = \frac{1}{(x + 2)^2}\left[(x + 2)\frac{d}{dx}(x + 1) - (x + 1)\frac{d}{dx}(x + 2)\right].$$

Simplifying,

$$\frac{dy}{dx} = \frac{1}{(x + 2)^2}[(x + 2) - (x + 1)] = \frac{1}{(x + 2)^2}.$$

2. $\dfrac{dy}{dx} = \dfrac{1}{(1 + x^2)^2}\left[(1 + x^2)\dfrac{d}{dx}(1) - 1\dfrac{d}{dx}(1 + x^2)\right] = \dfrac{-2x}{(1 + x^2)^2}$

3. $\dfrac{dy}{dx} = \dfrac{1}{(1 + x^3)^2}[(1 + x^3)(1 + 3x^2) - (x + x^3)(3x^2)]$

$$= \frac{1 + 3x^2 - 2x^3}{(1 + x^3)^2}.$$

How do we differentiate the function $y = (x + x^2)^{1/2}$? Of course, we can always return to the definition of the derivative as a limit, but there is a much easier method. If we define a new function $u = x + x^2$, then we observe that $y = u^{1/2}$ and the derivative of y with respect to u is $dy/du = \frac{1}{2}u^{-1/2}$. The derivative of u with respect to x is $du/dx = 1 + 2x$. Finally, the derivative of y with respect to x is

$$\frac{dy}{dx} = \frac{dy}{du}\frac{du}{dx} = \frac{1}{2}u^{-1/2}(1 + 2x) = \frac{1 + 2x}{2(x + x^2)^{1/2}}.$$

More generally, suppose that $y = f(u)$ and $u = g(x)$, where $f(u)$ and $g(x)$ are differentiable functions. Then $y = f(g(x))$ and the derivative of y with respect to x is

$$\frac{dy}{dx} = \frac{dy}{du}\frac{du}{dx} = f'(g(x))g'(x). \tag{B8}$$

This is the *chain rule of differentiation.* It expresses a very simple idea. For example, if y is growing three times as fast as u and u is growing three times as fast as x, then y must be growing nine times as fast as x.

Example B2.3   Use the chain rule to calculate the derivatives of the following functions.

1. $y = (1 + x^2)^6$.

2. $y = (x + x^3)^{-1}$.

3. $y = \left(x - \dfrac{1}{x}\right)^4$.

4. $y = (1 + 2x)^{1/3}$.

Solution:

1. If we define $u = (1 + x^2)$, then $y = u^6$ and the derivative of y with respect to x is

$$\frac{dy}{dx} = \frac{dy}{du}\frac{du}{dx} = 6u^5(2x) = 12x(1 + x^2)^5.$$

2. Define $u = x + x^3$. Then, $y = u^{-1}$ and

$$\frac{dy}{dx} = \frac{dy}{du}\frac{du}{dx} = \frac{-1}{u^2}(1 + 3x^2) = \frac{-(1 + 3x^2)}{(x + x^3)^2}.$$

3. If $u = x - 1/x$, then $y = u^4$ and

$$\frac{dy}{dx} = \frac{dy}{du}\frac{du}{dx} = 4u^3\left(1 + \frac{1}{x^2}\right) = 4\left(x - \frac{1}{x}\right)^3\left(1 + \frac{1}{x^2}\right).$$

4. Define $y = u^{1/3}$ with $u = 1 + 2x$. Then

$$\frac{dy}{dx} = \frac{1}{3}u^{-2/3}\frac{du}{dx} = \frac{2}{3}(1 + 2x)^{-2/3}.$$

The product rule, the quotient rule, and the chain rule greatly expand the number of functions whose derivatives can be calculated without too much work. In all the examples we have considered, y has been given as an *explicit function* of x, $y = f(x)$. However, we do encounter problems in which y is an *implicit function* of x, where it may not be possible to determine y as an explicit function. As an example, the equation $1/(1 + x) + \sqrt{1 + y^2} = 5(y/x)^{1/3}$

defines y as an implicit function of x, and it is not possible to solve for y as an explicit function of x. The problem of calculating the derivative dy/dx is more difficult but can be solved by the method of *implicit differentiation*.

To illustrate this method, suppose that y is defined as a function of x by the equation $x^2 + y^2 = 1$. To determine dy/dx, we differentiate this equation with respect to x. By the chain rule,

$$\frac{d}{dx}y^2 = \frac{d}{dy}y^2\frac{dy}{dx} = 2y\frac{dy}{dx}.$$

Using this result, we obtain

$$\frac{d}{dx}(x^2 + y^2) = 2x + 2y\frac{dy}{dx} = \frac{d}{dx}(1) = 0.$$

We conclude that

$$\frac{dy}{dx} = \frac{-x}{y} \qquad \text{if } x^2 + y^2 = 1.$$

The method of implicit differentiation is illustrated by the following example and by problems at the end of this section.

Example B2.4 Use the method of implicit differentiation to calculate dy/dx in the following equations.
  1. $xy = 1$.   2. $x^2y^5 + y = 1 + x$.   3. $x\sqrt{1 + y} + y^3 = 2 + x^2$.

Solution:
  1. Using the product rule to differentiate the left side, we have

$$\frac{d}{dx}(xy) = x\frac{dy}{dx} + y = \frac{d}{dx}(1) = 0.$$

  Therefore

$$\frac{dy}{dx} = -\frac{y}{x}.$$

  2. $\frac{d}{dx}(x^2y^5) + \frac{dy}{dx} = x^2\frac{d}{dx}y^5 + y^5(2x) + \frac{dy}{dx} = 5x^2y^4\frac{dy}{dx} + 2xy^5 + \frac{dy}{dx}.$

Therefore,

$$(5x^2y^4 + 1)\frac{dy}{dx} + 2xy^5 = \frac{d}{dx}(1 + x) = 1.$$

We conclude that

$$\frac{dy}{dx} = \frac{1 - 2xy^5}{1 + 5x^2y^4}.$$

3.

$$\frac{d}{dx}(x\sqrt{1 + y}) = x\frac{d}{dx}(\sqrt{1 + y}) + \sqrt{1 + y}$$

$$= \frac{x}{2\sqrt{1 + y}}\frac{dy}{dx} + \sqrt{1 + y}.$$

Differentiating both sides of the equation yields

$$\frac{x}{2\sqrt{1 + y}}\frac{dy}{dx} + \sqrt{1 + y} + 3y^2\frac{dy}{dx} = 2x,$$

and we can then solve to obtain

$$\frac{dy}{dx} = \frac{2x - \sqrt{1 + y}}{\dfrac{x}{2\sqrt{1 + y}} + 3y^2}.$$

Finally, in this section, we define the *higher derivatives* of a function of x. The derivative $dy/dx = f'(x)$ of a function $y = f(x)$ is a function of x whose derivative is defined whenever $\lim_{\Delta x \to 0} [f(x + \Delta x) - f(x)]/\Delta x$ is defined. The derivative of $f'(x)$ is called the *second derivative* of $f(x)$. The two standard notations for the second derivative of $y = f(x)$ are d^2y/dx^2 and $f''(x)$. The *third derivative* of $y = f(x)$ is defined to be the derivative of $f''(x)$. The third derivative is written d^3y/dx^3 or $f'''(x)$. The *n*th *derivative*, written d^ny/dx^n or $f^{(n)}(x)$, is obtained by differentiating $y = f(x)$ n times.

Example B2.5   Calculate the second and third derivatives of the following functions.

1. $y = x^4$.   2. $y = x^2$.   3. $y = \dfrac{1}{x + 1}$.   4. $y = x^{1/2}$.

Solution:

1. $\dfrac{dy}{dx} = 4x^3, \quad \dfrac{d^2y}{dx^2} = \dfrac{d}{dx}(4x^3) = 12x^2, \quad \dfrac{d^3y}{dx^3} = 24x.$

2. $\dfrac{dy}{dx} = 2x, \quad \dfrac{d^2y}{dx^2} = 2, \quad \dfrac{d^3y}{dx^3} = 0.$

3. $\dfrac{dy}{dx} = \dfrac{-1}{(x+1)^2}, \quad \dfrac{d^2y}{dx^2} = \dfrac{2}{(x+1)^3}, \quad \dfrac{d^3y}{dx^3} = \dfrac{-6}{(x+1)^4}.$

4. $\dfrac{dy}{dx} = \dfrac{1}{2}x^{-1/2}, \quad \dfrac{d^2y}{dx^2} = -\dfrac{1}{4}x^{-3/2}, \quad \dfrac{d^3y}{dx^3} = \dfrac{3}{8}x^{-5/2}.$

Problems B2

1. Use the product rule of differentiation to calculate the derivatives of the following functions.
   (a) $(x^2 + 3)(x + 5)$.
   (b) $(x^2 + 2x + 1)(x + 1)$.
   (c) $(1 + x^2)(1 + x^4)$.
   (d) $(x^{1/2} + x^{-1/2})(x^{3/2} + x^{-3/2})$.
   (e) $x \sin x$.
   (f) $(x^2 + 1) \sin x$.

2. Use the quotient rule of differentiation to calculate the derivatives of the following functions.
   (a) $\dfrac{1 + x^2}{1 - x^2}$.
   (b) $\dfrac{1}{1 + x + x^2}$.
   (c) $\dfrac{1 + 2x}{1 - 5x}$.

   (d) $\dfrac{\sin x}{1 + \sin x}$.
   (e) $\dfrac{\sin x + \cos x}{1 + x^2}$.
   (f) $\dfrac{\sin x}{\cos x}$.

3. Use the chain rule of differentiation to calculate the derivatives of the following functions.
   (a) $(2x + 3)^7$.
   (b) $(x^2 + x^3 + x^4)^{3/2}$.
   (c) $\sin(\sqrt{x})$.
   (d) $\sin(1 + x^2)$.
   (e) $\cos(1 + x + x^2)$.
   (f) $(1 + x^{1/2})^{1/2}$.

4. Given that $d(\sin x)/dx = \cos x$, differentiate implicitly the equation $\sin^2 x + \cos^2 x = 1$ to prove that $d(\cos x)/dx = -\sin x$.

5. *Derivatives of the Trigonometric Functions.* Given that $d(\sin x)/dx = \cos x$ and $d(\cos x)/dx = -\sin x$, prove the following results.

   (a) $\dfrac{d}{dx}(\tan x) = \dfrac{1}{\cos^2 x} = \sec^2 x.$

   (b) $\dfrac{d}{dx}(\cot x) = \dfrac{-1}{\sin^2 x} = -\csc^2 x.$

(c) $\dfrac{d}{dx}(\sec x) = \dfrac{\sin x}{\cos^2 x} = \tan x \sec x,$

(d) $\dfrac{d}{dx}(\csc x) = \dfrac{-\cos x}{\sin^2 x} = -\cot x \csc x.$

6. *Derivatives of the Inverse Trigonometric Functions.* To differentiate the inverse sine function $y = \sin^{-1} x$, write $x = \sin y$ and use implicit differentiation to find

$$1 = \cos y \frac{dy}{dx} \quad \text{and} \quad \frac{dy}{dx} = \frac{1}{\cos y} = \frac{1}{\sqrt{1 - \sin^2 y}} = \frac{1}{\sqrt{1 - x^2}}.$$

Prove the following results.

(a) $\dfrac{d}{dx}(\sin^{-1} x) = \dfrac{1}{\sqrt{1 - x^2}}.$

(b) $\dfrac{d}{dx}(\cos^{-1} x) = \dfrac{-1}{\sqrt{1 - x^2}}.$

(c) $\dfrac{d}{dx}(\tan^{-1} x) = \dfrac{1}{1 + x^2}.$

7. Calculate the derivatives of the following functions.
   (a) $(\sin x)^3.$ (b) $(\sin x + \cos x)^2.$ (c) $\sin x(1 + \tan x).$
   (d) $2 \sin 3x + 4 \cos 3x.$ (e) $x \sin x.$ (f) $x^2 \cos 2x.$

8. Calculate the derivatives of the following functions.
   (a) $\sin^{-1} 2x.$ (b) $\sin^{-1} x^2.$ (c) $(\cos^{-1} x)^2.$
   (d) $\cos (\sin^{-1} x).$ (e) $\sin (\cos^{-1} x).$ (f) $\tan^{-1} 4x.$

B3 Maxima, Minima, and Curve Plotting

One very useful way to study the properties of a function is to draw a graph of the function. A typical function $y = f(x)$ is graphed in Figure B3. Consider the three points $(x_0, f(x_0))$, $(x_1, f(x_1))$, and $(x_2, f(x_2))$ on the curve. At $(x_0, f(x_0))$, the slope of the tangent line to the curve $y = f(x)$ is positive. If x increases from x_0 by a small amount, then $y = f(x)$ also increases. We conclude that $f(x)$ is an *increasing function in any interval in which its derivative is positive.* At $(x_2, f(x_2))$, the slope of the tangent line is negative. If x increases from x_2 by a small amount, then $y = f(x)$ decreases. Therefore, $f(x)$ is a *decreasing function in any interval in which its derivative is negative.*

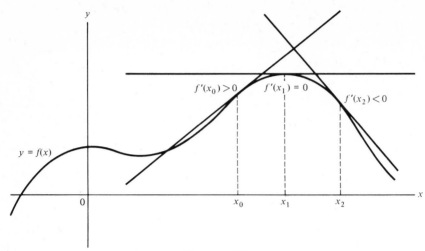

Figure B3

As x increases from x_0 to x_2, the derivative $f'(x)$ goes from positive to negative values. If the function $f'(x)$ is continuous, there must be a value of x between x_0 and x_2 where $f'(x) = 0$. In Figure B3, this is the point $(x_1, f(x_1))$, where $f'(x_1) = 0$. Such a point is called a *critical point*.

Example B3.1   For the following functions, determine the intervals in which the function is increasing and where it is decreasing. Determine all critical points.

   1. $f(x) = 3x$.   2. $f(x) = x^2$.   3. $f(x) = x^3$.

Solution:

   1. If $f(x) = 3x$, $f'(x) = 3 > 0$ and $f(x)$ is an increasing function for all x. [See Figure B4(1).] There are no critical points.

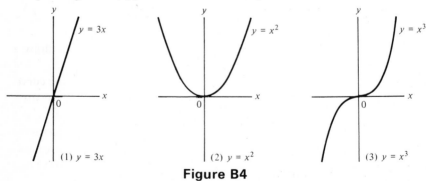

Figure B4

2. The derivative $f'(x) = 2x$ is negative for $x < 0$ and positive for $x > 0$. Therefore, $f(x)$ is increasing for $x > 0$ and decreasing for $x < 0$. The point $(0, 0)$ is the only critical point. The properties of $f(x)$ are obvious from Figure B4(2).

3. $f'(x) = 3x^2$. The slope is positive for all x except $x = 0$. The point $(0, 0)$ is the only critical point. The function $y = x^3$ is an increasing function of x in every interval. [See Figure B4(3).]

We have seen that a great deal of information about the curve $y = f(x)$ is contained in the derivative $dy/dx = f'(x)$. The second derivative d^2y/dx^2 is also very helpful in studying the curve $y = f(x)$. In Figure B3, the curve $y = f(x)$ has a *local maximum* at the point $(x_1, f(x_1))$. This means that, near $x = x_1$, the largest value of $f(x)$ is $f(x_1)$. If x is close to x_1, then $f(x) < f(x_1)$. The reason that this maximum is called a local maximum is that $f(x)$ may have larger values for x not close to x_1.

In Figure B3, as x increases from x_0 to x_2 along the curve, the slope $f'(x)$ decreases from positive to negative values. This is the condition that must be satisfied in order to have a local maximum between x_0 and x_2. We have a *local minimum* between the points x_0 and x_2 if the slope $f'(x)$ increases from negative to positive values as x increases from x_0 to x_2; that is, if $f'(x)$ is increasing near the point x_1. This occurs if $f''(x)$, the derivative of $f'(x)$, is positive at the point x_1. Therefore, the point $(x_1, f(x_1))$ is a local minimum if $f'(x_1) = 0$ and $f''(x_1) > 0$ [since this implies that $f'(x)$ is an increasing function near $x = x_1$]. The point $(x_1, f(x_1))$ is a local maximum if $f'(x_1) = 0$ and $f''(x_1) < 0$ [since this implies that the slope $f'(x)$ is a decreasing function near $x = x_1$].

Example B3.2   Find the local maxima and minima of the curve $y = f(x) = x^4 - 8x^3 + 22x^2 - 24x + 17$.

Solution: Differentiating, $f'(x) = 4x^3 - 24x^2 + 44x - 24 = 4(x - 1) \times (x - 2)(x - 3)$ and $f''(x) = 12x^2 - 48x + 44 = 4(3x^2 - 12x + 11)$. The critical points occur when $x = 1$, $x = 2$, and $x = 3$. At $x = 1$, $f''(1) = 8 > 0$, so that the point $(1, 8)$ is a local minimum of the curve $y = f(x)$. At $x = 2$, $f''(2) = -4 < 0$, so that there is a local maximum at the point $(2, 9)$. At $x = 3$, $f''(3) = 8 > 0$, which indicates that $y = f(x)$ has a local minimum at the point $(3, 8)$. This information can be used to simplify the problem of graphing the curve $y = f(x)$ (Figure B5).

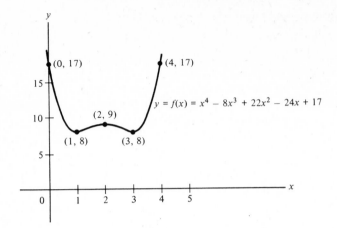

$y = f(x) = x^4 - 8x^3 + 22x^2 - 24x + 17$

Figure B5

Example B3.3   The reaction of a person to a drug can be increased blood pressure, decreased body temperature, changed pulse, or other physiological change. The strength of the reaction depends on the amount of the drug administered. Suppose that x is the amount of the drug administered and that the reaction is $y = f(x) = x^2(a - x)$, where a is a positive constant. For what value of x is the reaction a maximum?

Solution: To determine the maximum, we calculate $f'(x) = 2ax - 3x^2$ and $f''(x) = 2a - 6x$. The critical points are $x = 0$ and $x = 2a/3$. When $x = 2a/3$, $f''(2a/3) = 2a - 6(2a/3) = -2a < 0$. Therefore, $x = 2a/3$ is the dose level that produces the maximum reaction.

Example B3.4   A population of 1000 bacteria is introduced to a nutrient medium. The population grows according to the equation $p(t) = 1000 + 1000t/(100 + t^2)$, where t is measured in hours. Determine the maximum size of this population.

Solution: To determine the maximum, we calculate

$$p'(t) = \frac{(100 + t^2)1000 - (1000t)(2t)}{(100 + t^2)^2} = \frac{1000(100 - t^2)}{(100 + t^2)^2}.$$

The critical points occur at $t^2 = 100$ or $t = \pm 10$. The point $t = +10$ is a maximum. This can be verified by showing that $p''(10) < 0$. The maximum size of the population is $p(10) = 1000 + (10{,}000/200) = 1050$, reached after 10 hours of growth.

We have defined the critical points of the function $y = f(x)$ to be the points where $dy/dx = f'(x) = 0$. The *inflection points* of $y = f(x)$ are defined to be the points where $d^2y/dx^2 = f''(x) = 0$ and the second derivative changes sign. Determining the inflection points can be very useful in graphing the function. The slope of the curve reaches a local maximum (or minimum) at inflection points.

Example B3.5   Determine the inflection points of the curve $y = 1/(1 + x^2)$.

Solution: Differentiating,

$$\frac{dy}{dx} = \frac{-2x}{(1 + x^2)^2} \quad \text{and} \quad \frac{d^2y}{dx^2} = \frac{6x^2 - 2}{(1 + x^2)^3}.$$

The only critical point occurs at $x = 0$, $y = 1$. The inflection points are determined by the equation $d^2y/dx^2 = (6x^2 - 2)/(1 + x^2)^3 = 0$. The two inflection points are $(1/\sqrt{3}, 3/4)$ and $(-1/\sqrt{3}, 3/4)$. Figure B6 gives the graph of $1/(1 + x^2)$.

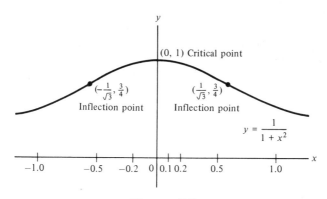

Figure B6

Problems B3

1. For the following functions $y = f(x)$, determine $f'(x)$ and $f''(x)$. Determine the intervals on which $f(x)$ is an increasing function. Determine the critical points and the inflection points. Graph the curves of these functions.

(a) $y = f(x) = 3x^4$.   (b) $y = 4x^2 + 4x + 1$.   (c) $y = x(x + 1)(x + 2)$.

(d) $y = \dfrac{4}{4 + x^2}$.   (e) $y = 1 + x^{3/2} + x^{5/2}$.   (f) $y = \dfrac{1 + x}{1 + x^2}$.

2. Determine the local maxima and minima of the following functions.

(a) $y = 1 + 2x^2 - 3x^4$.   (b) $y = (x - 1)(x + 1)(x - 2)$.

(c) $y = \dfrac{1}{1 + 2x^2 + x^4}$.   (d) $y = \dfrac{\sin x}{2 - \sin x}$.

(e) $y = \dfrac{1}{1 + \sin^2 x}$.   (f) $y = (x + 1) \sin x$.

3. The growth rate y of a population x is given by $y = (1/1,000)x(100 - x)$ when time is measured in days. For what population size is this growth rate a maximum? What are the stable populations, that is, the populations for which the growth rate is zero?

B4 Exponentials and Logarithms

Two functions of fundamental importance in the applications of calculus are the exponential function $f(x) = a^x$ and the logarithm function $g(x) = \log_a x$ defined for any positive real number a. If x is a positive integer m, then $f(m) = a^m = a \cdot a \cdot \ldots \cdot a$ (m times). If $x = 1/n$, where n is a positive integer, we define $f(1/n) = a^{1/n} = n$th root of a. If $x = m/n$, where m and n are positive integers, we define $f(m/n) = a^{m/n} = (a^{1/n})^m =$ the mth power of the nth root of a. If $x = -m/n$, where m and n are positive integers, then $f(-m/n) = a^{-m/n} = 1/a^{m/n}$.

This defines the exponential function $f(x) = a^x$ for all rational numbers x. How do we define $a^{\sqrt{2}}$ or a^π? The obvious way to do this is to recall that $\sqrt{2} = 1.4142\ldots$ and $\pi = 3.1416\ldots$. Then, $\sqrt{2}$ can be approximated by the rational number $14,142/10,000$ and we can calculate this power of a. The value of $a^{\sqrt{2}}$ is defined as the limit of the numbers a^1, $a^{14/10}$, $a^{141/100}$, $a^{1414/1000}$, $\ldots$. This can be done for every irrational number. To summarize the properties of the exponential function a^x, we have, for any pair of real numbers x and y:

1. $a^x > 0$.   2. $a^{-x} = 1/a^x$.   3. $a^{x+y} = a^x a^y$.   4. $a^{x-y} = a^x a^{-y}$.
5. $a^0 = a^{x-x} = a^x/a^x = 1$.   6. $(a^x)^y = a^{xy}$.
7. If $a > 1$, a^x is an increasing function of x for all x.
8. If $0 < a < 1$, a^x is a decreasing function of x for all x.

Figure B7 gives the graph of $f(x) = 2^x$.

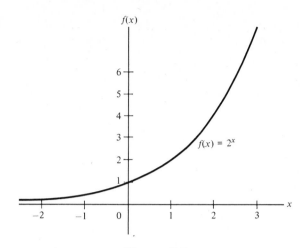

Figure B7

Example B4.1   Calculate (1) $9^{3/2}$, (2) $4^{-5/2}$, (3) $8^{2/3}$.

Solution:

1. $9^{3/2} = (9^{1/2})^3 = 3^3 = 27.$
2. $4^{-5/2} = (4^{1/2})^{-5} = 2^{-5} = \frac{1}{32}.$
3. $8^{2/3} = (8^{1/3})^2 = 2^2 = 4.$

If $y = a^x$, then we may ask a question: Which value of x corresponds to a given y? The solution of this problem is written $x = \log_a y$; that is, x *is the*

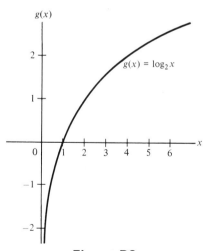

Figure B8

logarithm of y to the base a. By definition, $y = a^{\log_a y}$. The function $g(y) = \log_a y$ is defined only for $y > 0$. The function $g(x) = \log_2 x$ is graphed in Figure B8.

The properties of the logarithm follow from the properties of the exponential function. To summarize these properties, suppose that $a > 0$ $(a \neq 1)$ and that x and y are any positive numbers.

1. $\log_a xy = \log_a x + \log_a y$.
2. $\log_a (x/y) = \log_a x - \log_a y$.
3. $\log_a 1 = 0$ and $\log_a a = 1$.
4. $\log_a x^y = y \log_a x$.
5. If $a > 1$, $\log_a x$ is an increasing function of x for $x > 0$.
6. If $0 < a < 1$, $\log_a x$ is a decreasing function of x for $x > 0$.

To prove the first identity, for example, we know that $a^{\log_a xy} = xy = a^{\log_a x}a^{\log_a y} = a^{\log_a x + \log_a y}$. This implies that $\log_a xy = \log_a x + \log_a y$. To prove 4, we have $a^{\log_a x^y} = x^y = (a^{\log_a x})^y = a^{y\log_a x}$.

The number a in the definition of $\log_a x$ is called the *base* of the logarithms. For ordinary calculations, the most convenient base is $a = 10$. Logarithms to the base 10 are called *common logarithms*. The common logarithms have the great advantage that, if $\log_{10} x$ is known for x between 1 and 10, then it is very easy to determine $\log_{10} x$ for any x. As an example, $\log_{10} 137 = \log_{10} (1.37)(100) = 2 + \log_{10} 1.37$ since $\log_{10} 100 = 2$. Similarly, $\log_{10} (.137) = -1 + \log_{10} (1.37)$.

For reasons that will soon be apparent, the most convenient base for logarithms in calculus is a certain irrational number $e \approx 2.71828183\ldots$ defined as $e = \lim_{n \to \infty} (1 + 1/n)^n$. To study this limit, let us calculate $(1 + 1/n)^n$ for some increasing values of n. As n increases, $(1 + 1/n)^n$ approaches a limiting number defined as e.

n	1	2	3	10	1000	10,000
$(1 + 1/n)^n$	2	2.25	$2.3704\ldots$	$2.5937\ldots$	$2.7171\ldots$	$2.7182\ldots$

To see why this number e plays such an important role, let us calculate the derivative of $y = \log_a x$. For this function,

$$\frac{dy}{dx} = \lim_{\Delta x \to 0} \frac{\log_a (x + \Delta x) - \log_a x}{\Delta x} = \lim_{\Delta x \to 0} \frac{1}{\Delta x} \log_a \left(\frac{x + \Delta x}{x}\right).$$

But

$$\frac{1}{\Delta x} \log_a \left(\frac{x + \Delta x}{x} \right) = \frac{x}{x \, \Delta x} \log_a \left(1 + \frac{\Delta x}{x} \right) = \frac{1}{x} \log_a \left(1 + \frac{\Delta x}{x} \right)^{x/\Delta x}.$$

(We have multiplied and divided by x.) Define $u = x/\Delta x$. Then, as Δx tends to zero, u tends to infinity. We have proved

$$\frac{d}{dx}(\log_a x) = \frac{1}{x} \lim_{u \to \infty} \log_a \left(1 + \frac{1}{u} \right)^u.$$

But we have defined $e = \lim_{u \to \infty} (1 + 1/u)^u$, and we have the result

$$\frac{d}{dx}(\log_a x) = \frac{1}{x} \log_a e. \tag{B9}$$

This formula is particularly simple if $a = e$. In this case,

$$\frac{d}{dx}(\log_e x) = \frac{1}{x}, \tag{B10}$$

since $\log_e e = 1$. For this reason, logarithms to the base e are called *natural logarithms*, and they will be used whenever logarithms occur in calculus applications.

To differentiate the exponential function $y = a^x$, we write $x = \log_a y$ and differentiate this implicit equation for y as a function of x. Therefore,

$$\frac{dx}{dx} = \frac{d}{dy}(\log_a y) \frac{dy}{dx} \quad \text{or} \quad 1 = \frac{1}{y}(\log_a e) \frac{dy}{dx} \quad \text{and} \quad \frac{dy}{dx} = \frac{y}{\log_a e}.$$

We conclude that

$$\frac{d}{dx}(a^x) = \frac{a^x}{\log_a e}. \tag{B11}$$

In the special case of natural logarithms ($a = e$),

$$\frac{d}{dx}(e^x) = e^x. \tag{B12}$$

We have proved that the function e^x is equal to its own derivative.

   To conclude this section, let us calculate the derivatives of the functions $y = \log_e f(x)$ and $y = e^{g(x)}$, where $f(x)$ and $g(x)$ are differentiable functions and $f(x) > 0$. If we set $u = f(x)$ and $v = g(x)$, then, using the chain rule,

$$\frac{dy}{dx} = \frac{dy}{du}\frac{du}{dx} \quad \text{and} \quad \frac{dy}{dx} = \frac{dy}{dv}\frac{dv}{dx},$$

respectively. We conclude that

$$\frac{d}{dx}(\log_e f(x)) = \frac{1}{f(x)}f'(x) \tag{B13}$$

$$\frac{d}{dx}(e^{g(x)}) = e^{g(x)}g'(x). \tag{B14}$$

These formulas for the derivatives are illustrated in the following example.

Example B4.2   Calculate the derivatives of the following functions.

1. e^{-5x}.   2. e^{-x^2}.   3. $\log_e(1 + x^2)$.   4. $\log_e(1 + e^x)$.

Solution:

1. $\dfrac{d}{dx}(e^{-5x}) = e^{-5x}\dfrac{d}{dx}(-5x) = -5e^{-5x}$.

2. $\dfrac{d}{dx}(e^{-x^2}) = e^{-x^2}\dfrac{d}{dx}(-x^2) = -2xe^{-x^2}$.

3. $\dfrac{d}{dx}\log_e(1 + x^2) = \dfrac{1}{1 + x^2}\dfrac{d}{dx}(1 + x^2) = \dfrac{2x}{1 + x^2}$.

4. $\dfrac{d}{dx}(\log_e(1 + e^x)) = \dfrac{1}{1 + e^x}\dfrac{d}{dx}(1 + e^x) = \dfrac{e^x}{1 + e^x}$.

Example B4.3   Determine the maximum and minimum values of the function $f(x) = xe^{-x^2}$ defined for $-\infty < x < \infty$.

Solution:

$$f'(x) = \frac{d}{dx}(xe^{-x^2}) = e^{-x^2} + x\frac{d}{dx}e^{-x^2}$$

$$= e^{-x^2} + x(-2xe^{-x^2}).$$

Therefore, $f'(x) = (1 - 2x^2)e^{-x^2}$. The critical points occur at $x = 1/\sqrt{2}$ and $x = -1/\sqrt{2}$. The function $f(x)$ has a maximum value of $(1/\sqrt{2})e^{-1/2}$, when $x = 1/\sqrt{2}$, and a minimum value of $(-1/\sqrt{2})e^{-1/2}$, when $x = -1/\sqrt{2}$.

Problems B4

1. Calculate the derivatives of the following functions.
   (a) $y = xe^x$.   (b) $x^2 e^{-x}$.   (c) $e^x \sin x$.
   (d) $e^{3x} \cos 3x$.   (e) $e^{-3x} \sin 2x$.   (f) $x^2 e^{-x^2}$.

2. Determine the critical points and inflection points of the following functions and plot their curves.
   (a) $f(x) = e^{-x^2/2}$.   (b) $f(x) = xe^{-x^2}$.

3. An antibacterial agent added to a bacteria population causes the population to decrease. If the population t minutes after the addition of the agent is $p(t) = p(0)2^{-t/3}$, determine the rate of change of the population at time t. If the initial population $p(0) = 10^6$, after what period of time has the population declined to 10^3 individuals?

4. Calculate the derivatives of the following functions.
   (a) $y = 2^x$.   (b) $y = (1.05)^x$.   (c) $y = x^2 3^x$.

5. Yeast is growing in a sugar solution in such a way that the weight of the yeast increases by 3 per cent every hour. If the initial weight is 1 gram, the weight after t hours is $w(t) = (1.03)^t$. Determine the rate of change of $w(t)$ at (a) $t = 1$ hour, (b) $t = 2$ hours, (c) $t = 5$ hours.

6. For a patient undergoing glucose infusion, the amount of glucose in his bloodstream at time t (measured in hours) is $C(t) = 10 - 8e^{-t}$ when measured in appropriate units. Plot $C(t)$ as a function of time for $t \geq 0$. Determine $\lim_{t \to \infty} C(t)$, the equilibrium amount of glucose in the bloodstream.

7. Calculate the derivatives of the following functions.
   (a) $y = e^{(x+1/x)}$.   (b) $y = e^{\sin x}$.   (c) $y = e^{e^x}$.

8. A bacteria population grows from an initial population of 1000 to a population $p(t)$ at time t (measured in days) according to the equation

$$p(t) = \frac{1000e^t}{1 + \frac{1}{10}(e^t - 1)}.$$

Determine the growth rate $p'(t)$. When is the growth rate a maximum? Determine $\lim_{t \to \infty} p(t)$, the equilibrium population.

9. The reactions as a function of time (measured in hours) to two drugs are $r_1(t) = te^{-t}$ and $r_2(t) = t^2 e^{-t}$. Which drug has the larger maximum reaction? Which drug is slower in its action?

10. Newton's law of cooling states that the temperature difference between an object and its surrounding medium decreases exponentially with time. If $T(t)$ is the temperature of the object at time t and T_m is the temperature of the surrounding medium, then $T(t) - T_m = ke^{-at}$, where k and a are constants.
    (a) Evaluate k in terms of $T(0)$, the initial temperature of the object.
    (b) If $T(0) = 10$, $T_m = 5$ and $a = 1$, plot a graph of the function $T(t)$.

11. X-radiation has been used to destroy viruses on tobacco plants. As the radiation dosage is increased, the number of surviving viruses decreases exponentially. If $p(R)$ is the proportion of viruses that survive the radiation dosage R (measured in roentgens), then $p(R) = e^{-aR}$, where a is a constant characteristic of the virus.
    (a) In terms of a, determine the roentgen dosage necessary to reduce the number of viruses by 50 per cent.
    (b) What is the roentgen dosage necessary to kill 90 per cent of the viruses?

12. A radioactive tracer drug loses half its radioactivity in 2 days. Assuming an exponential decay of the radioactivity, what proportion is lost after 1 day? After 4 days? After 8 days?

13. A drug is injected into the bloodstream at time $t = 0$. The concentration of this drug in the bloodstream at time t is described by $x(t) = c(e^{-at} - e^{-bt})$, where a, b, and c are positive constants with $a < b$.
    (a) Prove that $x(0) = 0$ and that $x(t) > 0$ for $t > 0$.
    (b) What is the maximum value of $x(t)$ and when does it occur?
    (c) Draw a graph of the concentration $x(t)$ as a function of time when $a = 1, b = 2$, and $c = 1$.

14. The oxygen consumption of yearling salmon increases exponentially with speed of swimming. Define $C(v)$ to be the oxygen consumption per hour of a yearling salmon swimming at the speed v feet per second. If $C(0) = 100$ and $C(3) = 800$ (measured in appropriate units), determine $C(1)$ and $C(2)$.

15. The intensity of light passing through a liquid decreases exponentially with distance in the liquid. This means that, if $I(x)$ is the intensity of light x feet from the surface of the liquid, then $I(x) = I(0)e^{-ax}$ where a is a positive constant characteristic of the liquid. Suppose that $a = 0.5$ in a certain body of water and that a certain water plant requires light of

intensity at least one tenth of the surface intensity, $I(0)$. What is the maximum depth of water in which this plant can grow?

B5 Partial Differentiation

In the previous sections, we have studied the derivatives of functions $y = f(x)$ of one variable x. However, in many applications, we are interested in functions of several variables, say $y = f(x, t)$ or $y = f(x, t, u, v)$. For example, the volume of a cube is the product of its length, height, and width. The reproduction rate of a bacteria population is a function of the food supply and the temperature. The increase in heart rate that results from the injection of a drug is a function of the amount of the drug injected, the body weight of the patient, the time that has passed since the injection, and other factors.

For simplicity, we will study a function $y = f(x, t)$ of two variables x and t. For this function, we can define two first derivatives. We may be interested in the changes in y produced by changes in x when t is held constant. Similarly, we can consider the changes in y produced by changes in t when x is held constant. This idea leads to the following definition.

Definition B5.1   Partial Derivatives   *If $y = f(x, t)$, the partial derivative of y with respect to x is*

$$\frac{\partial y}{\partial x} = \lim_{\Delta x \to 0} \frac{f(x + \Delta x, t) - f(x, t)}{\Delta x}.$$

The partial derivative of y with respect to t is

$$\frac{\partial y}{\partial t} = \lim_{\Delta t \to 0} \frac{f(x, t + \Delta t) - f(x, t)}{\Delta t}.$$

Calculating partial derivatives is not any more difficult than calculating ordinary derivatives. If $y = (x + t)e^{-t}$, then, to calculate $\partial y/\partial x$, we simply think of t as a constant and, to calculate $\partial y/\partial t$, we think of x as a constant. Therefore $\partial y/\partial x = e^{-t}$ and

$$\frac{\partial y}{\partial t} = \frac{\partial}{\partial t}(xe^{-t}) + \frac{\partial}{\partial t}(te^{-t}) = -xe^{-t} + e^{-t} - te^{-t} = e^{-t} - (x + t)e^{-t}.$$

Example B5.1   Calculate the partial derivatives $\partial y/\partial x$ and $\partial y/\partial t$ of the following functions.

(a) $y = x^2 + t^2$.   (b) $y = e^{-xt}$.   (c) $y = x^2 t^3 + t^4$.   (d) $y = \sin (x - t)$.

Solution:
(a) $\partial y/\partial x = 2x$, $\partial y/\partial t = 2t$.
(b) $\partial y/\partial x = -te^{-xt}$, $\partial y/\partial t = -xe^{-xt}$.
(c) $\partial y/\partial x = 2xt^3$, $\partial y/\partial t = 3x^2 t^2 + 4t^3$.
(d) $\partial y/\partial x = \cos (x - t)$, $\partial y/\partial t = -\cos (x - t)$.

Example B5.2   The reaction to the injection of x units of a drug is $y = x^2(a - x)te^{-t}$, when measured t hours after the injection. Calculate $\partial y/\partial x$ and $\partial y/\partial t$. For a given dosage of the drug, when is the reaction a maximum?

Solution: In this example,

$$\frac{\partial y}{\partial x} = te^{-t} \frac{\partial}{\partial x}(x^2(a - x)) = te^{-t}(2ax - 3x^2).$$

Similarly,

$$\frac{\partial y}{\partial t} = x^2(a - x) \frac{\partial}{\partial t}(te^{-t}) = x^2(a - x)e^{-t}(1 - t).$$

For a given dosage x of the drug, the reaction y is a maximum when $\partial y/\partial t = 0$. The maximum occurs when $t = 1$, that is, 1 hour after the injection of the drug.

There is an obvious definition of the *higher partial derivatives*. If $y = f(x, t)$, we define the *second partial derivative* $\partial^2 y/\partial x^2$ of y with respect to x as the partial derivative of $\partial y/\partial x$ with respect to x. This means that $\partial^2 y/\partial x^2 = \partial(\partial y/\partial x)/\partial x$. Similarly, $\partial^2 y/\partial t^2 = \partial(\partial y/\partial t)/\partial t$. The *mixed partial derivative* $\partial^2 y/\partial x \partial t$ is defined by $\partial^2 y/\partial x \partial t = \partial(\partial y/\partial t)/\partial x$. For example, if $y = x^2 t^3$, then

$$\frac{\partial^2 y}{\partial x^2} = \frac{\partial}{\partial x}\left(\frac{\partial y}{\partial x}\right) = \frac{\partial}{\partial x}(2xt^3) = 2t^3, \quad \frac{\partial^2 y}{\partial t^2} = \frac{\partial}{\partial t}\left(\frac{\partial y}{\partial t}\right) = \frac{\partial}{\partial t}(3x^2 t^2) = 6x^2 t,$$

$$\frac{\partial^2 y}{\partial x \partial t} = \frac{\partial}{\partial x}\left(\frac{\partial y}{\partial t}\right) = \frac{\partial}{\partial x}(3x^2 t^2) = 6xt^2.$$

Problems B5

1. Calculate the partial derivatives $\partial y/\partial x$ and $\partial y/\partial t$ of the following functions.
   (a) $y = tx^2 + xt^2$. (b) $y = (x + t) \sin (x + t)$. (c) $y = 1 + 2x^2t^2$.
   (d) $y = \sin x \cos t$. (e) $y = xe^{xt}$. (f) $y = e^{(x+t)} \sin (xt)$.

2. The volume of a fixed amount of gas varies proportionally with the temperature and inversely with the pressure. Therefore, $V = k(T/P)$, where k is a constant and V, T, and P are the volume, temperature, and pressure. Calculate $\partial V/\partial T$ and $\partial V/\partial P$. Prove that $k(\partial V/\partial P) + V(\partial V/\partial T) = 0$.

3. Two drugs are used simultaneously as a treatment for a certain disease. The reaction R (measured in appropriate units) to x units of the first drug and y units of the second drug is $R(x, y) = x^2y^2(a - x)(b - y)$. For a fixed amount x of the first drug, what amount y of the second drug produces the maximum reaction?

4. Calculate the partial derivatives $\partial^2 y/\partial x^2$, $\partial^2 y/\partial x\,\partial t$, and $\partial^2 y/\partial t^2$ of the following functions.
   (a) $y = x^2t^2$. (b) $y = \sin (xt)$. (c) $y = e^{(x+t)}$. (d) $y = xte^{xt}$.
   (e) $y = (x + 1)e^{(t+1)}$. (f) $y = x^2 + t^2 \sin^{-1} x$.

5. The reaction $R(x, t)$ to x units of a drug t hours after the drug has been administered is given by $R(x, t) = x^2(a - x)t^2e^{-t}$. For what amount x is the reaction as large as possible? When does the maximum reaction occur?

6. Prove that $y = f(x, t) = (1/\sqrt{t})e^{-x^2/t}$ satisfies the partial differential equation $4(\partial y/\partial t) = \partial^2 y/\partial x^2$.

<table>
<tr><td>Integral Calculus</td><td>**C**
Appendix</td></tr>
</table>

C1 The Antiderivative

In Appendix B, we studied the problem of determining the derivative $f'(x)$ of a given function $f(x)$. In this appendix, we study the inverse problem. Given a function $f(x)$, can we determine a function $F(x)$ with the property that the derivative $F'(x)$ is equal to $f(x)$? Any function $F(x)$ that has this property will be called an *antiderivative* or *indefinite integral* of the given function $f(x)$.

Example C1.1 Determine antiderivatives of the following functions.
  1. $f(x) = x^2$.   2. $f(x) = 1 + x$.   3. $f(x) = \cos x$.

Solution:
  1. We must determine a function $F(x)$ with the property that $F'(x) = f(x) = x^2$. It is easy to verify that $F(x) = \frac{1}{3}x^3$ is such a function. Another choice could be $F(x) = \frac{1}{3}x^3 + 1$.
  2. If $F(x) = x + x^2/2$, then $F'(x) = 1 + x = f(x)$.
  3. If $F(x) = \sin x$, then $F'(x) = \cos x = f(x)$.

439

To determine an antiderivative of a given function $f(x)$, we must determine a function $y = F(x)$ which satisfies $dy/dx = f(x)$. Symbolically, the relation between y and x is written $dy = f(x)\,dx$. The conventional notation for the solution of this problem is

$$y = F(x) = \int f(x)\,dx, \qquad (C1)$$

where the symbol $\int$ is called the *integral sign*. If $dy/dx = f(x)$, we write $\int dy = \int f(x)\,dx$. In the second integral, $f(x)$ is said to be the *integrand*.

Example C1.2   Determine an antiderivative or indefinite integral of the function $f(x) = x^3$.

Solution: The solution is written $y = F(x) = \int f(x)\,dx = \int x^3\,dx$. We recall that x^4 has the derivative $4x^3$ and, therefore, $\frac{1}{4}x^4$ has the derivative x^3. This implies that one antiderivative is $F(x) = x^4/4$. However, for any constant c, $F(x) = (x^4/4) + c$ is also an antiderivative, since $dc/dx = 0$. We conclude that $\int x^3\,dx = (x^4/4) + c$, where c is an arbitrary constant.

As with derivatives, we must now learn to calculate as efficiently as possible the antiderivatives of the functions that occur in the applications of calculus. Unfortunately, this is in general a difficult problem. Some integrals that are not difficult to evaluate are given in the following list.

1. $\int dx = x + c.$
2. $\int x^n\,dx = x^{n+1}/(n + 1) + c$ for any number $n \neq 1.$
3. $\int (1/x)\,dx = \log_e |x| + c.$
4. $\int \sin x\,dx = -\cos x + c.$
5. $\int \cos x\,dx = \sin x + c.$
6. $\int e^{ax}\,dx = (1/a)e^{ax} + c.$
7. $\int af(x)\,dx = a\int f(x)\,dx.$
8. $\int (f(x) + g(x))\,dx = \int f(x)\,dx + \int g(x)\,dx.$
9. $\int F'(f(x))f'(x)\,dx = F(f(x)) + c.$

To prove any one of these results, we must prove that the derivative of the right side is equal to the integrand of the left side. For example, in 6,

$$\frac{d}{dx}\left(\frac{1}{a}e^{ax} + c\right) = e^{ax}.$$

The last result follows from the chain rule of differentiation. To prove integral 9, we know that the derivative of $F(f(x)) + c$ is

$$\frac{d}{dx}(F(f(x)) + c) = \frac{dF}{du}\frac{du}{dx}$$

if we define $u = f(x)$. Therefore,

$$\frac{d}{dx}(F(f(x)) + c) = F'(u)\frac{df}{dx} = F'(f(x))f'(x).$$

This is the result.

We must now use these results and develop skills in recognizing functions which can be integrated.

Example C1.3   Determine the indefinite integrals of the following functions.
1. $f(x) = x^3 + 4x^5$.   2. $(1 + x^2)^5 2x$.
3. $x^3(1 + x^4)^{1/2}$.   4. $\sin 5x$.
5. xe^{-x^2}.   6. $\sin x \cos x$.

Solution:
1. $\int f(x)\,dx = \int (x^3 + 4x^5)\,dx = \int x^3\,dx + 4\int x^5\,dx = (x^4/4) + \frac{4}{6}x^6 + c$.
   Simplifying, we have $\int (x^3 + 4x^5)\,dx = (x^4/4) + \frac{2}{3}x^6 + c$, where c is an arbitrary constant.
2. To evaluate $\int (1 + x^2)^5 2x\,dx$, we define $u = 1 + x^2$. Then $du/dx = 2x$ and

$$\int (1 + x^2)^5 2x\,dx = \int u^5\,du = (u^6/6) + c = \frac{(1 + x^2)^6}{6} + c.$$

   This calculation is a special case of integral 9 in our list.
3. To calculate $\int x^3(1 + x^4)^{1/2}\,dx$, define $u = 1 + x^4$. Then $du/dx = 4x^3$ and

$$\int x^3(1 + x^4)^{1/2}\,dx = \int \frac{u^{1/2}}{4}\,du = \frac{1}{4}\left(\frac{2}{3}\right)u^{3/2} + c = \frac{(1 + x^4)^{3/2}}{6} + c.$$

   This result can be verified by differentiating.

4. Define $u = 5x$. Then

$$\int \sin 5x \, dx = \int \frac{\sin u}{5} \, du = \frac{-\cos u}{5} + c = -\frac{\cos 5x}{5} + c.$$

5. Define $u = -x^2$. Then

$$\int xe^{-x^2} \, dx = -\tfrac{1}{2} \int e^u \, du = -\tfrac{1}{2}e^u + c = -\tfrac{1}{2}e^{-x^2} + c.$$

6. Define $u = \sin x$. Then $du = \cos x \, dx$ and

$$\int \sin x \cos x \, dx = \int u \, du = (u^2/2) + c = \frac{\sin^2 x}{2} + c.$$

This integral could also be evaluated using the substitution $v = \cos x$. In this case, $dv = -\sin x \, dx$ and

$$\int \sin x \cos x \, dx = -\int v \, dv = \frac{-v^2}{2} + k = \frac{-\cos^2 x}{2} + k,$$

where k is an arbitrary constant. These two answers are the same, since $\sin^2 x + \cos^2 x = 1$.

Example C1.4 Suppose that a curve in the xy plane passes through the point $(0, 2)$ and is such that the slope of the tangent at (x, y) on the curve is $x^3 - 3x^2 + 2$. Determine the curve.

Solution: The slope at the point (x, y) on the curve is $dy/dx = x^3 - 3x^2 + 2$. Integrating, we find $y = (x^4/4) - x^3 + 2x + c$, where c is an arbitrary constant. Since the curve passes through the point $(0, 2)$, we must have $c = 2$ ($y = 2$ when $x = 0$). The equation of the curve is $y = (x^4/4) - x^3 + 2x + 2$.

To conclude this section, we describe a very useful method to evaluate certain types of integrals. The method is based on the product rule of differentiation. If $u(x)$ and $v(x)$ are differentiable functions of x, then

$$\frac{d}{dx}(u(x)v(x)) = u(x)\frac{dv}{dx} + v(x)\frac{du}{dx}.$$

Integrating on both sides, we have

$$\int \frac{d}{dx}(u(x)v(x))\,dx = \int u(x)\frac{dv}{dx}\,dx + \int v(x)\frac{du}{dx}\,dx.$$

But the integrand of the integral on the left is the derivative of $u(x)v(x)$. Therefore,

$$u(x)v(x) = \int u(x)\frac{dv}{dx}\,dx + \int v(x)\frac{du}{dx}\,dx.$$

This result is usually written in the form

$$\int u\,dv = uv - \int v\,du. \qquad\qquad (C2)$$

This is the *integration by parts method*. It can be used to evaluate integrals of the form $\int u\,dv$ when the integral $\int v\,du$ is known. The wide range of applications of this method is illustrated in the following example.

Example C1.5 Evaluate the integrals of the following functions by the integration by parts method.
  1. $f(x) = x\cos x$.   2. $\log_e x$.   3. xe^x.   4. $e^x\sin x$.

Solution:
  1. To evaluate $\int x\cos x\,dx$, we define $u = x$ and $dv = \cos x\,dx$. Then $du = dx$ and $v = \sin x$. Using the integration by parts method, we find that

$$\int x\cos x\,dx = \int u\,dv = uv - \int v\,du = x\sin x - \int \sin x\,dx.$$

  We have already evaluated this last integral. The solution is $\int x\cos x\,dx = x\sin x + \cos x + c$, where c is an arbitrary constant. This solution can be verified by differentiation.
  2. To evaluate $\int \log_e x\,dx$ by the integration by parts method, we define $u = \log_e x$ and $dv = dx$. Then, $du = (1/x)\,dx$ and $v = x$. Therefore, $\int \log_e x\,dx = \int u\,dv = uv - \int v\,du = x\log_e x - \int x(1/x)\,dx = x\log_e x - x + c$.
  3. Choose $u = x$ and $dv = e^x\,dx$. Then, $du = dx$ and $v = e^x$. Integrating by parts, we have $\int xe^x\,dx = xe^x - \int e^x\,dx = xe^x - e^x + c$.

4. To evaluate $\int e^x \sin x \, dx$, define $u = e^x$ and $dv = \sin x \, dx$. Then $du = e^x \, dx$ and $v = -\cos x$. Therefore, $\int e^x \sin x \, dx = -e^x \cos x + \int e^x \cos x \, dx$. To evaluate $\int e^x \cos x \, dx$, we again use integration by parts, defining $u_1 = e^x$ and $dv_1 = \cos x \, dx$. Then $du_1 = e^x \, dx$, $v_1 = \sin x$, and $\int e^x \cos x \, dx = e^x \sin x - \int e^x \sin x \, dx$. It seems that we have made no progress in evaluating the required integral. However, this is not the case. We have found

$$\int e^x \sin x \, dx = -e^x \cos x + \int e^x \cos x \, dx = -e^x \cos x$$
$$+ e^x \sin x - \int e^x \sin x \, dx.$$

Therefore, $2\int e^x \sin x \, dx = e^x(\sin x - \cos x)$. The required integral is $\int e^x \sin x \, dx = (e^x/2)(\sin x - \cos x) + c$, where c is an arbitrary constant.

Integrating by parts is a very simple method to apply once we have chosen u and v appropriately. Recognizing those integrals that can be evaluated by this method is a skill which can only be acquired by trying many examples.

Problems C1

1. Determine antiderivatives of the following functions.

(a) $f(x) = x^{10}$.   (b) $f(x) = 1 + x^2 + x^4$.   (c) $f(x) = \dfrac{1}{x^3}$.

(d) $\dfrac{1}{(x+1)^3}$.   (e) $1 + 2x + x^2$.   (f) $x + \dfrac{1}{x}$.

2. Determine antiderivatives of the following functions.

(a) $\cos 3x$.   (b) $x \sin(x^2)$.   (c) $x^4(1 + x^5)^3$.

(d) $x^2 e^{-x^3}$.   (e) $\sin^2 x \cos x$.   (f) $\sin x(\cos x + \cos^2 x)$.

3. Evaluate the integrals of the following functions by the integration by parts method.

(a) $x \sin x$.   (b) $x \log_e x$.   (c) xe^{-x}.

(d) $e^x \cos x$.   (e) $x^2 \sin x$.   (f) $x \sin 5x$.

4. Determine antiderivatives of the following functions.

(a) $\dfrac{1}{1+x}$.   (b) $\dfrac{1}{1+x^2}$.   (c) $\dfrac{1}{\sqrt{1-x^2}}$.

(d) $\dfrac{1}{(1+x)^2}$.   (e) $e^{3x} \cos 2x$.   (f) $\dfrac{x}{\sqrt{1-x^2}}$.

5. Determine a curve $y = f(x)$, which passes through the point $(1, 1)$ such that the slope at the point $(x, f(x))$ is $x^2 + 1$.

6. Determine a curve $y = f(x)$, which passes through the point $(0, 5)$ such that the slope at the point $(x, f(x))$ is $\cos x$.

7. A population of insects grows from an initial size of 10,000 to a population $p(t)$ after a time t (measured in days). If the growth rate at time t is $p'(t) = t + t^2$, determine the population after (a) 1 day, (b) 5 days, (c) 10 days.

C2 The Definite Integral

In this section, we study a remarkable connection between the problem of integrating $f(x)$ and the problem of calculating the area under the curve $y = f(x)$. In Figure C1, the area under the curve $y = f(x)$ between $x = a$ and $x = b$ is drawn. We assume that $f(x) > 0$ for x between a and b.

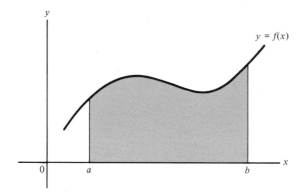

Figure C1. Area Under the Curve $y = f(x)$

If $y = f(x)$ is the constant function $y = c$, then the area under the curve between a and b is $c(b - a)$. This is the area of a rectangle with height c and base $(b - a)$. We will calculate the area under a general curve $y = f(x)$ by approximating the area by a sum of rectangles.

We begin by dividing the interval $a < x < b$ into n subintervals, each of length $(b - a)/n$. Define

$$x_0 = a, \quad x_1 = a + \frac{b - a}{n}, x_2 = a + \frac{2(b - a)}{n},$$

$$x_3 = a + \frac{3(b - a)}{n}, \dots, x_{n-1} = a + \frac{(n - 1)(b - a)}{n}, x_n = b.$$

This division of the interval $a < x < b$ is shown in Figure C2.

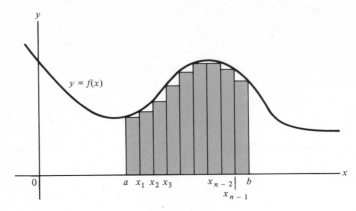

Figure C2

We approximate the area under the curve between $x = a$ and $x = b$ by adding up the areas of n rectangles. This area will be denoted by A_n to indicate that it does depend on the number of rectangles. We define A_n by the formula

$$A_n = f(x_0)\frac{b-a}{n} + f(x_1)\frac{b-a}{n} + \cdots + f(x_{n-2})\frac{b-a}{n} + f(x_{n-1})\frac{b-a}{n}$$

$$= \sum_{i=0}^{n-1} f(x_i)\frac{b-a}{n}.$$

As an example, consider the function $y = f(x) = x$ on the interval $1 < x < 2$. The area under this curve can be calculated by elementary methods to be

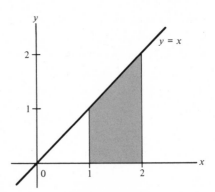

Figure C3

$\frac{3}{2}$ (see Figure C3). We define

$$x_0 = 1, \quad x_1 = 1 + \frac{1}{n}, \quad x_2 = 1 + \frac{2}{n}, \quad \ldots, \quad x_{n-1} = 1 + \frac{n-1}{n}, \quad x_n = 2.$$

Then,

$$A_n = \sum_{i=0}^{n-1} f(x_i)\frac{b-a}{n} = \sum_{i=0}^{n-1}\left(1 + \frac{i}{n}\right)\frac{1}{n} = \frac{1}{n}\sum_{i=0}^{n-1}\left(1 + \frac{i}{n}\right).$$

We evaluate

$$\sum_{i=0}^{n-1}\left(1 + \frac{i}{n}\right) = \sum_{i=0}^{n-1} 1 + \frac{1}{n}\sum_{i=0}^{n-1} i = n + \frac{n(n-1)}{2n} = n + \frac{n-1}{2} = \frac{3n}{2} - \frac{1}{2}.$$

Therefore,

$$A_n = \frac{1}{n}\left(\frac{3n}{2} - \frac{1}{2}\right) = \frac{3}{2} - \frac{1}{2n}.$$

We have used the fact that

$$\sum_{i=0}^{n-1} i = \frac{n(n-1)}{2},$$

which is proved in Appendix F. The area of the n rectangles is $\frac{3}{2} - \frac{1}{2}n$. For n large, this is a very good approximation to the area under the curve $y = x$ between $x = 1$ and $x = 2$.

Let us return to the general problem of calculating the area between $x = a$ and $x = b$ under the curve $y = f(x)$. Our definition of this area will be $A = \lim_{n \to \infty} A_n$, where A_n is our approximation to the area made up of n rectangles. This means that we use more and more rectangles of smaller and smaller width to improve our approximation to the exact area.

Suppose that $F(x) = \int f(x)\,dx$ is any antiderivative of $f(x)$. Then, $F'(x) = f(x)$ and, if Δx is a small change in x,

$$\frac{F(x + \Delta x) - F(x)}{\Delta x} \approx f(x) \qquad \text{or} \qquad F(x + \Delta x) - F(x) \approx f(x)\,\Delta x,$$

since

$$\lim_{\Delta x \to 0} \frac{F(x + \Delta x) - F(x)}{\Delta x} = F'(x) = f(x).$$

In calculating A_n, we add n terms of the form $f(x_i)[(b - a)/n]$. If $\Delta x = (b - a)/n$, then

$$F\left(x_i + \frac{b - a}{n}\right) - F(x_i) \approx f(x_i)\frac{b - a}{n},$$

and this approximation becomes very good for n large. If we recall that $x_{i+1} = x_i + [(b - a)/n]$, we have the result

$$F(x_{i+1}) - F(x_i) \approx f(x_i)\frac{b - a}{n}.$$

Our formula for A_n becomes

$$A_n = \sum_{i=0}^{n-1} f(x_i)\frac{b - a}{n} \approx \sum_{i=0}^{n-1} [F(x_{i+1}) - F(x_i)]$$

or

$$A_n \approx [F(x_1) - F(x_0)] + [F(x_2) - F(x_1)] + [F(x_3) - F(x_2)] + \cdots$$
$$+ [F(x_n) - F(x_{n-1})].$$

After canceling and remembering that $x_0 = a$ and $x_n = b$, we have the result that

$$A_n \approx F(b) - F(a).$$

Since our approximation for A improves as n increases, we conclude that

$$A = \lim_{n \to \infty} A_n = F(b) - F(a).$$

By this argument, we have seen that the area under the curve $y = f(x)$ between $x = a$ and $x = b$ is simply the difference $F(b) - F(a)$, where $F(x)$ is any antiderivative of $f(x)$. Because of the result, we use the notation $\int_a^b f(x)\,dx = F(b) - F(a)$ to represent the area under the curve from $x = a$ to $x = b$. This is the *definite integral* of $f(x)$ from $x = a$ to $x = b$. For convenience we write

$$\int_a^b f(x)\,dx = F(x)\Big|_a^b = F(b) - F(a).$$

This is read as "the definite integral of $f(x)$ from a to b is $F(x)$ at b minus $F(x)$ at a."

Example C2.1   Calculate the following definite integrals.

1. $\displaystyle\int_0^1 x^2\,dx.$ 2. $\displaystyle\int_1^2 x^3\,dx.$ 3. $\displaystyle\int_0^{2\pi} \sin x\,dx.$

Solution:

1. An antiderivative of $f(x) = x^2$ is $F(x) = x^3/3$. Therefore,

$$\int_0^1 x^2\,dx = \frac{x^3}{3}\Big|_0^1 = \tfrac{1}{3} - 0 = \tfrac{1}{3}.$$

Note that this result is independent of the antiderivative we choose. If $F(x) = (x^3/3) + c$ for some constant c, $F(1) - F(0) = (\tfrac{1}{3} + c) - (0 + c) = \tfrac{1}{3}$. The arbitrary constant c cancels.

2. $\displaystyle\int_1^2 x^3\,dx = \frac{x^4}{4}\Big|_1^2 = \frac{2^4}{4} - \frac{1^4}{4} = 4 - \frac{1}{4} = \frac{15}{4}.$

3. $\displaystyle\int_0^{2\pi} \sin x\,dx = -\cos x\Big|_0^{2\pi} = -\cos 2\pi + \cos 0 = -1 + 1 = 0.$

To understand this answer, we draw the graph of $y = \sin x$ from $x = 0$ to $x = 2\pi$ (Figure C4). The definite integral from 0 to 2π is the sum of two

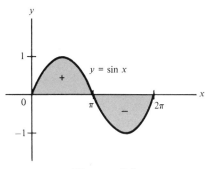

Figure C4

integrals, one from 0 to π and the other from π to 2π. When sin x is positive $(0 < x < \pi)$, there is a positive contribution to the integral. When sin x is negative $(\pi < x < 2\pi)$, the contribution to the integral is negative. These two contributions cancel, and we find $\int_0^{2\pi} \sin x \, dx = 0$.

The definite integral has the following three properties.

1. $\int_a^a f(x)\, dx = 0$.

2. $\int_a^b f(x)\, dx = -\int_b^a f(x)\, dx$.

3. $\int_a^b f(x)\, dx = \int_a^c f(x)\, dx + \int_c^b f(x)\, dx$.

These are immediate consequences of the identity $\int_a^b f(x)\, dx = F(b) - F(a)$.

Finally, in this section, we consider *improper integrals* defined as integrals $\int_a^b f(x)\, dx$, where either $f(x)$ is infinite for some x in the interval $a \le x \le b$ or one (or both) of the end points $x = a$ and $x = b$ is infinite. The integrals $\int_0^\infty e^{-x}\, dx$, $\int_0^1 \log_e x\, dx$, $\int_{-\infty}^\infty e^x\, dx$, and $\int_{-\infty}^\infty 1/(1 + x^2)\, dx$ are improper integrals. It is sometimes possible to evelute improper integrals as finite expressions. For example, we will interpret $\int_0^\infty e^{-x}\, dx$ to mean $\lim_{b \to \infty} \int_0^b e^{-x}\, dx$. But

$$\int_0^b e^{-x}\, dx = -e^{-x}\Big|_0^b = 1 - e^{-b} \quad \text{and} \quad \lim_{b \to \infty}(1 - e^{-b}) = 1.$$

Therefore $\int_0^\infty e^{-x}\, dx = 1$. This improper integral is said to be *convergent*. On the other hand,

$$\int_0^\infty e^x\, dx = \lim_{b \to \infty}\int_0^b e^x\, dx = \lim_{b \to \infty}(e^b - 1) = \infty.$$

This is a *divergent* improper integral.

Example C2.2   Evaluate (if possible) the following improper integrals.

1. $\displaystyle\int_0^1 \frac{1}{\sqrt{x}}\, dx.$ 2. $\displaystyle\int_0^1 \frac{1}{x^2}\, dx.$ 3. $\displaystyle\int_0^\infty xe^{-x^2}\, dx.$

Solution:

1. This integral is improper, since $1/\sqrt{x}$ is infinite at $x = 0$. We define

$$\int_0^1 \frac{1}{\sqrt{x}}\, dx = \lim_{a \to 0}\int_a^1 \frac{1}{\sqrt{x}}\, dx,$$

where $a > 0$. But

$$\int_a^1 \frac{1}{\sqrt{x}}\, dx = 2x^{1/2}\Big|_a^1 = 2 - 2\sqrt{a} \quad \text{and} \quad \lim_{a \to 0} (2 - 2\sqrt{a}) = 2.$$

Therefore, $\int_0^1 (1/\sqrt{x})\, dx = 2$, and this is a convergent improper integral.

2. $\displaystyle\int_0^1 \frac{1}{x^2}\, dx = \lim_{a \to 0} \int_a^1 \frac{1}{x^2}\, dx = \lim_{a \to 0} \left(-\frac{1}{x}\Big|_a^1\right) = \lim_{a \to 0} \left(\frac{1}{a} - 1\right) = \infty.$

This is a divergent improper integral.

3. $\displaystyle\int_0^\infty xe^{-x^2}\, dx = \lim_{b \to \infty} \int_0^b xe^{-x^2}\, dx = \lim_{b \to \infty} \frac{-e^{-x^2}}{2}\Big|_0^b$

$$= \lim_{b \to \infty} \tfrac{1}{2}(1 - e^{-b^2}) = \tfrac{1}{2}.$$

This is a convergent improper integral.

Problems C2

1. Evaluate the following definite integrals.
   (a) $\int_1^5 3x^2\, dx.$ (b) $\int_0^2 (1 + x^2)\, dx.$ (c) $\int_0^1 (1 + x + x^2)\, dx.$
2. Evaluate the following definite integrals.
   (a) $\int_4^7 (1 + e^x)\, dx.$ (b) $\int_3^5 (3e^{-3x} + 4e^{-4x})\, dx.$ (c) $\int_0^1 xe^{-x}\, dx.$
3. Evaluate the following definite integrals.
   (a) $\int_0^{\pi/4} \sin 4x\, dx.$ (b) $\int_{\pi/2}^\pi \sin x \cos x\, dx.$ (c) $\int_{-\pi/3}^{\pi/3} x \sin x\, dx.$
4. Use integration by parts to evaluate the following definite integrals.
   (a) $\int_0^1 xe^{3x}\, dx.$ (b) $\int_0^5 x^3 e^{-x^2}\, dx.$ (c) $\int_0^{\pi/4} x^2 \sin x\, dx.$
5. Evaluate the following improper integrals.

   (a) $\displaystyle\int_{-\infty}^\infty \frac{1}{1 + x^2}\, dx.$ (b) $\displaystyle\int_{-1}^1 \frac{1}{\sqrt{1 - x^2}}\, dx.$ (c) $\displaystyle\int_0^\infty \frac{1}{1 + x}\, dx.$

   [*Hint:* For parts (a) and (b), see Problem B2.6).]
6. The reaction to a given dose of a drug at time t hours after administration is given by $r(t) = te^{-t^2}$ (measured in appropriate units). Why is it reasonable to define the *total reaction* as the area under the curve $y = r(t)$ from $t = 0$ to $t = \infty$? Evaluate the total reaction to the given dose of the drug.
7. Continuing Problem 6, suppose that the reaction at time t is $r(t) = 1/(1 + t^2)$. Evaluate the total reaction to the given dose of the drug.

8. In Problems 6 and 7, define the total reaction up to time T as the area under the curve $y = r(t)$ from $t = 0$ to $t = T$. In both problems, evaluate the total reaction up to time T.

9. The rate of change of the concentration $C(t)$ at time t of a radioactive tracer drug is $C'(t) = 2^{-t}$, where t is measured in hours. If the initial concentration is 1 microgram per liter, determine the concentration at time t. (*Hint*: $2^{-t} = e^{-t \log_e 2}$.)

C3 Numerical Integration

It frequently happens that we must evaluate a definite integral $\int_a^b f(x)\,dx$, when we do not know an antiderivative of the function $f(x)$. For example, we cannot evaluate $\int_0^1 e^{x^2}\,dx$, since we do not know a function $F(x)$ whose derivative is $F'(x) = e^{x^2}$. However, we can estimate the value of this integral if we remember that it is equal to the area under the curve $y = e^{x^2}$ from $x = 0$ to $x = 1$. We could estimate the area very simply by drawing this curve as accurately as possible on graph paper and adding up the area of the squares under the curve.

In the previous section, we approximated the area under the curve $y = f(x)$ from $x = a$ to $x = b$ as a sum of n rectangles (see Figure C2). The rectangles were obtained by dividing the interval $a < x < b$ into n subintervals, each of length $(b - a)/n$. We defined

$$x_0 = a, x_1 = a + \frac{b-a}{n}, x_2 = a + \frac{2(b-a)}{n}, \ldots, x_n = b,$$

and we derived the estimate

$$\int_a^b f(x)\,dx \approx \sum_{i=0}^{n-1} f(x_i)\frac{b-a}{n}.$$

To improve this estimate, we will introduce a different approximation to the area as a sum of trapezoids. (A trapezoid is a quadrilateral that has two parallel sides.) This approximation method is called the *trapezoidal rule*. In Figure C5, the area under $y = f(x)$ from $x = a$ to $x = b$ is estimated as a sum of trapezoids. We estimate the area from x_i to x_{i+1} by the trapezoid drawn. The area of this trapezoid is the area of the rectangle plus the area

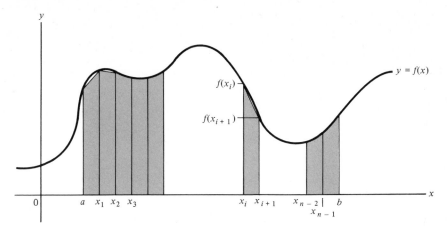

Figure C5

of the triangle, or

$$\frac{b-a}{n}f(x_{i+1}) + \frac{b-a}{2n}[f(x_i) - f(x_{i+1})] = \frac{b-a}{2n}[f(x_i) + f(x_{i+1})].$$

In Figure C5, we have drawn $f(x_i) > f(x_{i+1})$. The reader should verify that we obtain exactly the same estimate when $f(x_i) \le f(x_{i+1})$.

The total area under the curve is approximated by the sum of the areas of the individual trapezoids. Denoting by B_n the approximation by n trapezoids, we have

$$
\begin{aligned}
B_n &= \sum_{i=0}^{n-1} \frac{b-a}{2n}[f(x_i) + f(x_{i+1})] \\
&= \frac{b-a}{2n}[f(x_0) + f(x_1) + f(x_1) + f(x_2) + \cdots + f(x_{n-1}) + f(x_n)] \\
&= \frac{b-a}{2n}[f(x_0) + 2f(x_1) + 2f(x_2) + \cdots + 2f(x_{n-1}) + f(x_n)].
\end{aligned}
$$

This gives the estimate

$$\int_a^b f(x)\,dx \approx \frac{b-a}{2n}[f(x_0) + 2f(x_1) + \cdots + 2f(x_{n-1}) + f(x_n)].$$

We now have two estimates for the definite integral. We illustrate the accuracy of these estimates by the following examples.

Example C3.1   Estimate $\int_0^1 dx/(1 + x^2)$ by dividing $0 < x < 1$ into four subintervals.

Solution: In this example, $n = 4$, $a = 0$, and $b = 1$. We have the division points $a = x_0 = 0$, $x_1 = \frac{1}{4}$, $x_2 = \frac{1}{2}$, $x_3 = \frac{3}{4}$, and $x_4 = b = 1$. At these points $f(x) = 1/(1 + x^2)$ has values $f(x_0) = f(0) = 1$, $f(x_1) = \frac{16}{17}$, $f(x_2) = \frac{4}{5}$, $f(x_3 = \frac{16}{25}$, and $f(x_4) = f(1) = \frac{1}{2}$. Our first approximation to the definite integral is

$$\int_a^b f(x)\, dx \approx \frac{b - a}{n} \sum_{i=0}^{n-1} f(x_i) = \tfrac{1}{4}(1 + \tfrac{16}{17} + \tfrac{4}{5} + \tfrac{16}{25})$$

$$= \tfrac{1437}{1700} \approx .8453.$$

The second approximation is

$$\int_a^b f(x)\, dx = \frac{b - a}{2n}[f(x_0) + 2f(x_1) + 2f(x_2) + 2f(x_3) + f(x_4)]$$

$$= \tfrac{1}{8}(1 + \tfrac{32}{17} + \tfrac{8}{5} + \tfrac{32}{25} + \tfrac{1}{2}) \approx .7828.$$

We can check these approximations by comparing them with the exact value. We know

$$\int_0^1 \frac{dx}{1 + x^2} = \tan^{-1} x \Big|_0^1 = \frac{\pi}{4} \approx .7854.$$

We observe that the second approximation is very accurate, considering that we have used only four intervals. The accuracy of both estimates could be improved considerably by using more subintervals.

Example C3.2   Estimate $\int_1^2 dx/x$ by dividing the interval $1 < x < 2$ into four subintervals.

Solution: We have $n = 4$, $a = 1$, and $b = 2$. The end points of the sub-intervals are $a = x_0 = 1$, $x_1 = \frac{5}{4}$, $x_2 = \frac{6}{4} = \frac{3}{2}$, $x_3 = \frac{7}{4}$, and $x_4 = b = 2$. At these points, $f(x_0) = 1$, $f(x_1) = \frac{4}{5}$, $f(x_2) = \frac{2}{3}$, $f(x_3) = \frac{4}{7}$, and $f(x_4) = \frac{1}{2}$. The two estimates for the definite integral are

$$\int_1^2 \frac{dx}{x} = \tfrac{1}{4}(1 + \tfrac{4}{5} + \tfrac{2}{3} + \tfrac{4}{7}) = \tfrac{319}{420} \approx .7595$$

and

$$\int_1^2 \frac{dx}{x} = \tfrac{1}{8}(1 + \tfrac{8}{5} + \tfrac{4}{3} + \tfrac{8}{7} + \tfrac{1}{2}) \approx .6970.$$

Again, we can compare these estimates to the exact value of the definite integral, since $\int_1^2 dx/x = \log_e x|_1^2 = \log_e 2 \approx .6931$. The approximation obtained by the trapezoidal rule using four intervals is accurate to within 1 per cent of the correct value. This degree of accuracy is sufficient for many applications.

There are other methods that give better accuracy with the same number of intervals (or fewer). These would be used when a high degree of accuracy is called for. These methods, as well as the two methods discussed above, are easily programmed on a computer.

Problems C3

1. Estimate $\int_0^1 dx/(1 + x^2)$ by dividing the interval $0 < x < 1$ into eight subintervals. Compare the estimates with the exact value.

2. Use the trapezoidal rule with $n = 4$ to estimate $\int_1^3 dx/x$. Compare your estimate with the exact value.

3. Verify that the trapezoidal rule gives the exact value for the integral $\int_a^b f(x)\, dx$ if $f(x) = cx + d$, where c and d are constants.

4. Estimate $\int_0^2 x^2\, dx$ using the trapezoidal rule with $n = 8$. What is the exact value of this integral?

5. Use the trapezoidal rule with four subintervals to estimate $\int_0^\pi \sin x\, dx$. What is the exact value of this integral?

6. Estimate $\int_0^1 e^{x^2}\, dx$ using the trapezoidal rule with $n = 5$ and $n = 10$.

The Exponential Function

The exponential function is usually introduced in calculus and has been discussed in Appendix B. However, there are many noncalculus applications of this function, such as the Poisson distribution in probability (Section 2.9). For this reason, we collect a number of the noncalculus properties of the exponential function in this appendix.

Definition D.1   Exponential Function   *The exponential function* $\exp(x)$ *is defined by the following infinite sum for every real number* x.

$$\exp(x) = 1 + \frac{x}{1!} + \frac{x^2}{2!} + \frac{x^3}{3!} + \cdots + \frac{x^n}{n!} + \cdots = \sum_{k=0}^{\infty} \frac{x^k}{k!}. \qquad \text{(D1)}$$

For example, $\exp(0) = 1 + 0 + 0 + \cdots = 1$ and $\exp(1) = 1 + \frac{1}{1!} + \frac{1}{2!} + \frac{1}{3!}$ $+ \cdots = 1 + 1 + .5 + .1667 + .0417 + .0083 + .0014 + .0002 + \cdots \approx$ 2.7183.

The number $\exp(1)$ is denoted by the symbol e. A more accurate estimate is $e \approx 2.71828183$. It can be proved that $\exp(x) = e^x$. In other words, the exponential function is a power function. We will not prove this result in general, but we will verify the identity $\exp(x + y) = \exp(x)\exp(y)$. We have

$$\exp(x)\exp(y) = \sum_{k=0}^{\infty} \frac{x^k}{k!} \sum_{l=0}^{\infty} \frac{y^l}{l!} = \sum_{k=0}^{\infty} \sum_{l=0}^{\infty} \frac{x^k y^l}{k!l!}.$$

In this double infinite sum, we group terms with $k + l = r$ and then sum over r. Therefore,

$$\exp(x)\exp(y) = \sum_{r=0}^{\infty} \sum_{k=0}^{r} \frac{x^k y^{r-k}}{k!(r-k)!} = \sum_{r=0}^{\infty} \frac{1}{r!} \sum_{k=0}^{r} \frac{r!}{k!(r-k)!} x^k y^{r-k}.$$

But

$$\sum_{k=0}^{r} \frac{r!}{k!(r-k)!} x^k y^{r-k} = (x + y)^r$$

by the binomial theorem. We conclude that

$$\exp(x)\exp(y) = \sum_{r=0}^{\infty} \frac{1}{r!}(x + y)^r = \exp(x + y). \tag{D2}$$

This proves, for example, that $\exp(2) = \exp(1)\exp(1) = e^2$. In general, we have $\exp(x) = e^x$.

We have defined $\exp(x)$ as an infinite sum. If we expand $(1 + x/n)^n$ by the binomial theorem, we have a finite sum containing $n + 1$ terms. As n becomes large, this finite sum approaches the infinite sum $\exp(x)$. This is written

$$\exp(x) = e^x = \lim_{n \to \infty} \left(1 + \frac{x}{n}\right)^n. \tag{D3}$$

By the binomial theorem,

$$\left(1 + \frac{x}{n}\right)^n = \sum_{k=0}^{n} \binom{n}{k}\left(\frac{x}{n}\right)^k.$$

The coefficient of x^k is

$$\binom{n}{k}\frac{1}{n^k} = \frac{n}{n}\frac{n-1}{n}\frac{n-2}{n}\cdots\frac{n-k+1}{n}\frac{1}{k!}$$

$$= \left(1 - \frac{1}{n}\right)\left(1 - \frac{2}{n}\right)\cdots\left(1 - \frac{k-1}{n}\right)\frac{1}{k!}.$$

If k is fixed and n increases, this coefficient approaches $1/k!$. This is true for all k, and therefore

$$\lim_{n\to\infty}\left(1 + \frac{x}{n}\right)^n = 1 + x + \frac{x^2}{2!} + \frac{x^3}{3!} + \cdots + \frac{x^k}{k!} + \cdots = \exp x.$$

This is the result.

Finally, we will mention a remarkable connection between the exponential function and the trigonometric functions $\sin x$ and $\cos x$. If $i = \sqrt{-1}$ (see Appendix E) and x is measured in radians, then we have *Euler's formula*,

$$e^{ix} = \cos x + i \sin x. \tag{D4}$$

For example, $e^{i0} = \cos 0 + i \sin 0$ or $1 = 1$ and $e^{i\pi} = \cos \pi + i \sin \pi = -1$. As a check, note that $e^{ix}e^{-ix} = e^0 = 1$ and $(\cos x + i \sin x)(\cos x - i \sin x) = \cos^2 x + \sin^2 x = 1$. A rigorous proof of this identity requires considerable familiarity with calculus and will not be given here. Some applications of Euler's formula are given in Chapters 6 and 7 and in Appendix E.

Problems D

1. Using the definition of $\exp(x)$ as an infinite series, calculate
   (a) $\exp(.05)$.   (b) $\exp(.5)$.   (c) $\exp(-.1)$.   (d) $\exp(-1)$.
2. Using the result $\exp(x + y) = \exp(x)\exp(y)$, prove that
   (a) $\exp(3) = e^3$.   (b) $\exp(5) = e^5$.   (c) $\exp(-1) = e^{-1}$.
   (d) $\exp(-4) = e^{-4}$.   (e) $\exp(\frac{1}{2}) = e^{1/2}$.   (f) $\exp(\frac{3}{2}) = e^{3/2}$.
3. Euler's formula $e^{ix} = \cos x + i \sin x$ can be used to prove identities relating the trigonometric functions. Using the fact that $e^{2ix} = e^{ix}e^{ix}$, prove that $\cos 2x = \cos^2 x - \sin^2 x$ and $\sin 2x = 2 \sin x \cos x$. Use a similar argument to prove that $\cos 3x = \cos^3 x - 3 \cos x \sin^2 x$ and $\sin 3x = 3 \cos^2 x \sin x - \sin^3 x$.

4. Prove that $\cos(x + y) = \cos x \cos y - \sin x \sin y$ and $\sin(x + y) = \sin x \cos y + \cos x \sin y$ by using Euler's formula and the identity $e^{ix}e^{iy} = e^{i(x+y)}$.

Additional problems involving the exponential function can be found in Section B4. Problems 10, 11, 12, 14, and 15 do not require calculus.

Complex Numbers

Appendix

The problem of determining the roots of quadratic equations (finding the numbers x such that $ax^2 + bx + c = 0$, where $a \neq 0$ and b and c are real numbers) has the solutions

$$x = \frac{-b \pm \sqrt{b^2 - 4ac}}{2a}.$$

These roots are real numbers only if $b^2 - 4ac \geq 0$. It is very useful in many mathematical problems to use this formula for the roots, even when $b^2 - 4ac < 0$.

To handle this situation, we introduce the "imaginary" number $i = \sqrt{-1}$. Then, if $b^2 - 4ac < 0$,

$$\sqrt{b^2 - 4ac} = \sqrt{(-1)(4ac - b^2)} = i\sqrt{4ac - b^2}.$$

Defining $\alpha = -b/2a$ and $\beta = \sqrt{(4ac - b^2)}/2a$, we find that the roots of the quadratic equation are $\alpha + i\beta$ and $\alpha - i\beta$, where α and β are real numbers. A *complex number* z is any combination of the form $z = \alpha + i\beta$, where α and

β are real numbers. If $\beta = 0$, z is a real number. If $\alpha = 0$, $\beta \neq 0$, z is a *pure imaginary number*. We say that α is the *real part* of z and β is the *imaginary part* of z, written $\alpha = $ Re (z) and $\beta = $ Im (z). The ordinary rules for multiplication are valid for complex numbers.

The complex numbers can be represented as points in the complex plane, with Re (z) plotted along the x axis and Im (z) plotted along the y axis. This is shown in Figure E1.

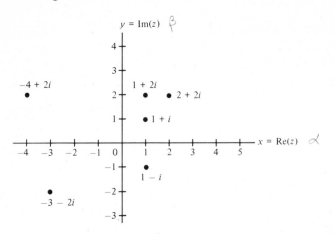

Figure E1. The Complex Plane

We associate to any complex number $z = \alpha + i\beta$ (Figure E2) the distance r from the point z to the origin O and the angle θ between the positive x axis and the line joining O to z. From the diagram, $r = \sqrt{\alpha^2 + \beta^2}$ (Pythagoras) and $\tan \theta = \beta/\alpha$ or $\theta = \tan^{-1}(\beta/\alpha)$ (where θ is measured in radians). We define r to be the *absolute value* of z, written $|z|$, and θ to be the *argument* of z, written arg z. It is clear from Figure E2 that $\sin \theta = \beta/r$ and $\cos \theta = \alpha/r$

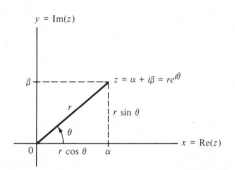

Figure E2. Polar Form of a Complex Number

or $\alpha = r \cos \theta$ and $\beta = r \sin \theta$. Therefore $z = \alpha + i\beta = r(\cos \theta + i \sin \theta)$. Using Euler's formula (Appendix D), we obtain $e^{i\theta} = \cos \theta + i \sin \theta$ and $z = re^{i\theta}$. This is the *polar form* of the complex number z.

Example E1   Determine the polar forms of the following complex numbers.
    1. i.   2. $1 + i$.   3. $1 - i$.   4. -1.

Solution: From Figure E1, $\arg (i) = \pi/2$, $\arg (1 + i) = \pi/4$, $\arg (1 - i) = -\pi/4$, and $\arg (-1) = \pi$. Also, $|i| = \sqrt{0^2 + 1^2} = 1$, $|1 + i| = \sqrt{2}$, $|1 - i| = \sqrt{2}$, and $|-1| = 1$. The required polar forms are

1. $i = e^{i\pi/2}$.   2. $1 + i = \sqrt{2}e^{i\pi/4}$.   3. $1 - i = \sqrt{2}e^{-i\pi/4}$.
4. $-1 = e^{i\pi}$.

The polar form is very useful in problems involving powers of complex numbers. If $z = re^{i\theta}$, then $z^n = (re^{i\theta})^n = r^n e^{in\theta} = r^n(\cos n\theta + i \sin n\theta)$. When $r = 1$, we have the *De Moivre formula*

$$(\cos \theta + i \sin \theta)^n = \cos n\theta + i \sin n\theta.$$

Example E2   Determine $(1 + i)^3$ by the De Moivre formula.

Solution: The polar form of $z = 1 + i$ is $\sqrt{2}e^{i\pi/4}$ with $r = \sqrt{2}$ and $\theta = \pi/4$. Therefore,

$$z^3 = (1 + i)^3 = (\sqrt{2})^3 e^{3i\pi/4} = 2\sqrt{2}\left(\cos \frac{3\pi}{4} + i \sin \frac{3\pi}{4}\right)$$

$$= 2\sqrt{2}\left(-\frac{\sqrt{2}}{2} + i\frac{\sqrt{2}}{2}\right) = -2 + 2i.$$

This can be checked by using the binomial theorem; $(1 + i)^3 = 1 + 3i + 3i^2 + i^3 = 1 + 3i - 3 - i = -2 + 2i$. If it seems that the second calculation is easier than the first, we suggest that the reader try to calculate $(1 + i)^{20}$ by the binomial theorem. By the De Moivre formula,

$$(1 + i)^{20} = (\sqrt{2})^{20} e^{20\pi i/4} = 2^{10}(\cos 5\pi + i \sin 5\pi) = -1024.$$

Problems E

1. Determine the polar forms of the following complex numbers.

(a) $3i$.   (b) $4 + 4i$.   (c) $4 + 3i$.   (d) $4 - 2i$.   (e) $1 + \sqrt{3}i$.

2. Write the following polar forms in the form $\alpha + i\beta$.

(a) $3e^{i\pi}$.   (b) $\frac{5}{2}e^{i\pi/3}$.   (c) $e^{5i\pi}$.   (d) $2e^{-(i\pi/2)}$.   (e) $5e^{3i\pi/4}$.

3. Use the De Moivre formula to calculate

(a) $(4 + 4i)^3$.   (b) $(1 + \sqrt{3}i)^4$.   (c) $(\sqrt{3} + i)^5$.

4. Derive expressions for $\cos 4\theta$ and $\sin 4\theta$ by comparing the De Moivre formula and the binomial expansion of $(\cos \theta + i \sin \theta)^4$.

5. Using De Moivre's formula, derive expressions for $\cos n\theta$ and $\sin n\theta$ as sums of powers of $\cos \theta$ and $\sin \theta$.

6. Suppose that λ_1 and λ_2 are the two roots of the quadratic equation with real coefficients $ax^2 + bx + c = 0$. Prove that $\lambda_1 + \lambda_2 = -(b/a)$ and $\lambda_1 \lambda_2 = c/a$.

Mathematical Induction

Mathematical induction is the long name given to a simple logical principle that can be used to prove mathematical statements of a certain type. Suppose that we have a series of statements that can be put in correspondence with the positive integers. For example, the statement that the sum of the first n integers is $n(n + 1)/2$ is, in fact, a series of statements, one for each positive integer.

To prove statements of this type by mathematical induction, we reason as follows. Suppose that we can verify that the statement is true for the first value of n (usually $n = 1$). Suppose further that, if we assume the statement is true for $n = k$, then we can verify that it is true for $n = k + 1$. We then know that the statement must be true for $n = 2$ and therefore for $n = 3$, and so on. We conclude that the statement is true for all n.

The application of the principle of mathematical induction is illustrated by the following examples.

Example F1   Prove that the sum of the first n integers is $n(n + 1)/2$.

Solution: Let us define $S_n = 1 + 2 + 3 + \cdots + n$, where n is a positive integer. We must prove that $S_n = n(n + 1)/2$. First observe that, for $n = 1$, the statement is true. Assume that it is true for $n = k$; that is, assume that $S_k = k(k + 1)/2$. Then consider

$$S_{k+1} = S_k + (k + 1) = \frac{k(k + 1)}{2} + (k + 1) = \frac{(k + 1)(k + 2)}{2}.$$

Therefore, S_{k+1} has the required formula. This completes the induction argument and the statement is true for all n.

Example F2   Prove that the sum of the squares of the first n integers is $n(n + 1)(2n + 1)/6$.

Solution: Define $T_n = 1^2 + 2^2 + 3^2 + \cdots + n^2$, where n is a positive integer. We will prove that $T_n = n(n + 1)(2n + 1)/6$. It is easy to verify that the statement is true for $n = 1$. Now assume that the statement is true for $n = k$; that is, assume that $T_k = k(k + 1)(2k + 1)/6$. Then

$$T_{k+1} = T_k + (k + 1)^2 = \frac{k(k + 1)(2k + 1)}{6} + (k + 1)^2$$

$$= \frac{k + 1}{6}[k(2k + 1) + 6(k + 1)] = \frac{k + 1}{6}(2k^2 + 7k + 6)$$

$$= \frac{(k + 1)(k + 2)(2k + 3)}{6}.$$

Therefore, T_{k+1} has the same formula as that for T_k with k replaced by $k + 1$. This completes the induction proof.

Example F3   Prove that

$$\tfrac{1}{2} + (\tfrac{1}{2})^2 + (\tfrac{1}{2})^3 + \cdots + (\tfrac{1}{2})^n = 1 - (\tfrac{1}{2})^n.$$

Solution: Define

$$R_n = \tfrac{1}{2} + (\tfrac{1}{2})^2 + (\tfrac{1}{2})^3 + \cdots + (\tfrac{1}{2})^n,$$

where n is a positive integer. We will prove by induction that $R_n = 1 - (\tfrac{1}{2})^n$. First, we observe that $R_1 = \tfrac{1}{2} = 1 - (\tfrac{1}{2})^1$. This proves the identity for $n = 1$.

If we assume that the identity is valid for $n = k$, that is, $R_k = 1 - (\frac{1}{2})^k$, then

$$R_{k+1} = R_k + (\tfrac{1}{2})^{k+1} = 1 - (\tfrac{1}{2})^k + (\tfrac{1}{2})^{k+1} = 1 - (\tfrac{1}{2})^{k+1}.$$

Therefore, R_{k+1} has the required form, and the result is proved.

Problems F

1. By mathematical induction, prove that

   (a) $(s + t) + (s + 2t) + (s + 3t) + \cdots + (s + nt) = n\left(s + \dfrac{n+1}{2}t\right).$

   (b) $(s + t + u) + (s + 2t + 4u) + \cdots + (s + nt + n^2u)$

   $$= n\left[s + \dfrac{n+1}{2}t + \dfrac{(n+1)(2n+1)}{6}u\right].$$

2. By mathematical induction, prove that (if $x \neq 1$)

   (a) $1 + x + x^2 + \cdots + x^n = \dfrac{1 - x^{n+1}}{1 - x}.$

   (b) $1 + 2x + 3x^2 + \cdots + nx^{n-1} = \dfrac{1 - (n+1)x^n + nx^{n+1}}{(1 - x)^2}.$

3. Prove that the number of subsets of a set S containing n elements is 2^n if the empty set is counted as a subset. Use mathematical induction.

4. A sequence of numbers $q_0, q_1, q_2, \ldots, q_n, \ldots$ satisfies the recurrence relation $q_n = q_{n-1}/(1 + q_{n-1})$. Prove by induction that $q_n = q_0/(1 + nq_0)$. (See Section 9.4 and Problem 9.4.10 for a biological interpretation.)

5. In a diploid organism such as man, genes occur in pairs on paired chromosomes (see Section 9.3). A gene locus with one allele, A, produces one genotype, AA. A gene locus with two alleles, A_1 and A_2, produces three genotypes, A_1A_1, A_1A_2, and A_2A_2. The genotypes A_1A_2 and A_2A_1 are identical.

   (a) What are the genotypes determined by a gene locus with three alleles, A_1, A_2, and A_3? By a gene locus with four alleles, A_1, A_2, A_3, and A_4?

   (b) Prove that a gene locus with n alleles $A_1, A_2, \ldots, A_n$ produces $(n^2 + n)/2$ genotypes.

6. Suppose that $A_1, A_2, \ldots, A_n$ are events in a finite probability space S. Prove by mathematical induction that $P(A_1 \cup A_2 \cup \cdots \cup A_n) \leq P(A_1) + P(A_2) + \cdots + P(A_n)$.

Tables

Table I Exponential Functions

x	e^x	e^{-x}	x	e^x	e^{-x}	x	e^x	e^{-x}	x	e^x	e^{-x}
0.00	1.0000	1.0000	0.85	2.3396	0.4274	2.4	11.023	0.0907	4.0	54.598	0.0183
0.05	1.0513	0.9512	0.90	2.4596	0.4066	2.5	12.182	0.0821	4.1	60.340	0.0166
0.10	1.1052	0.9048	0.95	2.5857	0.3867	2.6	13.464	0.0743	4.2	66.686	0.0150
0.15	1.1618	0.8607	1.0	2.7183	0.3679	2.7	14.880	0.0672	4.3	73.700	0.0136
0.20	1.2214	0.8187	1.1	3.0042	0.3329	2.8	16.445	0.0608	4.4	81.451	0.0123
0.25	1.2840	0.7788	1.2	3.3201	0.3012	2.9	18.174	0.0550	4.5	90.017	0.0111
0.30	1.3499	0.7408	1.3	3.6693	0.2725	3.0	20.086	0.0498	4.6	99.484	0.0101
0.35	1.4191	0.7047	1.4	4.0552	0.2466	3.1	22.198	0.0450	4.7	109.95	0.0091
0.40	1.4918	0.6703	1.5	4.4817	0.2231	3.2	24.533	0.0408	4.8	121.51	0.0082
0.45	1.5683	0.6376	1.6	4.9530	0.2019	3.3	27.113	0.0369	4.9	134.29	0.0074
0.50	1.6487	0.6065	1.7	5.4739	0.1827	3.4	29.964	0.0334	5.0	148.41	0.0067
0.55	1.7333	0.5769	1.8	6.0496	0.1653	3.5	33.115	0.0302	6.0	403.43	0.0025
0.60	1.8221	0.5488	1.9	6.6859	0.1496	3.6	36.598	0.0273	7.0	1096.6	0.0009
0.65	1.9155	0.5220	2.0	7.3891	0.1353	3.7	40.447	0.0247	8.0	2981.0	0.0003
0.70	2.0138	0.4966	2.1	8.1662	0.1225	3.8	44.701	0.0224	9.0	8103.1	0.0001
0.75	2.1170	0.4724	2.2	9.0250	0.1108	3.9	49.402	0.0202	10.0	22026.	0.0000
0.80	2.2255	0.4493	2.3	9.9742	0.1003						

Table II Natural Logarithms

N	0.0	0.1	0.2	0.3	0.4	0.5	0.6	0.7	0.8	0.9
1	0.0000	0.0953	0.1823	0.2624	0.3365	0.4055	0.4700	0.5306	0.5878	0.6419
2	0.6931	0.7419	0.7885	0.8329	0.8755	0.9163	0.9555	0.9933	1.0296	1.0647
3	1.0986	1.1314	1.1632	1.1939	1.2238	1.2528	1.2809	1.3083	1.3350	1.3610
4	1.3863	1.4110	1.4351	1.4586	1.4816	1.5041	1.5261	1.5476	1.5686	1.5892
5	1.6094	1.6292	1.6487	1.6677	1.6864	1.7047	1.7228	1.7405	1.7579	1.7750
6	1.7918	1.8083	1.8245	1.8405	1.8563	1.8718	1.8871	1.9021	1.9169	1.9315
7	1.9459	1.9601	1.9741	1.9879	2.0015	2.0149	2.0281	2.0412	2.0541	2.0669
8	2.0794	2.0919	2.1041	2.1163	2.1282	2.1401	2.1518	2.1633	2.1748	2.1861
9	2.1972	2.2083	2.2192	2.2300	2.2407	2.2513	2.2618	2.2721	2.2824	2.2925

Table III Common Logarithms

n	0	1	2	3	4	5	6	7	8	9
1.0	.0000	.0043	.0086	.0128	.0170	.0212	.0253	.0294	.0334	.0374
1.1	.0414	.0453	.0492	.0531	.0569	.0607	.0645	.0682	.0719	.0755
1.2	.0792	.0828	.0864	.0899	.0934	.0969	.1004	.1038	.1072	.1106
1.3	.1139	.1173	.1206	.1239	.1271	.1303	.1335	.1367	.1399	.1430
1.4	.1461	.1492	.1523	.1553	.1584	.1614	.1644	.1673	.1703	.1732
1.5	.1761	.1790	.1818	.1847	.1875	.1903	.1931	.1959	.1987	.2014
1.6	.2041	.2068	.2095	.2122	.2148	.2175	.2201	.2227	.2253	.2279
1.7	.2304	.2330	.2355	.2380	.2405	.2430	.2455	.2480	.2504	.2529
1.8	.2553	.2577	.2601	.2625	.2648	.2672	.2695	.2718	.2742	.2765
1.9	.2788	.2810	.2833	.2856	.2878	.2900	.2923	.2945	.2967	.2989
2.0	.3010	.3032	.3054	.3075	.3096	.3118	.3139	.3160	.3181	.3201
2.1	.3222	.3243	.3263	.3284	.3304	.3324	.3345	.3365	.3385	.3404
2.2	.3424	.3444	.3464	.3483	.3502	.3522	.3541	.3560	.3579	.3598
2.3	.3617	.3636	.3655	.3674	.3692	.3711	.3729	.3747	.3766	.3784
2.4	.3802	.3820	.3838	.3856	.3874	.3892	.3909	.3927	.3945	.3962
2.5	.3979	.3997	.4014	.4031	.4048	.4065	.4082	.4099	.4116	.4133
2.6	.4150	.4166	.4183	.4200	.4216	.4232	.4249	.4265	.4281	.4298
2.7	.4314	.4330	.4346	.4362	.4378	.4393	.4409	.4425	.4440	.4456
2.8	.4472	.4487	.4502	.4518	.4533	.4548	.4564	.4579	.4594	.4609
2.9	.4624	.4639	.4654	.4669	.4683	.4698	.4713	.4728	.4742	.4757
3.0	.4771	.4786	.4800	.4814	.4829	.4843	.4857	.4871	.4886	.4900
3.1	.4914	.4928	.4942	.4955	.4969	.4983	.4997	.5011	.5024	.5038
3.2	.5051	.5065	.5079	.5092	.5105	.5119	.5132	.5145	.5159	.5172
3.3	.5185	.5198	.5211	.5224	.5237	.5250	.5263	.5276	.5289	.5302
3.4	.5315	.5328	.5340	.5353	.5366	.5378	.5391	.5403	.5416	.5428
3.5	.5441	.5453	.5465	.5478	.5490	.5502	.5514	.5527	.5539	.5551
3.6	.5563	.5575	.5587	.5599	.5611	.5623	.5635	.5647	.5658	.5670
3.7	.5682	.5694	.5705	.5717	.5729	.5740	.5752	.5763	.5775	.5786
3.8	.5798	.5809	.5821	.5832	.5843	.5855	.5866	.5877	.5888	.5899
3.9	.5911	.5922	.5933	.5944	.5955	.5966	.5977	.5988	.5999	.6010
4.0	.6021	.6031	.6042	.6053	.6064	.6075	.6085	.6096	.6107	.6117
4.1	.6128	.6138	.6149	.6160	.6170	.6180	.6191	.6201	.6212	.6222
4.2	.6232	.6243	.6253	.6263	.6274	.6284	.6294	.6304	.6314	.6325
4.3	.6335	.6345	.6355	.6365	.6375	.6385	.6395	.6405	.6415	.6425
4.4	.6435	.6444	.6454	.6464	.6474	.6484	.6493	.6503	.6513	.6522
4.5	.6532	.6542	.6551	.6561	.6571	.6580	.6590	.6599	.6609	.6618
4.6	.6628	.6637	.6646	.6656	.6665	.6675	.6684	.6693	.6702	.6712
4.7	.6721	.6730	.6739	.6749	.6758	.6767	.6776	.6785	.6794	.6803
4.8	.6812	.6821	.6830	.6839	.6848	.6857	.6866	.6875	.6884	.6893
4.9	.6902	.6911	.6920	.6928	.6937	.6946	.6955	.6964	.6972	.6981
5.0	.6990	.6998	.7007	.7016	.7024	.7033	.7042	.7050	.7059	.7067
5.1	.7076	.7084	.7093	.7101	.7110	.7118	.7126	.7135	.7143	.7152
5.2	.7160	.7168	.7177	.7185	.7193	.7202	.7210	.7218	.7226	.7235
5.3	.7243	.7251	.7259	.7267	.7275	.7284	.7292	.7300	.7308	.7316
5.4	.7324	.7332	.7340	.7348	.7356	.7364	.7372	.7380	.7388	.7396

Table III Common Logarithms (Cont.)

n	0	1	2	3	4	5	6	7	8	9
5.5	.7404	.7412	.7419	.7427	.7435	.7443	.7451	.7459	.7466	.7474
5.6	.7482	.7490	.7497	.7505	.7513	.7520	.7528	.7536	.7543	.7551
5.7	.7559	.7566	.7574	.7582	.7589	.7597	.7604	.7612	.7619	.7627
5.8	.7634	.7642	.7649	.7657	.7664	.7672	.7679	.7686	.7694	.7701
5.9	.7709	.7716	.7723	.7731	.7738	.7745	.7752	.7760	.7767	.7774
6.0	.7782	.7789	.7796	.7803	.7810	.7818	.7825	.7832	.7839	.7846
6.1	.7853	.7860	.7868	.7875	.7882	.7889	.7896	.7903	.7910	.7917
6.2	.7924	.7931	.7938	.7945	.7952	.7959	.7966	.7973	.7980	.7987
6.3	.7993	.8000	.8007	.8014	.8021	.8028	.8035	.8041	.8048	.8055
6.4	.8062	.8069	.8075	.8082	.8089	.8096	.8102	.8109	.8116	.8122
6.5	.8129	.8136	.8142	.8149	.8156	.8162	.8169	.8176	.8182	.8189
6.6	.8195	.8202	.8209	.8215	.8222	.8228	.8235	.8241	.8248	.8254
6.7	.8261	.8267	.8274	.8280	.8287	.8293	.8299	.8306	.8312	.8319
6.8	.8325	.8331	.8338	.8344	.8351	.8357	.8363	.8370	.8376	.8382
6.9	.8388	.8395	.8401	.8407	.8414	.8420	.8426	.8432	.8439	.8445
7.0	.8451	.8457	.8463	.8470	.8476	.8482	.8488	.8494	.8500	.8506
7.1	.8513	.8519	.8525	.8531	.8537	.8543	.8549	.8555	.8561	.8567
7.2	.8573	.8579	.8585	.8591	.8597	.8603	.8609	.8615	.8621	.8627
7.3	.8633	.8639	.8645	.8651	.8657	.8663	.8669	.8675	.8681	.8686
7.4	.8692	.8698	.8704	.8710	.8716	.8722	.8727	.8733	.8739	.8745
7.5	.8751	.8756	.8762	.8768	.8774	.8779	.8785	.8791	.8797	.8802
7.6	.8808	.8814	.8820	.8825	.8831	.8837	.8842	.8848	.8854	.8859
7.7	.8865	.8871	.8876	.8882	.8887	.8893	.8899	.8904	.8910	.8915
7.8	.8921	.8927	.8932	.8938	.8943	.8949	.8954	.8960	.8965	.8971
7.9	.8976	.8982	.8987	.8993	.8998	.9004	.9009	.9015	.9020	.9025
8.0	.9031	.9036	.9042	.9047	.9053	.9058	.9063	.9069	.9074	.9079
8.1	.9085	.9090	.9096	.9101	.9106	.9112	.9117	.9122	.9128	.9133
8.2	.9138	.9143	.9149	.9154	.9159	.9165	.9170	.9175	.9180	.9186
8.3	.9191	.9196	.9201	.9206	.9212	.9217	.9222	.9227	.9232	.9238
8.4	.9243	.9248	.9253	.9258	.9263	.9269	.9274	.9279	.9284	.9289
8.5	.9294	.9299	.9304	.9309	.9315	.9320	.9325	.9330	.9335	.9340
8.6	.9345	.9350	.9355	.9360	.9365	.9370	.9375	.9380	.9385	.9390
8.7	.9395	.9400	.9405	.9410	.9415	.9420	.9425	.9430	.9435	.9440
8.8	.9445	.9450	.9455	.9460	.9465	.9469	.9474	.9479	.9484	.9489
8.9	.9494	.9499	.9504	.9509	.9513	.9518	.9523	.9528	.9533	.9538
9.0	.9542	.9547	.9552	.9557	.9562	.9566	.9571	.9576	.9581	.9586
9.1	.9590	.9595	.9600	.9605	.9609	:9614	.9619	.9624	.9628	.9633
9.2	.9638	.9643	.9647	.9652	.9657	.9661	.9666	.9671	.9675	.9680
9.3	.9685	.9689	.9694	.9699	.9703	.9708	.9713	.9717	.9722	.9727
9.4	.9731	.9736	.9741	.9745	.9750	.9754	.9759	.9763	.9768	.9773
9.5	.9777	.9782	.9786	.9791	.9795	.9800	.9805	.9809	.9814	.9818
9.6	.9823	.9827	.9832	.9836	.9841	.9845	.9850	.9854	.9859	.9863
9.7	.9868	.9872	.9877	.9881	.9886	.9890	.9894	.9899	.9903	.9908
9.8	.9912	.9917	.9921	.9926	.9930	.9934	.9939	.9943	.9948	.9952
9.9	.9956	.9961	.9965	.9969	.9974	.9978	.9983	.9987	.9991	.9996

Table IV Unit Normal Distribution

$$F(x) = \int_{-\infty}^{x} \frac{1}{\sqrt{2\pi}} \; e^{-t^2/2} \; dt$$

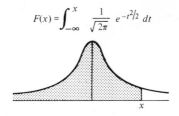

x	.00	.01	.02	.03	.04	.05	.06	.07	.08	.09
.0	.5000	.5040	.5080	.5120	.5160	.5199	.5239	.5279	.5319	.5359
.1	.5398	.5438	.5478	.5517	.5557	.5596	.5636	.5675	.5714	.5753
.2	.5793	.5832	.5871	.5910	.5948	.5987	.6026	.6064	.6103	.6141
.3	.6179	.6217	.6255	.6293	.6331	.6368	.6406	.6443	.6480	.6517
.4	.6554	.6591	.6628	.6664	.6700	.6736	.6772	.6808	.6844	.6879
.5	.6915	.6950	.6985	.7019	.7054	.7088	.7123	.7157	.7190	.7224
.6	.7257	.7291	.7324	.7357	.7389	.7422	.7454	.7486	.7517	.7549
.7	.7580	.7611	.7642	.7673	.7704	.7734	.7764	.7794	.7823	.7852
.8	.7881	.7910	.7939	.7967	.7995	.8023	.8051	.8078	.8106	.8133
.9	.8159	.8186	.8212	.8238	.8264	.8289	.8315	.8340	.8365	.8389
1.0	.8413	.8438	.8461	.8485	.8508	.8531	.8554	.8577	.8599	.8621
1.1	.8643	.8665	.8686	.8708	.8729	.8749	.8770	.8790	.8810	.8830
1.2	.8849	.8869	.8888	.8907	.8925	.8944	.8962	.8980	.8997	.9015
1.3	.9032	.9049	.9066	.9082	.9099	.9115	.9131	.9147	.9162	.9177
1.4	.9192	.9207	.9222	.9236	.9251	.9265	.9279	.9292	.9306	.9319
1.5	.9332	.9345	.9357	.9370	.9382	.9394	.9406	.9418	.9429	.9441
1.6	.9452	.9463	.9474	.9484	.9495	.9505	.9515	.9525	.9535	.9545
1.7	.9554	.9564	.9573	.9582	.9591	.9599	.9608	.9616	.9625	.9633
1.8	.9641	.9649	.9656	.9664	.9671	.9678	.9686	.9693	.9699	.9706
1.9	.9713	.9719	.9726	.9732	.9738	.9744	.9750	.9756	.9761	.9767
2.0	.9772	.9778	.9783	.9788	.9793	.9798	.9803	.9808	.9812	.9817
2.1	.9821	.9826	.9830	.9834	.9838	.9842	.9846	.9850	.9854	.9857
2.2	.9861	.9864	.9868	.9871	.9875	.9878	.9881	.9884	.9887	.9890
2.3	.9893	.9896	.9898	.9901	.9904	.9906	.9909	.9911	.9913	.9916
2.4	.9918	.9920	.9922	.9925	.9927	.9929	.9931	.9932	.9934	.9936
2.5	.9938	.9940	.9941	.9943	.9945	.9946	.9948	.9949	.9951	.9952
2.6	.9953	.9955	.9956	.9957	.9959	.9960	.9961	.9962	.9963	.9964
2.7	.9965	.9966	.9967	.9968	.9969	.9970	.9971	.9972	.9973	.9974
2.8	.9974	.9975	.9976	.9977	.9977	.9978	.9979	.9979	.9980	.9981
2.9	.9981	.9982	.9982	.9983	.9984	.9984	.9985	.9985	.9986	.9986
3.0	.9987	.9987	.9987	.9988	.9988	.9989	.9989	.9989	.9990	.9990
3.1	.9990	.9991	.9991	.9991	.9992	.9992	.9992	.9992	.9993	.9993
3.2	.9993	.9993	.9994	.9994	.9994	.9994	.9994	.9995	.9995	.9995
3.3	.9995	.9995	.9995	.9996	.9996	.9996	.9996	.9996	.9996	.9997
3.4	.9997	.9997	.9997	.9997	.9997	.9997	.9997	.9997	.9997	.9998

Table V Trigonometric Functions

Degrees	Radians	Sin	Tan	Cot	Cos		
0	0	0	0	—	1.000	1.5708	90
1	0.0175	0.0175	0.0175	57.290	0.9998	1.5533	89
2	0.0349	0.0349	0.0349	28.636	0.9994	1.5359	88
3	0.0524	0.0523	0.0524	19.081	0.9986	1.5184	87
4	0.0698	0.0698	0.0699	14.301	0.9976	1.5010	86
5	0.0873	0.0872	0.0875	11.430	0.9962	1.4835	85
6	0.1047	0.1045	0.1051	9.5144	0.9945	1.4661	84
7	0.1222	0.1219	0.1228	8.1443	0.9925	1.4486	83
8	0.1396	0.1392	0.1405	7.1154	0.9903	1.4312	82
9	0.1571	0.1564	0.1584	6.3138	0.9877	1.4137	81
10	0.1745	0.1736	0.1763	5.6713	0.9848	1.3963	80
11	0.1920	0.1908	0.1944	5.1446	0.9816	1.3788	79
12	0.2094	0.2079	0.2126	4.7046	0.9781	1.3614	78
13	0.2269	0.2250	0.2309	4.3315	0.9744	1.3439	77
14	0.2443	0.2419	0.2493	4.0108	0.9703	1.3265	76
15	0.2618	0.2588	0.2679	3.7321	0.9659	1.3090	75
16	0.2793	0.2756	0.2867	3.4874	0.9613	1.2915	74
17	0.2967	0.2924	0.3057	3.2709	0.9563	1.2741	73
18	0.3142	0.3090	0.3249	3.0777	0.9511	1.2566	72
19	0.3316	0.3256	0.3443	2.9042	0.9455	1.2392	71
20	0.3491	0.3420	0.3640	2.7475	0.9397	1.2217	70
21	0.3665	0.3584	0.3839	2.6051	0.9336	1.2043	69
22	0.3840	0.3746	0.4040	2.4751	0.9272	1.1868	68
23	0.4014	0.3907	0.4245	2.3559	0.9205	1.1694	67
24	0.4189	0.4067	0.4452	2.2460	0.9135	1.1519	66
25	0.4363	0.4226	0.4663	2.1445	0.9063	1.1345	65
26	0.4538	0.4384	0.4877	2.0503	0.8988	1.1170	64
27	0.4712	0.4540	0.5095	1.9626	0.8910	1.0996	63
28	0.4887	0.4695	0.5317	1.8807	0.8829	1.0821	62
29	0.5061	0.4848	0.5543	1.8040	0.8746	1.0647	61
30	0.5236	0.5000	0.5774	1.7321	0.8660	1.0472	60
31	0.5411	0.5150	0.6009	1.6643	0.8572	1.0297	59
32	0.5585	0.5299	0.6249	1.6003	0.8480	1.0123	58
33	0.5760	0.5446	0.6494	1.5399	0.8387	0.9948	57
34	0.5934	0.5592	0.6745	1.4826	0.8290	0.9774	56
35	0.6109	0.5736	0.7002	1.4281	0.8192	0.9599	55
36	0.6283	0.5878	0.7265	1.3764	0.8090	0.9425	54
37	0.6458	0.6018	0.7536	1.3270	0.7986	0.9250	53
38	0.6632	0.6157	0.7813	1.2799	0.7880	0.9076	52
39	0.6807	0.6293	0.8098	1.2349	0.7771	0.8901	51
40	0.6981	0.6428	0.8391	1.1918	0.7660	0.8727	50
41	0.7156	0.6561	0.8693	1.1504	0.7547	0.8552	49
42	0.7330	0.6691	0.9004	1.1106	0.7431	0.8378	48
43	0.7505	0.6820	0.9325	1.0724	0.7314	0.8203	47
44	0.7679	0.6947	0.9657	1.0355	0.7193	0.8029	46
45	0.7854	0.7071	1.0000	1.0000	0.7071	0.7854	45
		Cos	Cot	Tan	Sin	Radians	Degrees

Table VI Integrals

1. $\int [f(x) + g(x)] \, dx = \int f(x) \, dx + \int g(x) \, dx.$

2. $\int af(x) \, dx = a \int f(x) \, dx.$

3. $\int x^n \, dx = \frac{1}{n+1} x^{n+1} + c \quad (n \neq -1).$

4. $\int \frac{1}{x} \, dx = \log_e |x| + c.$

5. $\int \sin x \, dx = -\cos x + c.$

6. $\int \cos x \, dx = \sin x + c.$

7. $\int e^{ax} \, dx = \frac{1}{a} e^{ax} + c.$

8. $\int \frac{1}{a^2 + x^2} \, dx = \frac{1}{a} \tan^{-1} \frac{x}{a} + c.$

9. $\int \frac{1}{\sqrt{a^2 - x^2}} \, dx = \sin^{-1} \frac{x}{a} + c.$

10. $\int xc^{ax} \, dx = \frac{1}{a^2} (ax - 1) e^{ax} + c.$

11. $\int e^{ax} \sin bx \, dx = \frac{1}{a^2 + b^2} (a \sin bx - b \cos bx) e^{ax} + c.$

12. $\int e^{ax} \cos bx \, dx = \frac{1}{a^2 + b^2} (a \cos bx + b \sin bx) e^{ax} + c.$

13. $\int f(g(x)) g'(x) \, dx = \int f(u) \, du.$

14. $\int f(x)g'(x) \, dx = f(x)g(x) - \int g(x)f'(x) \, dx.$

Answers to Selected Problems

Chapter 1

1.1 **2.** $S_2 = S_6$, $S_1 \subset S_3$, $S_1 \subset S_5$, $S_4 \subset S_3$.

  4. $\varnothing$, s_1, s_2, s_3, $s_1 s_2$, $s_1 s_3$, $s_2 s_3$, $s_1 s_2 s_3$.

  6. $D_1 D_2$, $D_1 D_3$, $D_1 D_4$, $D_2 D_3$, $D_2 D_4$, $D_3 D_4$, $D_1 D_2 D_3$, $D_1 D_2 D_4$, $D_1 D_3 D_4$, $D_2 D_3 D_4$, $D_1 D_2 D_3 D_4$.

1.2 **1.** $A \cup B = \{\text{adults}\}$, $B \cup C = \{\text{females and boys}\}$, $(A \cup B) \cup C = A \cup (B \cup C) = \{\text{people}\}$.

  4. (a) $A \cap B = \{\text{positive integers divisible by } 6\} = \{6, 12, 18....\}$.
  (b) $D = \varnothing$.

  6. (a) 20. (b) 350. (c) 500.

  7. (c) 1000.

1.3 **3.** Yes, no.

  5. (a) $D(f) = \mathbf{R} \setminus \{1, -1\}$, $R(f) = \mathbf{R} \setminus \{y : 0 \le y < 1\}$.
  (b) $D(g) = \mathbf{R}$, $R(g) = \mathbf{R}^+ \setminus \{0\}$.
  (c) $D(h) = \mathbf{R}$, $R(h) = \{y : 0 \le y \le 2\}$.

  8. Yes, yes.

  9. R_4 is a function.

1.4   **1.** 120, 60, 60.

2. 125, 100, 5.

3. (a) 8.   (b) 10.   (c) 1.

4. (a) 30.   (b) 21.

5. (a) 1260.   (b) 10,080.   (c) 90,720.

6. $5! = 120$.

7. $\begin{pmatrix} 12 \\ 4,4,4 \end{pmatrix}$

8. No, 4.

9. 3.48×10^{20}.

10. 64.

11. (a) 16.   (b) 10!.

12. 256.

13. 5400.

14. No.

1.5   **1.** (a) 35.   (b) 35.

2. $\begin{pmatrix} 20 \\ 10 \end{pmatrix}\begin{pmatrix} 10 \\ 5 \end{pmatrix}\begin{pmatrix} 5 \\ 5 \end{pmatrix} = \begin{pmatrix} 20 \\ 10, 5, 5 \end{pmatrix}$.

3. $\begin{pmatrix} 10 \\ 4 \end{pmatrix}\begin{pmatrix} 10 \\ 7 \end{pmatrix}\begin{pmatrix} 10 \\ 5 \end{pmatrix}$.

4. (a) 210.   (b) 120.

7. None.

8. 6.

9. $\begin{pmatrix} 10 \\ 3 \end{pmatrix}\begin{pmatrix} 8 \\ 2 \end{pmatrix}$.

10. (a) $\begin{pmatrix} n \\ 2 \end{pmatrix}$.   (b) $\begin{pmatrix} n \\ 2 \end{pmatrix}^{k}$.

11. $\begin{pmatrix} 4 \\ 2 \end{pmatrix} + \begin{pmatrix} 4 \\ 2 \end{pmatrix}^{2} + \begin{pmatrix} 4 \\ 2 \end{pmatrix}^{3} + \begin{pmatrix} 4 \\ 2 \end{pmatrix}^{4}$.

12. (a) $A_1A_1, A_1A_2, A_1A_3, A_2A_3, A_2A_2, A_3A_3$.

14. 6^n.

15. (a) 27.   (b) 8.

16. (a) $\begin{pmatrix} 20 \\ 10 \end{pmatrix}$.

(b) $\begin{pmatrix} 10 \\ 6 \end{pmatrix}\begin{pmatrix} 10 \\ 4 \end{pmatrix} + \begin{pmatrix} 10 \\ 7 \end{pmatrix}\begin{pmatrix} 10 \\ 3 \end{pmatrix} + \begin{pmatrix} 10 \\ 8 \end{pmatrix}\begin{pmatrix} 10 \\ 2 \end{pmatrix} + \begin{pmatrix} 10 \\ 9 \end{pmatrix}\begin{pmatrix} 10 \\ 1 \end{pmatrix} + \begin{pmatrix} 10 \\ 10 \end{pmatrix}\begin{pmatrix} 10 \\ 0 \end{pmatrix}$.

(c) $\dbinom{10}{8}\dbinom{10}{2} + \dbinom{10}{9}\dbinom{10}{1} + \dbinom{10}{10}\dbinom{10}{0}$.

1.6   **1.** (a) 35.   (b) 35.   (c) 3.   (d) 4.

      2. (a) 21.   (b) 0.   (c) 3.   (d) 2.

      3. (a) $x^4 + 8x^3y + 24x^2y^2 + 32xy^3 + 16y^4$.

      4. (a) $\dbinom{9}{2,3,4}$.   (b) $-\dbinom{9}{2,3,4}$.   (c) $\dbinom{15}{6,2,3,4}$.

      8. $\dbinom{6}{4} + \dbinom{6}{5} + \dbinom{6}{6} = 22$.

      11. $\dbinom{n+m-1}{n}$, $\dbinom{14}{10} = 1001$.

      12. (a) $1.00985\ldots$ (b) $3.02985\ldots$

      13. (a) $\frac{5}{4}$.   (b) $\frac{4}{3}$.   (c) 4.   (d) 20.

      15. (a) $2 - \left(\frac{1}{2}\right)^9$.   (b) 121.   (c) $\frac{1}{2}\left(3 - \frac{1}{3^{20}}\right)$.   (d) 1.111111.

      16. (a) $(1.10)^{1/6} = 1.016\ldots$

      17. $w(1) \approx 159.1$.

Chapter 2

2.2   **1.** (a) $\frac{1}{52}$.   (b) $\frac{1}{10}$.   (c) $\dfrac{1}{\dbinom{30}{10}}$.   (d) $\dfrac{1}{10^7}$.

      4. (a) $\frac{33}{100}$.   (b) $\frac{6}{100}$.

      5. (a) $\dfrac{1}{20!}$.   (b) 0.

      6. (b) $P(WWW) = \dfrac{\dbinom{6}{3}}{\dbinom{10}{3}}$, $P(WWB) = \dfrac{\dbinom{6}{2}\dbinom{4}{1}}{\dbinom{10}{3}}$,

            $P(WBB) = \dfrac{\dbinom{6}{1}\dbinom{4}{2}}{\dbinom{10}{3}}$, $P(BBB) = \dfrac{\dbinom{4}{3}}{\dbinom{10}{3}}$.

7. (a) $\binom{52}{5}$.   (b) $\binom{47}{5}$.

8. $\dfrac{\binom{15}{5}}{\binom{20}{5}}, \dfrac{\binom{15}{4}\binom{5}{1}}{\binom{20}{5}}, \dfrac{1}{\binom{20}{5}}$.

9. (a) $S = \{0, 1, 2, \ldots, 250\}$.   (b) Probably not.
    (c) $\{51, 52, \ldots, 250\}$.   (d) $\bar{B} = \{0, 1, 2, \ldots, 39\}, A \cup B = S,$
                                                       $A \cap B = \{41, 42, \ldots, 50\}$.

10. (a) $\dfrac{1}{\binom{20}{10}}$.

    (b) $\dfrac{1}{\binom{20}{10}}\left[\binom{10}{6}\binom{10}{4} + \binom{10}{7}\binom{10}{3} + \binom{10}{8}\binom{10}{2}\right.$

                           $\left. + \binom{10}{9}\binom{10}{1} + \binom{10}{10}\binom{10}{0}\right]$.

11. (a) $(\tfrac{1}{5})^4$.   (b) $(\tfrac{1}{5})^3$.

2.3   **2.** (a) $\tfrac{29}{30}$.   (b) $\tfrac{1}{30}$.
       3. (a) $P(A_1) = .4, P(A_1 \cap B_1) = .4, P(A_1 \cap B_2) = 0$.
          (b) $P(B_1) = .6, P(B_2) = .4, P(A_1 \cup B_2) = .8$.
          (c) 1.
       4. (a) $P(W) = \tfrac{1}{8}, P(E) = \tfrac{3}{40}$.
          (b) $P(W \cap E) = \tfrac{1}{40}$.
          (c) $P(E \cap \bar{W}) = \tfrac{1}{20}, P(\bar{E} \cap W) = \tfrac{1}{10}$.
       5. (a) $\binom{9}{3}$.   (b) $P(A_1) = P(A_2) = P(A_3) = \tfrac{4}{9}$.   (d) $\tfrac{1}{6}, \tfrac{1}{21}, \tfrac{37}{42}$.
       6. (a) $\tfrac{4}{17}$.   (b) $\tfrac{16}{17}$.

2.4   **1.** (a) $\tfrac{3}{8}$.   (b) $\tfrac{3}{7}$.   (c) $\tfrac{1}{2}$.
       2. $\tfrac{17}{32}$.
       3. (a) 4.   (b) 3.
       4. (a) $\tfrac{5}{16}$.   (b) $\tfrac{1}{32}$.
       5. $\tfrac{1}{2}, \tfrac{1}{70}$.
       6. (a) $P(A) = P(B) = P(C) = \tfrac{1}{2}, P(D) = \tfrac{1}{4}, P(E) = \tfrac{7}{8}$.
          (b) No.
          (c) No.

7. Yes.

8. (a) $(.98)^6$. (b) $(.02)^6$. (c) $\binom{6}{5}(.98)^5(.02)$.

9. $\frac{1}{4}, \frac{1}{4}$.

10. $\frac{1}{9}, \frac{7}{18}$.

11. (a) $1 - (\frac{5}{6})^4$. (b) $\frac{2}{5}$. (c) $\dfrac{1 - 2(\frac{5}{6})^6}{1 - (\frac{5}{6})^7}$.

12. (a) $1 - \frac{1}{4} - \frac{3}{4}(\frac{1}{4}) - (\frac{3}{4})^2\,\frac{1}{4} - (\frac{3}{4})^3\,\frac{1}{4}$. (b) $\frac{3}{4}(\frac{1}{4}) = \frac{3}{16}$.

(c) $\frac{3}{4}(\frac{1}{4}) + (\frac{3}{4})^3\,\frac{1}{4} + (\frac{3}{4})^5\,\frac{1}{4} + \cdots = \dfrac{\frac{3}{16}}{1 - \frac{9}{16}} = \frac{3}{7}$.

13. (a) No. (b) $\frac{1}{30}$.

14. (a) $\frac{31}{72}$. (b) $\frac{6}{31}$.

2.5 **1.** $\frac{15}{19}$.

2. $\frac{3}{143}$.

3. $\frac{133}{201} \approx .66$, $\frac{49}{82} \approx .60$.

4. .095.

5. $\frac{3}{8}$.

6. .085.

7. $\frac{5}{44}, \frac{12}{44}, \frac{27}{44}$.

8. $\frac{85}{86} \approx 0.988$.

9. .40, .30.

10. $\frac{22}{57}$.

11. $\frac{1}{4}, \frac{3}{16}, \frac{9}{16}$.

12. (a) $\frac{5}{17} \approx .294$. (b) $\frac{63}{85} \approx .741$.

13. $\frac{5}{11}$.

14. $\frac{48}{79} \approx .61$.

15. $\frac{9}{14} \approx .64$.

16. $\frac{15}{16} = .9375$.

17. (a) .70. (b) .226.

18. 0.1 per cent.

19. $\frac{20}{719} \approx .028$.

2.6 **1.** (a) $(.05)^5$. (b) $(.95)^5$. (c) $5(.95)^4(.05)$.

2. $(.5)^7, \binom{7}{4, 2, 1}(.5)^4(.25)^3$.

3. (a) $\binom{10}{5, 5, 0}(.4)^{10}$. (b) $\binom{10}{4, 4, 2}(.4)^8(.2)^2$.

(c) $\binom{10}{3, 3, 4}(.4)^6(.2)^4$.

4. 230.

5. $(.95)^{10} \approx .599$.

6. $(.92)^5$.

7. $\binom{5}{3,0,2}(.65)^3(.25)^2$.

8. .001536.

9. (a) $(.5)^9$. (b) $\binom{9}{5,3,1}(.5)^5(.4)^3(.1)$. (c) $\binom{9}{3,3,3}(.5)^3(.4)^3(.1)^3$.

10. (a) $(.75)^6$. (b) $(.25)^6$. (c) $\binom{6}{4}(.75)^4(.25)^2$.

(d) $\binom{6}{4}(.75)^4(.25)^2 + \binom{6}{5}(.75)^5(.25) + \binom{6}{6}(.75)^6$.

11. (a) $\binom{n}{n/2}(\frac{1}{2})^n$. (b) $\frac{1}{2},\frac{1}{2} - \frac{1}{2}\binom{n}{n/2}(\frac{1}{2})^n$.

12. $\frac{1}{2},\frac{1}{2} - \frac{1}{2^{15}}\binom{14}{7}$.

13. (a) 4. (b) 4.

14. (a) $(.5)^{12}$. (b) $\binom{12}{6,6}(.5)^6(.3)^6$.

(c) $\binom{12}{4,4,4}(.5)^4(.3)^4(.2)^4$.

15. (a) 21. (b) $(.8)^5$.

16. $\binom{6}{2}(.2)^2(.8)^4, 1 - (.8)^6, 1 - (.2)^6 - 6(.2)^5(.8)$.

17. (a) $\frac{r_i}{t_i}, \frac{d_i}{t_i}$. (b) $\binom{t_i}{r_i}\left(\frac{M_i}{N}\right)^{r_i}\left(1 - \frac{M_i}{N}\right)^{t_i - r_i}$.

18. $\frac{1}{32}$.

19. $(\frac{1}{2})^{20}$.

2.7 **1.** (a) $R_X = \{-1, 0, 1\}$. (b) $f(-1) = f(0) = f(1) = \frac{1}{3}$,

$$f(x) = 0 \text{ elsewhere.}$$

(c) $F(x) = \begin{cases} 0 & \text{for } x < -1, \\ \frac{1}{3} & \text{for } -1 \le x < 0, \\ \frac{2}{3} & \text{for } 0 \le x < 1, \\ 1 & \text{for } 1 \le x. \end{cases}$

2. (a) $S = \{1, 2, \ldots, 10\}$.   (b) $R_X = \{0, 1, 2\}$.
  (c) $f(0) = \frac{4}{10}$, $f(1) = \frac{4}{10}$, $f(2) = \frac{2}{10}$

$$F(x) = \begin{cases} 0 & \text{for } x < 0, \\ \frac{4}{10} & \text{for } 0 \le x < 1, \\ \frac{8}{10} & \text{for } 1 \le x < 2, \\ 1 & \text{for } 2 \le x. \end{cases}$$

3. (a) $R_X = \{0, 1, 2, 3, 4\}$.   (b) $f(0) = (\frac{1}{2})^4$, $f(1) = 4(\frac{1}{2})^4, \ldots$.
4. $R_X = \{0, 1, 2, 3, 4, 5\}$.

  (b) $f(k) = P(X = k) = \binom{5}{k}(\frac{1}{6})^k(\frac{5}{6})^{5-k}$ for $k = 0, 1, 2, \ldots, 5$.

5. (a) $R_X = \{0, 1, 2, 3, \ldots, 1000\}$.

6. $R_X = \{0, 1, 2, 3, 4, 5\}$, $f(k) = P(X = k) = \binom{5}{4}(\frac{1}{4})^k(\frac{3}{4})^{5-k}$ for

  $k = 0, 1, 2, \ldots, 5$.

7. $R_X = \{0, 1, 2, \ldots, 8\}$, $f(k) = P(X = k) = \binom{8}{k}(.8)^k(.2)^{8-k}$ for

  $k = 0, 1, 2, \ldots, 8$.

8. (a) $R_X = \{0, 1, 2, \ldots, 20\}$.
9. (a) $R_X = \{0, 1, 2, 3, 4\}$, $f(k) = P(X = k) = \frac{1}{5}$ for $k = 0, 1, 2, 3, 4$.
  (b) $P(X \ge 3) = \frac{2}{5}$, $P(X > 0) = \frac{4}{5}$.
10. (a) $R_X = \{1, 2, 3, \ldots\}$, $R_Y = \{0, 1, 2, 3, \ldots\}$.
  (b) $P(X = 2) = (1 - p)^2 p$, $P(Y = 1) = (1 - p)p$, $P(Y > 2) = 1 - p - (1 - p)p - (1 - p)^3 p$.
  (d) Fair only for $p = 0$.
11. $R_X = \{1, 2, 3, \ldots, 25\}$, $P(X = 10) = (\frac{2}{3})^9 \frac{1}{3}$.
12. (a) $(.6)^4$.   (b) $R_X = \{4, 5, 6, \ldots\}$, $P(X = k) = \binom{k-1}{3}(.6)^4(.4)^{k-4}$

    for $k = 4, 5, 6, \ldots$.
13. (a) $\frac{1}{625}$.   (b) $P(X < 26) = \frac{26}{50}$, $P(22 < X < 28) = \frac{119}{625}$,
               $P(X > 45) = \frac{2}{125}$.   (c) 42.
14. $(1 - p)^2 p$, $(1 - p)^6 p$.

2.8   1. (a) $R_X = \{0, 1, 5\}$, $f(0) = \frac{1}{2}$, $f(1) = \frac{3}{20}$, $f(5) = \frac{7}{20}$, and
        $f(x) = 0$ otherwise.   (b) $E(X) = 1.9$, var $(X) = 5.29$, $\sigma = 2.3$.
  2. (a) $E(X) = 2.5$, var $(X) = 1.25$, $\sigma \approx 1.12$.
    (b) $R_Y = \{-5, -3, -1, 3, 5\}$, $E(Y) = 0$, var $(Y) = 5$, $\sigma \approx 2.24$.
  3. $E(X) = \frac{2}{3}$, var $(X) = \frac{5}{9}$, $\sigma \approx .745$.
  4. (a) $P(0 \text{ per cent}) = (\frac{2}{3})^5 \approx .13$.
    (b) $E(X) = 33.3$ per cent, $\sigma \approx 21$ per cent.

5. $E(X) = 75$, var $(X) = \frac{900}{16}, \sigma = 7.5$.

7. $E(X) \approx \frac{3}{2}, \sigma \approx .866$.

11. $E(X) = 500, \sigma \approx 316$.

12. $E(X) = 7.5, f(k) = \binom{15}{k}(\frac{1}{2})^{15}$ for $k = 0, 1, 2, \ldots, 15$.

13. (a) .3.   (b) .76.   (c) 2.33.

14. (a) .1.   (b) .037.

15. $f(k) = P(X = k) = \frac{1}{10}$ for $k = 1, 2, \ldots, 10$ and $E(X) = 5.5$.

2.9   **1.** $E(X) = 4, P(k) = \binom{20,000}{k}(.0002)^k(.9998)^{20,000-k} \approx \frac{4^k}{k!}e^{-4}$,

$P(0) \approx .0183$.

3. $E(X) = 5, P(X > 2) = 1 - e^{-5} - 5e^{-5} - \dfrac{25e^{-5}}{2} \approx .875$.

4. .271.

5. $E(X) = 1, P(k) = \dfrac{e^{-1}}{k!}, .735$.

6. $e^{-N/2500} = \dfrac{275}{2500}, N \approx 5518$.

7. (a) 301.   (b) 337.

8. (a) 147.   (b) 147.   (c) 32.

9. (a) e^{-10}.   (b) $1 - e^{-10}$.

10. (a) 135,000.   (b) 270,000.   (c) 145,000.

11. (a) .050.   (b) .100.

12. Poisson with $\mu = 1$ and $\mu = 10$.

13. (a) .050.   (b) .353.

14. $E(X) = 10$, Poisson with $\mu = 10$.

15. .368, .080.

17. (a) 8.   (b) 8.   (c) .4.   (d) $e^{-.4} \approx .67$.

18. $e^{-2}e^{-1.2} \approx .041, \dfrac{2^2e^{-2}}{2!} \dfrac{(1.2)^2 e^{-1.2}}{2!} \approx .059$.

19. (a) .168, .066.   (b) .916.   (c) Some medical treatments require very large numbers of X-rays.

21. 2218.

Chapter 3

3.1   **1.** $(1, 1, 2) = (1, 1, 1) + (0, 0, 1)$; no.

2. $c_1 = 2, c_2 = 1$.

4. (a) $(8, -1, -1)$. (b) $(0, -11, 29)$. (c) $\sqrt{62}$. (d) $\sqrt{26}$. (e) $\sqrt{38}$. (f) -25.

5. (a) $3(2, 1) - (3, 2) + 3(1, -1) = (0, 0)$.

8. (a) $\dfrac{11}{5\sqrt{5}}$. (b) $\dfrac{2}{\sqrt{5}}$. (c) $\dfrac{12}{13}$. (d) $\dfrac{-5}{\sqrt{26}}$.

13. (a), (c), (d).

14. $(x_1, x_2, x_3) = x_1(1, 0, 0) + x_2(0, 1, 0) + x_3(0, 0, 1)$.

15. (a) -2. (b) 1. (c) $\dfrac{-5}{2}$.

3.2 **1.**
$$AB = \begin{pmatrix} 1 & 1 & 0 \\ 0 & 1 & 0 \\ 0 & 0 & -1 \end{pmatrix}, \; BA = \begin{pmatrix} 1 & 1 & 3 \\ 0 & 1 & 2 \\ 0 & 0 & -1 \end{pmatrix}.$$

2.
$$A^2 = \begin{pmatrix} 0 & 0 & 15 \\ 0 & 0 & 0 \\ 0 & 0 & 0 \end{pmatrix}, \; A^3 = A^4 = A^{100} = \mathbf{0}.$$

3.
$$A^2 = \begin{pmatrix} 0 & 2 \\ -2 & 0 \end{pmatrix}, \; A^3 = \begin{pmatrix} -2 & 2 \\ -2 & -2 \end{pmatrix}, \; A^4 = \begin{pmatrix} -4 & 0 \\ 0 & -4 \end{pmatrix}.$$

5.
$$A^{-1} = \begin{pmatrix} \frac{1}{2} & 0 & 0 \\ 0 & \frac{1}{3} & 0 \\ 0 & 0 & 1 \end{pmatrix}, \; B^{-1} = \begin{pmatrix} \frac{1}{5} & 0 & 0 & 0 \\ 0 & \frac{1}{4} & 0 & 0 \\ 0 & 0 & -\frac{1}{3} & 0 \\ 0 & 0 & 0 & \frac{1}{4} \end{pmatrix}.$$

6. (b) $A = \begin{pmatrix} 3 & 0 & 0 \\ 0 & -2 & 0 \\ 0 & 0 & 1 \end{pmatrix}$.

8. $k = 3$.

10.
$$A = \begin{pmatrix} 1 & 2 & 1 \\ 3 & 2 & -1 \\ 5 & 0 & 2 \end{pmatrix}.$$

12. (a) $\begin{pmatrix} 1 & 5 \\ 2 & 7 \end{pmatrix}$ (b) $\begin{pmatrix} 1 & -1 & 2 \\ -1 & 2 & 2 \\ 0 & 1 & -1 \end{pmatrix}$. (c) $\begin{pmatrix} 1 & -1 \\ 2 & -2 \\ 3 & -3 \\ 4 & -4 \end{pmatrix}$.

14. (a), (d) symmetric; (c), (f) skew-symmetric.

18.
$$AA^t = \begin{pmatrix} 5 & 3 \\ 3 & 10 \end{pmatrix}, A^tA = \begin{pmatrix} 10 & -2 & 3 \\ -2 & 4 & 0 \\ 3 & 0 & 1 \end{pmatrix}.$$

20. (a) Average individuals of each species consume one-half unit of each of the other two species per day.

21. The ith component of $A\mathbf{r}$ represents the daily energy intake for an average predator of the ith species.

3.3
  1. (a) $(62, 40, -94)$.   (b) $(1, 3, 5)$.

  2. $(5000, 2000, 3000)$.

  3. $k = 0$.

  4. System consistent for all values of λ.

  5. (a) $(1, 3, 1)$.

  6. $x_1 = 10{,}000 + x_3, x_2 = 10{,}000 - 2x_3, 0 \le x_3 \le 5000$.

  7. (a) -5. (b) 0. (c) 1. (d) 4. (e) 4.

  8. (b), (d).

  9. (a) $(1, 1, 1)$.

  10. (a) $\mathbf{x} + k\mathbf{w}$ for any constant k.

  11. (a) $x_1 = \dfrac{20}{3} + x_3, x_2 = \dfrac{-2}{3}$.

  12. $(2, 4, 6)$.

  13. $g = \dfrac{72 - 2r}{7}, m = \dfrac{96 - 5r}{7}, 0 \le r \le \frac{96}{5}$ hours

  15. (a) $g = m = r = 8$ hours.   (b) $g = \dfrac{1200}{100 + Q}$ hours.

  17. $(\frac{1}{4}, \frac{1}{8}, \frac{1}{8}, \frac{1}{8}, \frac{3}{8})$.

3.4
  1. (a) $\begin{pmatrix} 1 & -2 & 1 & 0 \\ 0 & 1 & -2 & 1 \\ 0 & 0 & 1 & -2 \\ 0 & 0 & 0 & 1 \end{pmatrix}$.   (c) $\begin{pmatrix} -51 & -29 & 52 \\ 7 & 4 & -7 \\ 9 & 5 & -9 \end{pmatrix}$.

  2. (a) $(\frac{5}{2}, 3, \frac{7}{2})$.   (b) $(4, -1, 3)$.

  4. (a) $\begin{pmatrix} \frac{3}{11} & -\frac{1}{11} \\ -\frac{1}{22} & \frac{2}{11} \end{pmatrix}$.   (b) No inverse.   (c) $\begin{pmatrix} 7 & -2 \\ -3 & 1 \end{pmatrix}$.

6. (a) $\begin{pmatrix} 1 & -1 \\ 0 & 1 \end{pmatrix}$. (b) $\begin{pmatrix} 1 & -1 & 0 \\ 0 & 1 & -1 \\ 0 & 0 & 1 \end{pmatrix}$.

8. (b) $\begin{pmatrix} 1 & 0 & 0 \\ 0 & \frac{1}{5} & -\frac{3}{5} \\ 0 & 0 & \frac{1}{2} \end{pmatrix}$.

13. (a), (b), (c).

3.5 **1.** (a) -14. (b) 0. (c) 0. (d) 2.
2. (a) $(1, 3)$. (b) $(1, -1, 1)$.
3. $(3, 6, -3)$.
4. (a), (b), (d) independent.
5. (a) No nontrivial solution.
16. (a) 30. (b) 0. (c) 28. (d) 12.
17. $(5, -1, -\frac{5}{2})$.

3.6 **4.** $\mathbf{z} = \mathbf{y} + k\mathbf{x}$ for any constant k.

5. (a) $\lambda = 1, 4; \begin{pmatrix} 1 \\ -1 \end{pmatrix}, \begin{pmatrix} 1 \\ 2 \end{pmatrix}$. (c) $\lambda = 1, 6; \begin{pmatrix} 1 \\ -1 \end{pmatrix}, \begin{pmatrix} 3 \\ 2 \end{pmatrix}$.

6. (a) $\lambda = \pm i = \pm\sqrt{-1}; \begin{pmatrix} i \\ 1 \end{pmatrix}, \begin{pmatrix} 1 \\ i \end{pmatrix}$.

(b) $\lambda = 2 \pm i, \begin{pmatrix} 1 - i \\ 1 \end{pmatrix}, \begin{pmatrix} 1 + i \\ 1 \end{pmatrix}$.

7. $\lambda = -1, 3, 4; \begin{pmatrix} 2 \\ 1 \\ -6 \end{pmatrix}, \begin{pmatrix} 2 \\ 3 \\ 2 \end{pmatrix}, \begin{pmatrix} 1 \\ 3 \\ 2 \end{pmatrix}$.

9. (a) $\lambda = 1, \begin{pmatrix} 1 \\ 0 \end{pmatrix}$, (b) $\lambda = 3, \begin{pmatrix} 1 \\ 0 \end{pmatrix}, \begin{pmatrix} 0 \\ 1 \end{pmatrix}$. (c) $\lambda = -1, 2, \begin{pmatrix} 3 \\ -1 \end{pmatrix}, \begin{pmatrix} 0 \\ 1 \end{pmatrix}$.

10. (a) $\lambda = 2, 1; \begin{pmatrix} 1 \\ 0 \\ 0 \end{pmatrix}, \begin{pmatrix} -4 \\ 1 \\ 1 \end{pmatrix}$.

11. (a) $\lambda = 9, 4, 1, -3; \begin{pmatrix} 1 \\ 0 \\ 0 \\ 0 \end{pmatrix}, \begin{pmatrix} 0 \\ 1 \\ 0 \\ 0 \end{pmatrix}, \begin{pmatrix} 0 \\ 0 \\ 1 \\ 0 \end{pmatrix}, \begin{pmatrix} 0 \\ 0 \\ 0 \\ 1 \end{pmatrix}$.

13. (a)
$$\mathbf{n}(1) = \begin{pmatrix} 105 \\ 110 \\ 180 \end{pmatrix}, \mathbf{n}(2) = \begin{pmatrix} 115.5 \\ 128 \\ 144 \end{pmatrix}.$$
(c) After 10 time periods.

Chapter 4

4.1   **1.** $4x_1 + 2x_2 = 1200.$
2. $x_1 = 520, x_2 = 240.$
3. 200, 400.
5. 260, 120.
6. 520, 240.
7. 200, 400.
8. 20, 40, 40.
9. There is no maximum.

4.2   **2.** (a), (b) Nonempty convex. (c) Empty. (d) and (e) Nonempty, nonconvex.
3. (a) $a_1x_1 + a_2x_2 + a_3x_3 = a_1.$
(b) $a_1x_1 + a_1x_2 + a_3x_3 = a_1.$
(c) $x_1 + x_2 - x_3 = 1.$
4. (a) $a_1x_1 + a_2x_2 + a_3x_3 + a_1x_4 = a_1 + a_2 + a_3.$
(b) $a_1x_1 + a_2x_2 + a_3x_3 + a_1x_4 = 0.$
(c) $x_2 - x_3 = 0.$
6. (a) $15, \frac{8}{3}.$ (b) 6, 1. (c) $6, \frac{4}{3}.$
7. Add twice the first inequality to the second inequality.
8. (a) $1 \le f(x) \le 3.$ (b) $f(x) = 3.$ (c) $1 \le f(x) \le 6.$
9. 17, 1.
10. 25, 0.
12. $370, $220.
13. (a)
$$A = \begin{pmatrix} 1 & 2 \\ -1 & 0 \\ 0 & -1 \end{pmatrix}, \mathbf{b} = \begin{pmatrix} 5 \\ 0 \\ 0 \end{pmatrix}.$$
(b)
$$A = \begin{pmatrix} 1 & -2 \\ 1 & 1 \\ 1 & 3 \end{pmatrix}, \mathbf{b} = \begin{pmatrix} 4 \\ 2 \\ 3 \end{pmatrix}.$$
15. $(3, 1), (1, 3), (\sqrt{5}, \sqrt{5}).$
16. $(8\sqrt{2/5}, 4\sqrt{2/5}), (4\sqrt{2/5}, 8\sqrt{2/5}), (4, 4).$

4.3   **1.** (a) 2,0. (b) 8.5, 0. (c) 51, $-9.$
2. (a) $(1, 2), (-3, 4), (5, -4).$
3. (a) $(0, \frac{9}{2}, \frac{1}{2}), (1, 4, 0), (0, 0, 0), (0, 0, 2), (3, 0, 0), (0, 5, 0).$
4. (a) $\frac{11}{2}, -3.$
5. $x_1 \ge 1, -3x_1 + 2x_2 \le 5, x_1 + 2x_2 \le 17, 5x_1 - 4x_2 \le 1.$

6. (a) $17, 3$.   (b) $-4, -24$.   (c) $-4, -22$.   (d) $72, 12$.

7. (a) $(160, 0, 0)$.   (b) $(160, 0, 0)$.

8. (a) $(80, 40, 0)$.   (b) Any point on the line segment joining
$$(80, 40, 0) \text{ and } (\tfrac{320}{3}, 0, \tfrac{40}{3}).$$

9. $(9, 9, 6)$.

10. $(\tfrac{2}{3}, \tfrac{7}{3})$, 89 cents.

11. $(\tfrac{2}{3}, \tfrac{7}{3}, 0)$, 89 cents.

12. $(8000, 18{,}000, 4000)$.

4.4   **1.** $(250, 0, 500)$.

3. (a) Minimize $g = 5y_1 + 6y_2 + 4y_3$ subject to

$$y_1 + y_2 + 2y_3 \geq 1,$$
$$y_1 - 2y_2 - y_3 \geq 1,$$
$$y_1 + 2y_2 + y_3 \geq -3,$$
$$y_1, y_2, y_3 \geq 0.$$

(b) Minimize $g = y_1$ subject to $y_1 \geq 1, y_1 \geq -1, y_1 \geq 1, y_1 \geq -1$ and $y_1 \geq 0$. (Obvious minimum is $g = 1$.)

(c) Maximize $f = x_1 + x_2$ subject to

$$x_1 + 2x_2 \leq 1,$$
$$x_1 - x_2 \leq \tfrac{1}{2},$$
$$2x_1 - x_2 \leq 1,$$
$$x_1, x_2 \geq 0.$$

4. The dual problem is to minimize $g(\mathbf{y}) = \mathbf{b} \cdot \mathbf{y}$ subject to $A^t \mathbf{y} \geq \mathbf{c}$ and $\mathbf{y} \geq \mathbf{0}$.

(a) $A^t = \begin{pmatrix} 1 & 2 \\ 1 & 3 \end{pmatrix}.$ (b) $A^t = \begin{pmatrix} 1 & 4 & -2 \\ 0 & -1 & 1 \end{pmatrix}.$

(c) $A^t = \begin{pmatrix} 1 & 0 & 2 \\ -1 & 2 & 3 \end{pmatrix}.$

7. (a) Too many corner points.

(b) When the dual problem involves fewer variables.

(c) The solution of the dual problem determines the maximum value of f and the minimum value of g. This gives an equation that the components of any optimal solution must satisfy.

9. Minimize $g = 10,000y_1 + 12,000y_2 + 8000y_3$ subject to

$$\tfrac{1}{2}y_1 + \tfrac{1}{2}y_2 \qquad\qquad \geq .3,$$
$$\tfrac{1}{3}y_1 + \tfrac{1}{3}y_2 + \tfrac{1}{3}y_3 \geq .4,$$
$$\tfrac{1}{2}y_2 + \tfrac{1}{2}y_3 \geq .5,$$
$$y_1, y_2, y_3 \geq 0.$$

The minimum is $g = 11{,}600$ at $(.2, .4, .6)$.

10. If both problems are solved independently, the maximum of f will be found to be equal to the minimum of g. If this is not the case, a mistake has been made.

4.5   **1.** Maximize $f = x_1 + 2x_2 + 0 \cdot s_1 + 0 \cdot s_2$ subject to

$$x_1 + x_2 + s_1 \qquad\quad = 3,$$
$$2x_1 + x_2 \qquad + s_2 = 3,$$
$$x_1, x_2, s_1, s_2 \geq 0.$$

Slack variables are s_1 and s_2. The initial corner point is $x_1 = x_2 = 0$, $s_1 = s_2 = 3$. The solution is $f = 6$ at $x_1 = 0, x_2 = 3$.

2. Minimize $g = 3y_1 + 3y_2$ subject to

$$y_1 + 2y_2 \geq 1,$$
$$y_1 + y_2 \geq 2,$$
$$y_1, y_2 \geq 0.$$

The solution is $g = 6$ at $y_1 = 0, y_2 = 2$, or $y_1 = 2, y_2 = 0$.

4. Max $f = 4$ on the line segment joining $(4, 0, 0)$ and $(3, 0, 1)$.

5. Min $g = 4$ at $(1, 0)$.

6. (a)

x_1	x_2	x_3	s_1	s_2	s_3		
1	1	−1	1	0	0	1	s_1
1	−1	1	0	1	0	2	s_2
−1	1	1	0	0	1	3	s_3
1	2	3	0	0	0	f	

(b)

x_1	x_2	x_3	s_1	s_2	s_3	s_4		
1	1	2	1	0	0	0	5	s_1
2	1	1	0	1	0	0	7	s_2
2	-1	3	0	0	1	0	8	s_3
1	2	5	0	0	0	1	9	s_4
1	-1	1	0	0	0	0	f	

7. (a)

x_1	x_2	x_3	s_1	s_2	s_3		
1	2	4	1	0	0	1	s_1
-1	1	1	0	1	0	1	s_2
1	1	-4	0	0	1	1	s_3
0	2	3	0	0	0	f	

(b)

x_1	x_2	s_1	s_2	s_3		
2	1	1	0	0	1	s_1
1	1	0	1	0	2	s_2
1	-1	0	0	1	1	s_3
5	4	0	0	0	f	

8. (a)

②	-1	②	1	0	1
-2	0	3	0	1	2
1	1	1	0	0	f

(b)

①	①	①	1	0	1
-1	0	1	0	1	2
2	1	3	0	0	f

(c)

1	2	3	1	0	0	5
2	3	1	0	1	0	3
③	1	②	0	0	1	1
2	−1	3	0	0	0	f

9. (a) Unbounded solution, no terminal tableau.

(b) $f = 3$ at $(0, 0, 1)$; $g = 3$ at $(3, 0)$.

(c) $f = \frac{3}{2}$ at $(0, 0, \frac{1}{2})$; $g = \frac{3}{2}$ at $(0, 0, \frac{3}{2})$.

10. $f = 10$ at $(5, 0, 0)$.

11. $g = 10$ at $(2, 0)$.

12. $f = 16$ at $(8, 0, 0)$.

13. $g = 16$ at $(0, 0, 2)$.

14. $g = \frac{13}{2}$ at $(2, \frac{3}{2})$.

4.6   **1.** $f = 11$ at $(\frac{19}{5}, \frac{12}{5})$.

2. $f = 36$ at $(8, 4)$.

3. $f = 5$ at $(1, 4, 0)$.

4. $g = 11$ at $(2, 1)$.

5. $g = 36$ at $(0, 4, 1)$.

6. $f = 9$ at $(1, 0, 2)$.

Chapter 5

5.1   **1.** (b), (c).

2.

$$P^2 = \begin{pmatrix} \frac{3}{8} & \frac{5}{16} & \frac{5}{16} \\ \frac{5}{16} & \frac{3}{8} & \frac{5}{16} \\ \frac{5}{16} & \frac{5}{16} & \frac{3}{8} \end{pmatrix}, \; p_{11}^{(2)} = \frac{3}{8}, \; p_{12}^{(2)} = p_{13}^{(2)} = \frac{5}{16}.$$

3.

$$P = \begin{pmatrix} 0 & 0 & 0 & \frac{1}{2} & \frac{1}{2} \\ 0 & 0 & \frac{1}{2} & 0 & \frac{1}{2} \\ 0 & \frac{1}{2} & 0 & 0 & \frac{1}{2} \\ \frac{1}{2} & 0 & 0 & 0 & \frac{1}{2} \\ \frac{1}{4} & \frac{1}{4} & \frac{1}{4} & \frac{1}{4} & 0 \end{pmatrix}.$$

5. $(\frac{5}{13}, \frac{4}{13}, \frac{4}{13})$. Any multiple such as $(5, 4, 4)$ is a fixed vector.

6. (a) $\begin{pmatrix} 1 & 0 \\ \frac{3}{4} & \frac{1}{4} \end{pmatrix}.$ (b) $\begin{pmatrix} \frac{1}{2} & \frac{1}{2} \\ \frac{1}{2} & \frac{1}{2} \end{pmatrix}.$

7. $(0, 0, 1)$, yes.

8. $(0, 0, 1)$ or $(0, 1, 0)$.

9.
$$\lim_{n \to \infty} P^n = \begin{pmatrix} 0 & \frac{1}{2} & \frac{1}{2} \\ 0 & 1 & 0 \\ 0 & 0 & 1 \end{pmatrix}.$$

10.
$$\lim_{n \to \infty} P^n = \begin{pmatrix} 0 & 0 & 1 \\ 0 & 0 & 1 \\ 0 & 0 & 1 \end{pmatrix}.$$

11.
$$P = \begin{pmatrix} .98 & .02 \\ .30 & .70 \end{pmatrix}.$$

12. (a) $.3, (.7)(.3) = .21, .147$. (b) $\frac{7}{3}$.

13. (a) $.98$. (b) $.7$.

14.
$$P = \begin{pmatrix} .8 & .2 \\ 0 & 1 \end{pmatrix}.$$

15. (a) $.328$. (b) 4.

5.2 **2.** 50 per cent.

3.
$$P = \begin{pmatrix} \frac{1}{2} & \frac{1}{4} & \frac{1}{4} \\ \frac{1}{4} & \frac{1}{2} & \frac{1}{4} \\ \frac{1}{4} & \frac{1}{4} & \frac{1}{2} \end{pmatrix}, t = (\tfrac{1}{3}, \tfrac{1}{3}, \tfrac{1}{3}).$$

5.
$$P^n = \begin{pmatrix} \dfrac{1}{2} + \dfrac{1}{2 \cdot 7^n} & \dfrac{1}{2} - \dfrac{1}{2 \cdot 7^n} \\ \dfrac{1}{2} - \dfrac{1}{2 \cdot 7^n} & \dfrac{1}{2} + \dfrac{1}{2 \cdot 7^n} \end{pmatrix}.$$

6. (a) No. (b) $(\tfrac{1}{2}, 0, \tfrac{1}{2})$ or $(0, 1, 0)$.

(c)
$$P^m = \begin{pmatrix} 0 & 0 & 1 \\ 0 & 1 & 0 \\ 1 & 0 & 0 \end{pmatrix}(m \text{ odd}), \quad P^m = \begin{pmatrix} 1 & 0 & 0 \\ 0 & 1 & 0 \\ 0 & 0 & 1 \end{pmatrix}(m \text{ even}).$$

7. (a) Yes. (b) $(\tfrac{1}{3}, \tfrac{1}{3}, \tfrac{1}{3})$.

8. (a) $\mathbf{p}^{(2)} = (\tfrac{1}{4}, \tfrac{1}{2}, \tfrac{1}{4})$.

10. (a) $.15$. (b) $.216$.

5.3   **2.** None is absorbing.

3.
$$P = \begin{pmatrix} \frac{2}{3} & \frac{1}{3} & 0 \\ 0 & \frac{2}{3} & \frac{1}{3} \\ 0 & 0 & 1 \end{pmatrix}, \frac{4}{27}.$$

4. 6.

5.
$$P = \begin{pmatrix} \frac{1}{5} & \frac{4}{5} & 0 & 0 & 0 \\ 0 & \frac{1}{5} & \frac{4}{5} & 0 & 0 \\ 0 & 0 & \frac{1}{5} & \frac{4}{5} & 0 \\ 0 & 0 & 0 & \frac{1}{5} & \frac{4}{5} \\ 0 & 0 & 0 & 0 & 1 \end{pmatrix}.$$

6. 5.

8. $\frac{93}{49} \approx 1.90$.

10. (a) Rules favor team B.

5.4   **5.** $E(X) = -5$ cents, yes.

5.5   **3.** Play first row and second column.

4. Second row, first column.

5. (a) First row, second column. (b) First, first.
  (c) First, second.

6. (a) $p_0 = (\frac{3}{4}, \frac{1}{4})$, $q_0 = (\frac{1}{2}, \frac{1}{2})$.
  (b) $p_0 = (\frac{1}{2}, \frac{1}{2})$, $q_0 = (\frac{1}{2}, \frac{1}{2})$.

8. (a) $p_0 = (1 - t, t)$, $q_0 = (1 - t, t)$.
  (b) First row; second column if $t \leq \frac{1}{2}$ and first column if $t > \frac{1}{2}$.

11. (a), (b), (c).

13. 1000 acres I, 500 acres II.

14. 1500 acres II.

5.6   **1.** (a) $p_0 = (\frac{5}{7}, \frac{2}{7})$, $q_0 = (\frac{6}{7}, \frac{1}{7})$.
  (b) $p_0 = (\frac{2}{3}, \frac{1}{3})$, $q_0 = (\frac{1}{3}, \frac{2}{3})$.

2. (a) $p_0 = (\frac{3}{4}, \frac{1}{4}, 0)$, $q_0 = (\frac{1}{2}, \frac{1}{2}, 0)$, $v = \frac{1}{2}$.
  (b) $p_0 = (1, 0, 0)$, $q_0 = (0, 0, 1)$.

5. $p_0 = (1, 0, 0, 0)$, $q_0 = (0, 0, 0, 1)$.

Chapter 6

6.1   **1.** (a) $\lim_{n \to \infty} x_n = 100$. (b) $\lim_{n \to \infty} x_n = 100$.

2. (a) Third order. (b) Second order.

6. (a) $x_{n+2} = (1.25)x_n$. (b) Second order.
(c) $x_2 = 2000, x_4 = 2500$.

7. (a) $x_{n+1} = (1.04)x_n$. (b) First order.
(c) $x_1 = 5.20, x_2 = 5.41, x_3 = 5.62, x_4 = 5.85$.

6.2 **1.** (a) $x_n = x_0 + 2\left(1 - \dfrac{1}{2^n}\right)$. (b) $x_n = 3^n x_0$.

(c) $x_n = \dfrac{(n+4)(n+3)}{12} x_0$. (d) $x_n = 0, n \geq 1$.

(e) $x_n = (-1)^n x_0$.

2. (a) $x_n = 2^{-n}$. (b) $x_n = 1 + \dfrac{1 - e^{-n}}{1 - e^{-1}}$. (c) $x_n = n + 1$.

(d) $x_n = 2 \cdot 3^n - 1$.

3. (a) $cn^{-1/2}$. (b) $x_n = c\left(1 + \dfrac{1}{\sqrt{2}} + \dfrac{1}{\sqrt{3}} + \cdots + \dfrac{1}{\sqrt{n}}\right)$.

4. $p_n = \frac{2}{3} + \frac{1}{3}(-\frac{1}{2})^n$.

5. $w_{n+1} = (1 + a)w_n, (1 + a)^{30} = \frac{11}{4}, w_{15} \approx 124$ pounds.

6. $x_n = 9 + 2^n$.

8. $r_n = (.99)r_{n-1}, r_4 = (.99)^4 r_0 \approx (.96)r_0$.

9. Six years.

10. $p_0 = .3, p_n = (.3)(.7)^n$.

11. $p_n = \dfrac{1}{n!}p_0$.

15. .132, 6.

6.3 **1.** (a) $x_n = k_1\left(\dfrac{1 + \sqrt{5}}{2}\right)^n + k_2\left(\dfrac{1 - \sqrt{5}}{2}\right)^n$.

(b) $x_n = k_1 + k_2(-3)^n$. (c) $x_n = k_1(-2)^n + k_2 n(-2)^{n-1}$.
(d) $x_n = k_1 + k_2(-\frac{1}{3})^n$.

2. (a) $x_n = 2^n$. (b) $x_n = (-3)^n\left(1 - \dfrac{5n}{3}\right)$.

(c) $x_n = 2^{n/2}\left(\cos\dfrac{n\pi}{4} + \sin\dfrac{n\pi}{4}\right)$.

(d) $x_n = \frac{1}{2}\left(1 + \dfrac{3}{\sqrt{5}}\right)\left(\dfrac{1 + \sqrt{5}}{2}\right)^n + \frac{1}{2}\left(1 - \dfrac{3}{\sqrt{5}}\right)\left(\dfrac{1 - \sqrt{5}}{2}\right)^n$.

4. 16 years.

5. $x_{n+2} - 18x_{n+1} + 32x_n = 0.$

6. $x_{n+2} - 3x_{n+1} + \frac{9}{4}x_n = 0.$

7. $x_{n+2} + 16x_n = 0.$

8. (a) $x_n = \dfrac{1}{\sqrt{5}}\left(\dfrac{1+\sqrt{5}}{2}\right)^n - \dfrac{1}{\sqrt{5}}\left(\dfrac{1-\sqrt{5}}{2}\right)^n.$

    (b) $\displaystyle\lim_{n\to\infty}\dfrac{x_{n+1}}{x_n} = \dfrac{1+\sqrt{5}}{2}.$

10. Roots of auxiliary equation must have absolute value less than 1.

11. (a) $x_n = 2 - (-1)^n.$ (b) $x_n = 3 - 2^n.$

12. (a) $x_n = (-3)^n + 5n(-3)^{n-1}.$ (b) $x_n = -2(-1)^n - 6n(-1)^{n-1}.$

13. (a) $x_n = \dfrac{-1}{\sqrt{3}}\sin\dfrac{n\pi}{3} + \cos\dfrac{n\pi}{3}.$

    (b) $x_n = 4\left(\dfrac{1}{\sqrt{3}}\right)^{n-1}\sin\dfrac{n\pi}{6}.$

15. $x_n = 1000(1.6)^n + 1000(1.4)^n.$

6.4   **1.** (a) $x_n = n + k_0 + l_0(-1)^n.$ (b) $x_n = \frac{1}{14}(n - \frac{9}{14})^2 + k_0 + l_0(-6)^n.$
      (c) $x_n = -\frac{4}{7}(1.5)^n + k_0(2)^n + l_0(-2)^n.$

      (d) $x_n = \dfrac{3^n}{2} + k_1 + k_2(2)^n.$

  2. (a) $x_n = -\dfrac{1}{8} + \dfrac{3^n}{4} - \dfrac{(-3)^n}{8}.$

      (c) $x_n = \left(-\dfrac{19}{15}\right)(-1)^n + \left(\dfrac{64}{51}\right)\left(-\dfrac{1}{4}\right)^n + \dfrac{4^n}{85}.$

  3. $ar^2 + br + c \neq 0.$

  8. (c) $\frac{1}{14}3^n - \frac{1}{6}2^n.$

10. $x_n = (500 + 200\sqrt{5})\left(\dfrac{1+\sqrt{5}}{2}\right)^n$

         $+ (500 - 200\sqrt{5})\left(\dfrac{1-\sqrt{5}}{2}\right)^n + 1000.$

6.5   **1.** (a) $x_n = k_1 + k_2 3^n,\ y_n = k_2 3^n - k_1.$
      (c) $x_n = k_0 + (-1)^n l_0 + \frac{1}{3}2^n,\ y_n = -k_0 + (-1)^n l_0 + \frac{1}{3}2^n.$
  2. (a) $x_n = -50 + 150(3)^n,\ y_n = 150(3)^n + 50.$
  6. $x_n = (500)2^n + 500.$

Chapter 7

7.1 **1.** (a) $\lim_{t \to \infty} x(t) = 100$. (b) $\lim_{t \to \infty} x(t) = 100$.

2. (a) First order. (b) Second order. (c) Second order.
(d) Third order.

3. (a) $\left(\dfrac{dx}{dt}\right)^2 = 4x$. (b) $\dfrac{dx}{dt} = x$. (c) $\dfrac{dx}{dt} = x\left(1 + \dfrac{1}{t}\right)$.

6. (b) $x(t) = e^{2t}$.

7. $\dfrac{dx}{dt} = \dfrac{x}{5}$.

8. $\dfrac{dw}{dt} = \dfrac{w}{2}$.

7.2 **1.** (a) $x(t) = ce^{5t}$. (b) $x(t) = \sin t + c$.
(c) $x(t) = t^3 + 5t + c$. (d) $x(t) = ce^t - 1 - t$.

2. (a) $x(t) = e^{3t}$. (b) $x(t) = 1 + t + t^2 + \dfrac{t^3}{3}$.

(c) $x(t) = -\dfrac{e^{-2t}}{2} + \dfrac{3}{2}$. (d) $x(t) = 2e^t - 1$.

3. 124 pounds.

5. (a) $x(t) = 0$. (b) $x(t) = \dfrac{(t + 1)^2}{3} - \dfrac{1}{3(t + 1)}$.

(c) $x(t) = \dfrac{(1 + t)^4}{4} - \frac{1}{4}$. (d) $x(t) = \sin^2 t + \sin t$.

6. (a) $\dfrac{1}{k} \log_e 2 \approx \dfrac{.693}{k}$. (b) $\dfrac{1.386}{k}$.

7. (a) $T(1) \approx 95.1^\circ$, $T(10) \approx 60.7^\circ$.
(b) 13.9, 27.7 hours.

9. (a) $x(t) = \dfrac{3t^3}{2} + \dfrac{c}{t}$. (b) $x(t) = 1 + ce^{1/t}$.

12. (a) $x(t) = (1 - t + ce^{-t})^{-1}$. (b) $x(t) = \dfrac{-t^2}{t + c}$.

14. (a) $p(t) = p(0)e^{-\mu t}$, $q(t) = 1 - p(0)e^{-\mu t}$.

(b) $\dfrac{1}{\mu} \log_e \dfrac{5}{3} \approx \dfrac{.510}{\mu}$.

16. (a) $\dfrac{dS}{dt} = r - cS.$ (b) $S(t) = \dfrac{r}{c} + \left[S(0) - \dfrac{r}{c}\right]e^{-rt}.$

7.3   **1.** (a) $x = ke^{\sin t}.$ (b) $-\dfrac{1}{x} = t + \dfrac{t^3}{3} + c.$

    (c) $\tan^{-1} x = \dfrac{t^3}{3} + c.$

  2. (a) $x = e^{\sin t - \cos t + 1}.$ (b) $-\dfrac{1}{x} = t + \dfrac{t^3}{3} - 1.$

    (c) $\tan^{-1} x = \dfrac{t^3}{3} + \dfrac{\pi}{4}.$

  5. $x(t) = \dfrac{100e^{\beta t}}{1 - 1/1000 + \dfrac{1}{1000}e^{\beta t}},$ where

    $\beta = \log_e \dfrac{120 - 3/125}{100 - 3/125} \approx .18.$

  9. (b) $\lim\limits_{t \to \infty} c(t) = \sqrt{\dfrac{\omega}{r}}.$

  10. (a) $\dfrac{b - x}{a - x} = \dfrac{b}{a}e^{(b-a)rt}.$ (b) $\lim\limits_{t \to \infty} x(t) = a = 10.$

  13. (b) $x(t) = \left[x(0)^{1/3} + \dfrac{ct}{3}\right]^3.$ (c) .260.

  15. $x(t) = \dfrac{n(n + 1)e^{-(n+1)rt}}{1 + ne^{-(n+1)rt}},$ when $x = \dfrac{n + 1}{2}.$

7.4   **1.** (a) $x(t) = k_1e^{-t} + k_2e^{-2t}.$ (b) $x(t) = k_1e^t + k_2e^{-5t}.$

    (c) $x(t) = k_1 \cos\dfrac{t}{2} + k_2 \sin\dfrac{t}{2}.$ (d) $x(t) = k_1e^{-2t} + k_2te^{-2t}.$

  2. (a) $x(t) = \frac{4}{5}e^{2t} + \frac{1}{5}e^{-3t}.$

    (b) $e^{-t/2}\left(\cos\dfrac{\sqrt{3}}{2}t + \sqrt{3}\sin\dfrac{\sqrt{3}}{2}t\right).$

    (c) $x(t) = \cos 3t + \frac{1}{3}\sin 3t.$
    (d) $x(t) = e^{-t/2} + \frac{3}{2}te^{-t/2}.$

  3. (a) $x(t) = \sin t.$ (b) $x(t) = 1 - e^{-t}.$
    (c) $x(t) = te^{-t}.$ (d) $x(t) = e^{-t}\sin t.$

4. $x''(t) - 8x'(t) - 9x(t) = 0.$

5. $x''(t) - 4x'(t) + 4x(t) = 0.$

6. $x''(t) + 6x'(t) + 10x(t) = 0.$

9. (a) $x(t) = \frac{1}{2}e^t + \frac{1}{2}e^{-t}.$

10. (a) $x(t) = 4te^{2t} - e^{2t}.$

11. (a) $x(t) = e^{(-5/2)t}\left[2\cos\left(\frac{\sqrt{3}}{2}t\right) + \frac{10}{\sqrt{3}}\sin\left(\frac{\sqrt{3}}{2}t\right)\right].$

14. $x''(t) - (.18)x'(t) + (.008)x(t) = 0.$

7.5   **1.** (a) $x(t) = -t + k_1e^t + k_2e^{-t}.$

(b) $x(t) = \frac{t^2}{2}e^{-2t} + k_1e^{-2t} + k_2te^{-2t}.$

(c) $x(t) = \frac{-t\cos t}{2} + k_1\cos t + k_2\sin t.$

(d) $x(t) = t + 3 + k_1e^t + k_2e^{-t/3}.$

2. (a) $x(t) = \frac{3}{4} - \frac{3}{4}\cos 2t + \frac{1}{2}\sin 2t.$

(b) $x(t) = \frac{1}{3}e^t - \frac{1}{3}e^{-t/2}\cos\frac{\sqrt{3}}{2}t + \frac{1}{\sqrt{3}}e^{-t/2}\sin\frac{\sqrt{3}}{2}t.$

3. $ar^2 + br + c \neq 0.$

8. (a) $x(t) = \frac{1}{9} + \frac{e^{2t}}{25} + k_1e^{-3t} + k_2te^{-3t}.$

(b) $x(t) = -\frac{e^t}{6} - \frac{e^{2t}}{6} + k_1e^{4t} + k_2e^{-t}.$

9. (a) $x(t) = x(0) + x'(0)t - \frac{g}{2}t^2.$ (b) $\sqrt{2gh}$ vertically downward.

10. $x(t) = 1000 + 500\cos 2\pi t.$

7.6   **1.** (a) $x(t) = k_1 + k_2e^{2t}, y(t) = k_1 - k_2e^{2t}.$

(c) $x(t) = k_1e^t + k_2e^{-t} + te^{-t}, y(t) = -k_1e^t + k_2e^{-t} + te^{-t}.$

2. (a) $x(t) = 150 - 50e^{2t}, y(t) = 150 + 50e^{2t}.$

(c) $x(t) = -50e^t + 150e^{-t} + te^{-t}, y(t) = 50e^t + 150e^{-t} + te^{-t}.$

5. (b) $x(t) = te^{2t} + e^{2t}, y(t) = te^{2t}.$

7. (a) $x(t) = \left[x(0) - \frac{by(0)}{c+a}\right]e^{-at} + \frac{by(0)}{c+a}e^{ct}, y(t) = y(0)e^{ct}.$

Chapter 8

8.1 **1.** All are distribution functions.

2. (a) 0. (b) $\frac{1}{2} - \frac{1}{2}e^{-2} \approx .432$. (c) 1. (d) $\frac{1}{2}$.

3. (a) $k = 1$. (b) $P(X \leq 2) = F(2) = 1 - e^{-1} \approx .632$.

4. (a) $A = \frac{1}{2}$. (b) $P(X \geq -1) = 1 - F(-1) = 1 - \frac{1}{2}e^{-1} \approx .816$.

5. $\frac{10}{11} \approx .909, \frac{60}{61} \approx .984$.

6. $\frac{1}{64}$.

7. $\frac{9}{16}, \frac{6}{16}, \frac{1}{16}$.

8. (a) $0 \leq x \leq 250$. (c) .28, .49.

8.2 **1.** All are density functions.

2. (b) $0, \frac{1}{2}, .229, \frac{1}{4}, .316, \frac{1}{2}$.

(c) $E(X) = -1, 0, \dfrac{\pi}{2}, 0, 0, 0; \sigma^2 = 1, \frac{1}{12}, \dfrac{\pi^2}{4} - 2$, undefined, $2, \frac{1}{5}$.

3. (a) $k = 3$. (b) $P(X \leq 1) = 1 - e^{-3} \approx .950, P(X \geq 2) = e^{-6}$,
$$P(1 \leq X \leq 2) = e^{-3} - e^{-6}.$$

4. (a) $1 - e^{-1} \approx .632$. (b) $1 - e^{-2} \approx .865$.

6. (b) .594, .092, .314, .332. (c) $E(X) = 1, \sigma \approx .71$.

8. (a) $\dfrac{1}{\gamma}$ days. (b) .632. (c) .9933.

9. (a) 133.3 days. (b) .0225. (c) .23.

10. (a) $k = \frac{1}{36}$.

11. (a) $\frac{1}{2}\sqrt{\pi}$ years. (b) $e^{-4} \approx .0183$.

13. .223, .393, .632, .777.

14. .135.

8.3 **1.** (a) $Y = \dfrac{X - 5}{2}$.

(b) $P(3 \leq X) = P(Y \leq -1) = 1 - P(Y \geq 1) = .1587, .6915, .8664$.
(c) $a = 8.29, b = 9.65$.

2. (b) $21.08 \leq X \leq 28.92, 19.84 \leq X \leq 30.16$.

3. .440, .119.

4. .308, $26.08 \leq X \leq 33.92, 24.84 \leq X \leq 35.16$.

5. (a) .106. (b) .468. (c) .106.

6. (a) .274. (b) .345. (c) .341, .615.

7. .015.

8. .977, .999.

9. .955.

10. .955.

11. .159, .023.

12. .317.

13. .82.

14. .16.

8.4 **1.** (b) .9545, .9973.

2. (a) $P(45 < X < 105) \geq 1 - (\frac{1}{4})^2$. (b) 1500.

4. (a) 5. (b) 25.5.

5. (a) 18,000. (b) 3460.

7. .26 to .36 and .245 to .375, .43 to .53 and .415 to .545.

9. (a) .9772, .9772. (b) .0228.

11. (a) 80, 80.

13. (a) 1920. (b) 369.

14. .173 to .191, .170 to .194.

Chapter 9

9.2 **4.** $\frac{4}{7}$.

9.3 **5.** .495.

6. $\frac{6241}{6400}, \frac{158}{6400}, \frac{1}{6400}$.

7. Expand $(\frac{1}{4} + \frac{1}{4} + \frac{1}{4} + \frac{1}{4})^2$. For example, proportion of genotype $A_1 A_1$ is $\frac{1}{16}$.

8. $\dfrac{p_1^2 + p_2^2 + p_3^2 + \cdots + p_n^2}{1 - (p_1^2 + p_2^2 + \cdots + p_n^2)}$.

9.4 **2.** 50 generations.

5. Selection acts on genotypes.

7. No. The 95 per cent confidence interval is between .35 and .45 for the allele frequency. The 99 per cent confidence interval is between .33 and .47.

10. (b) 900 generations.

9.5 **1.** $x_1 = \dfrac{2k_1 - k_2}{3}, x_2 = \dfrac{2k_2 - k_1}{3}$.

6. $x_1 = \dfrac{2b_1 a_{22} - b_2(a_{12} + a_{21})}{4a_{11} a_{22} - (a_{12} + a_{21})^2}, x_2 = \dfrac{2b_2 a_{11} - b_1(a_{12} + a_{22})}{4a_{11} a_{22} - (a_{12} + a_{21})^2}$.

9. $x_1 = \dfrac{(b_1 - k_1)a_{22} - (b_2 - k_2)a_{12}}{a_{11} a_{22} - a_{21} a_{12}}$,

$x_2 = \dfrac{(b_2 - k_2)a_{11} - (b_1 - k_1)a_{21}}{a_{11} a_{22} - a_{21} a_{12}}$.

Appendix A

5. (a) $\dfrac{\pi}{12}$. (b) $\dfrac{2\pi}{5}$. (c) $-\dfrac{\pi}{2}$. (d) 4π. (e) $-\dfrac{3\pi}{4}$.

(f) $\dfrac{\pi}{90}$. (g) $\dfrac{7\pi}{6}$. (h) $\dfrac{7\pi}{4}$.

8. 5π inches.

9. 2.5π ft/sec.

10. $\cos^{-1}\frac{11}{14}$, $\cos^{-1}\frac{1}{2}$, $\cos^{-1}\frac{1}{7}$.

12. (a) $\sin\theta = -\dfrac{\sqrt{2}}{2}$, $\cos\theta = \dfrac{\sqrt{2}}{2}$. (b) $\sin\theta = -\dfrac{\sqrt{3}}{2}$, $\cos\theta = -\dfrac{1}{2}$.

(c) $\sin\theta = -\dfrac{1}{2}$, $\cos\theta = \dfrac{\sqrt{3}}{2}$. (d) $\sin\theta = 0$, $\cos\theta = 1$.

(e) $\sin\theta = 0$, $\cos\theta = -1$. (f) $\sin\theta = -\dfrac{\sqrt{3}}{2}$, $\cos\theta = \dfrac{1}{2}$.

(g) $\sin\theta = -1$, $\cos\theta = 0$. (h) $\sin\theta = -\dfrac{1}{2}$, $\cos\theta = -\dfrac{\sqrt{3}}{2}$.

13. (a) $\sin^{-1} x = 0$. (b) $\sin^{-1} x = \dfrac{\pi}{2}$. (c) $\sin^{-1} x = \dfrac{\pi}{6}$.

(d) $\sin^{-1} x = \dfrac{\pi}{3}$. (e) $\sin^{-1} x = -\dfrac{\pi}{2}$. (f) $\sin^{-1} x = -\dfrac{\pi}{3}$.

(These are the values of $\sin^{-1} x$ in the range $-\dfrac{\pi}{2} \le \theta \le \dfrac{\pi}{2}$.)

14. (a) $\csc\theta = \dfrac{1}{\sqrt{1 - \cos^2\theta}}$. (b) $\tan\theta = \dfrac{\sin\theta}{\sqrt{1 - \sin^2\theta}}$.

(c) $\cos\theta = \dfrac{\cot\theta}{\sqrt{\cot^2\theta + 1}}$.

15. $\sin\theta = \frac{4}{5}$, $\cos\theta = \frac{3}{5}$.

16. (a) $\frac{1}{2}\sqrt{2 - \sqrt{3}}$. (b) $\frac{1}{2}\sqrt{2 + \sqrt{2}}$. (c) $\sqrt{2} + 1$.

17. (a) $\frac{1}{4}(\sqrt{6} + \sqrt{2})$. (b) $\frac{1}{4}(\sqrt{6} + \sqrt{2})$. (c) $\frac{1}{4}(\sqrt{2} - \sqrt{6})$.

Appendix B

B1 **1.** (a) $6x$. (b) $1 + 2x$. (c) $\frac{1}{2}x^{-1/2} - \frac{1}{2}x^{-3/2}$.

2. (a) $\dfrac{1}{(1-x)^2}$. (b) $\dfrac{2}{(1-x)^3}$. (c) $\dfrac{-10x}{(1+x^2)^2}$.

3. $u + at$.

4. $p(0) = 1000, p'(t) = \dfrac{9000}{(1+t)^2}$.

5. (a) 127.25. (b) 3.975. (c) .2525. (d) $\frac{383}{192}$.

8. (a) 1.005. (b) 1.01.

9. 7.85×10^{-9} cm^3.

10. (a) 8000. (b) 0. (c) $-10{,}000$.

B2 **1.** (a) $3x^2 + 10x + 3$. (c) $6x^5 + 4x^3 + 2x$.
(e) $x \cos x + \sin x$.

2. (a) $\dfrac{4x}{(1-x^2)^2}$. (c) $\dfrac{7}{(1-5x)^2}$.

(e) $\dfrac{(x-1)^2 \cos x - (x+1)^2 \sin x}{(1+x^2)^2}$. (f) $\sec^2 x$.

3. (a) $14(2x + 3)^6$. (c) $\dfrac{\cos \sqrt{x}}{2\sqrt{x}}$. (e) $-(1 + 2x) \sin (1 + x + x^2)$.

7. (a) $3(\sin x)^2 \cos x$. (c) $\cos x + \sin x + \sec x \tan x$.
(f) $2x(\cos 2x - x \sin 2x)$.

8. (a) $\dfrac{2}{\sqrt{1 - 4x^2}}$. (c) $\dfrac{2 \cos^{-1} x}{\sqrt{1 - x^2}}$. (e) $\dfrac{-x}{\sqrt{1 - x^2}}$.

B3 **1.** (a) Increasing for $x > 0$. Critical point $(0, 0)$. No inflection points.
(b) Increasing for $x > -\frac{1}{2}$. Critical point $(-\frac{1}{2}, 0)$. No inflection points.
(c) Increasing on $x < -1 - \sqrt{3}/3$ and $x > -1 + \sqrt{3}/3$. Critical points $(-1 + 1/\sqrt{3}, -2/3\sqrt{3})$, $(-1 - 1/\sqrt{3}, 2/3\sqrt{3})$. Inflection point $(-1, 0)$.
(d) Increasing for $x < 0$. Critical point $(0, 1)$. No inflection points.
(e) Increasing for $x > 0$. Critical point $(0, 1)$. No inflection points.

2. (a) Local maxima $y = \frac{4}{3}$ at $x = \pm 1/\sqrt{3}$, local minimum $y = 1$ at $x = 0$.
(b) Local minimum at $x = \frac{2}{3} + \frac{1}{3}\sqrt{7}$, local maximum at $x = \frac{2}{3} - \frac{1}{3}\sqrt{7}$.
(c) Maximum at $x = 0$.
(d) Maximum at $x = \pi/2$. Minimum at $x = 3\pi/2$.

(e) Maxima at $x = 0, \pi$. Minima at $x = \pi/2, 3\pi/2$.

(f) Maximum at the point where $x + 1 = -\tan x$.

3. $50:0, 100$.

B4   1. (a) $e^x + xe^x$.   (c) $e^x (\sin x + \cos x)$.   (f) $2xe^{-x^2}(1 - x^2)$.

2. (a) Critical point at $x = 0$; inflection points at $x = \pm 1$.

(b) Critical points at $x = \pm 1/\sqrt{2}$; inflection points at $x = 0$, $\pm\sqrt{3/2}$.

3. 29.9 minutes.

4. (a) $2^x \log_e 2 \approx (.693)2^x$.   (c) $3^x(x^2 \log_e 3 + 2x)$.

5. (a) .0304.   (b) .0314.   (c) .0342.

6. 10.

7. (a) $e^{x+(1/x)}\left(1 - \dfrac{1}{x^2}\right)$.   (b) $e^{\sin x} \cos x$.   (c) $e^{e^x + x}$.

9. The second drug has a larger maximum reaction and is slower acting.

10. (a) $k = T(0) - T_m$.

11. (a) $\dfrac{.693}{a}$.   (b) $\dfrac{2.303}{a}$.

14. $C(1) = 100e^a$ and $C(2) = 100e^{2a}$, where $a = \log_e 2$.

B5   1. (a) $\dfrac{\partial y}{\partial x} = 2xt + t^2, \dfrac{\partial y}{\partial t} = 2xt + x^2$.

(b) $\dfrac{\partial y}{\partial x} = \sin (x + t) + (x + t)\cos (x + t) = \dfrac{\partial y}{\partial t}$.

(c) $\dfrac{\partial y}{\partial x} = 4xt^2, \dfrac{\partial y}{\partial t} = 4x^2 t$.

(e) $\dfrac{\partial y}{\partial x} = e^{xt}(1 + xt), \dfrac{\partial y}{\partial t} = x^2 e^{xt}$.

2. $\dfrac{\partial V}{\partial T} = \dfrac{k}{P}, \dfrac{\partial V}{\partial P} = \dfrac{-kT}{P^2}$.

3. $\dfrac{2b}{3}$.

4. (a) $\dfrac{\partial^2 y}{\partial x^2} = 2t^2, \dfrac{\partial^2 y}{\partial x \, \partial t} = 4xt, \dfrac{\partial^2 y}{\partial t^2} = 2x^2$.

(d) $\dfrac{\partial^2 y}{\partial x^2} = t^2 e^{xt}(2 + tx), \dfrac{\partial^2 y}{\partial x \, \partial t} = e^{xt}(t^2 x^2 + 3tx + 1)$,

$\dfrac{\partial^2 y}{\partial t^2} = x^2 e^{xt}(2 + tx)$.

5. $x = 2a/3$, 2 hours after administration.

Appendix C

C1 **1.** (a) $\frac{1}{11}x^{11} + c.$ (c) $-\frac{1}{2x^2} + c.$

(e) $x + x^2 + \frac{x^3}{3} + c.$ (f) $\frac{x^2}{2} + \log_e x + c.$

2. (a) $\frac{\sin 3x}{3} + c.$ (b) $\frac{-\cos (x^2)}{2} + c.$ (c) $\frac{(1 + x^5)^4}{20} + c.$

(d) $-\frac{1}{3}e^{-x^3} + c.$ (f) $-\frac{\cos^2 x}{2} - \frac{\cos^3 x}{3} + c.$

3. (a) $\sin x - x \cos x + c.$ (b) $\frac{x^2}{2}\log_e x - \frac{x^2}{4} + c.$

(c) $-xe^{-x} - e^{-x} + c.$ (d) $\frac{1}{2}e^x(\sin x + \cos x) + c.$

(e) $(2 - x^2) \cos x + 2x \sin x + c.$

(f) $-\frac{x \cos 5x}{5} + \frac{1}{25} \sin 5x + c.$

4. (a) $\log_e (1 + x) + c.$ (b) $\tan^{-1} x + c.$

(c) $\sin^{-1} x + c.$ (d) $-\frac{1}{1 + x} + c.$

(e) $\frac{1}{13}(2e^{3x} \sin 2x + 3e^{3x} \cos 2x) + c.$ (f) $-\sqrt{1 - x^2} + c.$

5. $y = \frac{x^3}{3} + x - \frac{1}{3}.$

6. $y = \sin x + 5.$

7. (a) 10,001. (b) 10,054. (c) 10,383.

C2 **1.** (a) 124. (b) $\frac{14}{3}$. (c) $\frac{11}{6}$.

2. (a) $3 + e^7 - e^4$. (b) $e^{-9} + e^{-12} - e^{-15} - e^{-20}$.

(c) $1 - \frac{2}{e}$.

3. (a) $\frac{1}{2}$. (b) $-\frac{1}{2}$. (c) $\sqrt{3}$.

4. (a) $\frac{1}{9}(1 + 2e^3)$. (b) $\frac{1}{2}(1 - 26e^{-25})$.

(c) $\frac{1}{32}(32\sqrt{2} + 8\sqrt{2\pi} - \pi^2\sqrt{2} - 64)$.

5. (a) π. (b) π. (c) Diverges.

6. $\frac{1}{2}$.

7. $\dfrac{\pi}{2}$.

9. $C(t) = 1 + \dfrac{1 - 2^{-t}}{\log_e 2}$.

C3 **1.** Estimates .813, .782; exact value $\pi/4 \approx .785$.
     2. Estimate 1.117; exact value $\log_e 3 \approx 1.099$.
     6. Integral is 1.463 to four significant figures of accuracy.

Appendix D

1. (a) 1.0513. (b) 1.6487. (c) .9048. (d) .3679.

Appendix E

1. (a) $3e^{i\pi/2}$. (b) $4\sqrt{2}e^{i\pi/4}$. (c) $5e^{i\theta}, \theta = \tan^{-1}\frac{3}{4}$.
  (d) $2\sqrt{5}e^{-i\pi/6}$. (e) $2e^{i\pi/3}$.

2. (a) -3. (b) $\dfrac{5}{4} + i\dfrac{5\sqrt{3}}{4}$. (c) -1. (d) $-2i$.

  (e) $-\dfrac{5}{\sqrt{2}} + \dfrac{5i}{\sqrt{2}}$.

3. (a) $-128 + 128i$. (b) $-8 - 8\sqrt{3}i$. (c) $-16\sqrt{3} + 16i$.
4. $\cos 4\theta = \cos^4 \theta - 6 \cos^2 \theta \sin^2 \theta + \sin^4 \theta$
  $\sin 4\theta = 4 \sin \theta \cos^3 \theta - 4 \cos \theta \sin^3 \theta$.

Index

EVOLUTION OVER SIX GENERATIONS OF SOME SIMPLE INITIAL CONFIGURATIONS